Advances in Fine Particles Processing

Advances in Fine Particles Processing

Proceedings of the International Symposium on Advances in Fine Particles Processing

Editors

John Hanna

Mineral Resources Institute
College of Engineering
University of Alabama
Tuscaloosa, Alabama, USA

and

Yosry A. Attia

Department of Materials Science and Engineering
The Ohio State University
Columbus, Ohio, USA

Elsevier
New York · Amsterdam · London

Elsevier Science Publishing Co., Inc.
655 Avenue of the Americas, New York, New York 10010

Sole distributors outside the United States and Canada:
Elsevier Science Publishers B.V.
P.O. Box 211, 1000 AE Amsterdam, The Netherlands

This book was printed on acid-free paper.

ISBN 0-444-01557-4

Current printing (last digit):
10 9 8 7 6 5 4 3 2 1

Manufactured in the United States of America

Contents

PART 4. SURFACE AND COLLOIDAL CHEMISTRY IN THE PROCESSING OF CLAYS

PART 5. PROCESSING OF FINE PARTICLES BY FLOCCULATION AND DISPERSION

PART 6. SEPARATION OF FINE PARTICLES BY FLOTATION

PART 7. FINE PARTICLE PROCESSING WITH MULTIPLE PHYSICAL, CHEMICAL AND BIOLOGICAL PHENOMENA

Foreword

Processing of fine particles has presented numerous challenges to scientists and engineers for many years. Considerable progress has already been made in meeting these challenges across various fields of applications around the world. Research on every aspect of fine particle processing has gained momentum in recent years, resulting in the development of new processes, improved products, and better understanding of the science and engineering fundamentals of fine particles. This symposium addressed the recent progress in fine particles processing, particularly in the production of minerals for chemicals, pigments and metal production, ceramic materials, and fossil fuels.

This book represents the edited proceedings of the International Symposium on Advances in Fine Particles Processing, where selected peer-reviewed papers describe current practices, review the state of the art and report original fundamental and applied research on fine particle production, sizing, characterization of the interface, fluid flow, and interparticle colloidal interactions, leading to dispersion, flocculation and flotation. Processing of fine particles by multi-chemical, physical and biological phenomena has also been addressed. Accordingly, the book consists of seven parts, with each part addressing a specific topic. Part One deals with production of fine particles by comminution methods where different milling practices, mathematic modeling and physical-chemical control methods are reported. Part Two covers particle flow properties in various fluids. Part Three addresses surface and colloidal phenomena in fine particle processing, while Part Four continues this topic but with emphasis on clay minerals. Part Five describes the roles of particle dispersion and flocculation, including the design of selective flocculants for processing of fine particles. Part Six shows the role of flotation in processing fine particles. In Part Seven, fine particle processing with multiple physical, chemical and biological phenomena is reported. This book will be of great interest and benefit to research scientists and engineers, graduate students and faculty, and all persons interested in fine particles processing.

This book was made possible through the cooperation and enthusiastic support of many colleagues and organizations, to whom we are indeed most grateful. We would particularly like to thank all of the peer-reviewers, the symposium speakers, authors of articles, and session chairmen for their contributions. We wish to acknowledge the support of Professor Teoman Ariman, President of the Fine Particle Society, for hosting and sponsoring this symposium. We wish also to acknowledge the efforts of Professor Carl Rampacek, Director Emeritus of the Mineral Resources Institute (MRI) of the University of Alabama, and Mr. R. S. Akins, Director of Mining, and Dr. Hassan El-Shall, Director of Beneficiation of the Florida Institute of Phosphate Research, for co-sponsoring this meeting. Sincere thanks are due to Dr. L. J. Morgan, Englehard Corporation, for her tireless efforts to promote the symposium. Special thanks are also due to Mr. I. J. Anazia, Dr. R. K. Mehta and all of the MRI staff for their valuable assistance throughout the meeting and in preparing this volume.

John Hanna and Yosry A. Attia

Editors

May 10, 1990

PART 1.

PRODUCTION OF FINE PARTICLES BY COMMINUTION

COMMINUTION ENERGY REDUCTION BY TWO - STAGE CLASSIFICATION

D. A. DAHLSTROM* and W.-P. KAM*
*Research Professor and Graduate Student, respectively, Metallurgical Engineering Department, University of Utah, Salt Lake City, Utah 84112

ABSTRACT

A previous computer study indicated that significant comminution energy savings could be experienced by use of two stage, counter-current cyclone classification. To further prove this important potential a pilot plant was constructed which permitted analysis of energy savings for two stage as compared to single stage classification. Because energy input to the ball mill was constant, savings potential are actually illustrated by the greater percentage of minus 400 mesh solids generated by two stage classification at the same circulating load and product production rate. It was shown that the following conclusions can be made:

1. Energy savings increase as the fineness of grind increases.

2. As recycle ratio decreases, energy savings increase.

3. Energy consumption savings ranged from 7 to over 40% to date.

4. For existing ball mills either capacity can be increased at the same grind or a finer grind can be produced at the same tonnage rate.

5. Overgrinding is reduced at the percent minus 400 mesh is always less than single stage at the same grind.

6. Pilot plant results to date exhibited 6 to 8 percentage points greater amount of minus 400 mesh solids by two stage classification at constant energy input and similar product production rate.

7. Two stage countercurrent classification yields a slightly lower product solids concentration.

INTRODUCTION

Comminution is undoubtedly one of the highest single categories in industrial energy consumption. About 1.3 percent of all electrical energy is consumed by crushing and grinding, most of which is found in the mineral and coal industries.[1] It is almost always the single largest energy consumer in the processing mill and many times the highest percentage of the capital and operating costs.

Advances in Fine Particles Processing
John Hanna and Yosry A. Attia, Editors

One study estimated that the classifier device design potential for energy reduction was 9-13% while the only other one in this high potential area was the comminution device design at 9-16%.[2] It is believed the reason is due to the reduction in overgrinding by better removal of fines below the desired size and the ability to impact more efficiently on the oversize.

Accordingly, two stage classification was considered as a possible improvement to classifier operation. In addition, hydrocyclones were employed as they are generally accepted in fine grinding (ball mill operation). Single stage cyclone classification results in only about 65% removal of the extreme fines (-400 mesh or -37 microns) in this circuit. This is caused by the fact that most cyclone classifiers closing the ball mill circuit are fed at 50-55wt % solids with an underflow concentration of around 60 to 70 wt %. It can be easily shown that at 54 wt % in the feed and 70% solids in the underflow with a 400% circulating load, 40.3% of the feed water will report to the underflow. Thus, at a minimum, 40.3% of the finest solids at less than about 10 microns will report to the underflow and thus recirculate to the ball mill. Anything coarser than that obviously will report to the underflow at even higher percentages and therefore lesser percentages of fines below liberation size are removed to the overflow per pass. Lower percent solids in the underflow will cause these percentages to increase also. Going to 200% circulating load decreases still further the recovery of fines below liberation size to the cyclone overflow.

If two stages of cyclone classification are employed, as much as 85% and even up to 90% of the -400 mesh can be removed by two stage cyclone classification per pass. While two stages will require another set of cyclones and a feed pump, this extra energy should be easily offset by energy savings with the ball mill.

The question logically asked is why should removing more fines below the liberation size per pass through the ball mill result in a reduction of energy per ton of product? The answer is hypothesized as follows:

1. Obviously, overgrinding is reduced as a higher percentage of the fine particles are not recycled and thus there should be less energy expended.

2. The fines are known to increase slurry viscosity as their percentage is increased. Thus, ball impact is decreased. It was very noticeable that the second stage cycle underflow was "very coarse" by comparison to the first stage and the settling velocity of the solids is greatly increased.

3. The weight of solids recycled to the ball mill at any circulating load is significantly reduced as compared to single stage classification operation which should decrease energy consumption per ton of product.
4. It is also believed that the solids above liberation size are more "exposed" to ball impact reducing energy consumption.

TWO STAGE FLOWSHEET SELECTION

While there are several two stage classification flowsheets, only two would reduce energy consumption. Figure 1 illustrates these two circuits. However, flowsheet B would result in excessive dilution of the product because fresh water is added to both ball mill discharge and first stage cyclone underflow. All input fresh water must end up in the single product stream coming from the overflows of both stages. While this would minimize ball

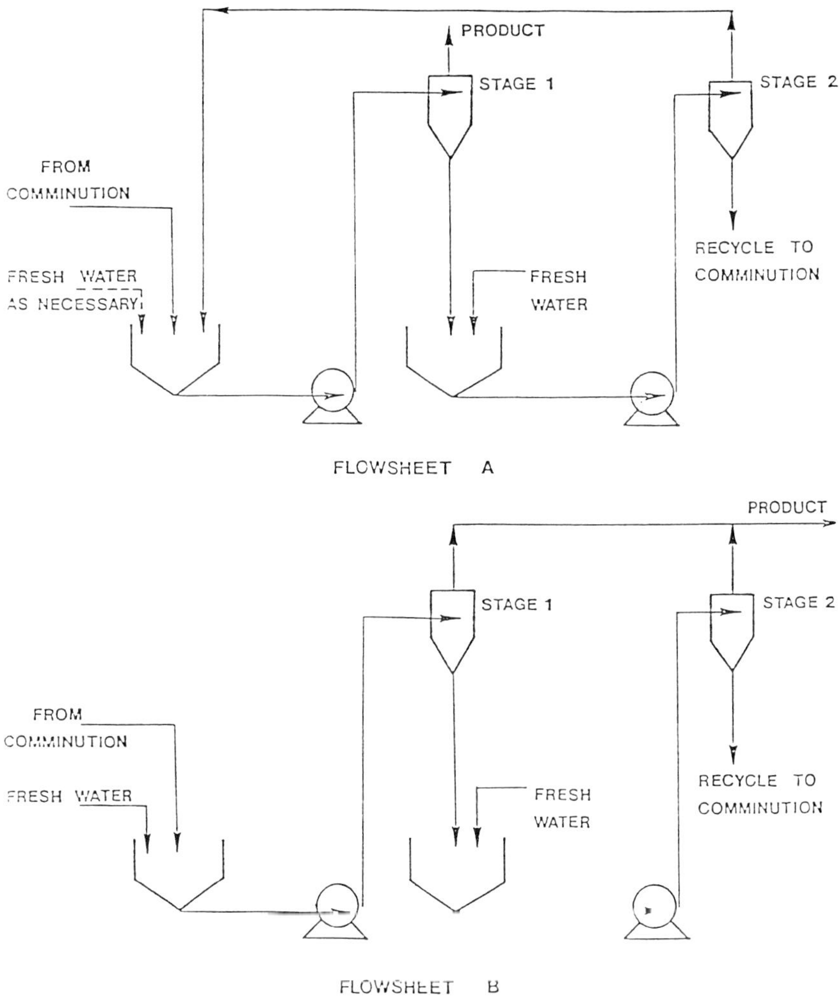

FIGURE 1
POSSIBLE ENERGY SAVING TWO STAGE CYCLONE CLASSIFICATION CIRCUITS

milling energy, it would also yield a very low product solids concentration. As most products will probably go either to flotation or leaching, this would either require thickening in the case of flotation or would result in an excessively dilute pregnant liquor after leaching.

Flowsheet A employs counter-current operation of the water as product is taken only from the first stage overflow. Second stage overflow is returned to the solids from ball milling and fresh water would only be added at this point as a control method to insure proper final product. size distribution.

In spite of the counter-current operation, some dilution of the final product will result with two-stage classification. Figure 2 is a plot of wt % solids in the product for both single stage and two stage cyclone classification.[3] In the latter case, the product is coming from the first stage cyclone overflow as illustrated in flowsheet A of Figure 1. For both cases, cyclone feed solids concentrations were maintained at 54 wt % solids and all cyclone underflows were at 75 wt % solids. It will be observed that if product solids concentration of 25 wt % is necessary, circulating loads would be a maximum of about 350% while single stage would be as much as 400% circulating load. As 350 % represents a reasonable and typical circulating load, it is felt that this small disadvantage may not be of any importance.

For the following computer studies, feed and underflow solids concentrations were maintained at the level of 54 and 75 wt % solids, respectively, as indicated previously. Accordingly, fresh water was added to the sump preceding the second stage cyclone while only a small amount of water was added as an adjustment to provide 54 wt % solids in the feed to the first stage.

COMPUTER STUDY OF SINGLE AND TWO STAGE CYCLONE CIRCUITS

There are several major factors that are very influential in their effect upon results in the classification of ball mill operation and grinding. These are listed in Table 1. Obviously, these factors cannot all be held constant . Table 2 gives those factors that were held constant while Table 3 list those that were variables. New feed size distribution was a typical rod mill discharge prior to the ball mill.

A computer program known as MODSIM was developed at the University of Utah for determining performance of grinding units with varying feed rate, size distribution and other factors.[4] A second program was developed including a classification circuit which employed the formulas developed by Plitt. The program is called SCALEMILL.[5] Employing this program and using the six variables plus assuming hydrocyclones dimensions, ball mill size, energy input, number of cyclones, and pressure drop required per stage, final product size distribution and circulating loads could be determined with relatively few iterations.

This permitted the development of correlations of energy consumption per metric ton of product as a function of product size with parameters of circulating load. All variable terms will also be determined for each set of conditions of Table 2 as well as a complete material balance. An example from

FIGURE 2

WEIGHT % SOLIDS vs CIRCULATING LOAD
SINGLE AND TWO STAGE CLASSIFICATION

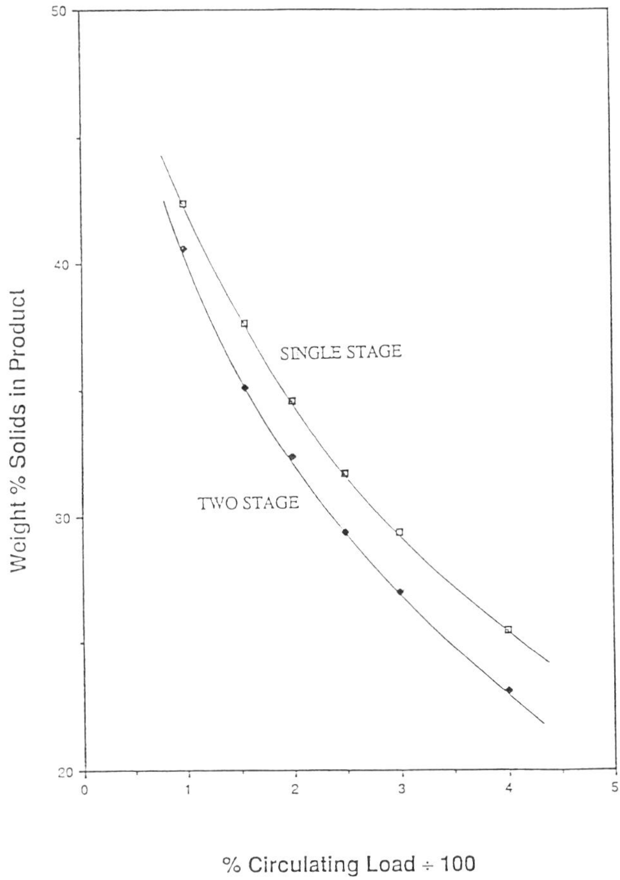

From Dahlstrom, D. A. and Kam, W. P., "Influence of the Classification Circuit on Energy Consumption in Comminution"

FIGURE 3

ENERGY CONSUMPTION vs WEIGHT % + 200 MESH
PER METRIC TON OF PRODUCT BASIS
PARAMETERS OF RECYCLE RATIO

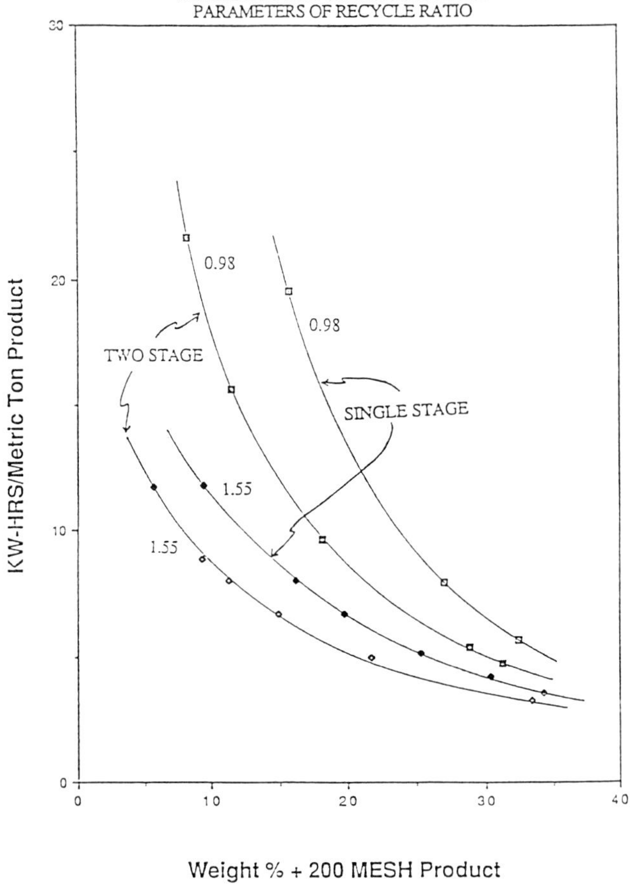

From Dahlstrom, D. A. and Kam, W. P.,
"Influence of the Classification Circuit on
Energy Consumption in Comminution"

Table 1

Influential Factors in Classification and Communition on Energy Consumption

1. Required grind for liberation
2. Energy input to the comminution unit
3. Selection and breakage functions
4. Size distribution of new feed to comminution.
5. Weight % solids in the comminution (ball mill) unit.
6. Solids specific gravity
7. Product and new feed solids rate
8. Feed solids concentration to the cyclones
9. Cyclone underflow solids concentration
10. Cyclone centrifugal force

Table 2

Fixed inputs in the Computer Study

1. Selection and breakage functions
2. New feed size distribution
3. Cyclone feed solids concentration (54 wt %)
4. Cyclone under flow solids concentration (75 wt%)
5. Cyclone flowsheet (flowsheet A, Figure 1 for two-stage)
6. Solids specific gravity
7. Product and new feed solids rate (300 metric tons/hour)
8. Weight % solids in the ball mill

Table 3

Variables Employed in Computer Studies

1. Ball mill power input
2. Ball mill size
3. Cyclone dimensions
   a. diameter
   b. number per stage
   c. inlet diameter
   d. vortex finder diameter
   e. apex diameter
   f. distance between vortex finder and apex
4. Number of cyclones per stage
5. Pressure drop across the cyclone stages
6. Product size distribution
7. Circulating load

an earlier paper, Figure 3 is a plot of energy consumption in kw-hours/metric ton as a function of the grind expressed as wt % plus 200 mesh product.[3] Parameters of recycle ratios (expressed as tons recycled per ton of new feed) of 0.98 and 1.55 are given for both single and two stage classification. Figure 4 is a similar correlation for the 200 mesh grind basis but with parameters of 3.0 and 4.0 recycle ratios.[3] It is very obvious in all cases that two stage classification always lies below the similar recycle ratio for single stage classification.

Comparisons can now be made from these results between energy consumption and any grind and recycle ratio for two stage and single stage classification. Figure 5 is a plot of percent reduction in energy consumption by two stage classification as compared to single stage with recycle ratios ranging from 0.98 to 4.0 based on plus 100 mesh grinds. Figure 6 is a similar plot for plus 200 mesh grinds with 5 recycle ratios from 0.98 to 4.0.[3]

Some very apparent observations can be drawn from these two figures as influenced by the two classification methods:

1. As recycle ratio decreases, energy savings increase dramatically.

2. As the percent plus the mesh size of grind decreases, energy savings increase.

3. Energy savings are as high as over 40% to a minimum of 7%.

4. In analyzing the product size distribution for all runs, it was noted that at the same grind, the -400 mesh solids were always less with two stage classification than single stage classification. Thus, overgrinding is reduced.[6]

Obviously, these are very significant energy savings but represent only computer analyses based upon a specific program. These values should be examined based on pilot studies.

PILOT PLANT RESULTS

A pilot plant was constructed in the Ore Dressing Laboratory of the Metallurgical Engineering Department at the University of Utah. Two inch diameter cyclones were contributed by the Krebs Engineers Co. A 30 inch diameter ball mill is installed in the laboratory and two progressive cavity type pumps were used for feeding the cyclones. A variable speed screw feeder was employed for controlling the raw feed rate and agitated sumps containing approximately five gallons were used ahead of the pumps. For two stage classification, flowsheet A of Figure I was employed. Only fresh water was added for dilution to the second stage sump, with that cyclone overflow returned to the first stage feed as the major dilution water. Only a relatively small amount of water was added to the ball mill feed and first stage sump to control solids concentration.

Relatively high purity commercial aragonite ore was available south of the Great Salt Lake in Northwest Utah. Table 4 gives the average screen size for the aragonite.

FIGURE 4

ENERGY CONSUMPTION vs WEIGHT % + 200 MESH
PER METRIC TON PRODUCT BASIS
PARAMETERS OF RECYCLE RATIO

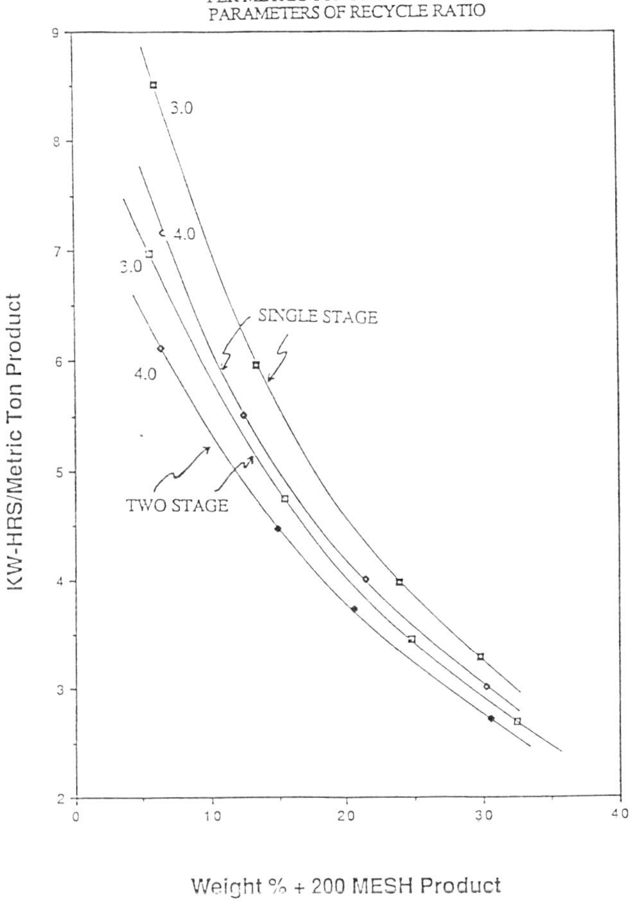

From Dahlstrom, D. A. and Kam, W. P.,
"Influence of the Classification Circuit on
Energy Consumption in Comminution"

FIGURE 5

% ENERGY CONSUMPTION REDUCTION
vs
WEIGHT % + 100 MESH
PARAMETERS OF RECYCLE RATIO

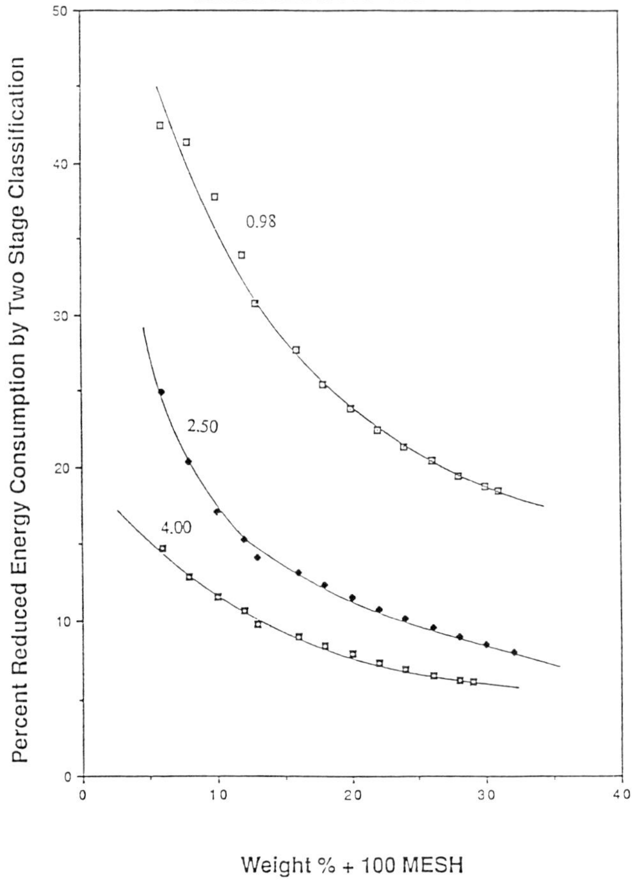

From Dahlstrom, D. A. and Kam, W. P., "Influence of the Classification Circuit on Energy Consumption in Comminution"

FIGURE 6

% ENERGY CONSUMPTION REDUCTION
VS
WEIGHT % + 100 MESH
PARAMETERS OF RECYCLE RATIO

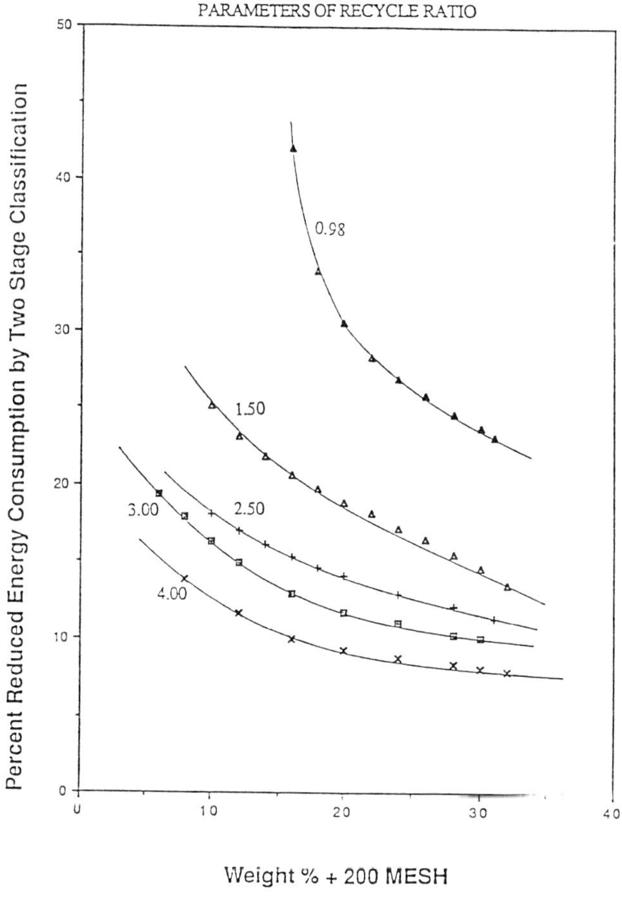

From Dahlstrom, D. A. and Kam, W. P., "Influence of the Classification Circuit on Energy Consumption in Comminution"

Table 4

Raw Aragonite Pilot Plant Feed

Tyler Mesh	Weight %	Cumulative %
+10	5.83	5.83
10x20	25.34	31.17
20x35	20.16	51.33
35x65	15.55	66.88
65x100	6.65	73.53
100x200	11.52	85.05
200x270	3.19	88.24
270x400	1.09	89.33
-400	10.67	100.00

With these solids, cyclone feed was always controlled at 54 wt % solids and underflows at 66 wt % solids plus or minus 2 percentage points. This meant that cyclone feed rates would range from about 3 to 7 gpm based on 100 to 400 percent circulating loads. This is why the very small diameter cyclones had to be employed.

No data was used unless the cyclone feeds and underflows were within the specified limits and reasonable material balances were obtained indicating equilibrium conditions had been experienced. The following were held constant in regards to the ball mill:

1. 35RPM speed
2. Volume % of feed and ball charge
3. Ball charge
4. Temperature
5. Feed solids concentration

Thus energy input was constant. Therefore, to determine if energy efficiency was increased by two stage classification in the pilot plant, the weight % solids of -400 mesh was determined on each run on the product from the first stage cyclone overflow. If two stage classification is more efficient, this would be indicated by an increase in the -400 mesh fraction at the same circulating load and solids processing rate as compared to single stage classification.

There is still insufficient data at the time this paper was written to sufficiently determine the greater amount of -400 mesh produced by two stage classification. However, there were enough runs accomplished to indicate that this result does occur.

Figure 7 is a plot of weight percent -400 mesh in the product as a function of the percent circulating load with parameters of production rate in terms of pounds of solids product per hour for both single and two-stage classification.

While it is readily apparent that there is insufficient data at present to determine the exact shape of parameters drawn in Figure 7 it is very evident that more -400 mesh solids were generated by two-stage classification. An average of eight percentage points more of -400 mesh were produced by two-stage classification at a production rate of 300 to 400 pounds of solids per hour and 6 percentage points at 400 to 500 pounds per hour. This represents a

FIGURE 7
WEIGHT % -400 MESH SOLIDS IN PRODUCT VERSUS
PERCENT CIRCULATING LOAD
PARAMETERS OF CLASSIFICATION TYPE AND PRODUCT RATE

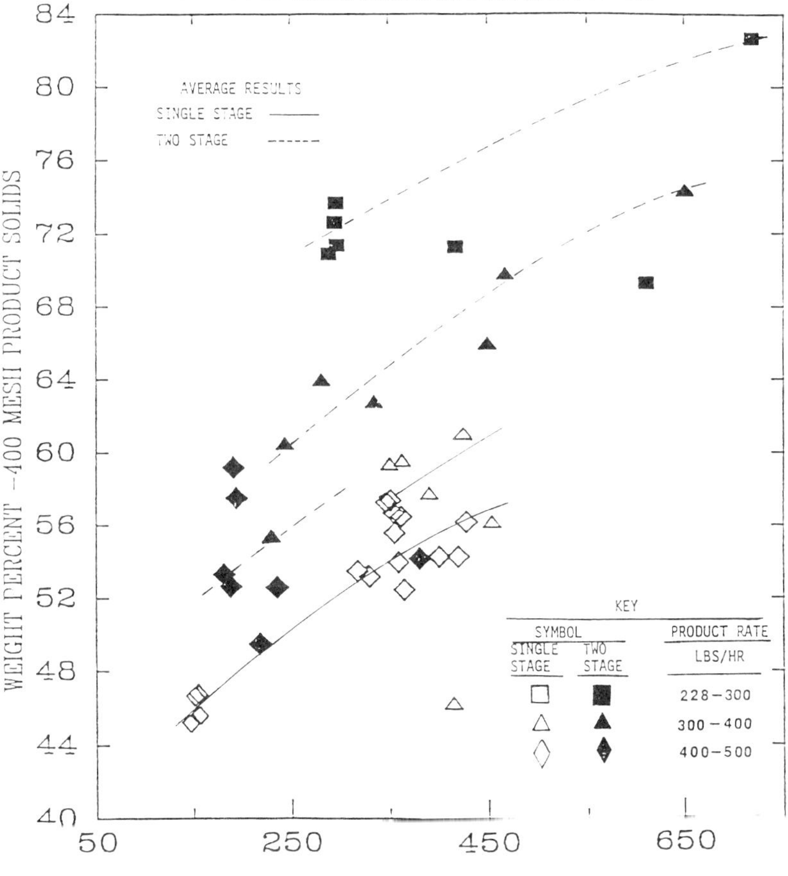

significant increase in energy efficiency. The actual value would have to be related to work index value for the solids for the two classification methods. However, it is expected that the percentage energy savings would be at least similar to those indicated in Figures 5 and 6.

CONCLUSIONS

Several conclusions can be drawn from the work completed to this point for comminution energy savings by two stage classification:

1. Energy savings will be a function of the grind and will increase as the grind becomes finer.
2. As recycle ratio increases energy savings, although significant, are reduced.
3. Energy consumption savings range from 7 to over 40 percent as a function of grind and recycle ratio.
4. For existing ball mills, two stage classification at the same circulating load could either produce more tonnage at the same grind or a finer grind at the same capacity when compared to single stage classification.
5. Overgrinding should also be achieved by two stage classification. In the computer analysis, the -400 mesh % was always lower at the same grind than single stage classification.
6. Pilot plant runs to date with constant energy input to the 30 inch diameter ball mill exhibited 6 to 8 percentage points greater -400 mesh solids by two stage classification with the larger percentage at 300 to 400 pounds of product solids per hour and the smaller number at 400 to 500 pounds of product solids per hour.
7. Two stage counter-current classification yields a slightly lower wt % solids product as compared to single stage.

REFERENCES

1. Pitt, C. H. and Wadsworth, M.E., "An Assessment of Energy Requirements in New Copper Processes," U. S. Dept. of Energy, Division of Industrial Energy Conservation, Final Report, 1980, Contract No. EM-78-5-07-1743.
2. Herbst J. A. (Chairman), "Comminution and Energy Consumption," U. S. Dept. of Energy, National Materials Advisory Board, Committee on Comminution and Energy Consumption, Publication NM+FB-364, National Academy Press, Washington, D.C.
3. Dahlstrom, D. A. and Kam, W.-P., "Influence of the Classification Circuit on Energy Consumption in Comminution," Annual Meeting, Society of Mining Engineers of AIME, February, 1988.

4. Herbst, J. A., Lo, Y. C. and Rajamani, K., "Population Balance Model Predictions of the Performance of Large-Diameter Mills," Minerals and Metallurgical Processing, Society of Mining Engineers of AIME, 1985.
5. Herbst, J. A., Schena, G. D., and Fu L. S., "Computer Aided Design of Comminution Circuits," Annual Meeting AIME, 1986, New Orleans, LA.
6. Dahlstrom, D. A. and Kam, W.-P., "Potential Energy Savings in Comminution by Two-Stage Classification," International Journal of Mineral Processing, Vol. 22, pages 239-250, 1988.

POWER REQUIREMENTS FOR ULTRAFINE GRINDING AND DRYING OF LOW-RANK COALS IN A FLUID-ENERGY MILL

C. W. BOUCHILLON,* W. G. STEELE,* J. D. BURNETT**
*Professors of Mechanical Engineering, Mississippi State University, Mississippi State, MS 39762; **Project Director Micro-Energy/Ergon, Inc., Vicksburg, MS 39180

ABSTRACT

An experimental evaluation of the power requirements for simultaneous ultrafine grinding and drying of low-rank coals in a fluid-energy mill was undertaken. Two different lignites and a subbituminous coal were used in the study. Ultrafine grinding tests were conducted with air nominally at ambient conditions and with steam at temperatures ranging from 225°F to 488°F. It was found that the power required for grinding and drying increased linearly with the grinding medium temperature.

INTRODUCTION

Low-rank coals have relatively high moisture levels and it is known that thermal treatment of the particles will result in a reduction of the equilibrium moisture of the coals. Therefore, a combined grinding/drying process such as the one investigated here is of practical interest.

Three coals were selected for use in this project - 1) a Martin Lake, Texas lignite, 2) a Beulah, North Dakota lignite, and 3) an Eagle Butte, Wyoming subbituminous coal. Ultrafine grinding tests were conducted with air nominally at ambient conditions and with steam at temperatures ranging from 225 to 488°F. The fluid-energy mill used in the tests was located at the development facility of the Micro-Energy Division of Ergon, Inc. in Vicksburg, MS.

Ultrafine grinding is typically defined as a grinding process which produces particles with a size distribution such that ninety-nine percent have a mean particle diameter of less than 44 microns in diameter on a volume basis. Research on ultrafinely ground coal has shown that when this product is burned, much less slagging occurs than with normally pulverized coal which is typically in the -200 mesh range [1-4].

The small ash particles in the ultrafine coal tend to follow the flow streams around the heat transfer surfaces rather than impacting the tubes and causing slagging. These ultrafine particles then pass through the plant and are caught in a baghouse. The benefit of such a fuel is that a solid fuel, either as a dry powder or in a slurry form, may be burned in a boiler which was originally designed for oil or gas without major boiler modifications. The tube spacing in oil and gas boilers is closer than that in slagging, coal fired boilers. These ultrafinely ground coal and lignite fuels, which are potentially less expensive than oil and gas, allow for a relatively inexpensive retrofit for using coal in existing power plants and boilers.

Until recently there had been only limited work done on ultrafinely ground lignites. In Australia, tests were conducted using ultrafinely ground lignite in a diesel engine [5]. The authors for this paper have reported results of studies on the ultrafine grinding of a Mississippi lignite [6-8]. Those investigations showed that the lignite, even with its high moisture content, could be ultrafinely ground in a fluid energy

Advances in Fine Particles Processing
John Hanna and Yosry A. Attia, Editors

mill to a product with a mean diameter based on population of 2 to 5 microns and a mean diameter based on volume of 7 to 25 microns. Simultaneous in-the-mill grinding and drying tests showed that some permanent drying of the lignite could be accomplished.

FLUID-ENERGY MILL

The mill used for the grinding tests was a recently patented jet-vortex mill which had significantly reduced erosion rates over those that normally occur in fluid-energy grinding. The Ergon, Micro-Energy mill, Fig. 1, is based on the patent of Taylor [9] and consists of a vertically oriented cylindrical chamber with nozzles arranged along the periphery near the base of the cylinder. The trajectory of the jets from the nozzles is towards a point slightly outward of the center line and slightly upward. This creates a vortex in the center of the vessel which has upward axial velocities near the center and downward velocities due to recirculation near the outer walls. The feed material may be fed into either the top or bottom with the most recent developmental model having bottom feed.

The convergent-divergent nozzles produce supersonic flows which provide rapid particle accelerations when particles are lifted into the issuing jet. The particles are then ground by mutual collision and attrition in the chamber. The smaller particles are more readily carried out of the overflow which may be fitted with a mechanical classifier to obtain better top-size control. The centrifugal effects on the larger particles cause them to be thrown outward and then gravity and the recirculating fluid causes them to be carried back into the nozzle flow region.

The jet-vortex mill has been further developed by Ergon/ Micro-Energy and has been used for processing 20 tons per hour of coal. The mill size can be scaled from one ton per hour up to these larger sizes. This mill has distinct advantages over other fluid energy mills in reduced erosion and higher production rates. Further development of this mill is in progress.

EXPERIMENTAL PROCEDURE

The facility in which the ultrafine grinding and drying of the samples of the low-rank coals was performed consisted of a fluid-energy mill with a design capacity of 2,000 lbm/hr of solid material. The grinding medium was either compressed air or steam. The air supply was capable of furnishing up to approximately 4,000 lbm/hr of air at approximately 100 psig. The steam supply was capable of furnishing up to approximately 4,000 lbm/hr of superheated steam at 150 psig and 750°F.

In order to evaluate the power requirements for the ultrafine grinding with simultaneous drying of low-rank coals, it was necessary to appropriately instrument the facility to obtain energy and mass balances on the fluid and the solids streams. A schematic diagram of the pilot plant and the location of the various instruments is shown in Fig. 2.

The instrumentation on the fluid stream consisted of pressure and temperature measurements upstream of the orifice in the supply line, pressure drop across the orifice, pressure and temperature at the nozzle ring of the pulverizer, pressure and temperature in the pulverizer, and pressure and temperature on the exhaust stream. The instrumentation required for the solids stream included a mass flow rate measurement, the temperature of the feed, and the temperature of the product. Commercial

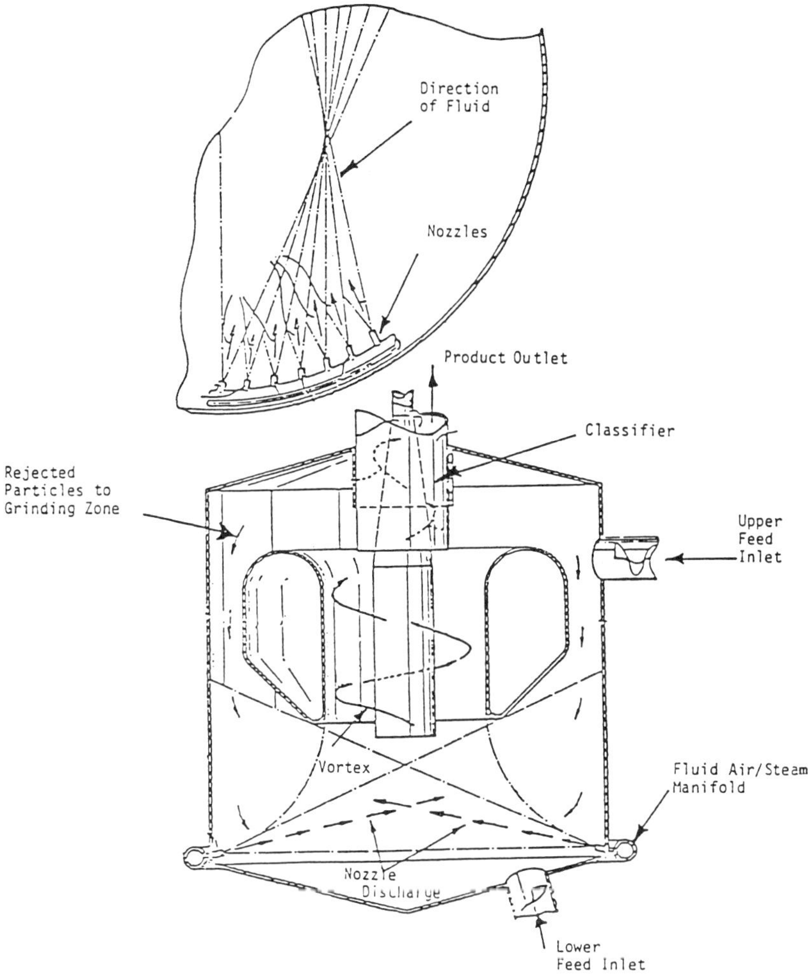

FIG. 1. Jet-vortex fluid energy mill (Ergon/Micro-Energy, Inc.).

instrumentation was available on the system and this was augmented by pressure transducers and thermocouples which were compatible with a computer based data acquisition system which allowed experimental data to be recorded throughout the experimental period.

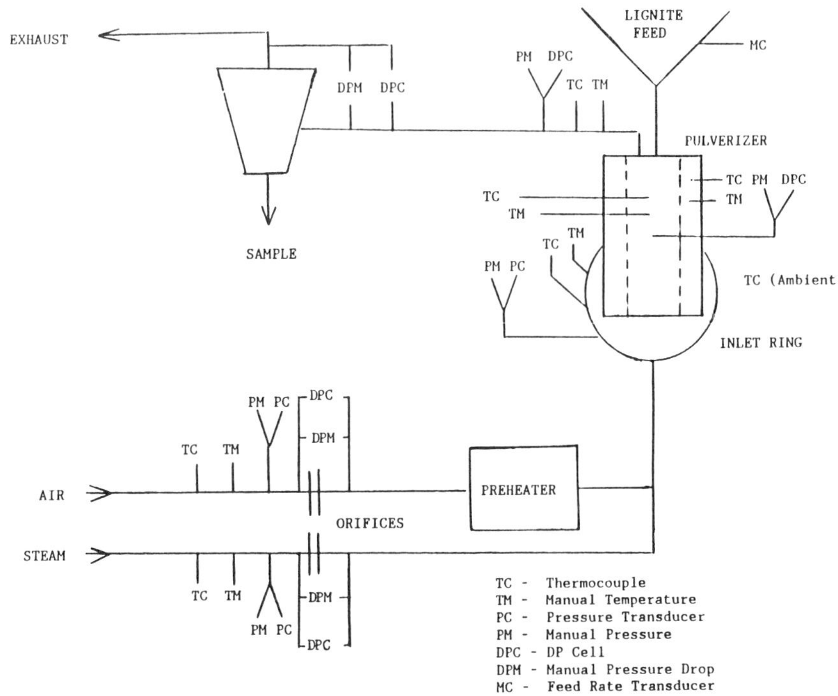

Fig. 2. Schematic diagram of the instrumentation on the fluid-energy mill.

The energy balance required that the heat losses from the system be determined. This was accomplished by running only the fluid stream through the system at various temperatures and calculating the appropriate heat loss from an energy balance on the system. Then a heat loss correlation was developed which was used in the data reduction program to ascertain the process energy requirements.

The speed of the belt (or screw in some tests) feeder to the pulverizer was used as a feed rate transducer and was calibrated prior to the conduct of the tests. Temperature measurements of the feed material were also taken during the tests.

Prior to the data runs all of the transducers were calibrated. The pressure transducers were calibrated against standard devices over their appropriate ranges. The thermocouples were calibrated against known reference temperatures.

Prior to each test the system was preheated by allowing fluid only to flow through the mill. Then the low-rank coal feed rate was adjusted to the desired value. Monitoring of the various temperatures and pressures using a microcomputer based data acquisition system allowed determination that a steady state condition had been reached in the system, after which data were recorded for use in the energy consumption determinations as well as for use in describing the conditions of operation for the process.

The runs were continued for approximately 30 minutes after steady state had been reached. The conditions and runs which were conducted during the experiments are presented in Table I.

TABLE I. Conditions of operation for the experimental studies.

Material	Grinding Medium	Mixture Temperature in Mill (°F)	Mill Inlet Temperature (°F)	Mill Inlet Pressure (psia)	Fluid Flow Rate (lbm/hr)
Texas Lignite	Air	116	320	69	1600
	Steam	225	689	93	1525
	Steam	310	688	112	1925
	Steam	350	700	123	2100
	Steam	400	708	135	2275
	Steam	488	730	191	*
North Dakota Lignite**	Air	90	128	110	1808
	Steam	240	639	62	1407
	Steam	275	653	120***	1857
	Steam	310	563	88	2265
	Steam	360	565	87	2259
Wyoming Subbituminous Coal**	Air	90	126	109	1798
	Steam	240	647	69	1496
	Steam	305	588	74	1667
	Steam	370	549	103	2387

* Flow orifice overranged.
** Fluid-energy mill internals were changed after the Texas lignite (and one North Dakota Lignite) tests causing a different system pressure drop.
*** Internals same as with Texas lignite.

Feedstock samples and product samples of the micropulverized lignite were taken during each run. These were collected and sealed in air tight cans for subsequent moisture determinations. The resultant product moisture for the various conditions of operation are shown in Fig. 3.

The particle size distribution was determined for each run using a Coulter Counter model TaII particle analyzer. The resultant average mean particle size based on volume of the products for the various coals is shown in Fig. 4. The two unusually high particle sizes for the Texas lignite were for conditions of operation with greater lignite flow rate than the design condition for the mill.

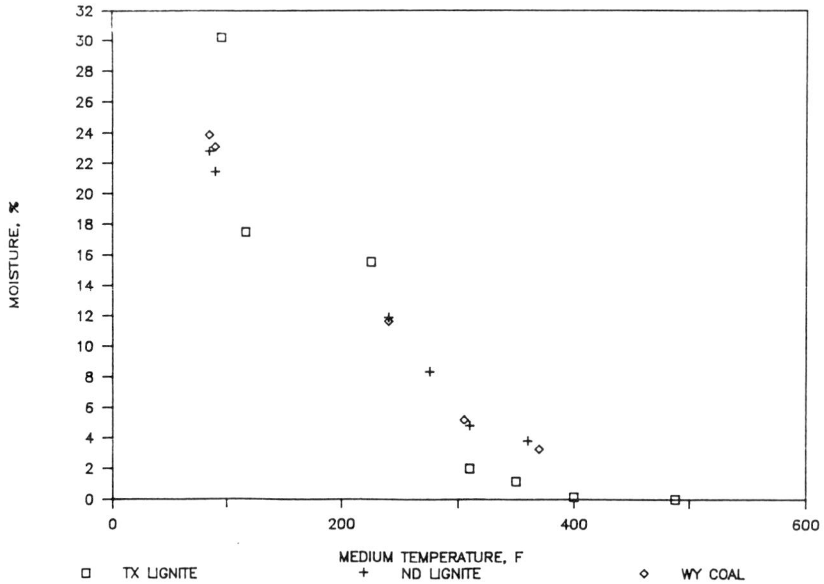

FIG. 3. Sample moisture as a function of in the mill temperature.

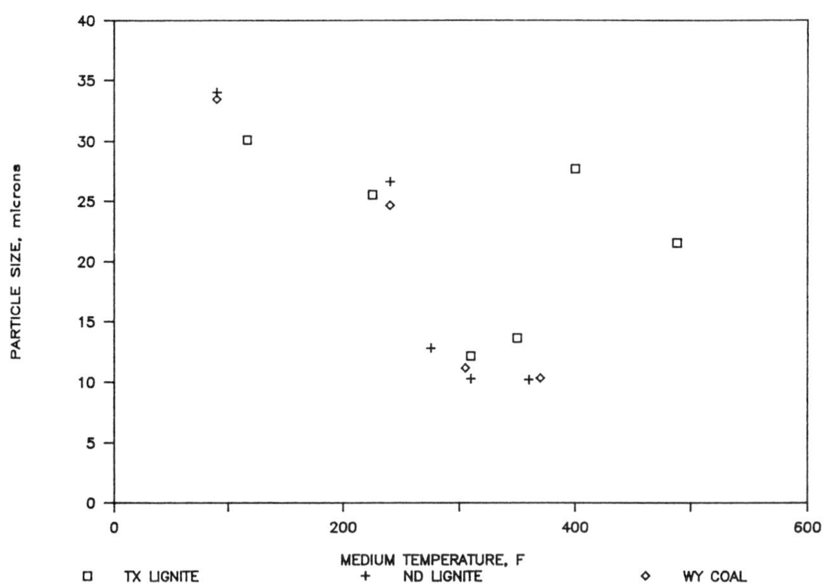

FIG. 4. Mean volume particle size as a function of in the mill temperature.

POWER REQUIREMENTS FOR ULTRAFINE GRINDING AND DRYING OF LIGNITE IN A FLUID-ENERGY MILL

During the tests on low-rank coals, data were collected in order to determine the energy consumption requirements for the ultrafine grinding. The flow rate of the air or steam grinding fluid was determined from the orifice plate pressure drop, the fluid density, and the orifice area and conversion constant. The density for the air runs was calculated from the ideal gas relationship using the measured temperature and pressure at the orifice plate. The density for the steam runs was calculated from a correlation for superheated steam specific volume [10] which is also a function of temperature and pressure. The lignite feed rate was determined from the calibration curve for the lignite mass flow rate versus feed belt (or screw) speed.

Samples of the feedstock and products for each run were analyzed for moisture in order to determine the amount of water driven out of the lignite. The flow rates of the product lignite (solid + final water in lignite) and the moisture removed from the lignite were determined by the following relationships:

$$\dot{m}_{LP} = \dot{m}_{LF} \left[\frac{1 - \eta_f}{1 - \eta_p}\right] \qquad (1)$$

and

$$\dot{m}_{WV} = \dot{m}_{LF} \left[\frac{\eta_f - \eta_p}{1 - \eta_p}\right] \qquad (2)$$

where $\dot{m}_{LF}$ is the lignite feed flow rate, $\dot{m}_{LP}$ is the lignite product flow rate, $\dot{m}_{WV}$ is the flow rate of water vapor driven out of the lignite, η_f is the moisture fraction of feed, and η_p is the moisture fraction of product.

An energy balance on the fluid-energy mill is illustrated in Fig. 5. The energy lost by the grinding fluid minus the heat loss from the system is equal to the energy gained by both the lignite and the water vapor driven out of the lignite:

$$\dot{m}_F(h_{F1} - h_{F2}) - Q_{HL} = \dot{m}_{LP}(h_{L2} - h_{L1}) + \dot{m}_{WV}(h_{WV2} - h_{WV1}) \qquad (3)$$

For the grinding fluid the loss of enthalpy was determined from

$$\Delta h_{fluid} = \int_{T_1}^{T_2} C_p dT \qquad (4)$$

where T_1 and T_2 are the inlet and outlet temperatures and Cp is the specific heat. For air Cp was taken as 0.24 Btu/lbm°R. For the superheated steam, a correlation for Cp was used [11] where

$$Cp_{steam} = \left[19.86 - \frac{597}{\sqrt{T}} - \frac{7500}{T}\right]/18.016 \qquad (5)$$

with T in this expression in °R. Over the range of temperatures used in the test, this correlation gave a Δh value within 1.0% of tabulated values [10].

For the lignite, the gain in enthalpy was determined from

$$\Delta h_{lignite} = Cp_L(T_2 - T_L) \qquad (6)$$

where T_L is the temperature of the feedstock (90 to 100°F) and Cp_L equals 0.37 Btu/lbm°R [12]. Here it was assumed that the ground lignite and grinding fluid leave at the same temperature. For the water vapor driven out of the lignite, the change in enthalpy was determined from

$$\Delta h_{\substack{water\\vapor}} = h_{\substack{saturated\ vapor\\at\ 672°R\ and\\14.7\ psia}} - h_{\substack{liquid\ water\\at\ T_L}} + \int_{672°R}^{T_2} Cp_{steam}dT \qquad (7)$$

where the saturated vapor enthalpy at 672°R and 14.7 psia is 1150.5 Btu/lbm and

$$h_{liquid\ water\ at\ T_L} = T_L - 32 \qquad (8)$$

The above expression has the units Btu/lbm with T_L in °F.

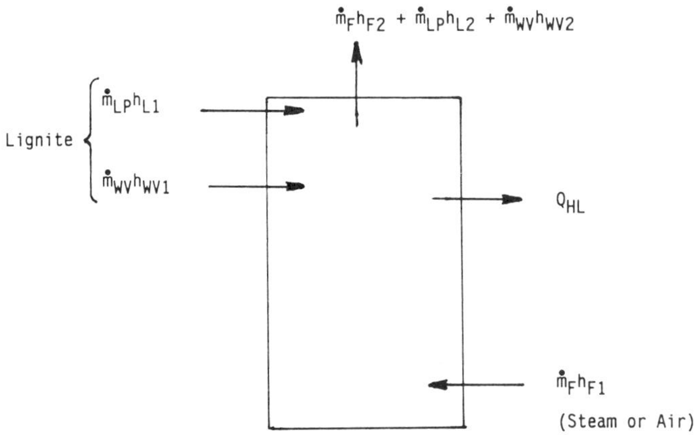

$\dot{m}_F$ = Grinding fluid flow rate
$\dot{m}_{LP}$ = Lignite product flow rate
$\dot{m}_{WV}$ = Flow rate of water vapor driven out of the lignite
h_F = Grinding fluid enthalpy
h_L = Lignite product enthalpy
h_{WV} = Enthalpy of water driven out of the lignite
Q_{HL} = Heat loss from system

FIG. 5. Energy balance on fluid-energy mill.

With the above expressions it was possible to determine the power required to dry and grind the lignite in the fluid-energy mill. It should be noted that there is no way to separate the grinding power from the drying power since the grinding process creates heat which goes into heating the lignite and the water vapor. The left-hand side of equation (3) is the power required to dry and grind the lignite as determined from measurements on the grinding fluid and from the heat loss. This quantity served as the primary calculation of the power requirements.

The grinding/drying power values for each low-rank coal are given in Tables II-IV. These results are expressed in KWH/Ton and are the ratio of the power requirements divided by the lignite feed flow rate. Also given in Tables II-IV are the energy-per-ton values determined from the lignite feed and product moistures and temperatures (the right hand side of equation (3)). These second calculations served as a balance check on the grinding/drying power measurements.

TABLE II. Power requirements to grind and dry Martin Lake Texas lignite in a fluid-energy mill.

Test Condition	Power Requirement/Lignite Feed Rate (Calculated from Fluid) KWH/Ton	(Calculated from Lignite) KWH/Ton	Percent Comparison
116°F A.G.*	24	--	--
225°F S.G.**	95	129	-37.8
310°F S.G.	208	217	- 4.3
350°F S.G.	212	227	- 7.1
400°F S.G.	229	237	- 3.5
488°F S.G.	***	259	--

* Air Ground ** Steam Ground
*** Fluid flow orifice pressure drop transducer was overranged.

TABLE III. Power requirements to grind and dry Beulah North Dakota lignite in a fluid-energy mill.

Test Condition	Power Requirement/Lignite Feed Rate (Calculated from Fluid) KWH/Ton	(Calculated from Lignite) KWH/Ton	Percent Comparison
90°F A.G.*	21	--	--
240°F S.G.**	119	104	12.6
275°F S.G.	191	135	29.3
310°F S.G.	156	166	-6.4
360°F S.G.	247	185	25.1

* Air Ground ** Steam Ground

TABLE IV. Power requirements to grind and dry Eagle Butte Wyoming subbituminous coal in a fluid-energy mill.

Test Condition	Power Requirement/Lignite Feed Rate (Calculated from Fluid) KWH/Ton	(Calculated from Lignite) KWH/Ton	Percent Comparison
90°F A.G.*	12.5	--	--
240°F S.G.**	113	105	7.1
305°F S.G.	132	171	-29.5
370°F S.G.	229	188	17.9

* Air Ground ** Steam Ground

The comparison between the KWH/ton calculated from the steam and from the lignite data varies from 4% to 39%. While this comparison is high in some cases it is intended only as a check on power requirement calculated from the fluid which was significantly more accurate than that calculated from the lignite. For the air ground condition, the exit was below saturation and no calculations could be made from the lignite data since the model used to calculate the change in enthalpy of the water removed from the lignite (equation (7)) assumes that a superheated steam exists at the exit.

The average grinding and drying power versus mill temperature is shown in Fig. 6 for all three low-rank coals. The trend is a linear increase in power required with grinding medium temperature. Grinding with ambient air used about 20 KWH/Ton. Steam grinding at about 500°F mill temperature required about 260 KWH/ton.

It should be stressed that these power data only account for the process inside the fluid-energy mill. They do not include the compressor/boiler efficiency, the transmission losses, and the other inefficiencies that are part of the entire grinding circuit. Also, at the higher temperature grinding conditions, there is still significant energy in the exit steam. For an economical process, this hot exhaust would have to be used to preheat the boiler water in order to reduce the overall system losses.

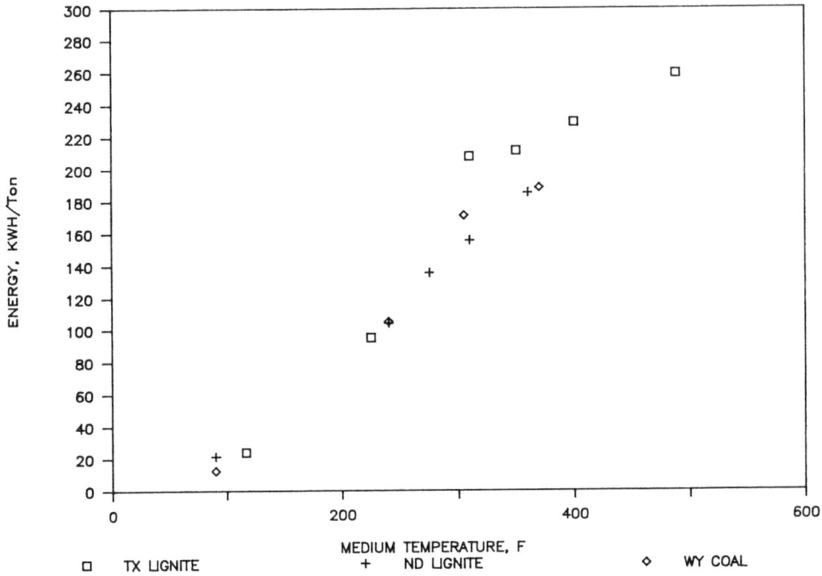

FIG 6. Process power required for simultaneous grinding/drying as a function of in the mill temperature.

Previous grinding power requirements data for the North Dakota lignite (Beulah Mine) were reported by Ellman, et. al. [13]. In that study, the pulverizer was a hammermill-type equipped with a classifier. The net mechanical power input to the pulverizer was determined for various in-the-mill drying temperatures ranging from about 70°F to 650°F. The power consumption value measured was the pulverization power requirements only. The report did not include sufficient data to determine the power requirements for the drying process. Therefore, a comparison of the power data from Ellman, et. al. [13] and the power data presented here can be made only for the air ground condition where minimum drying occurred.

For the hammermill tests [13], the net mechanical power consumption was about 40 KWH/Ton for ambient temperature sweeping air with a product fineness of about 59% passing a 200-mesh sieve. In the test results of the present study for the fluid-energy mill grinding, the power consumption was 21 KWH/Ton for an ambient temperature air grinding medium. The product fineness for the fluid-energy mill grinding with air was about 75% of the material less than 74 microns (200 mesh) on a volume basis. For the air grinding runs the product particle sizes were larger than those in the steam runs. (see Fig. 4.)

Care must be exercised in comparing the above power consumptions for the hammermill and the fluid-energy mill. For the hammermill, the reported power is that required to turn the pulverizer. For the fluid-energy mill, the power reported is the net power consumed in the mill which includes the mechanical work done on the lignite and the reheating of the fluid.

CONCLUSIONS

A jet-vortex type fluid-energy mill was successfully used to grind low-rank coals to ultrafine sizes with simultaneous permanent drying. The power required to grind and dry the coals was determined from temperature, pressure, and flow rate measurements made during the tests. The power requirements increased linearly with grinding medium temperature.

ACKNOWLEDGMENTS

The authors would like to acknowledge the support of the U.S. Department of Energy for this project (Contract No. AC21-84FC10622) with special thanks to Mr. Leland E. Paulson for his advice and assistance.

REFERENCES

1. A.E. Margulies, G.F. Moore, and R.J. West, "Ultrafine Pulverized Coal as an Alternate Fuel for Oil and Gas Fired Boilers," 9th Energy Technology Conference, Washington, DC, February, 1982.
2. R.J. West, G. Haider, T.A. Morris, A.E. Margulies, and G.F. Moore, "Potential of Micronized Coal as an Alternate Fuel in Oil- and Gas-Fired Boilers," International Symposium on Conversion to Solid Fuels, Newport Beach, CA, October 26-28, 1982.
3. L. Nall, "Coal/Steam Burning Status Report," Southeastern Electric Exchange 1984 Conference, Bal Harbour, FL, April 12, 1984.
4. J.L. Hartness, and M.M. Koeroghlian, "Dry Micronized Coal Applications Progress Report on a Utility Station Gas-Fired Boiler," 7th International Coal and Lignite Conference and Exhibition, Houston, TX, November 13-15, 1984.
5. J.J. Kowalczewski, P.D. Bandopandhayay, R.J. Downie, and W.R. Read, "Ultra Fine Grinding of Australian Brown Coal for Use in Diesel Engines," Fifth International Symposium on Coal Slurry Combustion and Technology, Tampa, FL, April 25-27, 1983.
6. W.G. Steele, C.W. Bouchillon, J.A. Clippard, and J.L. Hartness, "Evaluation of the Micropulverization of Mississippi Lignite," Twelfth Biennial International Meeting on Technology and Utilization of Low-Rank Coals, Grand Forks, ND, May 18-19, 1983.
7. C.W. Bouchillon, W.G. Steele, J.A. Clippard, and J.D. Burnett, "Evaluation of the Micropulverization, Drying and Beneficiation of Lignite," 6th International Coal and Lignite Utilization Exhibition and Conference, Houston, TX, November 15-17, 1983.

8. W.G. Steele, C.W. Bouchillon, and R.B. Ross, "Beneficiation Processes for Ultrafinely Ground Low Rank Coals," Thirteenth Biennial Lignite Symposium on Technology and Use of Low-Rank Coals, Bismarck, ND, May 21-23, 1985.
9. D.W. Taylor, "Comminution of Pulverulent Material by Fluid Energy," U. S. Patent Number 4,219,164, August 1984.
10. J.H. Keenan, and F.G. Keyes, Thermodynamic Properties of Steam, John Wiley and Sons, Inc., New York, 1936.
11. R.E. Sonntag, and G.J. Van Wylen, Introduction to Thermodynamics: Classical and Statistical, John Wiley and Sons, Inc., New York, 1971.
12. B. Stanmore, D.N. Baraia, and L.E. Paulson, "Steam Drying of Lignite--A Review of Processes and Performance," Report No. DOE/GFETC/RI-82/1, Grand Forks Energy Technology Center, Grand Forks, ND, 1982.
13. R.C. Ellman, J.W. Belter, and L. Dockter, "Effects of In-the-Mill Drying on Pulverization Characteristics of Lignite," Bureau of Mines Report of Investigations, 6074, U. S. Department of the Interior, Bureau of Mines, 1962.

PROBLEMS INHERENT IN USING THE POPULATION BALANCE MODEL FOR WET GRINDING IN BALL MILLS

T. P. MELOY*, M. C. WILLIAMS* AND P. C. KAPUR**
*Particle Analysis Center, WVU, Mgtn., WV 26505, USA
**IIT Kanpur, 208016, India

ABSTRACT

A stuffing matrix **S** is the PBM solution to any Inverse Problem for wet grinding with straight line product curves observed for some types of wet ball mills. This PBM solution assumes all particles are broken out of their own size class. In other words, if one starts the mill, no matter when it is stopped, every particle in every class is a new particle. This corresponds to no known physical situation. Dry grinding feed matrices can all be inverted, because the product size distributions all curve to the right in the larger sizes showing that not all the larger particles are broken in a finite length of time. For parallel straight line feed and product size distributions, the use of a simple function for the product distribution is proposed but not described.

INTRODUCTION

For a quarter of a century, the population balance model, PBM, for comminution in ball mills has been solved using matrix methods (Broadbent and Calcott 1960, Meloy and Gaudin 1962, Reid 1965, Mika 1970, Herbst and Mika 1970, Klimpel and Austin 1970, Austin 1971, Malgan and Fuerstenau 1976, Herbst et al. 1971, Fuerstenau et al. 1984). In theory, selection and breakage functions are created and combined into a mill matrix. This mill matrix is multiplied (on the right side) by the feed matrix which describes the particle size distribution entering the mill. This multiplication yields the product matrix which describes the size distribution of the particles exiting the mill. In practice one creates the mill matrix by measuring the size distribution of the feed and product, then calculating the mill matrix by various methods. This is referred to as solving the Inverse Problem. Underlying all numerical solutions to the Inverse Problem is the assumption that the mill matrix is independent either of time or the size distribution of the feed. As will be shown, for some wet grinding ball mill simulations, this assumption of linearity is not valid and the matrix methods of the PBM cannot be used.

In wet grinding, plots of the product from the mill (Fuerstenau & Sullivan, 1961; Agar & Charles, 1961) often turn out to be series of straight lines when plotted as a log cumulative finer, M(x), versus log particle size, x (See Figure 1). It is in this case, when the plot of the mill product is a straight line, that the linearity assumption and the matrix methods are both mathematically and physically incorrect. Since this type of straight line product is common in wet grinding, the phenomena is not a quirk, and, if the

Advances in Fine Particles Processing
John Hanna and Yosry A. Attia, Editors

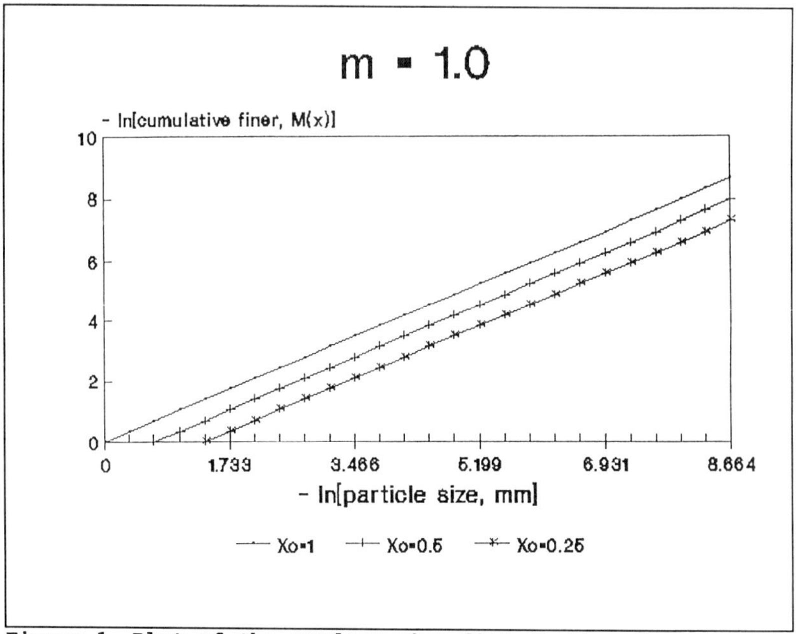

Figure 1 Plot of the product size distribution from a wet grinding mill showing the straight-line breakage at various times showing that the topsize (largest particle) in the mill changes from 1.00 to 0.25 millimeters (Slope m is 1).

grinding time was either more or less, one would still get a straight line product.

If, when sampling the product from a wet ball mill, there are two straight line plots for the product at two different grinding times, the larger size distribution of the earlier sampling time can be considered to be the feed for the smaller size product of the later sampling time. If one looks at the difference in the top size particles between any two straight lines, one observes that all the particles in the top size range have been completely ground away - all the top size particles are gone. This means the probability of a particle in this largest top size fraction being broken is identically one, not approximately one. No matter how close together the straight line plots for the mill product are, the probability that the largest particle will be broken is one. Put another way, the largest particle size is broken out of its class in an infinitely short time interval, that is, instantaneously. This is physically unrealistic. Moreover, if the largest particle is always broken instantaneously, the probabilities of a particles being broken are dependent on the size distribution in the mill, a violation of the assumption of linearity.

These observations are not trivial and may explain the difficulty experienced in developing a body of knowledge about selection and breakage functions that is transferable from mill to mill - the way heat transfer coefficients were developed in the early in the study of heat transport theory. This comminution information has not yet appeared in the literature.

USE OF THE PBM FOR WET GRINDING

The typical way to look at the problem of simulating wet ball mill operation is the PBM. When the plot of the product from a mill is a straight line, the method of solution is to use matrix algebra to find the mill matrix, **M**, from idealized feed, **F**, and product, **P**, matrices, that is, to solve the Inverse Problem. The relationship between **M**, **F** and **P** is:

$$1)\quad \mathbf{M}\,\mathbf{F} = \mathbf{P}$$

M is a sparse, square, n by n, steady-state, stochastic matrix; n is the number of particle size classes into which the particle feed has been arbitrarily divided. The top particle size is designated one and the bottom size is designated size n. **F** and **P** are n by one, column matrices. The elements of **M**, the m_{ij} of row i and column j, are all fractions, less than one. The value of m_{ij} is the probability of a particle in the j-th particle size class being broken into the i-th particle class. In the absence of any agglomeration phenomenon, m_{ij} is zero for all j > i. In addition, conservation of mass requires that:

$$2)\quad \sum_{i=1}^{n} m_{ij} = 1.0 \quad \text{for each } j$$

Hence, m_{nn} must be one.

A typical five by five, mill matrix is:

$$(3)\quad \mathbf{M} = \begin{matrix} m_{11} & m_{12} & m_{13} & m_{14} & m_{15} \\ m_{21} & m_{22} & m_{23} & m_{24} & m_{25} \\ m_{31} & m_{32} & m_{33} & m_{34} & m_{35} \\ m_{41} & m_{42} & m_{43} & m_{44} & m_{45} \\ m_{51} & m_{52} & m_{53} & m_{54} & m_{55} \end{matrix} = \begin{matrix} 0.30 & 0.00 & 0.00 & 0.0 & 0.0 \\ 0.30 & 0.30 & 0.00 & 0.0 & 0.0 \\ 0.15 & 0.30 & 0.40 & 0.0 & 0.0 \\ 0.15 & 0.20 & 0.40 & 0.5 & 0.0 \\ 0.10 & 0.20 & 0.20 & 0.5 & 1.0 \end{matrix}$$

A typical feed distribution with slope of unity is given by:

$$4)\quad \mathbf{F} = \begin{matrix} f_1 \\ f_2 \\ f_3 \\ f_4 \\ f_5 \end{matrix} = \begin{matrix} 0.516 \\ 0.258 \\ 0.129 \\ 0.064 \\ 0.033 \end{matrix}$$

For the case when **M** and **F** are known, the product matrix by matrix algebra is:

$$5)\quad \mathbf{P} = \mathbf{M}\,\mathbf{F} = \begin{matrix} p_1 \\ p_2 \\ p_3 \\ p_4 \\ p_5 \end{matrix} = \begin{matrix} 0.155 \\ 0.232 \\ 0.206 \\ 0.213 \\ 0.194 \end{matrix}$$

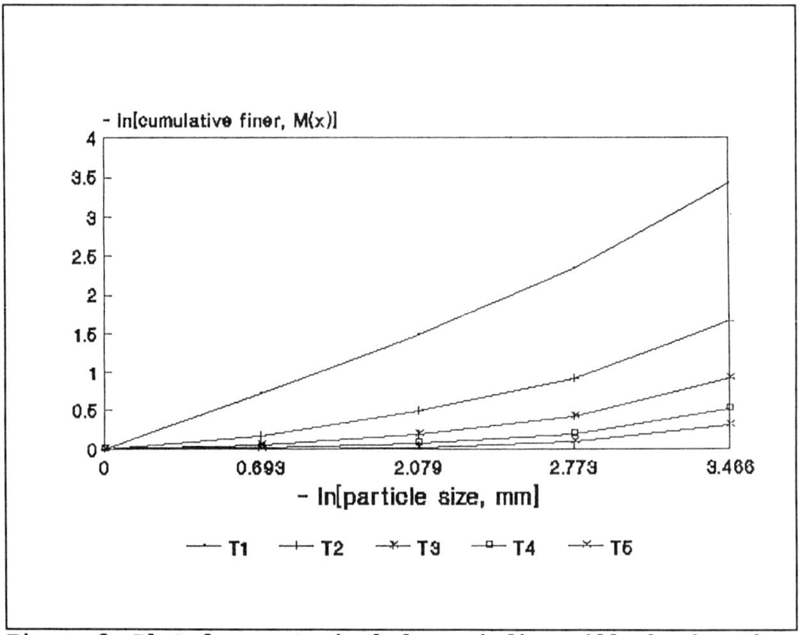

Figure 2 Plot from a typical dry grinding mill showing the breakage at various times during which the topsize in the mill does not change.

Figure 2 shows the mill matrix of Equation 3 operating of the feed five times in succession. Note that some product exits in the topsize at all times.

For the case that **P** and **F** are known, the problem (Inverse Problem) is to find **M**. For the case of wet grinding in a ball mill where all the top size is ground away during a small time interval, m_{11} and p_1 must both be zero. With these constraints, the number of unknowns in the mill matrix is reduced. The mill matrix becomes:

$$6) \quad \mathbf{M} = \begin{matrix} 0.0 & 0.0 & 0.0 & 0.0 & 0.0 \\ m_{21} & m_{22} & 0.0 & 0.0 & 0.0 \\ m_{31} & m_{32} & m_{33} & 0.0 & 0.0 \\ m_{41} & m_{42} & m_{43} & m_{44} & 0.00 \\ m_{51} & m_{52} & m_{53} & m_{54} & m_{55} \end{matrix}$$

There are fourteen unknowns in the mill matrix. By performing the matrix multiplication of **M** and **F**, four equations relating the p_j to the m_{ij} and f_j may be developed. They are:

$$7) \quad p_2 = \sum_{j=1}^{5} m_{2j} f_j$$

8) $p_3 = \sum_{j=1}^{5} m_{3j} f_j$

9) $p_4 = \sum_{j=1}^{5} m_{4j} f_j$

10) $p_5 = \sum_{j=1}^{5} m_{5j} f_j$

From the conservation of mass described in Equation 2, five additional equations can be developed. These are:

11) $\sum_{i=1}^{5} m_{i1} = 1.0$

12) $\sum_{i=1}^{5} m_{i2} = 1.0$

13) $\sum_{i=1}^{5} m_{i3} = 1.0$

14) $\sum_{i=1}^{5} m_{i4} = 1.0$

15) $\sum_{i=1}^{5} m_{i5} = 1.0$

The result is fourteen unknowns but only nine equations. The consequence is the solution is degenerate or underspecified, that is, there is **no unique solution** to the Inverse Problem unless additional information is provided or further assumptions are made.

The nonuniqueness of solution is not the only problem inherent in the solution of PBMs. An even more fundamental problem inherent in the PBM solution is the assumption that the elements of **M** are invariant with respect to time and composition. The problem is that m_{22}, m_{33} and m_{44} are changing during the wet grinding time. At some time during the wet grinding they must become zero. However, initially they are not zero. Not even m_{11} is initially zero. Thus, the m_{ij} are functions of the mill conditions such as mill loading and/or particle size distribution. A new mill matrix must be developed for **each** grinding time interval. This fact obviously contradicts the basis for the assumptions which are used to justify the constancy and interrelatedness of the breakage and selectivity functions used to generate mill matrices, such as the Arbiter-Bhrany relationship (1960).

It will become apparent now that no mill matrix PBM solution will be valid for the entire grinding time. Consider two possible PBM solution types, the two matrices, **R** and **S**:

$$16)\quad \mathbf{R} = \begin{matrix} 0.00 & 0.00 & 0.00 & 0.00 & 0.00 \\ 0.53 & 0.00 & 0.00 & 0.00 & 0.00 \\ 0.27 & 0.57 & 0.00 & 0.00 & 0.00 \\ 0.13 & 0.29 & 0.67 & 0.00 & 0.00 \\ 0.07 & 0.14 & 0.33 & 1.00 & 1.00 \end{matrix}$$

$$17)\quad \mathbf{S} = \begin{matrix} 0.00 & 0.00 & 0.00 & 0.00 & 0.00 \\ 1.00 & 0.00 & 0.00 & 0.00 & 0.00 \\ 0.00 & 1.00 & 0.00 & 0.00 & 0.00 \\ 0.00 & 0.00 & 1.00 & 0.00 & 0.00 \\ 0.00 & 0.00 & 0.00 & 1.00 & 0.00 \end{matrix}$$

The first solution, **R**, represents a legitimate mill matrix which breaks everything in the upper size range and the second, known as the stuffing matrix, yields straight line products for straight line feeds, but with the unrealistic result that the new surface area created during grinding is proportional to the square of the grinding time.

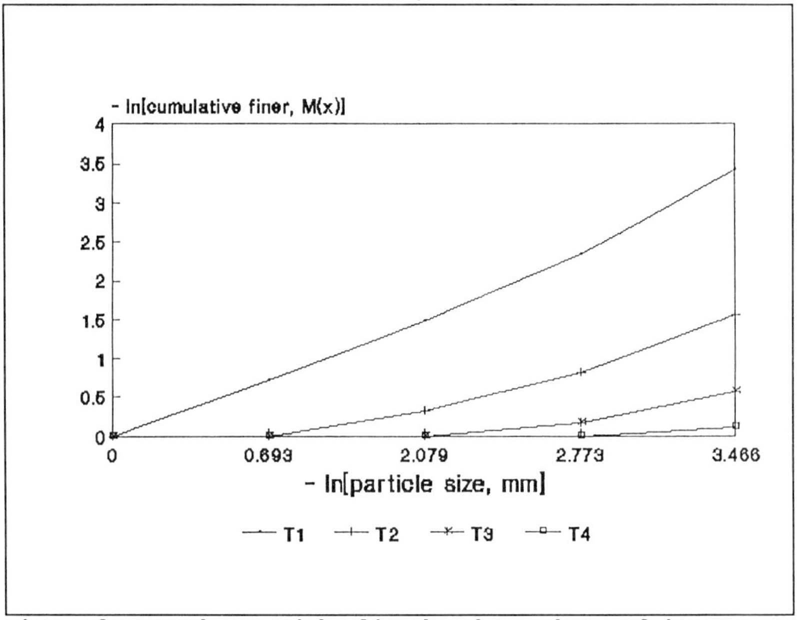

Figure 3 The straight-line breakage observed in wet grinding mills may be simulated using the Mill Matrix "R".

Refer to the first of these two matrices. Along the major diagonal, left to lower right, each term is zero except for the m_{55} position, corresponding to the physical fact that everything in the top size fraction is broken - a requirement if the product is to be a series of straight lines. This corresponds to m_{22}, m_{33} and m_{44} being zero. These particles are assumed to be broken into a size distribution that is a straight line of the same slope as the feed, the ideal case. (In the example used the slope is one.) When the mill matrix **R** is multiplied with the straight line feed, the product is curved to the right - not a straight line as required. See

Figure 3. When forming the mill matrix, regardless of what the breakage function looks like, the resulting mill matrix, when multiplied by a straight line feed matrix, results in a product size distribution curved to the right as long as the particles are broken into a continuous size distribution.

Consider now the matrix second solution, **S**. Once again the left to lower right diagonal is zero, except for the m_{55} position, signifying that all the largest particles are broken. The diagonal below the principle diagonal is all one's, meaning that the material in the size fraction above is being broken into the size fraction below, clearly physically impossible because particles always break into size distributions. When for a given grinding time this artificial matrix is post multiplied with the feed matrix, it yields the straight line product size distribution. See Figure 4. This matrix is the solution using the PBM approach. If this matrix is used as a mill matrix, then the average particle size is the original particle size divided by the grinding time, which means the new surface area created is proportional to the square of the grinding time - obviously desirable but physically impossible.

This means, both physically and mathematically, that the use of linearity to compute the mill matrix results in a matrix that: 1) is fallacious, 2) is a non unique curve fitted mill matrix (there are many matrices that will yield the same results), and 3) breaks down for variations in power

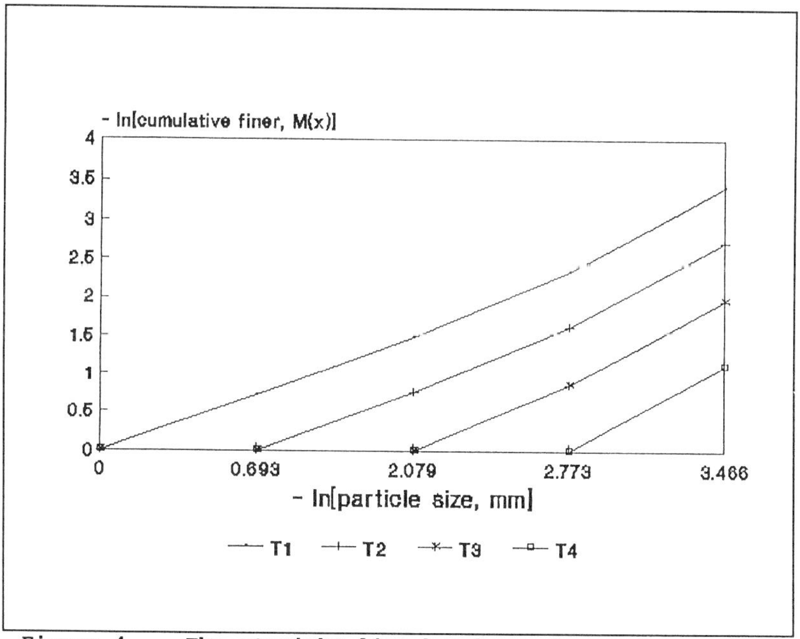

Figure 4 The straight-line breakage observed in wet grinding Mills may be simulated by using the Stuffing Matrix "S".

input, feed input, grinding time or any other operating variation. While dry grinding feed matrices can all be inverted - because the product size distributions all curve to the right in the larger sizes showing that not all the larger particles are broken in a finite length of time - for wet grinding with straight line product curves, realistic matrices for simulation of wet grinding **do not exist**.

It was while the first author was looking for a simple transfer function to represent the comminution unit operation for use in the development of circuit analysis techniques (Meloy 1983 and Williams and Meloy 1983) that the above described anomaly was discovered. For parallel straight line feed and product size distributions, a very simple transfer function can be developed. See Figure 5. (After further refinement, the equation will be published is a subsequent paper.) This new transfer fuction is too simple a solution to warrant the use of the matrices requiring hundreds of constants. In the attempt to determine what this simple transfer function meant, the above described problems with the matrix PBM approach were discovered.

In addition to aiding in discovering the problems inherent in PBM's, this simple transfer function is useful in its own

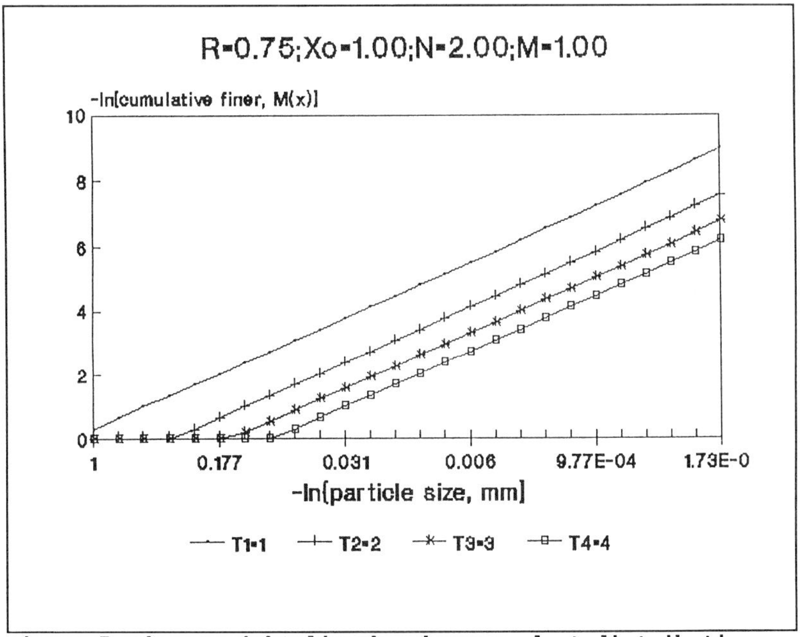

Figure 5 The straight-line breakage product distribution, observed in wet grinding mills, may be modeled using a simple equation.

right and may lead, not only to a simple method of computing products in wet ball mills, but it may also be directly related to energy or operating conditions required to achieve a degree of size reduction in a mill. In short, this serendipitous finding of the comminution transfer function may lead to a far more accurate method of analyzing ball mills and other types of grinding circuits with a simpler mathematical form.

CONCLUSION

However, unrealistic, the stuffing matrix **S** is obviously a PBM solution to any Inverse Problem of the wet grinding where the product size distribution is a straight or nearly straight line size. This PBM solution to the wet ball mill problem assumes, no matter how short the ginding time, that all particles are broken out of their own size class. In other words, even if one grinds for an infinitesimal short time, every particle in a given size class is a new particle; that is, no particles in any size range remain unbroken irrespective of how short the grinding time. This corresponds to **no known physical process.**

While all dry grinding feed matrices can be inverted - because the product size distributions all curve to the right in the larger sizes showing that not all the larger particles are broken in a finite length of time - for wet grinding with straight line product curves, realistic **matrices cannot be used for simulation.**

For parallel straight line feed and product size distributions, a simple transfer functions may be developed that simulates wet grinding and may be used for control stratagies. This new approach is too simple a solution to warrant the use of the matrices which require determining many constants.

REFERENCES

Agar, G., and Charles, R., 1962, "Size Distribution Shift in Grinding", AIME Trans., 223, pp. 390-395.

Arbiter, N., and Bhrany, U., 1960, "Product Size, Power and Capacity in Tumbling Mills", AIME Trans., 217, pp. 245-252.

Austin, L.G., "A Review Introduction to the Mathematical Descriptions of Grinding", Powder Technology, 5, pp. 1-17 (1971).

Broadbent, S.R. and Calcott, T.B., "Coal Breakage Processes II: A Matrix Representation of Breakage", J.INST. Fuel, 29, pp. 258-265 (1960).

Fuerstenau, D.W., Sullivan, D.A., 1961, Size Distribution and Energy Conservation in Grinding Mills " AIME Trans., 220, pp. 397 402.

Fuerstenau, D.W., K.S. Venkataraman, and M.C. Williams, "Simulation of Mill Dynamics of Grinding Mixtures Using PBMs," in Control'84, AIME, New York, 1984.

Herbst, J.A. and Mika, T.S., "Mathematical Simulation of Tumbling Mill Grinding: An Improved Method". Rudy, 18, pp.75-80 (1970).

Herbst, J.A., Gandy, G.A. and Mika, T.S., "On the Development and Use of Lumped Parameter Models for Continuous Open- and Closed-Circuit Grinding Systems," Inst.Min.Metall.Trans, 80, C.193-198 (1971).

Klimpel, R.R. and Austin, L.G., "Determination of Selection-for-Breakage Functions in the Batch Grinding Equations by Nonlinear Optimization," Ind.Eng.Chem.Fund., 9, pp.230-237 (1970).

Malghan, S.G. and Fuerstenau, D.W., "The Scale-up of Ball Mills Using Population Balance Models and Specific Power Input," in Zerkleinern.DECHEMA-Monogr., 79(II), No. 1586, pp. 613-30 (1976).

Meloy, T.P. and Gaudin, A.M., "Model and Comminution Distribution Equation for Repeated Fracture", Trans. AIME, 223, pp.243-50 (1962).

Meloy, T.P., "Analysis and Optimization of Mineral Processing and Coal Cleaning Circuits - Circuit Analysis", IJMP, 10, pp. 61-80 (1983).

Mika, T.S., Population Balance Models of a Continuous Grinding Mill as a Distributed Process, Diss., Univ. Calif., Berkeley, Ca., 424 pp. (1970).

Reid, K.J., "A Solution to the Batch Grinding Equation," Chem.Eng.Sci., 20, pp. 953-963 (1965).

Williams, M.C. and Meloy, T.P., "Dynamic Model of Flotation Cell Banks - Circuit Analysis", IJMP, 10, pp. 141-160 (1983).

CORRELATION OF ADSORPTION OF SURFACTANTS WITH FRACTURE AND GRINDING OF QUARTZ

HASSAN EL-SHALL* AND P. SOMASUNDARAN**
*Florida Institute of Phosphate Research, Bartow, FL; **Henry Krumb School of Mines, Columbia University, New York City, NY

ABSTRACT

Fracture of quartz by single impact as well as wet ball milling is studied in presence of different surfactants (dodecylammonium chloride), and inorganic electrolytes ($AlCl_3$, $CaCl_2$). The data show an increase or decrease in fracture efficiency depending on parameters such as pH, type and concentration of added surfactant. Adsorption of these chemicals as indicated by their effect on surface properties such as zeta potential, and pulp characteristics including flocculation and fluidity, is correlated with the ionic species distribution, and fracture and grinding results. Such correlation shows, in general, that fracture efficiency is higher under pH conditions where surface active species are dominant.

INTRODUCTION

Adsorption of surfactants on solids can lead to significant changes in surface and interfacial properties such as zeta potential, surface energy, surface hardness, friction, stability (flocculation/ dispersion) of solid suspensions, etc. Such effects, most probably will influence the performance of processes involving the surfactant/solid systems. For instance, grinding efficiency has been reported to decrease or increase due to addition of chemicals [1-10]. Investigators have attributed the obtained data to changes in surface energy of the solid [11], or to changes in surface hardness of the ground material [12]. Other reports [8-10] suggest changes in the flow properties of the pulp to be responsible for the observed effects.

Grinding is, however, an integral process involving several simultaneous subprocesses such as transport of particles to the grinding zone, application of different kinds of stresses on the particles, and transport of the ground products to the discharge end of the mill. Evidently, such subprocesses depend on mechanical properties of the solid e.g. hardness, tensile and compressive strength, etc., as well as pulp properties such as fluidity and state of flocculation or dispersion.

Adsorption of surfactants can alter the strength properties of the solids and this becomes clear by examining the minimum stress required for fracture as given by Griffith's Law [13]: $\sigma = \left(\frac{4E\gamma}{L} \right)^{\frac{1}{2}}$

where

E	=	Young's modulus
γ	=	Free energy of created surface
L	=	Crack length

Published 1990 by Elsevier Science Publishing Co., Inc.
Advances in Fine Particles Processing
John Hanna and Yosry A. Attia, Editors

Thus, any changes in surface energy and/or the crack length resulting from interaction of the solid with surfactants, are expected to modify the strength of the particles. Another important parameter that might influence the strength of materials is the radius of the crack tip. The sharper the crack tip the higher the localized stress concentration, and consequently, the strength of material is lower. It is important to note here that surfactants may react with the cracked surfaces leading probably to changes in the radius of the crack, and as a result, in the strength of the particles.

Griffith's Law describes the brittle fracture of solids. However, most solids have defects such as dislocations that can move under stress leading to a plastic behavior and consequently higher stresses are needed to produce fracture. Thus, if the dislocation movement is pinned due to adsorption of surfactants, the solid should become brittle and fracture occurs at lower levels of stresses.

Regarding transport of particles inside the grinding mill, surfactants can influence the friction characteristics of particles and the grinding media. In other words, chemicals may affect the flow characteristics of the mill contents including the fluidity of pulps containing high percent solids. In addition, adsorption of surfactants might lead to changes in flocculation/dispersion properties of the suspension and consequently, the breakage efficiency. For instance, fine particles might coagulate with the larger ones and/or coat the grinding media leading to decrease in breakage of the coarse particles due to the cushioning effect. In this paper, changes in solid and pulp properties due to surfactant(s) adsorption, are discussed and correlated.

EXPERIMENTAL

Details of the experimental procedure, equipment and materials are described in previous papers (3-7). Nevertheless, a summary is provided below.

Materials and Chemicals

Dodecylammonium chloride (DDACl), $AlCl_3$, and $CaCl_2$ were used as additives during testing fracture, zeta potential, flocculation/dispersion, and fluidity of suspensions containing highly crystalline quartz of 99.9% purity. KNO_3 or NaCl was used to adjust ionic strength, and NaOH or HCl was added to adjust the pH. $FeCl_3$ or $Fe(NO_3)$ was used to simulate the iron released from the mill during grinding.

Equipment and Methods

Grinding tests were carried out in 7.5 inch diameter and 8.0 in length stainless steel ball mill at two different operating conditions. Testing of amine was done under conditions of 45% mill volume of ball loading with 100% void filling with pulp of 67% solids by weight and the mill was running at 78% of the critical speed.

Grinding in $AlCl_3$ and $CaCl_2$ was performed in the same mill using 20% ball loading, 50% void filling, 65% pulp density, and 59% of the critical speed.

Fracture testing was done using a dropweight mill by allowing a hardened steel mass to fall from a fixed height on a monolayer of quartz immersed in the required solution whose volume and depth was calculated to be just enough to cover the largest particle. Flocculation was determined by monitoring the turbidity of the quartz suspension using a spectronic 20 spectrophotometer or Brinkman PC/600 Colorimeter. For this purpose, 0.3 gm samples of -400 mesh quartz were prepared in an agate mortar then mixed with 50 cc. of 10^{-4} mole/liter $Fe(NO_3)_3$ or $FeCl_3$ solution adjusted to the desired pH, ionic strength, and additive concentration to simulate the grinding environment. After tumbling the suspension for 10 minutes, a 6 ml. sample was poured into the spectronic 20 holding tube or the probe of the colorimeter was lowered to 1.0 inch level from the surface of the suspension to measure the percentage of light transmitted. All the measurements were made at a wave length of 620 nm. In each case, the instrument was calibrated with water for a reading of 100% transmission.

Zeta potential measurements were carried out on the samples used for turbidity measurements. A Zeta Meter was used for this purpose.

The pulp fluidity was measured by sensing the torque exerted on a six-blade turbine, four inch diameter impeller stirring 0.2 liters of quartz pulp (67.8% solids by weight).

RESULTS AND DISCUSSION

Adsorption at Quartz/Solution Interface

Adsorption of ions at solid/solution interface can lead to changes in interfacial properties such as zeta potential. In fact, specific adsorption might lead to reversal of the sign of zeta potential. In this study, changes in zeta potential of quartz due to addition of surfactants are used as indication of adsorption of surfactant species at the quartz/solution interface.

For instance, H^+ and OH^- are potential determining ions for silicates and oxides, and the reaction of quartz with these ions is shown as [14]:

at acidic pH:

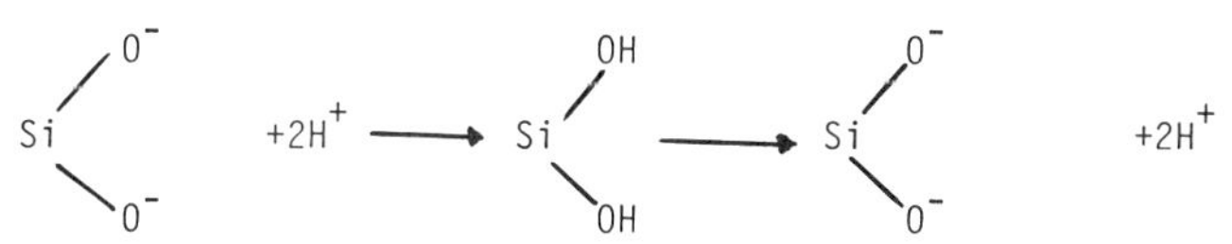

at alkaline pH

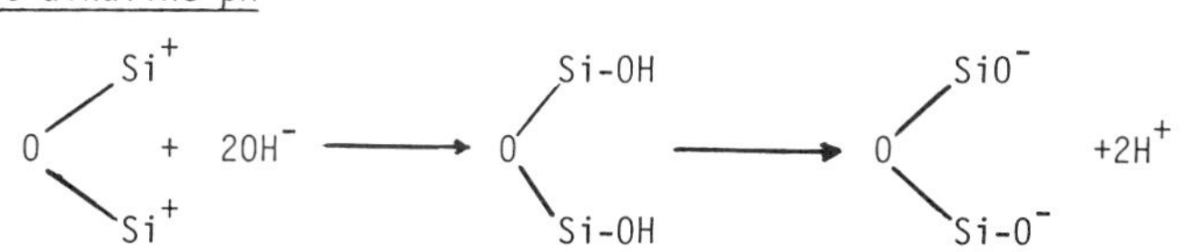

Thus, the zeta potential of quartz is expected to be of negative sign at all pH values. This can be clear by examining the experimental data in Figure 1 (curve #1). Addition of 10^{-4} mole/liter of ferric ions to simulate the iron released from the mill during grinding has resulted in reversal of the zeta potential sign from negative to positive then to negative again above pH 6.4 (curve #2).

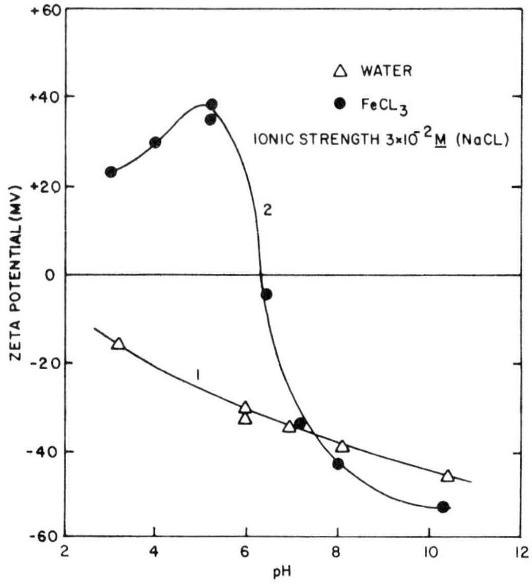

Figure 1. Effect of 10^{-4} M $FeCl_3$ on Zeta Potential of Quartz as a Function of pH.

The charge reversal from negative to positive in the acidic pH range can be attributed to adsorption of positively charged ferric hydroxy complexes (see Figure 2) on the negatively charged quartz surface.

Above pH 3.0, the positive charge is due to precipitation of ferric hydroxide which is positively charged in the acidic pH range and has a point of zero charge at about pH 6.7 [15].

Zeta potential of quartz was measured in solutions of different concentrations of $AlCl_3$ and containing 10^{-4} mole/liter $FeCl_3$ and the results are given in Figure 3. These data suggest that in 10^{-3} mole/liter $AlCl_3$ solutions (curve 2), quartz surface is positively charged in the pH range of 2 to 8.7 and becomes negative above pH 8.7. The reversal of the potential from negative to positive below pH 3.0 can be attributed to the adsorption of complex ferric ions as explained above. The positive charge observed in the pH range 3.0-8.7 is possibly due to adsorption of aluminum species e.g. $(Al(OH)^{++}$ and $Al(OH)2^{+})$, (see Figure 4) as well as iron complexes leading to the shift of the isoeletric point from 6.2 to 8.7. The negative potential obtained above pH 8.7 might be the result of the precipitation of aluminum hydroxide which has an isoelectric point at about pH 9.0.

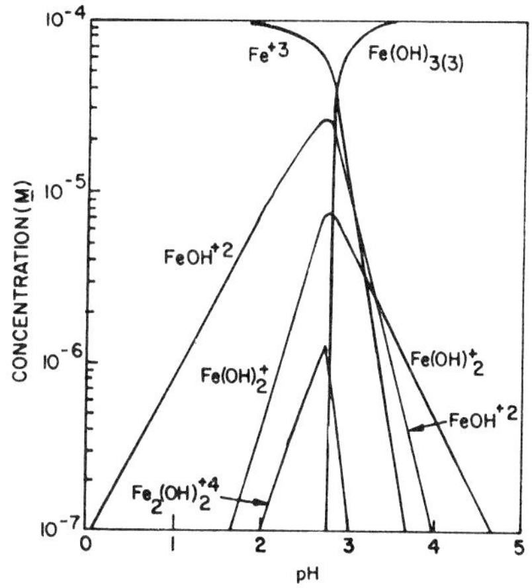

Figure 2. Logarithmic Concentration Diagram for 1×10^{-4} $\underline{M}$ $FeCl_3$ (15).

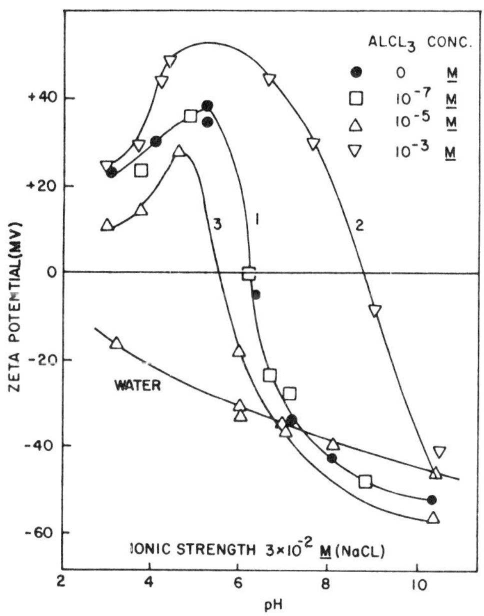

Figure 3. Effect of $AlCl_3$ additions on Zeta Potential of Quartz Slimes in 10^{-4} $\underline{M}$ $FeCl_3$ Solutions as a Function of pH.

At lower levels of $AlCl_3$ addition (10^{-5} mole/liter), zeta potential is found to be positive in the acidic pH range with isoelectric point at pH 5.5 (Figure 3, curve #3). This could be attributed to adsorption of the surface active species $AlOH^{++}$ which have a maximum in the stated pH range, (see Figure 4). Data obtained for the flocculation characteristics of quartz fines in 10^{-4} mole/liter $FeCl_3$ solution with and without $AlCl_3$ is given in Figure 5. Flocculation is indicated by an increase in the percent light transmitted. The results show that, at a given pH, quartz particles can be flocculated or dispersed by changing the level

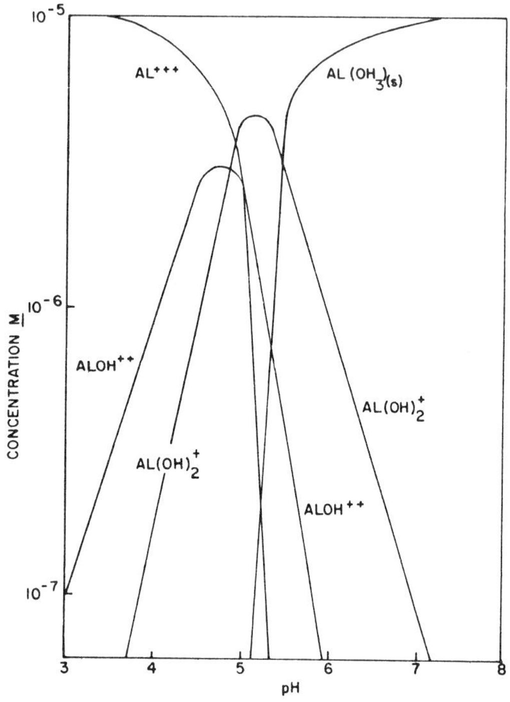

Figure 4. Logarithmic Concentration Diagram for 1×10^{-5} M Aluminum Salt.

of $AlCl_3$ concentration. Higher levels (10^{-3} mole/liter) are found to help flocculation in pH range of 8-10.5 (curve 2). On the other hand, in 10^{-5} mole/liter $AlCl_3$ solutions, the pulp is slightly dispersed.

The flocculation data given in Figure 5 and zeta potential results given in Figure 3 suggest a strong correlation between the zeta potential of quartz and its flocculation/dispersion status in various solutions. Significant flocculation is noticed when the zeta potential value is in the range of $\pm$ 15 mv.

Organic surfactants are also found to affect the zeta potential of quartz as shown in Figure 6.

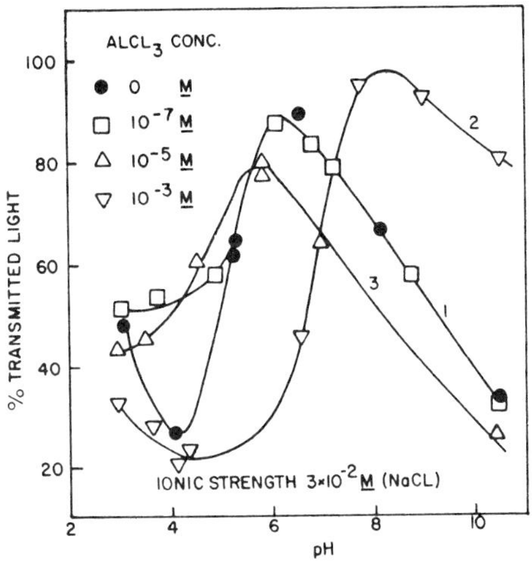

Figure 5. Effect of $AlCl_3$ Additions on the Turbidity of Quartz Slimes in 10^{-4} M $FeCl_3$ Solutions as a Function of pH.

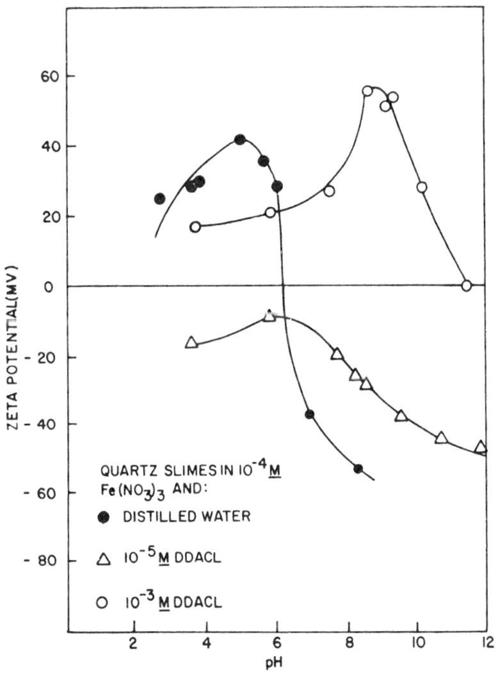

Figure 6. Effect of Dodecylammonium Chloride on Zeta Potential of Quartz in 10^{-4} M Fe $(NO_3)_3$ Solution.

The given results show a decrease in the negative value of zeta potential in presence of 10^{-5} mole/liter (DDACl), and a reversal of sign of zeta potential due to addition of higher levels of 10^{-3} mole/liter of amine. This is explained on the basis of the ionic species distribution of amine as a function of pH as given in Figure 7 which indicates that adsorption of cationic species of amine is responsible for the decrease in the negative zeta potential values and the reversal of it to positive value could be a result of multilayer adsorption of the cations. It should be noted that maximum positive value is obtained in the alkaline pH range where the ionomolecular complexes (amine-aminium ions) are predominant (see Figure 7).

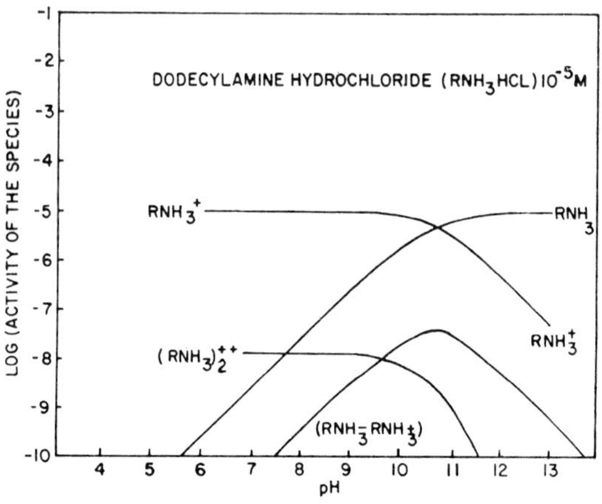

Figure 7. Amine Species Distribution Diagram as a Function of pH. Total amine = 10^{-5} moles/L (16).

Decrease in zeta potential in highly alkaline solutions could be due to decrease in adsorption of such highly surface active species resulting from the shown decrease in their concentration in that pH range.

The above data will be correlated later in this paper with fracture and grinding of quartz in surfactant solutions.

Fracture of Quartz in Surfactant Solutions

Gaudin-Schuhmann size distribution of quartz broken by a single impact in 10^{-5} mole/liter aluminum chloride solution at pH 3.0 is plotted in Figure 8 along with that obtained in water for the purpose of comparison. The data indicate more breakage of quartz in $AlCl_3$ solution than in water in the acidic pH range. Similar improvements in fracture of quartz were obtained when 10^{-5} $CaCl_2$ or dodecylammonium chloride were added at pH 10.5. It is interesting to mention that the above mentioned surfactants did not produce appreciable effects at other pH values as seen in Table 1.

TABLE 1. Change in Surface Area of Fracture Products of Quartz Produced by a Single Impact in Different Surfactant Solutions (3×10^{-5} M NaCl were added to control ionic strength).

Surfactant	pH		
	3.0	6.0	10.5
10^{-4} M $FeCl_3$	0.0	0.0	0.0
10^{-5} M $AlCl_3$	+4.6	+0.2	0.0
10^{-5} M $CaCl_2$	---	+1.2	+4.3
10^{-3} M Amine	+0.9	+1.0	+4.3

** control test, 10^{-4} M $FeCl_3$ was added to all tests to simulate iron released during grinding.

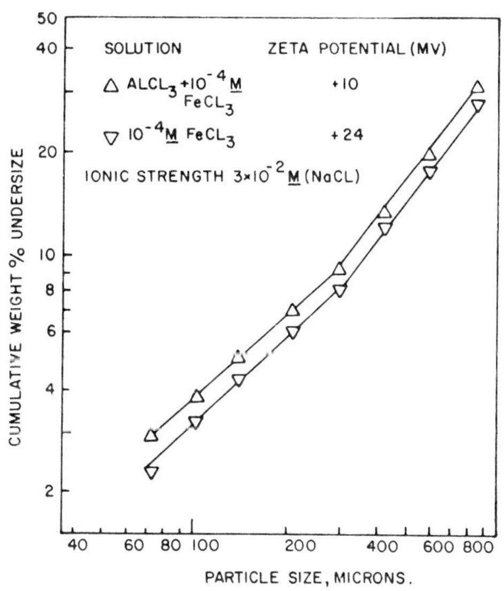

Figure 8. Effect of 10^{-5} M $AlCl_3$ at pH 3.0 on Size Distribution of Quartz Crushed by a Single Impact in a Drop-Weight Mill.

The data given in Figure 8 and in Table 1 show that the surfactants influence fracture of quartz in the pH region(s) where the highly surface active species (e.g. $AlOH^{++}$ in the acidic pH range, $CaOH^{+}$ and amine-aminium complex in the alkaline pH ranges) are present and adsorbing on quartz surface as discussed above.

<u>Grinding of Quartz in Surfactant Solutions</u>

The data for the effect of $AlCl_3$ additions at neutral pH values on grinding of quartz is shown in Figure 9 (curve #1). It can be seen that

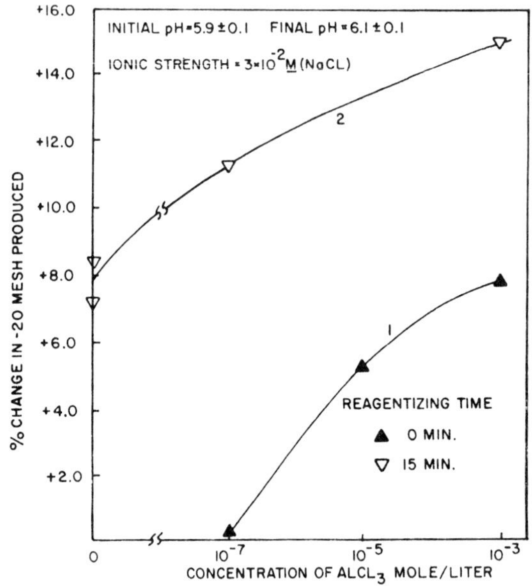

Figure 9. Effect of Addition of $AlCl_3$ on Wet Ball Milling of Quartz at pH 5.9.

the grinding of quartz is improved at all levels of addition of $AlCl_3$. It is interesting to note that further improvement in grinding can be obtained by tumbling the material (inside the mill) in the solutions before grinding as indicated by curve #2 in the same figures. This effect of tumbling could not be achieved when the material was scraped from the mill wall after tumbling. This suggests that distribution of the material on mill due to tumbling is the primary reason for the observed improvement.

Influence of different levels of addition of $CaCl_2$ on grinding of quartz at pH 5.9 is given in Figure 10. The data suggest a definite decrease in grinding at all levels of addition. However, the observed effect depends on $CaCl_2$ concentration up to 10^{-5} <u>M</u> and independent of it above that level within the experimental error. It should be mentioned here that this level of addition at other pH values indicated that $CaCl_2$ causes a detrimental effect in the acidic and neutral pH ranges with no measurable change noticed in alkaline pH solutions [5].

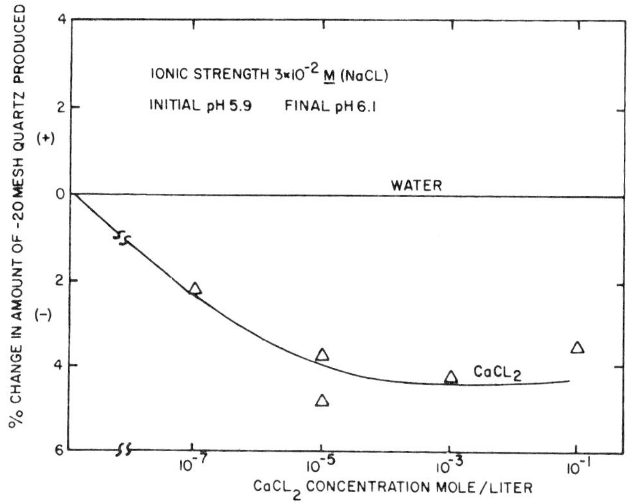

Figure 10. Effect of Different Additions of $CaCl_2$ on Wet Ball Milling of Quartz at pH 5.9.

Dodecylammonium chloride is found to influence grinding of quartz both beneficially and detrimentally depending on pH and additive concentration as shown in Figure 11. It can be seen that amine addition is beneficial except under acid conditions where detrimental effects are obtained at both amine levels. Also, in the neutral pH range, addition of lower levels (10^{-5} mole/liter) is found to lead to reduced grinding.

Since grinding consists of several simultaneous processes, correlation of effect of surfactants on such processes and the grinding results will be discussed next.

Correlation in Surfactant(s)/Quartz Systems

1. $AlCl_3$
   The changes obtained in properties of quartz suspensions due to the addition of $AlCl_3$ are summarized in Figure 12. The data suggest the following:

   (a) Improvement in grinding performance at pH 3.0 could be due to improvement in fracture of quartz as well as pulp fluidity;

   (b) Enhanced grinding at pH 5.9 is probably due to better dispersion of the pulp in the $AlCl_3$ solution; and

   (c) Absence of any effect at pH 10.5 can be attributed probably to the positive influence of pulp dispersion which has nullified the detrimental effect resulting from the decrease in pulp fluidity.

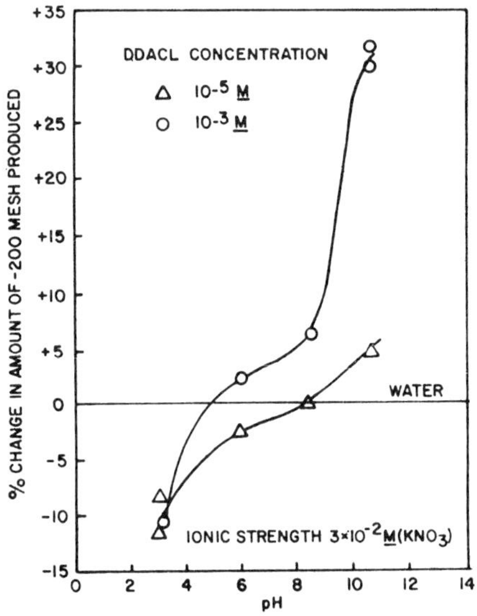

Figure 11. Effect of Amine on Amount of -200 Mesh Produced by Wet Ball Milling of Quartz.

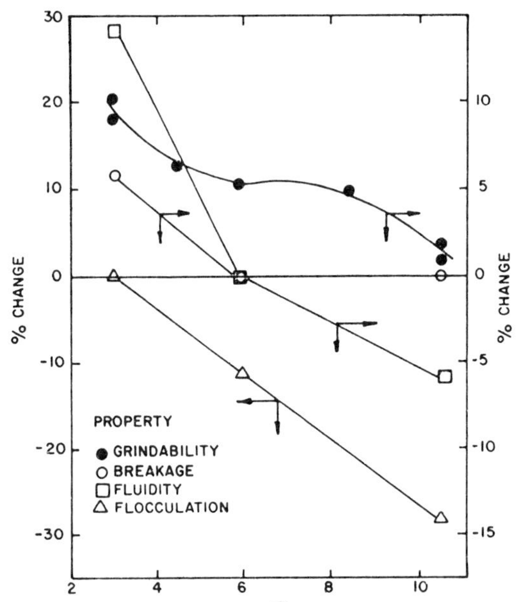

Figure 12. Change in Different Properties of Quartz Suspensions Due to 10^{-5}M $AlCl_3$ Addition as a Function of pH.

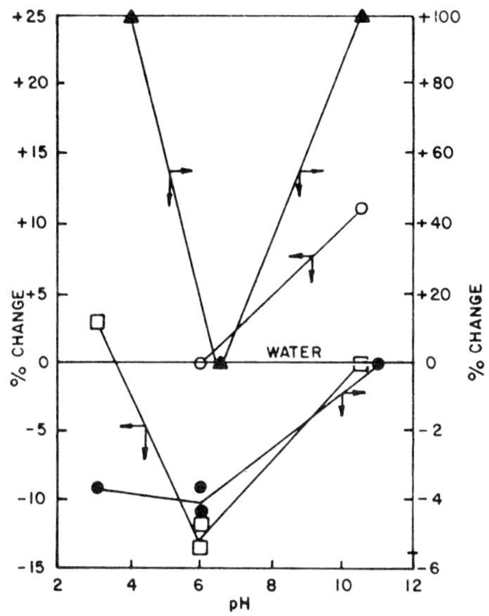

Figure 13. Change in Different Properties of Quartz Suspensions due to $10^{-5}\underline{M}$ $CaCl_2$ Addition.

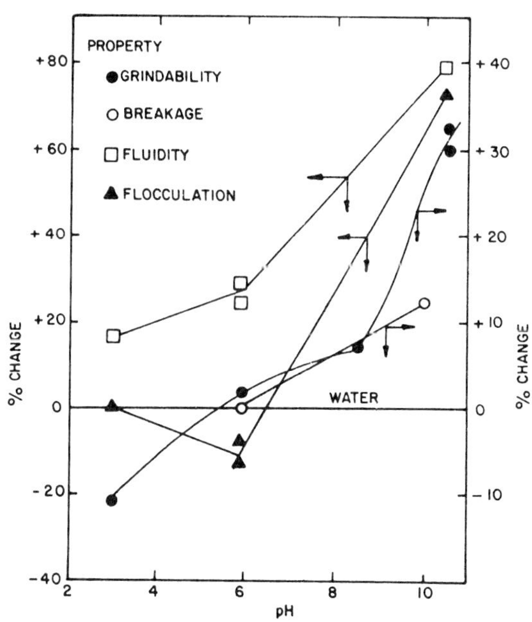

Figure 14. Change in Different Properties of Quartz Suspensions Due to $10^{-3}\underline{M}$ Amine Addition as a Function of pH.

2. $CaCl_2$
The results of tests conducted in $CaCl_2$ solutions are plotted in Figure 13. Correlation of the data obtained under different conditions suggest the following:

(a) The decrease in grinding performance in the acidic and neutral pH ranges could be due to the flocculation of the ground products inside the mill and decrease in pulp fluidity in the stated pH ranges.

(b) $CaCl_2$ enhances fracture of quartz at pH 10.5 at the same time increases the flocculation of the quartz fines. Thus, the absence of any effect on grinding in this pH range could be attributed to cancellation of the beneficial effect of fracture by detrimental effects of flocculation.

3. Dodecylammonium Chloride:

The changes in different properties of quartz suspension due to amine addition are shown in Figure 14. Correlation of these changes indicate the following:

(a) Improvement in grinding performance in the alkaline and neutral pH ranges is probably due to enhanced fracture and/or pulp fluidity.

(b) Reduced grinding efficiency in the acidic pH range cannot be correlated with the observed increase in pulp fluidity. In a previous paper [6], we have speculated that such detrimental effects could be due to increase in floc strength due to presence of amine. Nevertheless, more work is needed to prove such speculation.

CONCLUSION

It is clear from the above analysis that surfactants addition can be beneficial or detrimental to grinding depending on the effect of it on basic solid and suspension characteristics. It is possible to predict, under most conditions, the influence of additives on grinding from knowledge of their effect on fracture, pulp fluidity and flocculation. All of these properties can be affected by adsorption as well as consequent changes in interfacial properties.

ACKNOWLEDGMENT

The support of the American Institute of Steel Industries and Florida Institute of Phosphate Research is gratefully acknowledged.

REFERENCES

1. P. Somasundaran and I.J. Lin, "Effect of the Nature of the Environment on Comminution Processes," I and EC Processes Des. Div.,11 (1972), pp. 321-331.

2. H. El-Shall, S. Vindage, and P. Somasundaran, "Grinding of Quartz in Amine Solutions, Int. J. Mineral Proc. 6, (1979) pp. 105-117.

3. H. El-Shall, A. Gorken, and P. Somasundaran, "Effect of Chemical Additives on Grinding of Iron Ore Minerals," Proc. of the XII Int. Min. Proc. Congr. Warsaw (1974) V. 2, Part A, pp. 695-726.

4. H. El-Shall, A. Gorken, and P. Somasundaran, "Effect of Chemical Additives on Grinding," Dr.Eng.Sc., Thesis, Columbia University, New York, NY (1980) p. 288.

5. P. Somasundaran, and H. El-Shall, "Mechanochemical Effects in Ultrafine Grinding," in Ultrafine Grinding and Separation of Industrial Materials, S.G. Malghan, ed., AIME, (1983) pp. 21-35.

6. H. El-Shall and P. Somasundaran, "Mechanisms of Grinding Modification of Chemical Additives: Organic Reagents," Powder Technology, 38 (1984) pp. 267-273.

7. H. El-Shall, and P. Somasundaran, "Physico-Chemical Aspects of Grinding: A Review of Use of Additives," Powder Technology, 38 (1984) pp. 275-293.

8. R. Klimpel and W. Manfroy, "Chemical Grinding Aids for Increasing Throughput in the Wet Grinding of Ores," I & EC Proc. Des. & Dev., V. 17, (1978) pp. 518-523.

9. M. Katzer, R. Klimpel and J. Swell, "Examples of the Laboratory Characterization of Grinding Aids in the Wet Grinding of Ores," Mining Engineering, V. 33, No. 10 (1981) pp. 1471-1476.

10. R. Klimpel and L.G. Austin, "Chemical Additives for Wet Grinding of Minerals," Powder Technology, 31 (1982) No. 2 pp. 234-253.

11. P.A. Rehbinder, "Hardness Reduction Through Adsorption of Surface Active Agents," Physik, V. 72, (1931), pp. 191-205.

12. A.R.C. Westwood and D.L. Goldheim, J. Appl. Physic., 39 (1968), p. 3401.

13. A.A. Griffith, "The Phenomena of Rupture and Flow in Solids," Phil. Trans Roy Soc. London, Ser. A 221 (1920-1921) pp. 163-198.

14. A.M. Gaudin, and D.W. Fuerstenau, "Quartz Flotation with Cationic Collectors," Trans. Amer. Inst. Met. Eng. 202 (1955)

15. J.M.W. Mackenzie, Trans. AIME, (1966) pp. 82-87.

16. K. Ananthpadmanabhan, P. Somasundaran, and T.W. Healy, "Chemistry of Oleate and Amine Solutions in Relation to Flotation," Trans. AIME, V. 266 (1979) pp. 2003-2009.

17. "Communition and Energy Consumption," Report of the Committee on Comminution and Energy Consumption, National Acad. Press, Washington, DC (1981) p. 47.

18. J.N. Bultler, Ionic Equillibrium, Addison-Wesley, Reading, MA (1964) p. 287.

COMMINUTION AND ASH REDUCTION OF COAL PARTICLES

MASAAKI NAKAMURA, NAOKI ITO, YUKIO SAKURAI, and SHIGEKI TOYAMA
Department of Chemical Engineering, Nagoya University,
Furo-cho, Chikusa-ku, Nagoya 464-01, Japan

ABSTRACT

In the mineral processing industry, there are two purposes in crushing and grinding operation. Namely one is reduction of the particle size and the other is liberation of the target minerals from locked particles prior to their recovery. To predict the particle size distribution in such process, the population balance method composed of the selective and breakage functions is very useful.

In this study, Japanese coals were ground in a vibrating ball mill and the distributions of ash and combustibles as well as the size distribution were analyzed. In grinding these coals, combustibles were concentrated in the fine products, and then ash content of the fine products was lower than that of the coarse ones. A new breakage function for ash distribution in coal grinding process was proposed to interprete the concentration of ash into the coarse products, and solved simultaneously with the well-known selection and breakage functions which had been established to calculate the time-dependent size distribution of particles.

In addition the effect of ammonia treatment on coal comminution rate was tested and the effectiveness was observed in the initial stage of comminution.

INTRODUCTION

Coal utilization has been developed in the processing of gasification, liquefaction and slurry with oil or water as an alternative fuel of oil. In any case, a high amount of ash content prevents from increasing consumption of coal. To cope with this it becomes important to reduce the ash content before and/or after consuming coal.

There are many kinds of physical or chemical methods for coal cleaning. In general, coal consists of useful carbonaceous combustibles and useless inorganic minerals, and their grindability is differed each other. This provides an idea to classify the two contents by making use of the different grindability. Some solvents are capable of swelling and fragmentation of coal particles, and this results in enhancing classification of combustibles and ash. Some papers reported that the grindability of coal was increased by the treatment of liquefied ammonia [1] or ammonia gas [2,3].

In this study the ammonia gas treatment was applied before comminution of Japanese coals. The effect of this treatment on enhancement of coal comminution rate and ash redistribution in ground products were discussed. Then a new breakage function for ash redistribution in coal comminution was proposed to determine the cumulative ash distribution function based on the difference of grindability. The population balance equations for coal particles size and ash content were solved simultaneously with the well-known selection and breakage functions and a new breakage function proposed in this study for ash redistribution.

Advances in Fine Particles Processing
John Hanna and Yosry A. Attia, Editors

EXPERIMENT

Table 1 shows results of the proximate analysis for two Japanese coals, Miike and Taiheiyo coal, used in this study. Coal particles were sieved between 850 and 1000μm before grinding. And then they were exposed to ammonia gas under atmospheric pressure for 72 hours, which was a sufficient time to attain the equilibrium of absorbing ammonia gas.

A vibration mill of 0.120m in diameter and 0.120m in length was used, and contained coal particles of $0.160 \times 10^{-3}m^3$ with alumina or steel balls of 0.013m in diameter. The net volume of balls was 50% of the mill volume in all experiments. The amplitude and frequency of the vibration mill was kept constant at 3.5×10^{-3}m and 28.5Hz, respectively.

Particle size distribution was measured by sieves and granulometer CILAS HR750, Alcatel Co. Ash was analyzed by ignition method of JIS (Japanese Industry Standard) M8812.

TABLE I. Proximate analysis of coals (%).

Coal	Moisture	Volatile matter	Fixed carbon	Ash
Miike	0.7	37.6	37.1	24.6
Taiheiyo	2.4	44.0	38.6	15.0

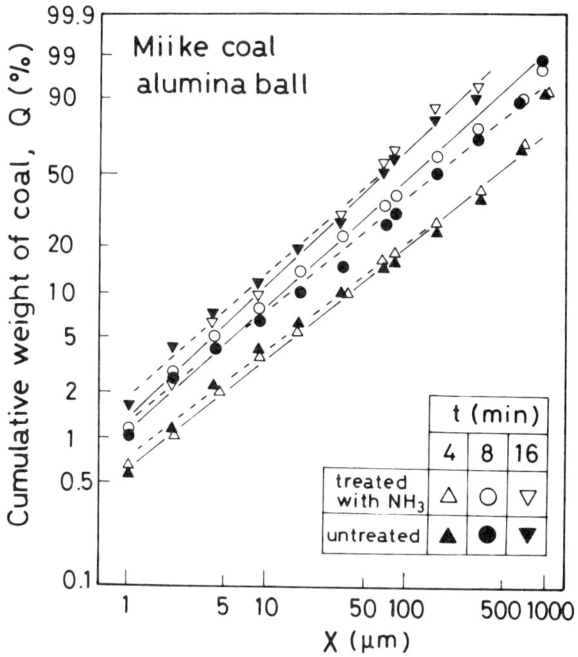

FIG. 1 Rosin-Rammler diagram of cumulative size distribution of Miike coal ground with alumina balls

RESULT AND DISCUSSION

Particle Size Distribution of Ground Products

The cumulative distribution of ground products under the particle size, x, is defined by

$$Q(x,t) = \int_0^x (\partial Q/\partial x)\,dx \tag{1}$$

$$Q(x_0,t) = 100 \tag{2}$$

Figure 1 and 2 show the Rosin-Rammler diagrams of Miike coal ground with alumina and steel balls, respectively. Remarkable difference in the size distribution between ammonia-treated and untreated coal was not observed. This is attributed to the strong impact force of the vibrating ball mill. However, the effect of ammonia treatment was observed at the beginning of comminution, and the communition rate of ammonia-treated coal was higher than that of untreated coal. It is noticeable that the treatment with ammonia gas was effective for the comminution of coarse particles. Tamai et.al. also reported that the effect of treatment with liquefied ammonia was observed in the ball mill comminution of coarse coal particles above 325 mesh (about 500μm) but not found under 250μm [1].

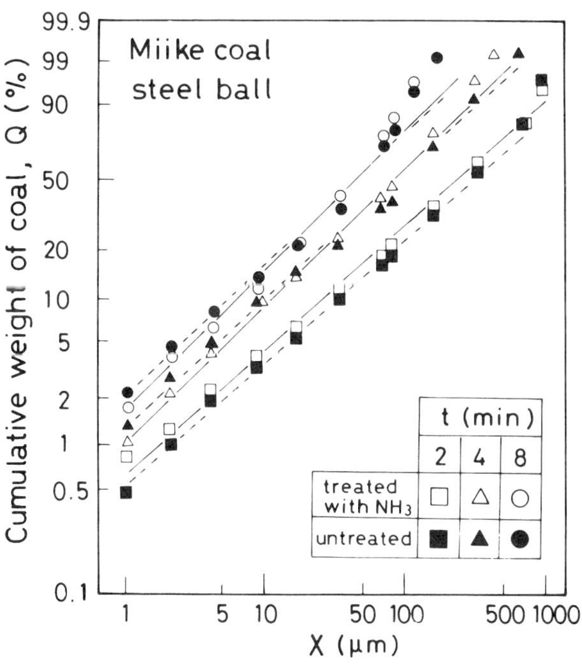

FIG. 2 Rosin-Rammler diagram of cumulative size distribution of Miike coal ground with steel balls

Ash Distribution of Ground Products

Figure 3 and 4 show the distribution of ash weight percentage, w, of ground products, which were sieved and then measured by the ignition method. Ash contents of the coarse particles were much higher than the initial ash content, w_0. On the other hand, those of the fine particles were less than w_0. These tendencies were observed not only in the comminution of ammonia-treated coal but also in that of untreated coal.

Cumulative Distributions of Ash and Combustibles

The size of ground products is usually distributed over a wide range, and the cumulative distribution of ash is defined.

$$Q_a(x,t) = \frac{\int_0^x w(\partial Q/\partial x)\,dx}{\int_0^{x_0} w(\partial Q/\partial x)\,dx} \times 100 = \int_0^x \left(\frac{w}{w_0}\right)\left(\frac{\partial Q}{\partial x}\right)dx \tag{3}$$

$$Q_a(x_0,t) = 100 \tag{4}$$

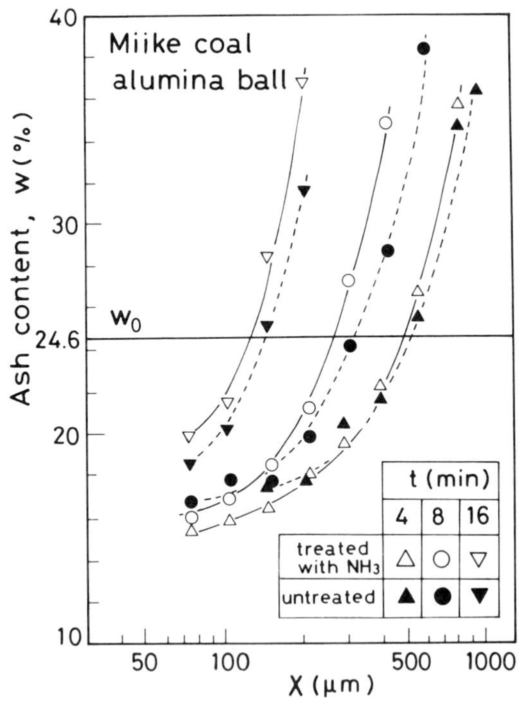

FIG. 3 Ash content distribution of Miike coal ground with alumina balls

The cumulative distribution of combustibles is defined by

$$Q_c(x,t) = \frac{\int_0^x (100 - w)(\partial Q/\partial x)dx}{\int_0^{x_0} (100 - w)(\partial Q/\partial x)dx} = \int_0^x (\frac{100 - w}{100 - w_0})(\frac{\partial Q}{\partial x})dx \tag{5}$$

$$Q_c(x_0,t) = 100 \tag{6}$$

The average ash content under the particle size, x, is given by

$$w_{av} = \frac{\int_0^x w(\frac{\partial Q}{\partial x})dx}{\int_0^x (\frac{\partial Q}{\partial x})dx} = \frac{w_0}{\frac{w_0}{100} + \frac{Q_c}{Q_a}(1 - \frac{w_0}{100})} \tag{7}$$

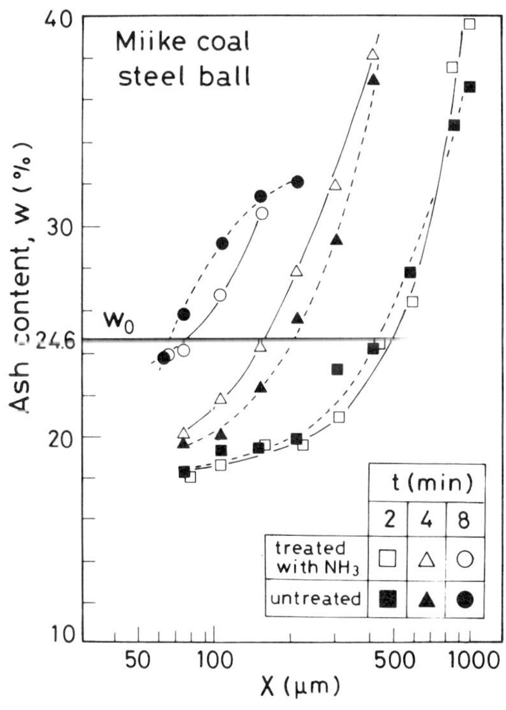

FIG. 4 Ash content distribution of Miike coal ground with steel balls

If $Q_c/Q_a = 1 + \delta$,

$$Z = \frac{w_0}{100} + \frac{Q_c}{Q_a}\left(1 - \frac{w_0}{100}\right) = 1 + \delta\left(1 - \frac{w_0}{100}\right) \tag{8}$$

Since $w_0 < 100$, $Z > 1$ if $\delta > 0$.
Therefore, if $Q_c/Q_a > 1$,

$$w_{av} < w_0 \tag{9}$$

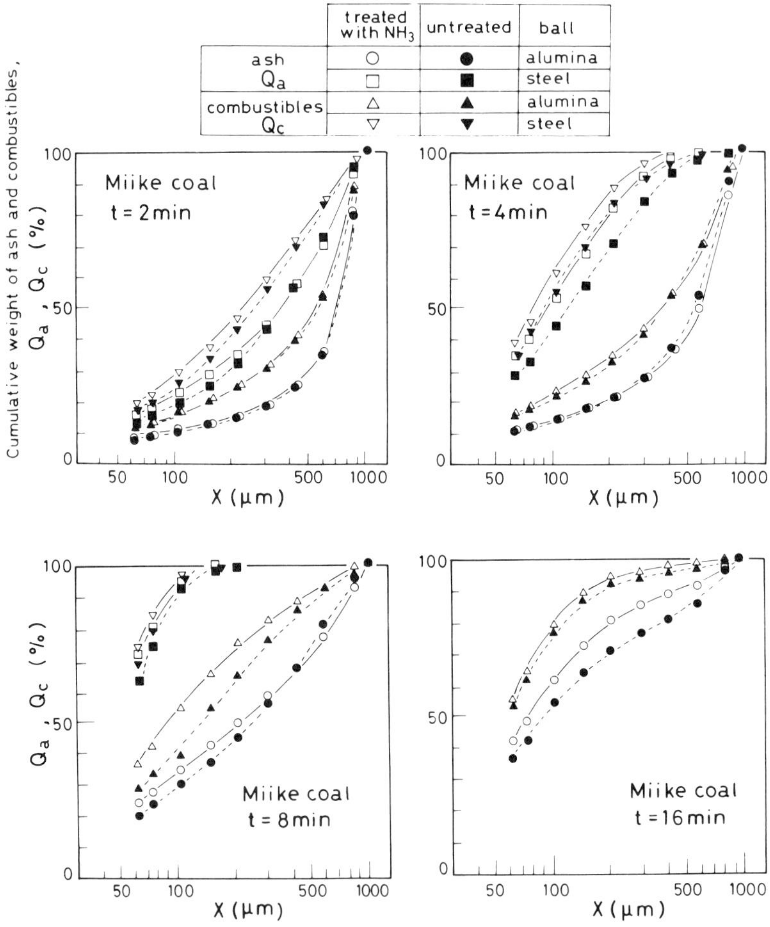

FIG. 5 Cumulative distributions of ash and combustibles in ground products of Miike coal

Figure 5 shows that the cumulative combustibles distribution function, Q_c, is higher at any particle size, x, than the cumulative ash distribution function, Q_a. A big ratio of Q_c to Q_a should be preferable to concentrate ash in the coarse products. Figure 6 shows that the difference, $Q_c - Q_a$, at the comminution time, t, had a maximum at some particle diameter, which shifted from the large particle size to the small one with the comminution time. The maximum value of $Q_c - Q_a$ was about 20 to 25%.

According to Fig. 5 and 6, the ground products are capable of classifying the fine particles of low ash content from the coarse ones of high ash content, if comminution is stopped at the most suitable time.

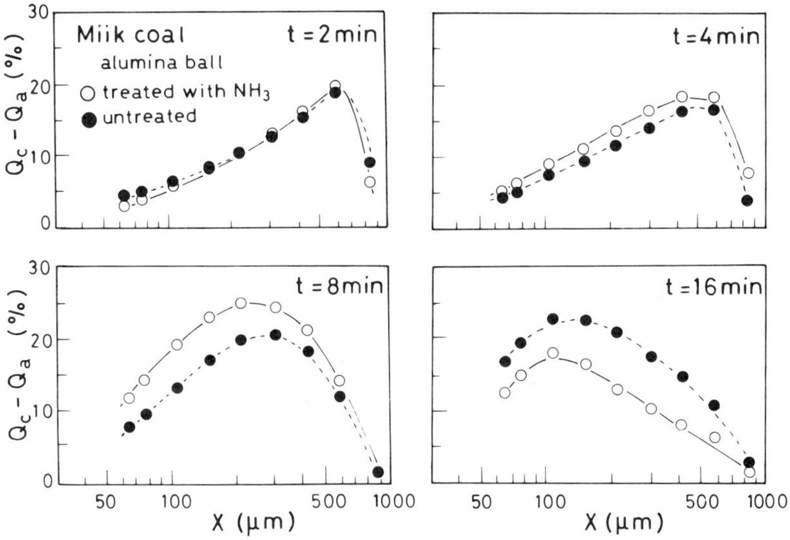

FIG. 6 Difference between cumulative distributions of ash and combustibles in ground products of Miike coal

Analysis of Comminution Process

The population balance method has been developed to simulate the comminution process. The selection function, S(x,t), and breakage function, B(x,y), are represented as Eq. (10) and (11) [4].

$$S(x,t) = Kx^n \tag{10}$$

$$B(x,y) = (x/y)^m \tag{11}$$

where y is the original size of particles selected for crushing, and K, m and n are constants determined by experiments. The cumulative function under the particle size, x, can be calculatd by Eq. (12).

$$\frac{\partial^2 Q(x,t)}{\partial t \, \partial x} = -S(x,t)\frac{\partial Q(x,t)}{\partial x} + \int_x^{x_m} S(y,t)\frac{\partial Q(y,t)}{\partial y}\,\frac{\partial B(x,y)}{\partial x}\,dy \tag{12}$$

The cumulative ash distribution function, $Q_a(x,t)$, were linear in Rosin-Rammler diagram. The distribution constant in Rosin-Rammler equation was obtained as shown in Fig. 7, and this was nearly equal to unity in the range briefly below 700μm.

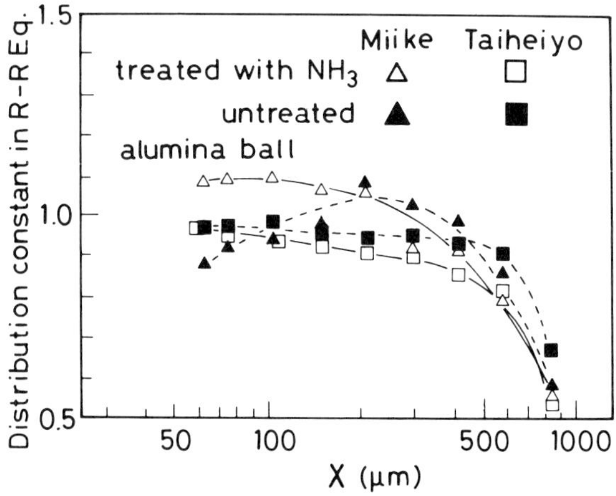

FIG. 7 Distribution constant in Rosin-Rammler equation for cumulative ash distribution of ground products

In addition it was confirmed that the combustibles concentrated into the fine products and ash did into the coarse ones during the comminution process of coal composed of soft carbonaceous combustibles and hard inorganic ash. On the basis of these experimental results, a breakage function, $A(x,y)$, for ash redistribution in coal comminution was assumed as Eq. (13) similar to the breakage function, $B(x,y)$.

$$A(x,y) = (x/y)^{\alpha} \tag{13}$$

where α is constant as well as m in Eq. (11). Since combustibles and ash coexist in coal particles and are crushed together, the selection function, $S(x,t)$, for prediction of ash redistribution was assumed to be the same relationship as Eq. (10).

Finally, the cumulative ash distribution funtion, $Q_a(x,t)$, can be calculated by Eq. (14) similar to Eq. (12).

$$\frac{\partial^2 Q_a(x,t)}{\partial t\, \partial x} = -S(x,t)\frac{\partial Q_a(x,t)}{\partial x} + \int_x^{x_m} S(y,t)\frac{\partial Q_a(y,t)}{\partial y}\,\frac{\partial A(x,y)}{\partial x}\,dy \tag{14}$$

Coal comminution process was simulated in two steps. In the first step, the parameter K, in the selection function, $S(x,t)$, was determined from curve fitting of the experimental data of $Q(x,t)$ to the calculated values. An example was shown in Fig. 8, where K was determined as 2.20×10^{-5} $s^{-1}\cdot\mu m^{-1}$ under on the assumption that Rosin-Rammler equation could be valid (m = n = 1.0). Both experimental and numerical results showed fairly good agreement at the beginning of comminution.

In the second step, parameter α in Eq. (13) was estimated to calculate the ash redistribution in ground products. Figure 9 shows the comparison of the experimental results with the numerical ones obtained by substituting α for 1.1. The ash content of ground products was determined by experiments for the coarser particles than 50 m in diameter. Now, it becomes clear that the simulation of both $Q(x,t)$ and $Q_a(x,t)$ are feasible at the initial stage of comminution when the ash concentration in the coarse products is enhanced.

CONCLUSION

Although the effect of ammonia treatment on coal comminution rate was observed only at the beginning of comminution, the ash redistribution of ground products was enhanced by ammonia treatment. On the basis of the ash concentration into the coarse products in coal comminution, a new breakage function for ash redistribution was formulated in the population balance method and the ash distribution was simulated. The numerical results showed fairly good agreement at the initial stage of comminution when the ash concentration into the coarse products was enhanced.

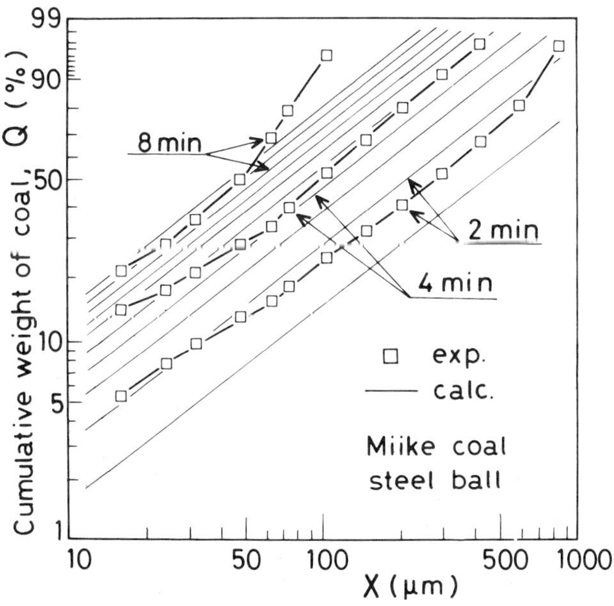

FIG. 8 Comparison between cumulative size distributions of experimental and numerical results

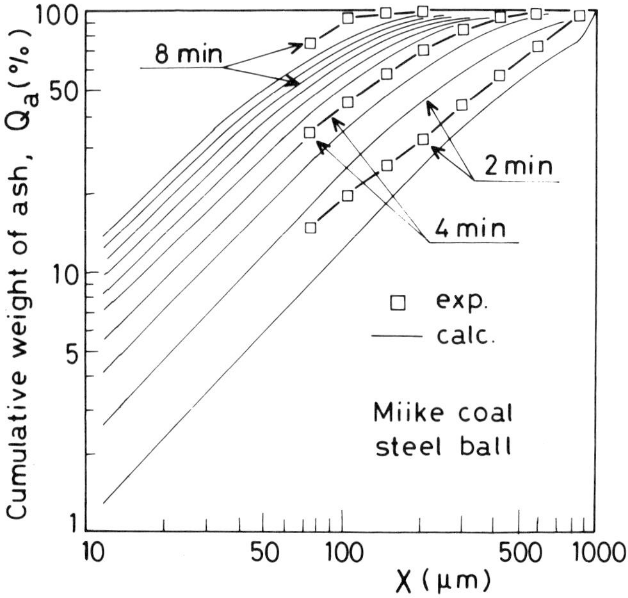

FIG. 9 Comparison between cumulative ash distributions of experimental and numerical results

NOMENCLATURE

$A(x,y)$	: breakage function for ash redistribution in coal comminution	[-]
$B(x,y)$	: breakage function	[-]
K	: rate constant of comminution	[$s^{-1} \cdot \mu m^{-n}$]
m	: power of breakage function	[-]
n	: power of selection function	[-]
$Q(x,t)$	: cumulative weight percentage of coal under particle size, x	[%]
$Q_a(x,t)$	: cumulative weight percentage of ash under particle size, x	[%]
$Q_c(x,t)$	: cumulative weight percentage of combustibles under particle size, x	[%]
$S(x,t)$	: selection function	[s^{-1}]
t	: comminution time	[s]
$w(x,t)$	: ash content of particle size, x	[%]
w_{av}	: average ash content under particle size, x	[%]
w_0	: initial ash content	[%]
$x(t)$	: particle size	[μm]
x_m	: maximum particle size	[μm]
x_0	: initial particle size	[μm]
y	: particle size (integral variable)	[μm]
α	: power of breakage function for ash redistribution	[-]

REFERENCES

1. Y. Tamai, A. Tomita and T. Takarada, Nippon Kagaku Kaishi, 951 (1980).
2. R.S. Datta, P.H. Howard and A. Hanchett, Feasibility Study of Pre-Combustion Coal Cleaning Using Chemical Comminution, Final Report, ERDA Contract No. 14-32-0001-1777 (1976).
3. P.L. Silveston, R.R. Hudgins, D.R. Spink, B. Smith and G. Mathieu, Proceedings 64th CIC Coal Symposium, 277 (1982).
4. Funtai-Kogaku Benran, p. 497, Nikkan Kogyo Shinbunsha (1986).

PART 2.

SIZING, MIXING AND FLOW PROPERTIES

RHEOLOGY OF CONCENTRATED SUSPENSIONS

TH F TADROS
ICI Agrochemicals, Jealotts Hill Research Station,
Bracknell, Berks. RG12 6EY, UK.

INTRODUCTION

The rheology of concentrated suspensions, namely their viscoelasticity, is determined by the balance of three forces; Brownian diffusion, hydrodynamic interaction and interparticle forces. These three contributions depend on the volume fraction of the suspension, the particle size and interparticle forces. Thus, various viscoelastic responses may be obtained depending on the time scale of the experiment and the structure of the system. If the time scale of the experiment is shorter than the relaxation time of the system, a predominantly elastic response is produced. This means that the system has a relatively high Deborah number D_e,

$$D_e = t_r/t_e \qquad (1)$$

where t_r is the relaxation time of the system and t_e the time scale of the rheological experiment.

Conversely, if the system has a small D_e (i.e. $D_e << 1$, a predominantly viscous response is produced. This is seldom the case with concentrated suspensions with high effective volume fractions. Thus, concentrated suspensions usually show non-Newtonian response ranging from predominantly elastic to a mixed viscoelastic system.

Four different systems may be distinguished: hard-sphere suspensions, electrostatically stabilised suspensions (referred to as soft interaction), sterically stabilised suspensions and unstable suspensions. The latter may be arbitrarily divided into weakly flocculated and strongly coagulated systems depending on the magnitude of the energies of interaction involved. When such interaction is of the order of few kT units (where k is the Boltzmann constant and T the absolute temperature) one usually refers to weak reversible flocculation. This is, for example, the case with suspensions flocculated in the secondary minimum and those flocculated by addition of free (non-adsorbing) polymer. In the latter case the energy of interaction may be of the order of tens of kT units depending on the volume fraction and polymer concentration. Strongly flocculated or coagulated suspensions are these involving large energies of interactions usually exceeding hundreds of kT units eg. primary minimum flocculation and incipient flocculation of sterically stabilised suspensions.

The above four systems increase in the order of the complexity of their rheology, with the hard-sphere system being the most simple and the flocculated or coagulated systems being the most complicated both experimentally and theoretically. For this reason, progress on the rheology of concentrated suspensions has been very slow and only in recent years has some progress been made. This is due to the development of modern rheological techniques which allows one to obtain information on the structure of the system. However, theoretical analysis of the rheological data is far from being quantitative and only general trends may be drawn. As we will see in the next section, viscoelastic measurements provide valuable information on the interactions in concentrated suspensions and in some cases it is possible to obtain the magnitude of the forces involved.

Advances in Fine Particles Processing
John Hanna and Yosry A. Attia, Editors

STABLE SYSTEMS WITH HARD-SPHERE INTERACTIONS

These are sometimes referred to as systems with neutral stability in which case both repulsion and attraction are screened. Such systems may be produced by screening double layer repulsion in electrostatically stabilised suspension, eg. polystyrene latex suspensions. The screening may be obtained by addition of moderate concentration of electrolyte (eg 10^{-3} mol dm^{-3} 1:1 electrolyte) or replacing water with a less polar solvent such as benzylalcohol (1,2). Under these conditions all interactions are weak relative to Brownian diffusion and therefore the main forces responsible for flow are hydrodynamic and Brownian. If the results of viscosity are plotted in dimensionless quantities, then all data for different particle sizes should fall on the same curve at any given volume fraction ϕ. This is illustrated in Fig. 1 which shows a plot of the reduced viscosity η_r ($\eta_r = \eta / \eta_0$ where η is the viscosity of the suspension and η_0 that of the medium) versus reduced shear rate. The latter is simply the ratio

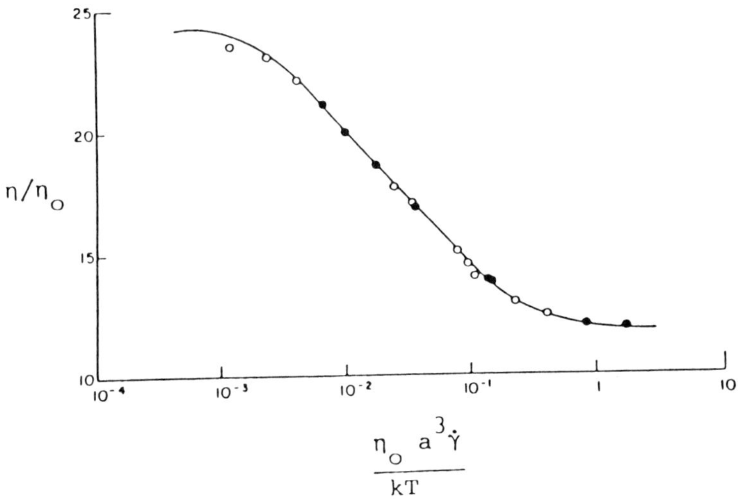

Fig. 1. Reduced viscosity versus reduced shear rate for hard-sphere suspensions at constant volume fraction ϕ = 0.4)

between the time scale of Brownian diffusion (which takes into account the particle size) and the time scale of the experiment ie. the reciprocal shear rate,

$$\gamma_{red} = t_r/(1/\dot{\gamma}) = t_r \dot{\gamma} = (6\pi \eta_o a^3 kT) \dot{\gamma} \qquad (2)$$

where a is the particle radius.

The η_{red} - $\dot{\gamma}_{red}$ curve shows two Newtonian regions at the low and high shear rate ranges with a shear thinning region at intermediate values. At low $\dot{\gamma}$, Brownian diffusion predominates over hydrodynamic flow and the suspension shows a high viscosity resulting from the random arrangement of particles. As the shear rate is increased beyond a certain limit, the particles arrange themselves in layers coincident with the plane of shear and the viscosity decreases with increase of applied shear rate (shear thinning). In the high shear rate regime, the hydrodynamic flow prevails over Brownian diffusion and the suspension shows a second Newtonian region but with much lower η_r than the value obtained in the low shear region.

A plot of η_r versus ϕ is shown in Fig. 2 and the data can be analysed using the Dougherty-Krieger equation (1,2), ie.

$$\eta_r = \left[1 - (\phi / \phi_o) \right]^{-[\eta]\ \phi p} \qquad (3)$$

where $[\eta]$ is the intrinsic viscosity, ie. the slope of the curve when $\phi \rightarrow 0$ and ϕ_p is the so called maximum packing fraction, which is equal to 0.64 for random packing and 0.74 for hexagonal close packing.

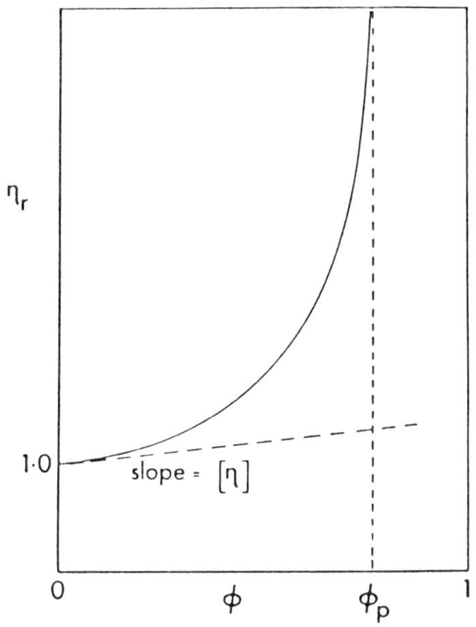

Fig. 2. η_r versus ϕ for a hard-sphere dispersion.

A theory for the rheology of hard-sphere suspension has been developed by Bachelor (3) who considered the balance between hydrodynamic and Brownian diffusion, ie.,

$$\eta/\eta_o = 1 + 2.5\phi + 6.2\ \phi^2 + O\ \phi^3 \qquad (4)$$

The first two terms on the right hand side of equation (3) represent the Einstein limit whereas the third term (6.2 ϕ^2) is the contribution from hydrodynamic interaction, the term in ϕ^3 represents higher order interactions. The experimental results could be fitted using equation (3) when $\phi < 0.2$ while retaining the first three terms only ie. considering the hydrodynamic interaction alone.

STABLE SYSTEMS WITH ELECTROSTATIC (SOFT) INTERACTIONS

These are systems with extended double layers, ie. at low electrolyte concentrations, whereby the interaction is dominated by double layer repulsion. The latter is determined by three main parameters, the surface potential ψ_o, the particle radius a and the electrolyte concentration C (which determines the double layer thickness $1/\kappa$). The best model for such system is polystyrene latex that can be prepared monodisperse with well defined surface charge. The latex can be cleaned (eg. by dialysis) and the electrolyte concentration can be accurately controlled.

Recently (4) we have studied the viscoelastic properties of concentrated polystyrene latex suspensions (with radius a = 700 nm) at various volume fractions and at two NaCl concentrations namely 10^{-5} and 10^{-3} mol dm^{-3}. The electrolyte range will give an order of magnitude reduction in double layer thickness ie. from 100 nm in 10^{-5} to 10 nm in 10^{-3} mol dm^{-3} NaCl. Viscoelasticity was investigated by dynamic (oscillatory) measurements. At each volume fraction and electrolyte concentration, measurements were made of the complex modulus G*, the storage modulus G' and loss modulus G" as a function of strain amplitude (at constant frequency) and frequency (at constant strain amplitude in the linear viscoelastic region). The complex modulus G* is simply the ratio between the stress and the strain amplitudes, ie.

$$G^* = \tau_0/\gamma_0 \qquad (5)$$

G' and G" are given by

$$G' = G^* \cos \delta \qquad (6)$$

$$G'' = G^* \sin \delta \qquad (7)$$

and

$$G^* = G' + i\, G'' \qquad (8)$$

where δ is the phase angle shift between stress and strain ($\delta = 2\pi\omega\Delta t$ where ω is the frequency in Hz and Δt is the time shift of the two sine waves) and i is a constant that is equal to $(-1)^{1/2}$.

Plots of G*, G' and G" versus ϕ at the two NaCl concentrations studied are shown in Fig. 3. The data are at a frequency of 1 Hz and at low strain amplitude (0.004 for 10^{-5} and 0.01 for 10^{-3} mol dm^{-3}, as close as possible to the linear region). The results show a number of interesting features: (a) a rapid increase in the modulii above a critical volume fraction, ϕ_{cr}, which is lower at the lower NaCl concentration; (b) at any given ϕ, the modulii decrease by orders of magnitude as the NaCl concentration is increased from 10^{-5} to 10^{-3} mol dm^{-3}; (c) at 10^{-5} mol dm^{-3} $G' \gg G''$ at all ϕ values studied and G' approaches G* at high ϕ. On the other hand, in 10^{-3} mol dm^{-3} $G' < G''$ within most of the range studied.

The above results of rheology reflect the interaction in such electrostatically stabilised suspensions. As mentioned above this interaction is governed by double layer repulsion which becomes very strong as soon as the double layers significantly overlap with each other. This overlap becomes significant as the particle surface to surface separation h becomes less than twice the double layer thickness ($1/\kappa$). In 10^{-5} mol dm^{-3} ($1/\kappa$) = 100 nm and hence strong interaction occurs as soon as h becomes less than 200 nm. In contrast, in 10^{-3} mol dm^{-3}, strong interaction only occurs when h <20 nm ($1/\kappa$ in this case is 10 nm). It is perhaps useful to define an effective volume fraction ϕ_{eff} for electrostatically stabilised suspensions that takes the double layer into account, ie.,

$$\phi_{eff} = \phi \left[1 + \frac{(1/\kappa)}{a}\right]^3 \qquad (9)$$

Thus, in the above case of 700 nm particles, $\phi_{eff} = 1.48\ \phi$ in 10^{-5} mol dm^{-3} and $\phi_{eff} = 1.04$ in 10^{-3} mol dm^{-3} NaCl. This explains the lower ϕ_{cr} in 10^{-5} mol dm^{-3} compared to 10^{-3} mol dm^{-3} NaCl. It should be mentioned that when ϕ_{eff} approaches close packing (0.64 for random and

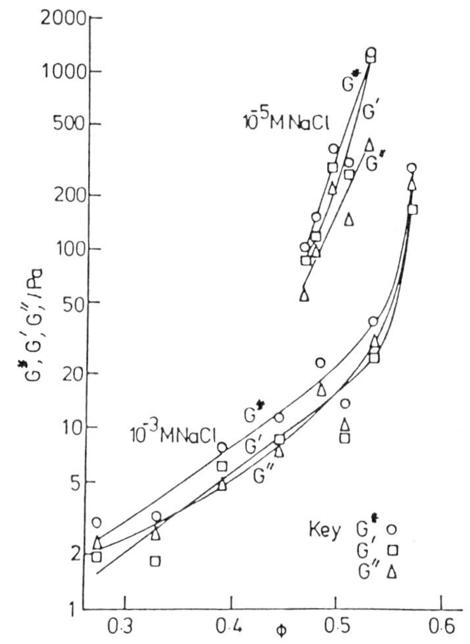

Fig. 3. Variation of G*, G' and G" with ϕ at two NaCl concentrations.

0.74 for hexagonal close packing) strong interaction occurs and the system becomes predominantly elastic. This is shown in the above system. In 10^{-5} mol dm^{-3}, the system becomes predominantly elastic at a ϕ value of 0.463 (see Fig. 3) which corresponds to ϕ_{eff} of 0.73 ie. double layer overlap is significant in this case. In contrast in 10^{-3} mol dm^{-3} NaCl, G" > G' at the highest ϕ value studied, namely 0.566. This corresponds to ϕ_{eff} of 0.59 which is lower than the maximum packing and hence double layer interaction is not very strong in this case. Clearly at any given ϕ, G' is much higher for 10^{-5} mol dm^{-3} NaCl when compared with the value at 10^{-3} mol dm^{-3} reflecting the much longer range of interaction in the first case.

Thus, viscoelastic measurements can be applied to follow interparticle interaction. The rheological characteristics of the system can be controlled by adjusting the properties of the dispersion. For example, if during preparation of a concentrated suspension, one needs to reduce the elasticity and viscosity of the system, a small addition of electrolyte could be beneficial. However, one has to be careful that the added electrolyte does not approach the critical coagulation concentration.

STERICALLY STABILISED SUSPENSIONS

These are exemplified by suspensions containing physically adsorbed or grafted polymer chains. Again strong repulsion occurs as soon as the adsorbed or grafted polymer layers interfere with each other eg. by interpenetration and/or compression. In this case the interaction is determined by the adsorbed layer thickness (which determines the range) and the chain-solvent interaction parameter χ (the Flory-Huggins interaction parameter). The interaction is repulsive when $\chi < 0.5$. Again, it is useful to define an effective volume fraction ϕ_{eff} that is given by the expression,

$$\phi_{eff} = \phi \left[1 + \left(\frac{\delta}{a} \right) \right]^3 \qquad (10)$$

As an illustration, Fig. 4 shows the variation of G*, G' and G" versus frequency ω (Hz) for polystyrene latex suspensions with grafted poly(ethylene oxide) (PEO) chains, at various ϕ values (5). The PEO molecular weight is 2000 giving an adsorbed layer thickness of the order of 20 nm (highly extended chains). At ϕ = 0.44, G" > G' and the suspension behaves as a viscous fluid. This is not surprising since at such volume fraction the interparticle separation distance is more than twice the adsorbed layer 2δ and the interaction between the PEO tails is relatively

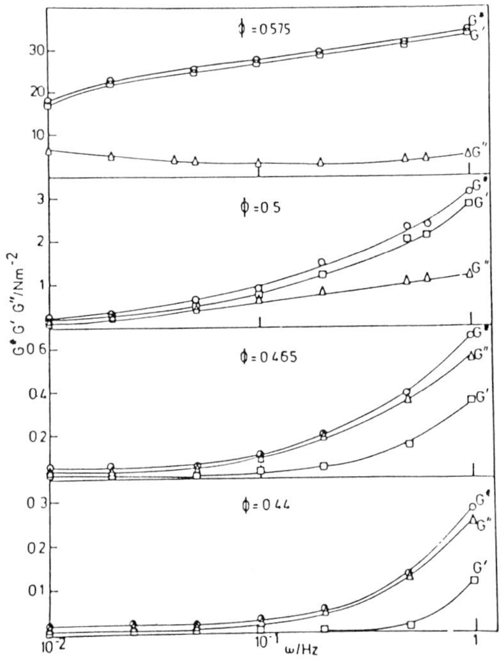

Fig. 4. Variation of G*, G' and G" with ω at various ϕ values for PEO grafted polystyrene suspensions.

weak. At ϕ = 0.465, G" is still larger than G' and the moduli values increase by about a factor of 2 compared to the values at ϕ = 0.44. At such volume fraction the interaction is still weak since the average particle separation is large compared to 2δ and relatively strong interactions take place between the particles. On further increase of ϕ to 0.575, G' becomes much larger than G" and it closely approaches G*. The suspension behaves as a near elastic body as a result of interpenetration and/or compression of the chains. Note that for such ϕ value, the interparticle separation is about 12 nm which is now significantly smaller than 2δ.

Fig. 5 shows plots of G*, G' and G" at 1 Hz versus ϕ. The cross overpoint at which G' becomes equal to G" denotes the start of the elastic interaction. Above that volume fraction, G' increases rapidly with ϕ and approaches G* very closely. Moreover, both G* and G' reach a very high value whereby the latex behaves as an elastic body.

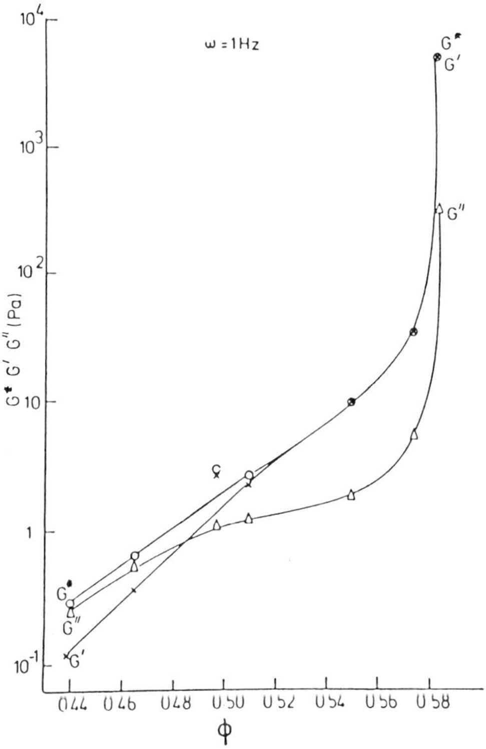

Fig. 5. Variation of G*, G' and G" (at 1 Hz) with ϕ for PEO grafted polystyrene latex suspensions.

Similar results were obtained (6) for physically adsorbed polymer layers eg. poly(vinyl alcohol) (PVA) (M_w = 45,000) on polystyrene latex suspensions. In this case, the physically adsorbed polymer layer forms trains, loops and tails. The latter are quite long giving a hydrodynamic thickness of 46 nm (obtained using photon correlation spectroscopy) (7). Fig. 6 shows the variation of G*, G' and G" (at ω = 1 Hz) with ϕ. It is clear that both G* and G' increase rapidly above ϕ = 0.53 whereas G" remains low over the whole volume fraction of the suspension. This range corresponds to an effective volume fraction of 0.64. That is the maximum random packing fraction. Thus, rheology curves reflect the strong steric interaction that occurs when the interparticle distance becomes smaller than 2δ.

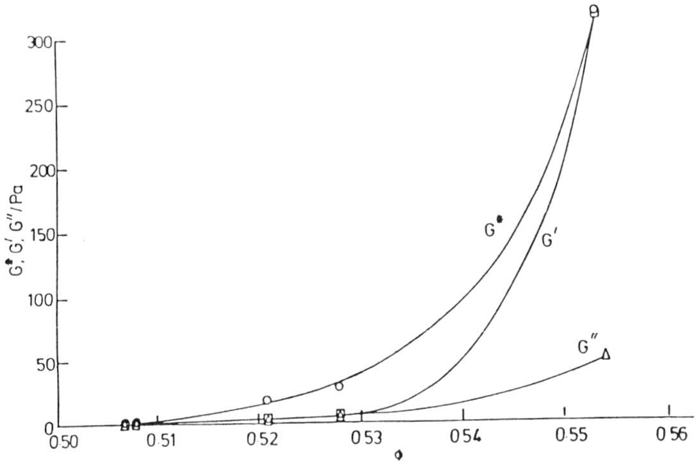

Fig. 6. Variation of G*, G' and G" (at ω = 1 Hz) with ϕ for polystyrene latex suspensions containing physicall adsorbed PVA layers.

FLOCCULATED AND COAGULATED SYSTEMS

The rheology of concentrated suspensions, in which the net force between the particles is attractive, is rather complex. This is due to the non-equilibrium nature of the structure at rest resulting from relatively weak Brownian motion (8). The systems pose difficult problems both from experimental and theoretical points of view. For that reason, advances on theories for rheology of flocculated or coagulated suspensions have been only slow and of a qualitative nature.

The steady state flow curve of a flocculated suspension is usually pseudoplastic; this is illustrated in Fig.7. The flow curve is characterised by three important parameters: (i) $\dot{\gamma}_{crit}$, ie. the shear rate above which the flow curve becomes linear; $\dot{\gamma}_{crit}$ is the value above which the shear causes collision to occur between flocs and a dynamic equilibrium is set up in which flocculi (the basic units from which flocs are found) are possibly transferred from one floc to another and may even be separated entirely for a short time, so that the floc radius of floc decreases with increase of shear rate but the ratio between floc volume and particles volume ie. ϕ_F/ϕ_p remains constant for all $\dot{\gamma} > \dot{\gamma}_{crit}$; (ii) τ_β the value of the stress obtained by extrapolating the linear portion of the flow curve to $\dot{\gamma}$ = 0; the residual stress arises from the residual effect of the interaction potential; (iii) the plastic (apparent) viscosity η_{pl} $(=(d\tau/d\dot{\gamma})_{\dot{\gamma}>\dot{\gamma}_{crit}}$, i.e. the gradient of the linear shear stress-shear rate flow curve. The plastic viscosity results from purely hydrodynamic effects.

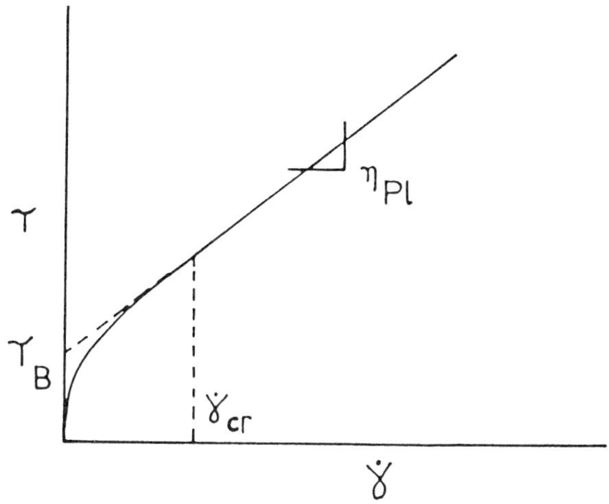

Fig. 7. Typical flow curve (pseudoplastic) of a flocculated suspension.

Flocculated suspensions may also show time dependent effects generally referred to as thixotropy. This is the continuous decrease of apparent viscosity with time under shear and subsequent recovery of the viscosity when the flow is discontinued (9), earlier referred to as sol $\rightleftharpoons$ gel transformation. This is usually the result of weak flocculation eg. in the secondary minimum resulting either in the formation of isolated large flocs or a single floc structure throughout the whole dispersion.

Several theories have been put forward to analyse the pseudoplastic flow curve of flocculated suspensions. One of the earliest theories, referred to as "impulse theory" was first proposed by Goodeve (10) and later extended by Gillespie (11). Goodeve (10) assumed that strictly hydrodynamic or Newtonian effects and interparticle interaction effects are simply additive, ie.

$$\tau = \tau_\beta + \eta_{app} \dot{\gamma} \quad (11)$$

where η_{app} is the viscosity that is descriptive of Newtonian effects and τ_β refers to particle interaction effects.

To calculate τ_β, Goodeve (10) proposed that, when shearing occurs, links between the particles in a flocculated structure would be stretched, broken and reformed and, that, during this process an impulse would be transmitted from a fast moving layer to a slower adjacent layer. Non-Newtonian behaviour would be due to the effect of shear on the number of links, the averge life of a link and any change in the size of interacting particles. Proceeding in this manner, Goodeve (10) derived the following expression for τ_β,

$$\tau_\beta = \frac{3\ \phi^2}{2\pi\ a3} E_A \quad (12)$$

where E_A is the total binding energy between two particles, ie. $E_A = n_L\ \varepsilon_A$, where n_L is the number of links and ε_A is the binding energy of each link.

The theory of Goodeve (10) was later extended by Hunter and coworkers (12) who introduced the elastic floc model concept. This floc consists of an open network of "girders" consisting of chains of particles, as illustrated in Fig.8. The floc undergoes extension and compression during

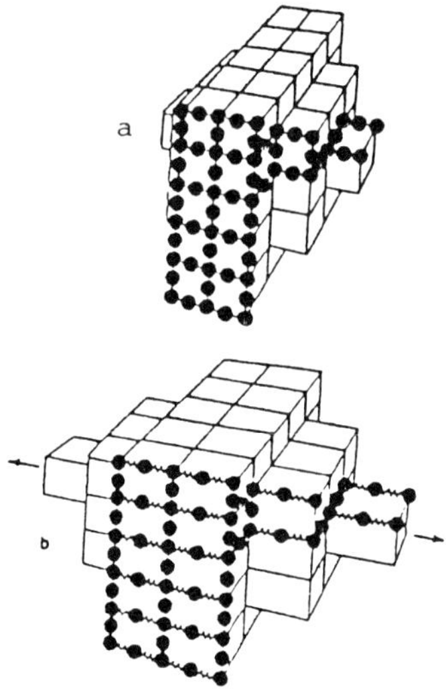

Fig. 8. Elastic floc according to Hunder *et al* (12). (a) Part of a floc with no applied stress; (b) same floc with applied stress.

rotation in shear flow and during mutual collision, resulting in the structure shown in Fig. 8b, in which the bonds are stretched by a small amount Δ (of the order of 0.1% of a particle diameter). This structure is used to explain the flow behaviour of coagulated suspensions as represented in Fig.7.

In order to calculate the Bingham yield value, τ_β, Hunter and coworkers (12) considered the energy dissipation during rupture of flocs, assumed simply to consist of doublets. Most of the energy dissipation was consumed in bond stretching which consisted of three main parts: energy required to overcome interparticle forces, energy required to overcome viscous drag during stretching of a floc (since each particle inside a floc is displaced by a small amount) and energy consumed in the internal movement of liquid within a floc (since during stretching the floc changes its shape and perhaps its volume). The third energy dissipation process was much larger than the other two which could be neglected. In this manner, Hunter and coworkers (12) derived the following expression for τ_β,

$$\tau_\beta = \alpha_o \; \beta \; \lambda \; \eta \; \dot{\gamma} \; (a^2_{floc}/a^3) \; \phi_s^2 \; \Delta \; C_{FP} \qquad (13)$$

where α_o is the collision frequency, β is a constant (= 27/5) and λ is a correction factor. Equation (13) predicts that τ_β increases with ϕ_s^2 and this has been found with many coagulated systems eg. kaolin suspensions.

It should be mentioned that the above theories are rather crude and based on several assumptions whose validity cannot be easily tested in practice. However, they provide useful information for analysis of the results. For better understanding of the rheology of flocculated and coagulated systems, it is essential to obtain results at low deformation, ie. without much perturbation of the flocculated structure. Such measurements have been recently carried out (13) and they constitute a starting point for understanding the structure of flocculated system. For convenience, it is perhaps useful to classify flocculated systems into various categories depending on the nature and magnitude of interaction. Three different cases will be considered and these are summarised below.

Weakly flocculated suspensions

These may be exemplified by the case where flocculation is induced by addition of free (non-adsorbing) polymer to a sterically stabilised suspension. An example of this case is where free PEO is added to a concentrated polystyrene latex suspension ($\phi = 0.3$) containing grafted PEO chains (14). The latex had a radius of 175 nm and a PEO grafted layer of $M_w = 2000$. Three different molecular weight free PEO were added, namely $M_w = 20000$, 35000 and 90000. The flocculation was investigated using three rheological techniques namely steady state shear stress-shear rate, shear wave and oscillatory measurements. As an illustration, Fig. 9 shows plots of the shear modulus G_∞, Bingham yield value τ_β and Casson (15) yield value τ_c versus ϕ_p, the volume fraction of the free polymer (with $M_w = 20000$). All rheological parameters show a rapid increase above a critical ϕ_p value which for that molecular weight is 0.02. Similar results were obtained with the higher M_w PEO samples, but ϕ_p decreases with increase of M_w. For example with $M_w = 35000$, $\phi_p = 0.01$ and with $M_w = 90000$, $\phi_p = 0.005$. This reduction with increase is M_w is expected from theoretical consideration since depletion interaction is a function of radius of gyration of free polymer.

The rheological parameters may be related to the interaction energy between the particles. For example, the Bingham yield value may be related to the energy of separation between the particles in floc, E_{sep}, by the equation (15),

$$\tau_\beta = N\, E_{sep} \tag{14}$$

where N is the total number of contacts between the particles that is given by the equation,

$$N = \frac{1}{2}\left(\frac{3\,\phi n}{4\,\pi a^3}\right) \tag{15}$$

where n is the average number of contacts in a floc, ie. the coordination number. Combining (14) and (15) one obtains

$$\tau_\beta = \frac{3\,\phi\, n\, E_{sep}}{8\,\phi\, a^3} \tag{16}$$

To calculate E_{sep} from τ_β one has to assume a value for n. The latter is probably of the order of 4 (since the floc structure is relatively open). E_{sep} may also be equated to G_{dep}, the energy minimum in the particle interaction produced by addition of a free polymer, ie. the depletion free energy. G_{dep} is related to the osmotic pressure of the free polymer solution. For example, Asakura and Oosawa (16) derived the following expression for G_{dep},

$$G_{dep} = -\,(3/2)\,\phi_2\,\beta\,x^2; \quad 0 < x < 1 \tag{17}$$

where ϕ_2 is the volume concentration of the polymer that is equal to $(4/3)\,\pi\,\Delta^3\,N_2/V$ with Δ being the depletion thickness, N_2 the total number of polymer molecules and V the total volume of the solution. β is equal to (a/Δ) whereas x is given by the following expression,

$$x = \left[\Delta - (h/2)\right] / \Delta \tag{18}$$

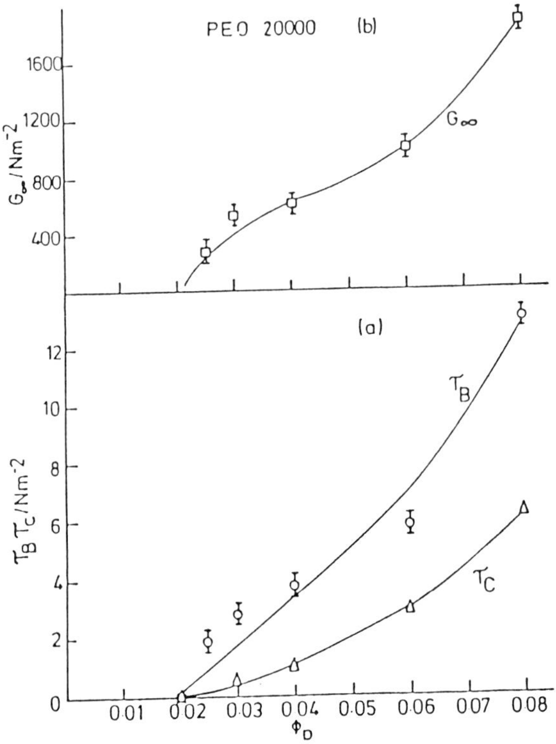

Fig. 9. Variation of G_∞, τ_β and τ_c with ϕ_p of PEO (M_w = 20000)

where h is the surface to surface separation. Clearly when h = 0, ie at the point where the polymer chains are "squeezed out", χ = 1.

A comparison of the values of E_{Sep} (assuming n = 4) obtained from τ_β and G_{dep} obtained using Asakura and Oosawa's theory (16) is given in Table I. Inspite of all the approximations and the assumptions made eg. for n and the hard-sphere nature of the interaction, the values of E_{Sep} calculated from τ_β are comparable to G_{dep} calculated using Asakura and Oosawa's model.

Flocculated sterically stabilised suspensions obtained by reduction of solvency (incipient flocculation).

Incipient flocculation usually occurs when the solvency of the medium for the chains is made worse than a θ-solvent, ie. $\chi > 0.5$. This may be produced by addition of high concentration of electrolyte and/or increase of temperature. As an illustration, Fig. 10 shows the variation of G*, G' and G" with Na_2SO_4 concentration for a polystyrene latex suspension (ϕ = 0.5) coated with PVA (M_w = 45000). It can be seen that the modulii (particularly G* and G') initially decrease with increase in electrolyte concentration (17). This initial decrease is due to reduction of the adsorbed layer thickness with increase in electrolyte concentration. Any reduction in solvency results in a reduction of the adsorbed layer thickness and hence ϕ_{eff} decreases with increase in $C_{Na_2SO_4}$. This results in a reduction in

TABLE I. Comparison of E_{Sep} (calculated from τ_β) and G_{dep} (calculated from Asakura and Oosawa's model)

ϕ_p	τ_β	E_{Sep}/kT	G_{dep}/kT
		(a) PEO, M_w = 20000	
0.025	2.0	18.2	25.3
0.03	2.8	25.4	30.3
0.04	3.8	34.6	40.5
0.06	5.8	52.8	60.7
0.08	13.1	111.2	80.9
		(b) PEO, M_w = 35000	
0.015	2.3	21.0	16.4
0.02	4.4	40.0	21.8
0.03	7.0	63.8	32.7
0.04	11.7	66.4	43.6
		(c) PEO, M_w = 90000	
0.01	1.2	11.0	12.3
0.015	2.8	24.5	18.4
0.02	4.4	40.0	24.5
0.025	5.9	53.8	30.6

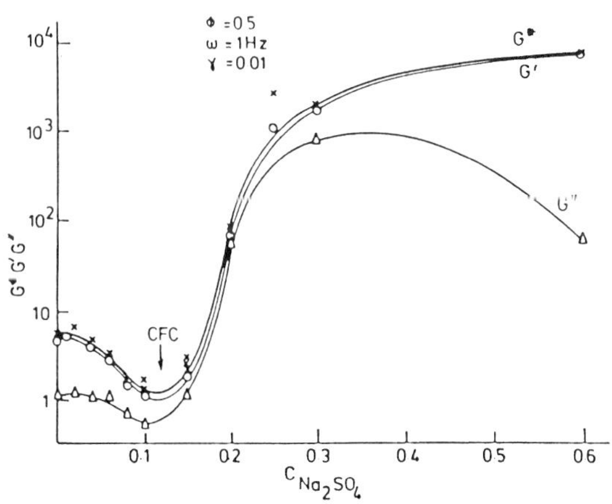

Fig. 10 Variation of G*, G' and G" with $C_{Na_2SO_4}$ for PVA coated polystyrene latex suspension (ϕ = 0.5)

the modulus. However, above a certain critical Na_2SO_4 concentration (0.15 mol dm^{-3}) G*, G', G" increase rapidly with increase in electrolyte concentration. This marks the onset of incipient flocculation. Thus, the critical flocculation concentration (CFC) for Na_2SO_4 is ∿0.15 mol dm^{-3} for the PVA coated polystyrene latex.

Incipient flocculation for sterically stabilised suspensions can also be produced by increasing the temperature at a given electrolyte concentration. This is illustrated in Fig. 11 which shows the results for a PVA coated polystyrene latex (ϕ = 0.5) in 0.15 mol dm^{-3} Na_2SO_4, ie. just below the CFC at 25°C. It can be seen that both G' and G" remain virtually constant (or

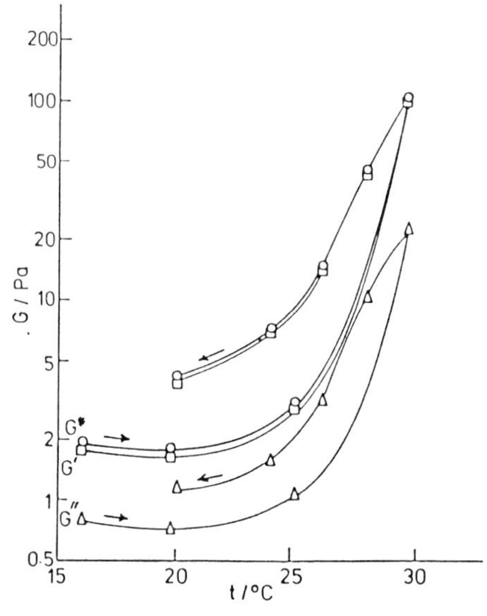

Fig. 11. Variation of G*, G' and G" with temp for a PVA coated polystyrene latex suspension (ϕ = 0.5) in the presence of 0.15 mol dm^{-3} Na_2SO_4.

slightly decrease) with increase in temperature until ∿ 25°C, above which there is a rapid increase in the modulii with further increase in temperature. Thus a critical flocculation temperature (CFT) of ∿ 25° C can be defined for this system at 0.15 mol dm^{-3} Na_2SO_4. After reaching 30°C, ie. well above the flocculation temperature, the suspension was cooled and G' and G" measured. It can be seen that the downward curve deviates significantly from the upward curve indicating that the flocculation is not completely reversible

Strongly Flocculated (coagulated) Suspensions.

These are systems produced by addition of electrolyte to an electrostatically stabilised suspension. An example is polystyrene latex suspension, where the flocculation is produced by addition of 0.2 mol dm^{-3}

NaCl (ie. well above the CFC). In this case coagulation into the primary minimum occurs. The structure of such coagulated systems becomes partially broken down above a critical strain (deformation) that depends on the volume fraction of the suspension (18). This is illustrated in Fig. 12 which shows a strain sweep for a coagulated suspension at various ϕ values and at ω = 1 Hz. It can be seen that G* and G' initially remain independent of the applied strain (the linear viscoelastic region) but above a critical strain

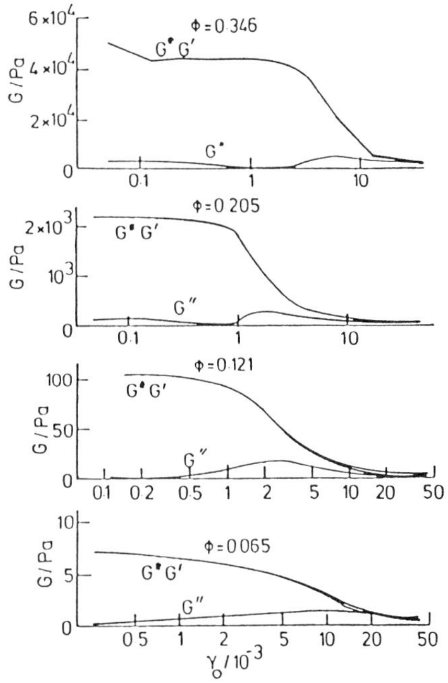

Fig. 12. Strain sweep for a coagulated polystyrene latex suspension at various ϕ values.

value, γ_{cr}, the modulii show a rapid reduction with applied strain. However, G" increases with increase of γ_o reaches a maximum and then drops to a minimum in the region of γ_{cr}. The region above γ_{cr} is the non-linear viscoelastic region whereby the flocculated structures start to become broken down with the applied shear. In the linear region, G* and G' are nearly independent of the applied frequency. A log-log plot of G' (in the linear region) versus ϕ is shown in Fig. 13. This linear plot can be represented by the following equation,

$$G' = 1.98 \times 10^7 \ \phi^{6.0} \tag{19}$$

The high power in the G' -ϕ relationship is indicative of a strongly flocculated structure.

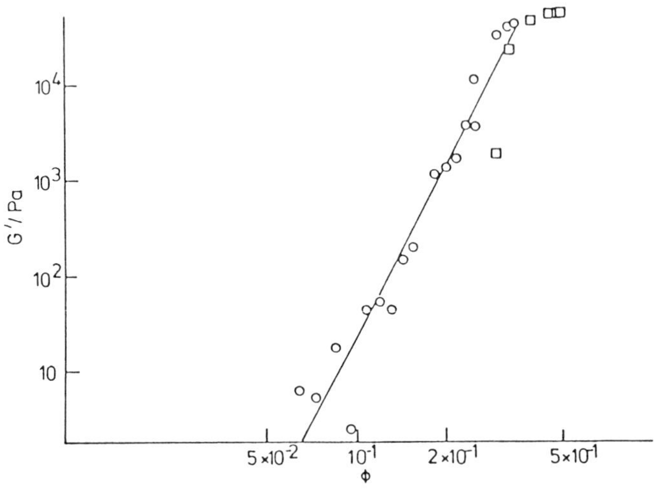

Fig. 13. Log-log plots of G' versus ϕ for a coagulated system.

It is also possible to obtain the cohesive energy in the flocculated structure from a knowledge of ϕ_{cr} and G' in the linear region (19). This is shown as follows. The cohesive energy is related to the stress in the flocculated structure σ by the following equation,

$$E_c = \int_0^{\gamma_c} \sigma \, d\gamma \tag{20}$$

Since $\sigma = \gamma_0 G'$

$$E_c = \int_0^{\gamma_c} \gamma_0 G' \, d\gamma = \frac{1}{2} \gamma_c^2 \quad G' \tag{21}$$

A log-log plot of E_c versus ϕ is shown in Fig. 14, which again can be by a power law, ie,

$$E_c = 1.02 \times 10^3 \quad \phi^{9.1} \tag{22}$$

Again, the high power in E_c - ϕ relationship is indicative of the strongly flocculated structure of the suspension. This is not surprising since flocculation is in the deep primary minimum in the energy-distance curve.

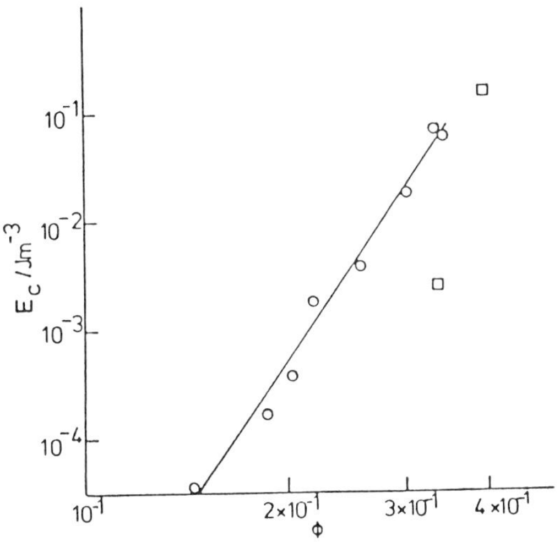

Fig. 14. Log-log plot of E_c versus ϕ.

REFERENCES

1. I.M. Krieger, Advances Colloid Interface Sci. 3, 111 (1972).
2. I.M. Krieger, I.M., Dougherty, M. Trans. Soc. Rheol. 3, 137 (1959)
3. G.K. Bachelor, J. Fluid Mech. 83, 97 (1977).
4. A. Hopkinson and Th.F. Tadros, to be published.
5. C. Prestidge and Th.F. Tadros, J. Colloid Interface Sci, 124, 660 (1988)
6. A. Hopkinson and Th.F. Tadros, to be published.
7. Th. Van Den Boomgaard, T.A. King, Th.F. Tadros, H. Tang and B. Vincent, J. Colloid Interface Sci. 66, 68 (1978).
8. W.C. Russel, J. Rheol. 24, 987 (1980).
9. J. Mewis, J. Non-Newtonian Fluid Mechanics, 6, 1 (1979).
10. C.F. Goodeve, Trans. Faraday Soc. 35, 342 (1939).
11. T. Gillespie, J. Colloid Sci. 15, 219 (1960).
12. B.A. Firth and R.J. Hunter, J. Colloid Interface Sci., 57, 248 (1976); 57, 257 (1976); 57, 266 (1957).
13. A. Hopkinson and Th.F. Tadros, to be published.
14. C. Prestidge and Th.F. Tadros, Colloids and Surfaces 31, 325 (1988).
15. P.F. Luckham, B. Vincent and Th.F. Tadros, Colloids and Surfaces, 6, 101 (1983).
16. Asakuras and F. Oosawa, J. Polym. Sci. 33, 245 (1958).
17. A. Hopkinson and Th.F. Tadros, to be published.
18. A. Hopkinson and Th.F. Tadros, to be published.
19. J.D.F. Ramsay, J. Colloid Interface Sci. 109, 441 (1986).

RHEOLOGICAL AND TRANSPORT ANALYSIS OF MICRONIZED COAL-WATER SUSPENSIONS PREPARED IN CONVENTIONAL AND HIGH-SPEED STIRRED BALL MILLS

Rajendra K. Mehta* and John A. Herbst**
*Mineral Resources Institute, The University of Alabama, Tuscaloosa, Alabama 35487;
**Control International Inc., University Research Park, 419 Wakara Way, Suite 101, Salt Lake City, Utah 84108

ABSTRACT

The purpose of this paper is to compare the rheological properties of micronized coal-water suspensions (CWS) prepared in different grinding devices. This hypothesis is based on the theory that the interaction between particulate phase and the nature of forces prevailing in the mill affects the particle size distribution and shape of particles . The grinding devices used were conventional tumbling ball mill and high speed stirred ball mill having pin and disc option. Flow properties of the suspensions were found to differ appreciably at desired fineness of grind (d_{80}). This was explained on the basis of packing density and morphology of particulate phase. Typically, the high-speed stirred ball mill produced broader particle size distribution yielding distribution modulus (DM) of 0.213 as opposed to 0.384 for the conventional mill. Rheological data collected on a typical distribution constructed based on the Farris analysis revealed that suspensions were most viscous for particles ground in the pin device followed by the disc device and the conventional mill. A simple theoritical analysis has been presented to estimate the shear rate in the vicinity of the tips for the stirred ball mill under typical operating condition for this type of application. This resulted in an estimate of 680 sec^{-1}.

Viscosity data was correlated to the shape factor of particles ground in different grinding devices. Finally, an analysis of pumping power requirement was carried out under typical fluid flow conditions utilizing rheological data which showed that suspensions ground in a stirred mill required 94.33 Hp/mile in comparison to the convention mill which required 84.13 Hp/mile.

NOMENCLATURE

CP	Centipoise (unit of viscosity)
C_v	Volume fraction solids in slurry
CWS	Coal-water suspension
d	Geometric mean size of a size interval
d_s	Smallest geometric mean particle size
d_l	Largest geometric mean particle size
D	Pipe diameter
DM	Distribution modulus
D_p	Maximum particle diameter
dp/dl	Pressure gradient along the length of pipe
d_{80}	80% passing size
dv/dx, γ	Shear rate
ELL-EL	Elliptical shape factor
$F_3(d)$	Cumulative weight fraction finer than size d
f	Friction factor
FORM-PE	Form shape factor
g	Acceleration due to gravity
g_c	Newton's constant
k	Coefficient of the power-law model

Advances in Fine Particles Processing
John Hanna and Yosry A. Attia, Editors

L	Length of stirrer shaft for pin device
mPa·sec	Milli-pascal sec (unit of viscosity)
n	Exponent of the power-law model
Nm	Newton-meter (unit of torque)
Pa	Pascal (unit of shear stress, dynes/cm^2)
P/L	Pumping power per unit length of pipe
RePL2	Reynolds number for steady-state laminar flow of pseudoplastic fluid
SBM	Stirred ball mill
T	Torque
TBM	Tumbling ball mill
v_θ	Velocity component in θ direction
v	Slurry velocity
YP_c	Critical yield point
ρ	Density
μ	Apparent viscosity
$\tau_{r\theta}$	Shear stress component in $r\theta z$ coordinate system
τ	Shear stress

INTRODUCTION

In order to obtain a structured and stable coal-water suspension, the ratio between hydrophobic and hydrophilic sections on the surface of coal must be optimum and correspond to the minimum amount of immobalized water mechanically retained in the cells of the structure [1]. Likewise the viscosity of the system must insure its hydraulic transport and spraying by nozzles for combustion purposes. Viscosity data on such suspensions are also important in order to design slurry transportation system [2], and to keep the suspension suspended in stirred storage tanks [3]. Jinescu [4] analyzed the viscosity data of many experiments and concluded that different type of behavior such as Newtonian, Viscoplastic, Dilatant, Bingham plastic etc; occurs at concentration of particulate phase which depends upon its shape, size distribution along with the interactive forces between particulate and continuous phase.

While the factors such as particle size, shape and packing density (narrow or broad particle size distribution) are important in the area of rheology and have been studied by researchers [5,6,7] but studies showing their variability with mill type have been neglected. It has been well accepted that the mechanism of comminution (i.e. impact-shear forces) changes from one extreme to another depending on mill type (i.e. impact forces in conventional to shear forces in stirred), it is very likely that the interaction between particulate phase and these forces will alter the nature of aforementioned factors subsequently affecting the rheological properties of suspensions.

Therefore, the objective of this study was to characterize and compare rheological and transport properties of coal-water suspensions (CWS) prepared in a conventional tumbling mill and high-speed stirred mill (Drais-Perl mill).

EXPERIMENTAL

Coal Type

The test work reported in this investigation was carried out using Illinois #6 coal supplied by Peabody Coal Company. Table I shows the proximate analysis and other relevant properties of coal feed. The coal feed was having a d_{80} of 600 microns.

Conventional Tumbling Ball Mill (TBM)

The conventional mill was a batch ball mill of stainless steel construction. This mill was 10 inches in diameter and 11.5 inches in length with eight square lifters. Wet grinding tests in this mill were carried out under standard conditions [8].

Stirred Ball Mill (SBM)

A stirred ball mill, supplied by Draiswerke, Inc., Allendale, New Jersey and

TABLE I. Analysis of -12 mesh Illinois #6 Coal Feed

Moisture	2.4%
Volatile Matter	45.6%
Ash	18.0%
Fixed Carbon	34.0%
Total Sulfur	3.8%
Hard Grove Grindability Index	55

especially designed to meet the requirements of research work was used. This device consists of a drive unit and two exchangeable grinding shells. The drive unit was equipped with an adjustable speed transmission in the range of 500-2500 rpm. Figure 1 shows a schematic drawing of shell #1, referred to as the disc device. The stirrer device for this option consists of six perforated discs that are mounted along the drive shaft. The net volume of the grinding chamber was 5.4 liters. Figure 2 shows another device, referred to as the pin device. This is an agitating device featuring ten annular rows of six stirrer pins each and can be coupled to the drive shaft. This shell was also equipped with the same type of pins which are radially fastened to the inner wall of the grinding chamber. Accordingly, stirrer pins plus stationary pins embedded in the mill shell formed a rotor-stator type of agitator. The net volume of the grinding chamber for this device was 11.1 liters. As in the disc device, holes along the uppermost part of the chamber permitted filling of media and coal.

Viscometer

The viscosity measurements on the coal suspension samples were made with a Haake Rotovisco RV12. The sensor system was a coaxial cylinder MVIIP . This profiled sensor was well suited for test substances which tend to slip on cup and rotor surfaces or which are likely to show a phase separation. The detail of this sensor system and instrument constants are given in Table II. A small representative sample of the suspension (55 cc) was used in each experiment. The rotor was run at different speeds and torque was read off. The detailed procedure of grinding and viscometric measurements is described elsewhere [9].

TABLE II. Sensor System Detail and Instrument Constants

Type	MVIIP
Inner Cylinder	
radius, R_i (cm)	1.84
height, h (cm)	6
Outer Cylinder (cup)	
radius, R_o (cm)	2.1
Sample Volume V(cm^3)	55
Instrument constants	
A (Pa/scale · grad)	3.76
M (min/sec)	0.88
G (mPa · s/scale . grad . min)	4274

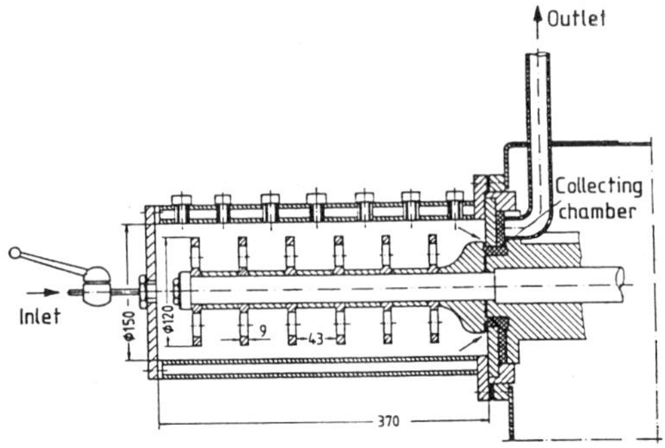

Fig. 1. Schematic diagram of the grinding chamber of the disc device

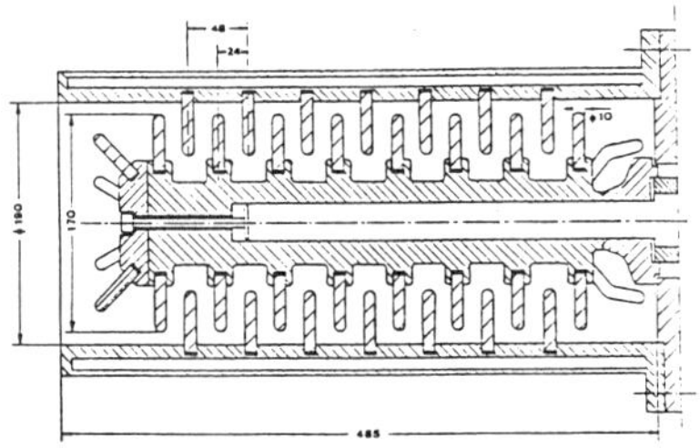

Fig. 2. Schematic diagram of the grinding chamber of the pin device.

DISCUSSION

Viscosity Evaluation

It was planned to study and compare the rheological properties of the suspensions prepared in two types of devices. Since the pin device was found to out perform the disc device for this type of application in terms of energy efficiency [10], direct comparison of conventional mill and pin device was of more interest to us. Therefore, based on the optimum conditions [9] found for grinding in conventional mill using grinding media of 0.5 inch and pin device at 786 rpm, two particle size distributions were simulated aiming at identical d_{80}=250 mesh (63 microns). This fineness was based on a typical grind of coal used in coal-water suspensions for industrial and utility purposes [11].

These distributions are plotted in Figure 3 along with calculated distribution moduli resulting from fitting finer part of the distributions by Gaudin-Schuhmann distribution. Samples were synthetically prepared at 50% coal by weight from particles ground in corresponding devices according to the simulated distribution. The rheological data are plotted in Figure 4. It is interesting to note that the behavior of suspensions from each device changes markedly with shear rate. At low shear rate, the viscosity of suspension prepared from particles ground in stirred mill is higher than that of suspensions prepared from particles ground in conventional mill, whereas the trend is opposite at higher shear rate. The higher viscosity of stirred mill suspension at low shear rate can be understood in terms of particle packing. It is clearly evident from the distribution moduli in Figure 4 that the SBM yields a relatively broader size distribution than TBM (in other words, the variance of particle size is higher for SBM). This allows the smaller particles to fit into the interstices formed by larger ones, therefore reducing the porosity of packing and thereby achieving a state of dense suspension. Two different approaches have been described in the literature for producing highly loaded suspensions: either to use fine and a relatively coarse size distribution [12] termed as bimodel size distribution or to use as relatively "broad" unimodel size distribution [13]. It is the second approach in the context of which this finding can be explained.

On the other hand, at high shear rate, the lower viscosity of stirred mill suspension is primarily due to shear forces. At high shear rates, shear orientation effects [14] will be more favorable for particles of SBM leading to larger shear thinning

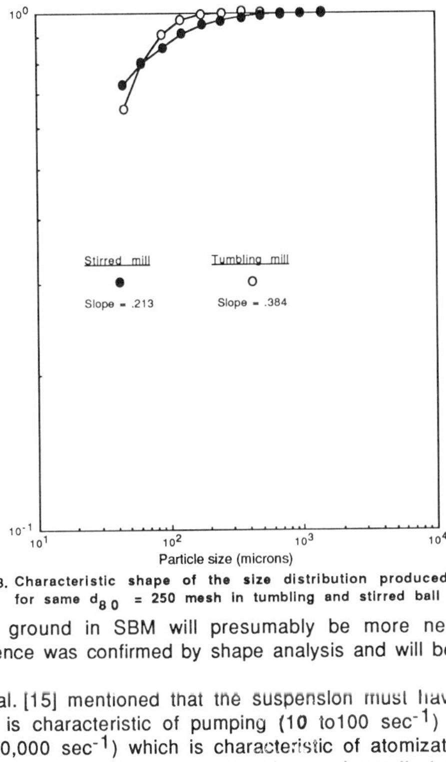

Fig. 3. Characteristic shape of the size distribution produced for same d_{80} = 250 mesh in tumbling and stirred ball mill.

effects. Particles ground in SBM will presumably be more needle-like or flaky, however, this inference was confirmed by shape analysis and will be addressed in the later section.

Morgan et al. [15] mentioned that the suspension must have low viscosity at shear rates which is characteristic of pumping (10 to100 sec^{-1}) and also at shear rates (10,000 to 30,000 sec^{-1}) which is characteristic of atomization. This may be because of less energy requirement in pumping and small droplet sizes during atomization. Based on this reasoning, suspension formulated in a SBM will be comparatively more atomizable, pourable, and suitable for pumping except at very low shear rates (<25 sec^{-1}). Hence, for SBM an effort was made to estimate shear rate in the vicinity of the tips of the stirrer and apparent viscosity of suspension under typical operating conditions from energy data.

To do this, we assumed the pin device stirred mill to be analogous to a Stormer viscometer [16] where the stirring shaft of the mill resembles a rotor and the grinding chamber resembles the outside stationary cylinder. Based upon the principles of fluid mechanics, the following set of boundary conditions and equations can be written for this system:

$$@\ r=KR \qquad v_\theta=\omega KR \quad 0<K<1 \quad (1)$$

$$@\ r=R \qquad v_\theta=0 \quad (2)$$

$$T=2\pi\cdot KR\cdot L\cdot KR\cdot \tau_{r\theta}|_{r=KR} \quad (3)$$

$$\tau_{r\theta}=2\mu r\,((\omega K^2R^2)/((1-K^2)r^3)) \quad (4)$$

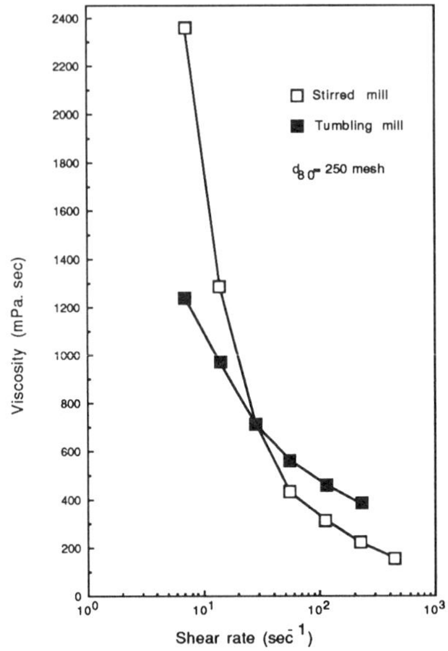

Fig. 4. Comparison of viscosity for samples obtained from two devices, measurement done at 50% coal by weight.

The pertinent size distribution shown in Figure 3 originating from SBM corresponds to a value of 16.6 Kwh/T and the observed value of torque and angular velocity corresponding to this experiment was 93 Nm and 70.58 sec^{-1} respectively. The constants for the pin device were as follows (see Figure 2):

KR=0.085 m; R=0.095 m; L=0.485 m

Substituting these values in Equation (3) yields:

$\tau_{r\theta}|_{r=KR}=4226.13$ Newton/m^2

Substituting this calculated value of $\tau_{r\theta}|_{r=KR}$ in Equation (4), we obtain

$\mu=6224$ P or 6224 mPa sec

An estimate of shear rate in the vicinity of the tip of the pin will thus be 42261.3/62.24=680 sec^{-1}.

It would have been interesting to verify the measured value of viscosity from the viscometer at 680 sec^{-1} with the predicted viscosity in the mill based on power draft data but unfortunately grinding and rheological measurement were not representative of the same sample. Nevertheless, this approach can be a useful tool to estimate in-situ viscosity in the mill.

EFFECT OF PARTICLE SIZE

One of the factor affecting the suspension viscosity is particle size. Samples were prepared form monosize particles (170 x 200 mesh and 325 x 400 mesh) ground in conventional mill and stirred mill, and viscosity was measured at 50% coal by weight. The data are plotted in Figure 5. Two observations can be made: firstly, as expected, the viscosity at all shear rates for finer paricles is higher irrespective of their origin (i.e. whether ground in SBM or TBM). Secondly, for both particle sizes suspension prepared from SBM ground particles is little more viscous at all shear rates, however, the difference is not appreciable. Interestingly, at low shear rates, the difference in viscosity is larger and the difference narrows down as shear rate is increased. This can be explained as follows: for any particle size, the viscosity contribution is due to the combination of two factors, partly due to packing considerations and partly due to physical properties of particles. At low shear rates, suspension of fine particles show viscosity values almost 4 times to that of coarse particles primarily because of packing considerations but at high shear rates viscosity values are almost the same indicating that packing effects do not contribute and the minor difference in viscosity values what we observe is probably because of the presence of more liberated mineral matter phase (hard) in finer particles.

STABILITY EVALUATION

From the viewpoint of pumping, suspensions of coarse paritcles having lower viscosity in the shear rate range of 1 to 100 sec^{-1} will be suitable provided such suspensions are nonsettling. The prameter which is an aid to determining suspension stability during transportation and storage is the yield point [17]. If the yield point is assumed to be the controlling parameter in suspension settling, an equation can be derived that results in a critical value of yield point to prevent settling. By equating the gravitational force equal to the opposing buoyant and drag forces [18],

$$YP_c > \frac{D_p g(\rho_c - \rho H_2O)}{3g_c} \qquad (5)$$

Where YP_c=critical yield point

D_p=maximum particle diameter

ρ_c=density of coal

ρH_2O=density of water

g, g_c=acceleration of gravity, Newton's constant

The standard test for transportation of materials require that fluid be tested at g_c=1.5. Using ρ_c=1.4 gm/cc and ρH_2O=1 gm/cc, D_p=75 microns, the above equation yields:

$$YP_c > 6.5 \text{ dynes/cm}^2 \text{ or } > 0.65 \text{ Pa}$$

The power-law fitting of the data of 170 X 250 mesh coal ground in TBM results in a value of 7 Pa, which is much higher than the critical value of 0.65 Pa; since rheological analysis of suspension prepared from 170 X 250 mesh coal ground in SBM and TBM is the same (see Figure 5), hence suspension prepared in either mill will be nonsettling. Figure 6 shows estimates of yield point for these two particle size fractions ground in TBM and similar plot is expected for paritcles ground in SBM.

EFFECT OF PARTICLE SHAPE

It has been reported in the literature [7] that particles having irregular shape yield higher suspension viscosity. In this context, it is safe to hypothesize that the characteristics of particles, particularly shape originating from different devices will be different due to the very nature of different breakage mechanisms and kind of forces prevalent. Also, sufficient but undetectable evidence in this regard was available when effect of particle size in the previous section was addressed possibly because particulate assembly was comprised of single size particles (Figure 5). Therefore, it was planned to confirm the effect of shape on viscosity by considering a particulate assembly where different sizes of particles are present. For this purpose, a typical distribution based on Farris analysis [19] was constructed from 4 different particle types. This distribution was prepared from narrowly sized particles in the size range of 180 X 0 microns. The procedure has been outlined in the Appendix. It was interesting to note that the viscosity of suspension formulated from particles ground in the pin device showed highest value, followed by particles ground in the disc device, tumbling mill and finally suspension prepared from particles of natural feed. Figure 7 shows the effect of particle shape on the viscosity of suspensions.

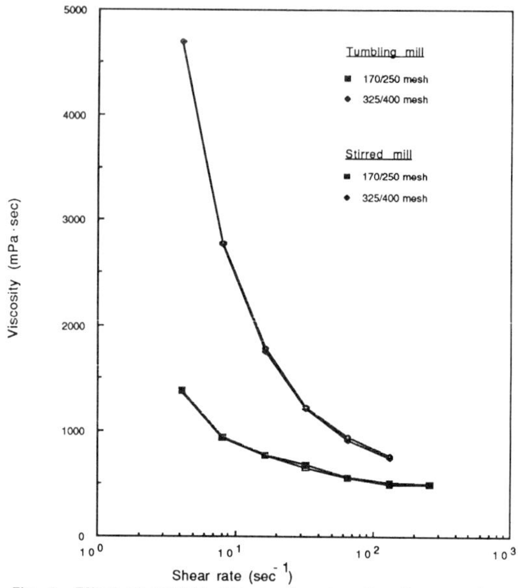

Fig. 5. Effect of particle size on the viscosity of suspension measured at 50% coal by weight for particles ground in tumbling and stirred ball mills.

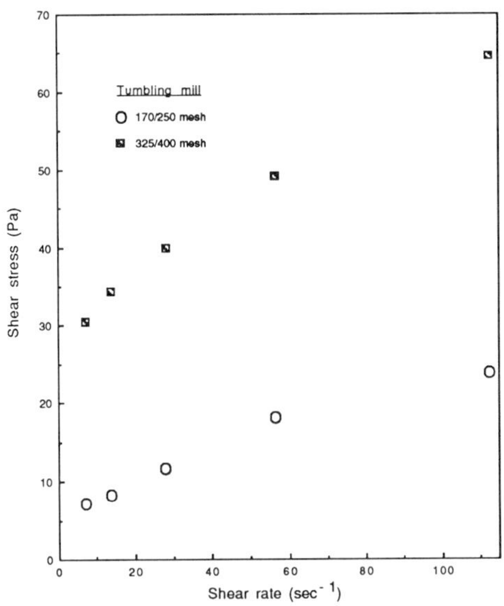

Fig. 6. Plot of shear stress and shear rate for the data plotted in Figure 7 to obtain an estimate of yield point for particles ground in tumbling mill.

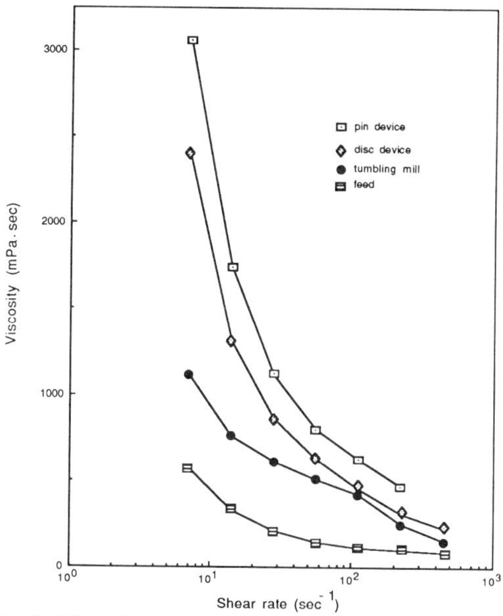

Fig. 7. Effect of particle shape on the viscosity of suspensions, based on Farris particle distribution, DM = 0.3 and C_v = 0.39

The difference in rheology of such suspensions can only be understood in terms of the morphological characteristics of particles. The irregularity in shape can be explained as merely resulting from impact and shear forces in grinding devices. Stirred mills produce more flaky or needle-like irregular particles than do conventional mills simply because intense shear action prevails in stirred mills. This was tested by mounting a single size-fraction of particles (115 X 170 mesh) ground in different devices and examining them under the IBAS image analyzer. Fifty different particles from each mounted specimen were picked and estimates of two shape factors (ELL-EL and FORM-PE) were determined for each particle. Quantitatively, a particle having a value of ELL-EL close to the highest value, 1, will be interpreted as having more like a spherical shape and a low value will mean that particles are elongated or needle-like (flakes). A sample mean of these shape factors for fifty particles ground in the pin device, disc device and tumbling mill was .5201, .5714 and .6297 respectively.

Qualitatively, particles from the pin device showed numerous surface serration, sharp corners and revealed irregularity. Also, particle entaglement was highest for these particles.

PUMPING CHARACTERISTICS

It is important to determine the power requirements to transport coal-water suspensions through a pipeline. The technical and economical feasibility of such mixtures will depend on energy consumption and rheological characteristics which in turn will dictate whether these suspensions are to be prepared in a conventional or in a stirred mill. Suspension prepared in these devices are compared in terms of the power requirements utilizing rheological data.

The pseudoplastic behavior of coal-water suspensions has been established in the literature and was reconfirmed by testing a variety of suspensions by fitting rheological data in a power-law type of relationship [9], (i.e. $\tau=k(dv/dx)^n$) or

$$\tau=k(\gamma)^n \qquad (6)$$

Based on the principles of fluid mechanics, the one dimensional momentum equation for "Poisuelle Flow" is given by

$$\tau=-(D/4)(dp/dl) \qquad (7)$$

where D is the pipe diameter and dp/dl is the pressure gradient along the length of the pipe. The underlying assumptions for this is fully developed flow and negligible end effects.

The friction factor and pressure drop terms are related as follows [19]:

$$dp/dl=(-2f\rho_c v^2)/D \qquad (8)$$

The calculation of friction factor for power-law pseudoplastic steady state laminar flow is given by [20]

$$f=16/RePL2$$

where $RePL2=(8D^n v^{2-n}\rho_c/k)(n/2 + 6n)^n$

Finally, the required pumping power per unit length is calculated by

$$P/L=(dp/dl)(volume/time)$$

Note that this analysis is valid for laminar flow, however, if the flow is turbulent, then the friction factor would simply be related to the Clapp power-law Reynolds number in another equation form [19].

Figure 8 shows the power-law model fit for suspensions formulated in a conventional and stirred mill for the same d_{80}=250 mesh. The parameters resulting from the fit are also shown on this plot. It is interesting to note that at low shear rate, which is characteristic of pumping, the viscostiy of suspension formulated in a conventional mill has a smaller value than that of the stirred mill. Therefore, one would anticipate conventional mill suspensions to require less pumping power. It was found that stirred mill suspensions required 10% more pumping power (94.33 HP/mile for stirred mill against 84.13 HP/mile for conventional mill). This calculation was based upon 2 million tons of coal per year to be transported at 50% coal by weight for a pipe diameter of 12 inches at an average velocity of suspensions to be 5 ft/sec [21]. From this analysis, it is worth mentioning that, if two hypothetical suspensions bearing the same consistency index (coefficient, k) but different flow behavior index (exponent, n)

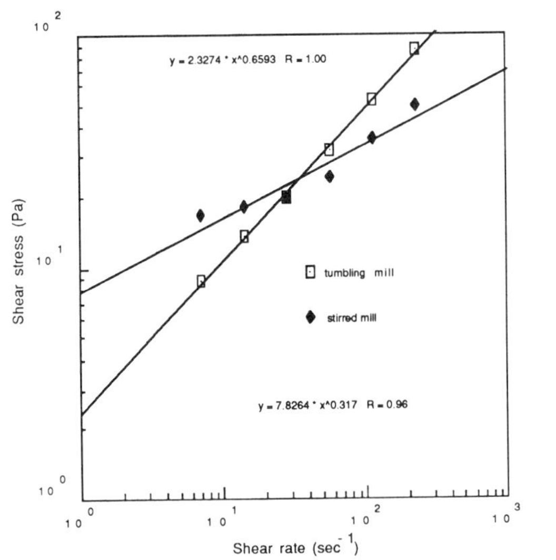

Fig. 8. Power law model fit for slurries prepared from particles ground in tumbling and stirred mill at same d_{80}= 250 mesh.

are compared with respect to power requirements, then the suspension bearing higher value of n will require more power to pump. This is particularly the case when suspensions prepared under diffierent operating conditions in one device are compared with respect to power requirements.

CONCLUSIONS

1. The viscosity data on coal-water suspensions prepared in conventional and stirred mills was successfully described by a power-law model. These suspensions were reconfirmed to be pseudoplastic in nature.
2. The comparison of suspension viscosity at the same d_{80} (-250 mesh) revealed that the viscosity of such suspension prepared in the stirred mill is higher than that of the conventional mill at low shear rates (<30 sec^{-1}) and smaller at higher shear rates (>30 sec^{-1}).
3. The stirred ball mill produced broader particle size distribution when compared to the conventional mill. For a product having fineness of d_{80} (-250 mesh), distribution moduli were found to be .384 and .213 for the conventional and pin-device stirred mill, respectively.
4. A simple theoritical analysis was presented to estimate the shear rate in the pin device stirred mill for this type of application under a typical operating condition and yielded a value of the order of 680 sec^{-1}.
5. Stability evaluation criterion showed that the suspension prepared in either device will be non-settling.
6. The effect of shape of particles on the rheology of suspensions was found to be very pronounced. Suspensions were most viscous if prepared from particles ground in the pin-device stirred mill, followed by those in disc-device stirred mill,then conventional mill, and finally natural coal feed. These results seemed to correlate well with the shape analysis done on particles ground in different grinding devices.
7. An analysis of pumping power requirements showed that stirred mill required 94.33 Hp/mile, as opposed to the tumbling mill which required 84.13 Hp/mile for suspension having the same d_{80}=250 mesh under typical fluid flow conditions.

ACKNOWLEDGEMENTS

The authors wish to acknowledge the financial support of the College of Mines and Mineral Industries at the University of Utah in the form of Utah Mineral Leasing funds and partial research support provided by Draiswerke, Inc. Also, one of the authors (RKM) wishes to thank The University of Utah Research Committee for the financial assistance in the form of Graduate Research Fellowships (1985-87), and Mr. Carl Rampacek, Director, Mineral Resources Institute of The University of Alabama for the financial support.

REFERENCES

1. Rukin, E.I., Groskaya, T.P., and Delyagin, G.N., 1976, "A Study of Aqueous Suspensions of Coal in the Presence of Surface-Active Agents", Khimiya Tverdogo Topliva, Vol. 10, No. 4, pp. 152-158.
2. Link, J.M., Laviangia, N.J., and Faddick, R.R., 1974, "The Economic Selection of a Slurry Pipeline", Hydrotransport, Vol. 3, May.
3. O'Hara, J.B., 1976, "Coal Liquifaction", H.C. Processing, Vol. 55, No. 11, p. 221.
4. Jinescu, V.V., 1974, "The Rheology of Suspensions", Inter. Chem. Eng., Vol. 14, p. 397.
5. Devaney, F.D., and Shelton, S.M., 1940, "Properties of Suspension Mediums for Float and Sink Concentration", U.S. Department of Interior, Bureau of Mines, Ri3469-R.

6. Eveson, J., 1954, "Viscosity of Suspensions", Ind. Eng. Chem., Vol. 46, p. 1146.
7. Aplan, F.F. and Spedden, H.R., 1965, "Viscosity Control in Heavy-Media Suspensions", Proceedings VII, International Mineral Processing Congress, Vol. 1, p. 103.
8. Siddique, M., 1972, "A Kinetic Approach to Ball Mill Scale-Up for Dry and Wet Systems", M.S. Thesis, University of Utah, Salt Lake City, Utah.
9. Mehta, R.K., 1987, "Characterization of Coal-Water Slurries Produced in a High-Speed Stirred Ball Mill", Ph.D. Dissertation, University of Utah, Salt Lake City, Utah.
10. Stehr, N., Mehta, R.K., and Herbst, J.A., 1987, "Comparison of Energy Requirements for Conventional and Stirred Ball Milling of Coal-Water Slurries", Coal Preparation, an International Journal, Vol. 4, pp. 209-226.
11. Sommer, T.M. and Funk, J.E., 1981, "Development of a High-Solids Coal-Water Mixture for Application as a Boiler Fuel", ASME/IEEE, Power Generation conference, St. Louis, October 4-8.
12. Mchale, E.T., Scheffee, R.S., and Rossmeissl, N.P., 1983, "Combustion of Coal-Water Slurry", Combustion and Flame, Vol. 45, pp. 121-135.
13. Funk, J.E., et al., 1981, "Preparation and Combustion of a High Solids Coal-Water Fuel CO-AL", DOE Workshop in Coal-Water Fuel Technology, Pittsburgh.
14. Schramm, Gebhard, 1985, Introduction to Practical Viscometry, Haake Viscometers, New York, p. 5.
15. Morgan, M.E., Heation, H.L., and Scheffee, R.S., 1985, "A Study of Yield Stress of CWF", Proceedings, U.S. Dept. of Energy, Pittsburgh Energy Technology Center, VIIth International Symposium on Coal Slurry Fuels Preparation and Utilization, May 21-24, New Orleans, Louisiana;
16. Bird, R.B., Stewart, W.E., and Lightfoot, E.N., 1978, Transport Phenomena, John Wiley & Sons, Inc., New York, p.
17. Charm, S. and McComis, W., 1965, "Determination of Yield Point for Transportation Systems", Food Technology, Vol. 19, p. 948.
18. Henderson, C.B. and Scheffee, R.S., 1983, "The Optimum Particle-Size Distribution of Coal for Coal-Water Slurries", Mini-Symposium, Coal Slurry Fuels, SME Annual Meeting, March, Atlanta.
19. Govier, G.W. and Aziz, K., 1972, The Flow of Complex Mixtures in Pipes, Van Nostrand-Reinhold, New York.
20. Davis, P.K., and Srivastava, P., 1982, "Rheological and Pumping Characteristics of Coal-Water Suspensions", Journal of Pipelines, Vol. 3, pp. 97-107.
21. Wasp, E.J., Kenny, J.P., and Gandhi, R.L., 1975/77, Solid Liquid Flow Slurry Pipeline Transportation, Series on Bulk Material Handling, Vol. 1, No. 4, Trans Tech Publications.

APPENDIX
PROCEDURE OUTLINING THE MEHTODOLOGY USED IN CONSTRUCTING THE FARRIS DISTRIBUTION

The Farris distribution is given by

$$F_3(d) = \frac{d^{DM} - d_s^{DM}}{d_l^{DM} - d_s^{DM}}$$

Where d is the geometric mean size of any interval

d_s is the smallest geometric mean particle size

d_l is the largest geometric mean particle size

$F_3(d)$ is the cummulative fraction finer than particle of size d

DM is the distribution modulus

For the particulate assembly of interest

$$d_l/d_s = 5.17$$

$$DM = \frac{\ln (1/(1-C_v))}{\ln (d_l/d_s)}$$

$$.3 = \frac{\ln (1/(1-C_v))}{\ln 5.17}$$

Hence C_v = .39 (volume fraction of solids)

$$.39 = \frac{25/1.4}{25/1.4 + x/1}$$

x = 28 cc of water

Mix 25 gms of coal according to the above distribution with 28 cc of water for the purpose of viscosity analysis.

VELOCITY OF VARIOUSLY SHAPED PARTICLES SETTLING IN NON-NEWTONIAN FLUIDS

MOACYR LARUCCIA, * CESAR SANTANA, * AND ERIC MAIDLA**
* Chemical Engineering Department; ** Petroleum Engineering Department, State University of Campinas (UNICAMP), C.P. 6122, 13081 Campinas, SP - Brazil

ABSTRACT

This research [1] concerns the development of a drag coefficient correlation for nonspherical particles settling in purely viscous non-Newtonian fluids. The dynamic interaction term between fluids and particles was studied using both the dimensional analysis and a large number of experimental data covering the laminar, transitional and turbulent flow regime to obtain a generalized correlation for the determination of the settling velocity valid for particles on a sphericity (ϕ) range from 0.5 to 1.

Unlike the previous published research in this area, this generalized correlation does not depend on a particular rheological model.

The developed correlation for the drag coefficient C_D assumes the form

$$C_D = \left\{ \left[\frac{24\ \Omega(\phi)}{Re_{gen}} \right]^m + \left[\chi(\phi) \right]^m \right\}^{1/m} \tag{1}$$

being the Reynolds number Re defined here as

$$Re = \frac{\rho\ v_t^2\ \Theta(\phi)}{\tau(\dot{\gamma})} \tag{2}$$

In equation (2), $\Theta(\phi)$ is a known form factor and $\tau(\dot{\gamma})$ is the shear stress correspondent to a shear rate $\dot{\gamma}$ related to the particle diameter d_p and to the settling velocity v_t by the following equation:

$$\dot{\gamma} = \frac{v_t}{d_p}\ \Theta(\phi) \tag{3}$$

In equation (1) the functions $\Omega(\phi)$ and $\chi(\phi)$ known from experiments considering the limit cases of laminar fully turbulent flow and the exponent m is determined from the data reduction using the Churchill's asymptotic method and an extensive data file from the literature.

A form for v_t can be obtained by combination of the above dimensionless numbers resulting

$$v_t = \left\{ \left[\frac{4}{3} g\ \frac{d_p\ (\rho_s - \rho)}{\chi(\phi)\ \rho} \right]^m - \left[\frac{24\ \tau(\dot{\gamma})}{\rho}\ \alpha(\phi) \right]^m \right\}^{\frac{1}{2m}} \tag{4}$$

Advances in Fine Particles Processing
John Hanna and Yosry A. Attia, Editors

The match of experimental data led to the following sphericity (ϕ) dependent parameters:

$$\chi(\phi) = e^{(4.69 - 5.53\,\phi)}$$

$$\alpha(\phi) = -\frac{(1.65 - 0.656\,\phi)\; e^{(5.53\,\phi - 4.69)}}{(3.45\,\phi^2 - 5.25\,\phi + 1.41)}$$

$$\Omega(\phi) = 1.65 - 0.656\,\phi$$

$$\Theta(\phi) = -3.45\,\phi^2 + 5.25\,\phi - 1.41$$

INTRODUCTION

The technological importance of the knowledge of the velocity for spherical and non-spherical particles settling in non-newtonian fluids is related to the techniques of roch cuttings transport by drilling fluids and to solid-liquid separation and processing in the chemical and mineral industry.

A review of the main developments in this subject is presented by Sample and Bourgoyne [2] and also by Meyer [3] and Peden & Luo [4], being the correlations used to predict the solid-liquid relative velocity valid for particular rheological models and showing very dissimilar results on a comparison between them.

In the present work we have the purpose of developing a drag-coefficient correlation for both spherical and non-spherical particles applicable for all flow regimes and, unlike the previous published research in this area, independent on particular rheological model for purely viscous non-newtonian fluids.

The validity of the resulting correlation is confirmed by a substantial number of experimental data obtained by several authors and also by data from our solid-liquid fluidization set-up designed to obtain the influence of particle concentration on the solid-liquid relative velocity and to obtain the settling velocity of particles at a vanishing solid concentration.

NON-SPHERICAL PARTICLE DYNAMICS AND NEWTONIAN FLUIDS

For a particle having a mass M with volume V and a density ρ_s moving in the gravitational field with a velocity v and submitted to a resistive force $\underset{\sim}{f}$ we have the motion equation and the drag coefficient C_D definition

$$M \frac{d\underset{\sim}{v}}{dt} = (\rho_s - \rho)\, V\underset{\sim}{g} + \underset{\sim}{f} \tag{5}$$

$$\underset{\sim}{f} = \frac{1}{2} A\, ||\underset{\sim}{u} - \underset{\sim}{v}||^2 \rho\, C_D \frac{\underset{\sim}{u} - \underset{\sim}{v}}{||\underset{\sim}{u} - \underset{\sim}{v}||} \tag{6}$$

Being d_p the diameter of a sphere with the same volume of the particle, we have for the terminal velocity v_t:

$$v_t^2 = \frac{4\,(\rho_s - \rho) g\, d_p}{3 \rho C_D} \tag{7}$$

where for isometric particles and Newtonian fluids we have,

$$C_D = F\,(Re,\ \phi) \tag{8}$$

$$Re = \frac{\rho\, d_p\, ||\underset{\sim}{u} - \underset{\sim}{v}||}{\mu} \tag{9}$$

some correlations for C_D and $C_D Re^2$ based on the experimental data of Pettyjohn and Christiansen [9] and containing the sphericity ϕ of the particles as a fundamental parameter were stablished by Massarani [4] and showed on TABLE I.

FLOW OF NON-NEWTONIAN FLUIDS IN THE VICINITY OF PARTICLES - CREEPING FLOW CASE

To obtain a Generalized Reynolds Number we can use the approach proposed by Massarani and Telles [6], based on adimensional analysis. Considering at first the movement of a particle in an infinite medium where the shear stress τ is the unique material function for the fluid and that the dimension and the form of the particle are characterized by d_p and ϕ, results for the resistive force $\underset{\sim}{f}$:

$$\underset{\sim}{f} = \Psi \left[\tau(\dot{\gamma}), \rho, d_p,\ \phi\ ,\ ||\underset{\sim}{u} - \underset{\sim}{v}|| \right] (\underset{\sim}{u} - \underset{\sim}{v}) \tag{10}$$

Being the shear rate $\dot{\gamma}$ a kinematic variable, we can suppose a dependence only on the geometry and the velocities

$$\dot{\gamma} = \Gamma\ (d_p,\ \phi\ , ||\underset{\sim}{u} - \underset{\sim}{v}||) \tag{11}$$

From the fundamental theorem of adimensional analysis result from equations (10) and (11):

$$\underset{\sim}{f} = \frac{\tau(\dot{\gamma})}{d_p ||\underset{\sim}{u} - \underset{\sim}{v}||} \left[\Psi_1 \frac{\rho ||\underset{\sim}{u} - \underset{\sim}{v}||^2}{\tau(\dot{\gamma})}\ ,\ \phi \right] (\underset{\sim}{u} - \underset{\sim}{v}) \tag{12}$$

$$\dot{\gamma} = \frac{||\underset{\sim}{u} - \underset{\sim}{v}||}{d_p}\, \Theta(\phi) \tag{13}$$

To analyze initially the creeping flow we can take the first term of the Taylor's series expansion of Ψ_1 and normalizing in order to retain the limit case of Stokes solution for newtonian fluids and sphericity equal to one:

TABLE I : CORRELATIONS DEVELOPED BY MASSARANI (REF.5)

FOR ISOMETRIC PARTICLES SETTLING IN NEWTONIAN FLUIDS

VARIABLE TO BE ESTIMATED	ASYMPTOT FOR $Re \leq 0,1$	ASYMPTOT FOR $Re \geq 10^3$	CORRELATION	n
C_D	$\dfrac{24}{K_1 Re}$	K_2	$\left\{\left[\dfrac{24}{K_1 Re}\right]^n + K_2^n\right\}^{1/n}$	$n=0,9$ for $0,6 \leq \phi \leq 0,9$ $n=3,15-2,50\,\phi$ for $0,9 \leq \phi \leq 1$
Re	$\dfrac{K_1 C_D Re^2}{24}$	$\left(\dfrac{C_D Re^2}{K_2}\right)^{0,5}$	$\dfrac{\dfrac{K_1}{24}\left(C_D Re^2\right)}{\left\{1 + \left[\left(\dfrac{K_1 K_2}{24}\right)^{0,5}\left(C_D Re^2\right)^{0,5}\right]^n\right\}^{1/n}}$	$n=1,3$ for $0,6 \leq \phi \leq 0,8$ $n=2,70-1,75\,\phi$ for $0,8 \leq \phi \leq 1$
Re	$\left(\dfrac{24}{K_1\, C_D/Re}\right)^{0,5}$	$\dfrac{K_2}{\left(C_D/Re\right)}$	$\left\{\left(\dfrac{24}{K_1\, C_D/Re}\right)^{n/2} + \left(\dfrac{K_2}{\left(C_D/Re\right)}\right)^n\right\}^{1/n}$	$n=1,5$ for $0,6 \leq \phi \leq 0,8$ $n=3,62-2,65\,\phi$ for $0,8 \leq \phi \leq 1$

$$K_1 = 0,843 \log_{10} \frac{\phi}{0,065} \qquad K_2 = 5,31 - 4,88\,\phi$$

$$\underset{\sim}{f} = \frac{18\ \tau(\dot{\gamma})}{d_p \| \underset{\sim}{u} - \underset{\sim}{v} \|} \ \frac{\Omega(\phi)}{\Theta(\phi)} \ (\underset{\sim}{u} - \underset{\sim}{v}) \tag{14}$$

where

$$\Omega(1) = 1$$

Measurements of the settling velocity v_t of particles leads to the determination of the sphericity functions Ω and Θ. In this way we have for example for newtonian fluids, from equations (5) and (14):

$$\Omega(\phi) = \frac{(\rho_s - \rho) g \, d_p^2}{18 \ \mu \ v_t} \tag{15}$$

And from determinations of settling velocities with fluids of known rheology $\tau(\dot{\gamma})$, we have also from eqs. (5), (13) and (14):

$$(\rho_s - \rho) \ g = \frac{18 \ \Omega(\phi) \ \tau(\dot{\gamma})}{d_p \ \Theta(\phi)} \tag{16}$$

$$\dot{\gamma} = \frac{v_t \ \Theta(\phi)}{d_p} \tag{17}$$

In FIGURES 1 and 2 are depicted the resulting curves for $\Omega(\phi)$ and $\Theta(\phi)$. The function $\Theta(\phi)$ was obtained from experiments with the particles

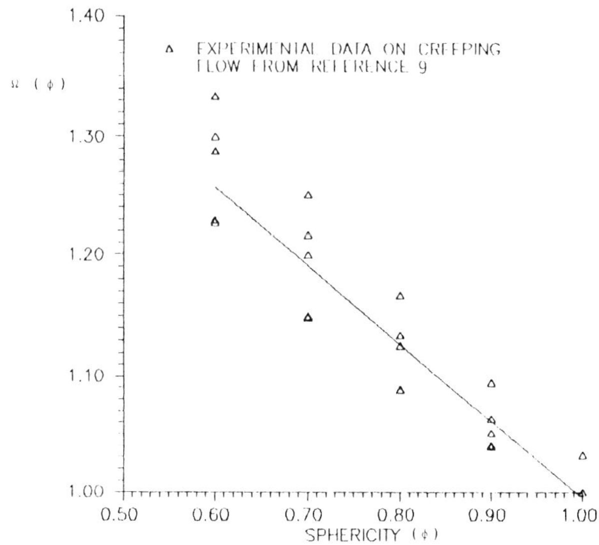

FIGURE 1. FUNCTION Ω (ϕ) FROM EXPERIMENTS WITH NEWTONIAN FLUIDS.

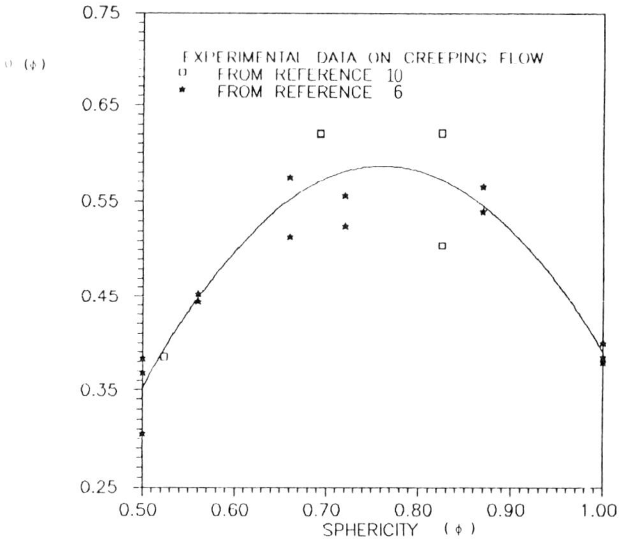

FIGURE 2. FUNCTION Θ (ϕ) FROM EXPERIMENTAL DATA WITH NON-NEWTONIAN FLUIDS.

and fluids with characteristics shown in TABLE II and TABLE III. The numerical fitting for $\Omega(\phi)$ and $\Theta(\phi)$ are:

$$\Omega(\phi) = 1.65 - 0.65\ \phi \tag{18}$$

$$\Theta(\phi) = -3.45\ \phi^2 + 5.25\ \phi - 1.41 \tag{19}$$

for $\qquad 0.5 \leqq \phi \leqq 1.0$

The dimensionless forms from the above analysis can be easily obtained from equation (2) and (8):

$$\frac{C_D}{\Omega(\phi)} = \frac{24}{Re_{gen}} \tag{20}$$

where

$$Re_{gen} = \frac{\rho \| \underset{\sim}{u} - \underset{\sim}{v} \|^2\ \Theta(\phi)}{\tau(\dot{\gamma})} \tag{21}$$

TABLE II: PARTICLE PROPERTIES, FROM REF. 6 AND 10

INVESTIGATOR	SHAPE	(equivalent sphere diameter) mm	SPHERICITY	DENSITY (g/cm^3)
Massarani |6|	DISK	4,65	0,50	2,70
Massarani |6|	DISK	3,87	0,56	2,70
Massarani |6|	DISK	2,97	0,66	2,70
Massarani |6|	DISK	2,54	0,72	2,70
Massarani |6|	DISK	2,58	0,87	2,65
Massarani |6|	SPHERICAL	1,50	1	8,02
Massarani |6|	SPHERICAL	1,50	1	3,89
Massarani |6|	SPHERICAL	2,40	1	10,95
Walker & Mayes |10|	DISK	7,26	0,524	2,83
Walker & Mayes |10|	DISK	5,77	0,825	2,68
Walker & Mayes |10|	DISK	4,57	0,693	2,68
Walker & Mayes |10|	DISK	18,31	0,693	1,38
Walker & Mayes |10|	DISK	23,09	0,825	1,38

TABLE III - FLUID PROPERTIES, FROM REF. 6 AND 10

INVESTIGATORS	TYPE	DENSITY (gm/cc)	SHEAR STRESS (dyne/cm²)	RANGE OF SHEAR RATE (SEC^{-1})
Massarani [6]	Polyacrylamine	1	$8.62\ \dot{\gamma}^{0.442}$	1.56 - 72.9
Massarani [6]	Polyacrylamine + Carbosymethyl Cellulose	1	$4.35\ \dot{\gamma}^{0.663}$	3.48 - 34.5
Massarani [6]	Hydrosyethyl Cellulose	1.01	$50.5\ \dot{\gamma}^{0.782}$	0.12 - 1.23
Walker and Mayes [10]	X.C.	1	$15.32\ \dot{\gamma}^{0.327}$	0.1 à 1,000.0
Walker & Mayes [10]	X.C. + C.M.C.	1	$13.02\ \dot{\gamma}^{0.399}$	0.1 à 1,000.0

EXTENSION FOR THE INTERMEDIATE AND TURBULENT REGIME

The Generalized Reynolds Number defined by eq. (14) and the form obtained in eq. (20) can be extended to higher values of Re by using an extensive data file from the literature (References 11 to 14) and the Churchill's asymptotic method [7] already used in paragraph 2, to obtain a C_D correlation in the form given by eq. (1) where $\chi(\phi)$ is a limit case for C_D at fully turbulent flow.

The non-linear least square method of Marquadt [8] coupled to the Churchill's method leaded to the following ϕ dependent parameters:

$$m(\phi) = 2.29 - 0.83\ \phi$$

$$\chi(\phi) = 108.7\ e^{(-5.53\ \phi)}$$

The function $\chi(\phi)$ and the correlation developed for C_D are depicted in FIGURES 3 and 4 where we have a comparison with many experimental data.

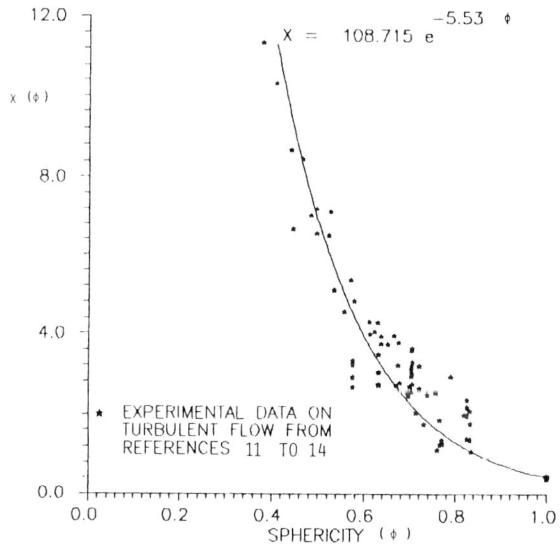

FIGURE 3. FITTING FOR FUNCTION X FROM EXPERIMENTS WITH NEWTONIAN AND NON-NEWTONIAN FLUIDS.

An useful form for v_t determination can be obtained by a combination of C_D and Re_{gen} resulting:

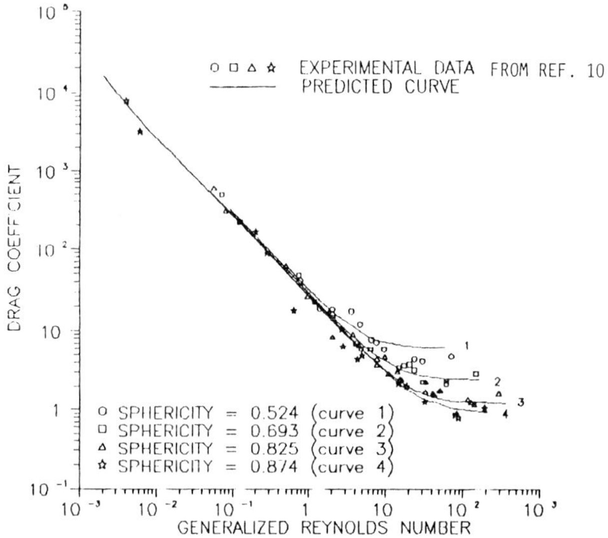

FIGURE 4. DRAG COEFFICIENT CURVES FOR SETTLING OF PARTICLES WITH SPHERICITY ϕ ON NON-NEWTONIAN FLUIDS.

$$v_t = \left\{ \left[\frac{4}{3} \; g \; \frac{d_p \; (\rho_s - \rho)}{\chi(\phi) \; \rho} \right]^m - \left[\frac{24 \; \tau(\dot{\gamma})}{\rho} \; \alpha(\phi) \right]^m \right\}^{\frac{1}{2m}} \tag{4}$$

where

$$\alpha(\phi) = - \frac{(1.65 - 0.656 \; \phi) \; e^{(5.53 \; \phi \; - \; 4.69)}}{3.45 \; \phi^2 - 5.25 \; \phi + 1.41}$$

From eq. (4) v_t can be determined using a numerical procedure to solve that implicit equation. FIGURES 5 to 9 shows the comparison between predicted and calculated v_t.

THE EFFECT OF PARTICLE CONCENTRATION

An experimental analysis for the effect of particle concentration on the settling velocity of a swarm of particles can be accomplished by making solid-liquid fluidization experiments.

In these simple experiments the measurement of the fluid superficial velocity and the height of the fluidized solids can substitute the measurement of the relative solid-liquid velocity and the average concentration, considering that the solid velocity is statiscally zero.

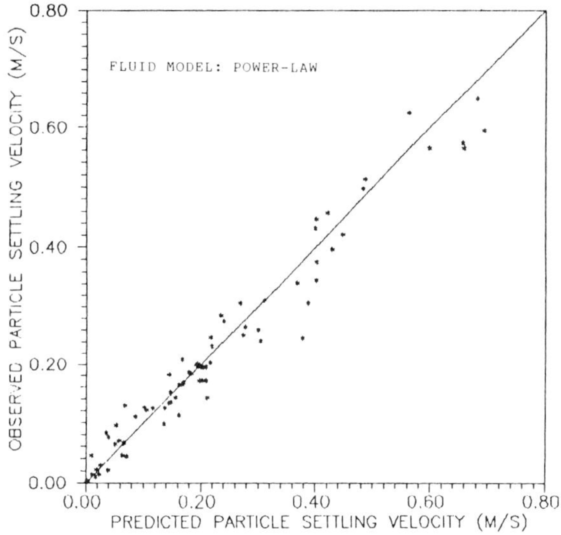

FIGURE 5. COMPARISON BETWEEN CALCULATED AND EXPERIMENTAL DATA FROM REFERENCE 10.

(Walker & Mayes, 1975)

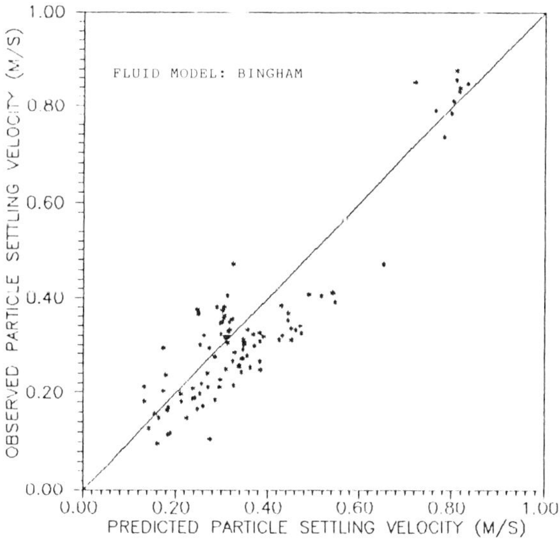

FIGURE 6. COMPARISON BETWEEN CALCULATED AND EXPERIMENTAL DATA FROM REFERENCE 11.

(Hall, Thompson & Nuss, 1950)

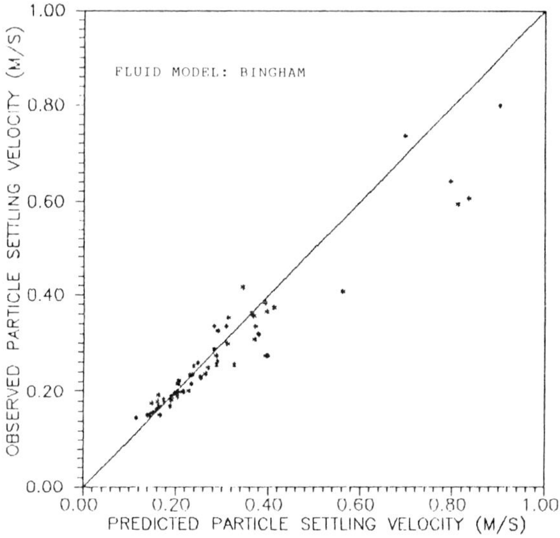

FIGURE 7. COMPARISON BETWEEN CALCULATED AND EXPERIMENTAL DATA FROM REFERENCE 12.
(Hopkin, 1967)

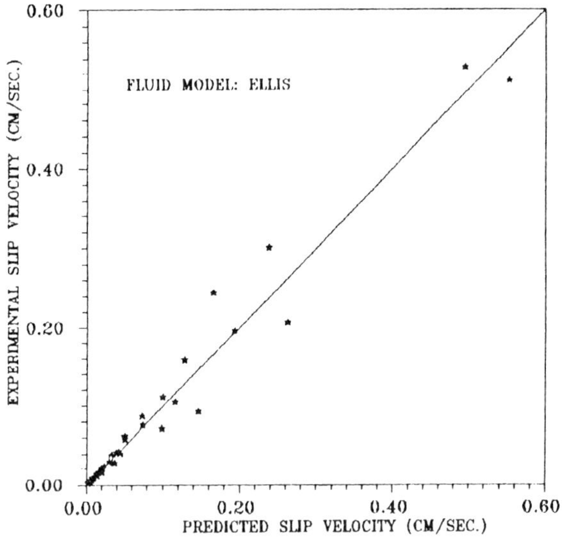

FIGURE 8. COMPARISON BETWEEN CALCULATED AND EXPERIMENTAL DATA FROM TURIAN REF. 15

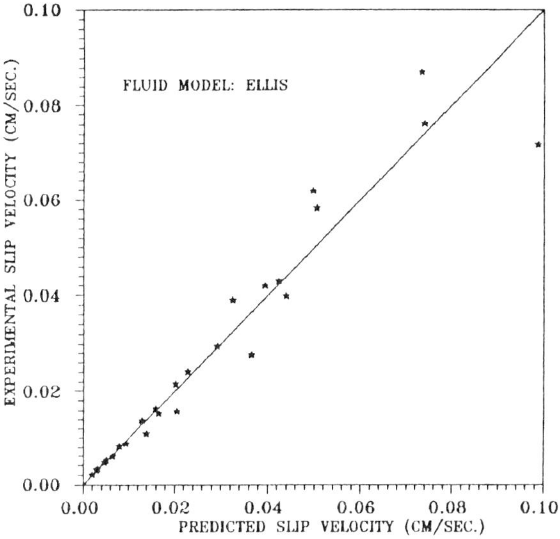

FIGURE 9. COMPARISON BETWEEN CALCULATED AND EXPERIMENTAL DATA FROM TURIAN REF. 15

In FIGURES 10 and 11 are shown the behavior of the relative velocity for several system porosities using carboxi-metil-cellulose aqueous solutions as fluidizing liquids. From these curves the settling velocity of particles in an infinite medium v_t can be obtained making the porosity ε

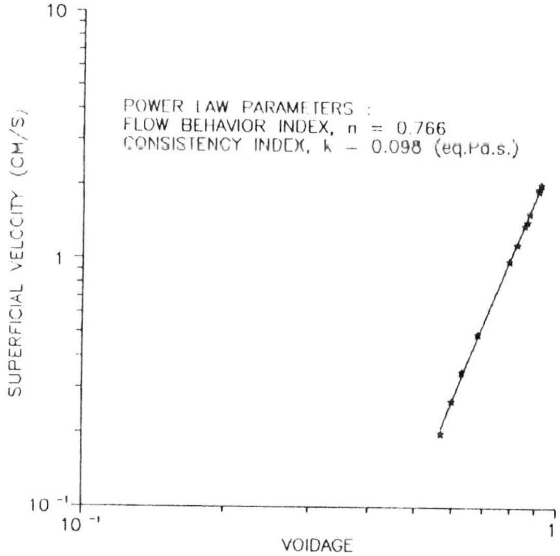

FIGURE 10. EXPERIMENTAL DATA OBTAINED FROM 0.66% WT. CMC AQUEOUS SOLUTION AND ELLIPTICAL DISK PARTICLES SHAPE OF NYLON WITH AVERAGE SPHERICITY OF 0.853.

equal to one.

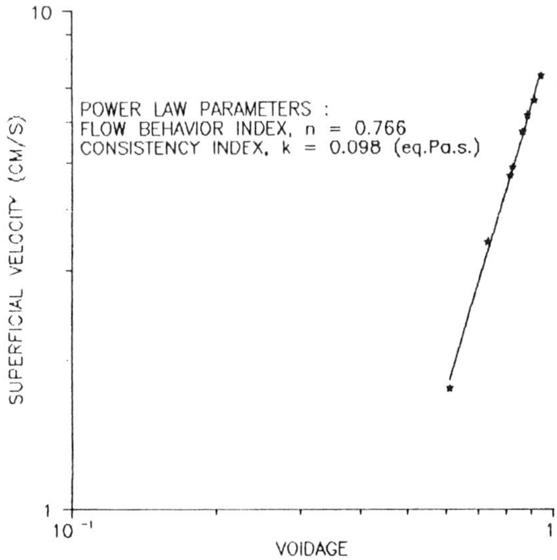

FIGURE 11. EXPERIMENTAL DATA OBTAINED FROM 0.66% WT. CMC AQUEOUS SOLUTION AND DISK PARTICLES SHAPE OF POLYETHYLENE.

The experimental data can be interpreted by an empirical equation in the form:

$$\frac{u_s}{v_t} = \varepsilon^i \qquad (22)$$

The exponent i is a function of the Generalized Reynolds Number Re_{gen}, as shown in FIGURE 12 and from the experimental data obtained in this work we can propose the form:

$$i = 4.693 \left[\frac{\rho\, v_t^2\, \Theta(\phi)}{\tau(\dot{\gamma})} \right]^{-0.111} \qquad (23)$$

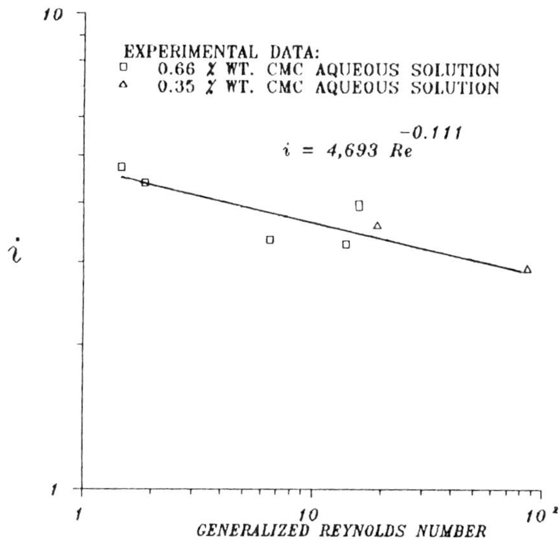

FIGURE 12. FITTING FOR PARAMETER i

CONCLUSIONS

A drag-coefficient correlation has been developed for both spherical and non-spherical particles applicable for settling in a purely viscous non-newtonian fluid as an infinite medium. Also the influence of particle concentration in the non-newtonian settling process has been investigated by using an experimental procedure based on the homogeneous fluidization of particles.

The validity of the results was confirmed by 275 experimental data of several authors and can be used, for instance, to predict the slip velocity between the fluid and particles on a cuttings transport system used in drilling of oil wells.

As the experimental data used here include solids larger than 0.6 mm in diameter it is necessary to extend the results to fines particles. Further research including visco elastic fluids are necessary to obtain a generalized Reynolds Number. To extend also the procedure to obtain the generalized Reynolds Number it is recommended for future work to include fluids with elastic properties.

REFERENCES

1. Laruccia, M.B., Estudo do Efeito da Forma e Concentração na Velocidade de Sedimentação de Partículas em Fluidos Não-Newtonianos, Tese de Mestrado. Dept. de Engenharia Química, UNICAMP, 1990.
2. Sample, K.J. and Bourgoyne, A.T., An Experimental Evaluation of Correlations used for Predicting Cutting Slip Velocity, SPE paper No. 6645, 1977.
3. Meyes, B.R., Generalized Drag Coefficient Applicable for all Flow Regimes, Oil and Gas Journal, May 26, pp. 71-77, 1986.

4. Peden, J.M. and Luo, Y., Settling Velocity of Variously Shaped Particles in Drilling and Fracturing Fluids, SPE Drilling Engineering, pp. 337-343, 1987.
5. Massarani, G., Some Aspects of Solid-Fluid Separation, Chapter 1 in Sepcial Topics in Particulate Systems, vol. 2, 1986.
6. Massarani, G. and Telles, A.C.S., Flow of Non-Newtonian Fluids in the Vicinity of Solid Particles, Proc. 3rd. Brazilian Symposium of Fluid Mechanics and Heat Transfer, May 1977.
7. Churchill, S.W., The Development of Theoretically Based Correlations for Heat and Mass Transfer, Latin American Journal of Heat and Mass Transfer, vol. 77, pp. 207-229, 1983.
8. Marquadt, D.W., An Algorithm for Least-Squares Estimation of Nonlinear Parameter, S.Soc. Indust. Math., 11, 2, June 1963.
9. Pettyjohn, E.S. and Christiansen, E.B., Effect of Particle Shape on Free-Settling Rates of Isometric Particles, Chem. Eng. Progress, 44, 2, 157, 1948.
10. Walker, R.E. and Mayes, T.M., Design of Muds for Carrying Capacity. Journal of Petroleum Technology, July, 1975.
11. Hall, H.N., Thompson, H. and Nuss, F., Ability of Drilling Mud to Lift Bit Cuttings. Trans. AIME 189, 35-46, 1950.
12. Hopkin, E.A., Factors Affecting Cuttings Removed During Rotary Drilling, Trans. AIME, vol. 240, p. 807, 1967.
13. Williams, C.E. and Bruce, G.H., Carrying Capacity of Drilling Muds. Trans. AIME 192, 111-120, 1951.
14. Sifferman, T.R. Meyers, G.M., Haden, E.L. and Wahl, H.A., Drill-Cutting Transport in full-scale Vertical Annuli. J.Pet.Tech. pp 1295-1302, November, 1974.
15. Turian, R.M., Ph.D. Thesis, Univ. Wisconsin, Madison, 1964.

NOMENCLATURE

A - Characteristic particle area
C_D - Drag coefficient
d_p - Particle Sauter mean diameter
$\underset{\sim}{f}$ - Resistive force
$\underset{\sim}{g}$ - Gravity acceleration
i - Exponent of Eq. (22)
M - Particle mass
m - Function of sphericity
$\underset{\sim}{u}$ - Local fluid velocity
u_s - Superficial fluid velocity
$\underset{\sim}{v}$ - Local particle velocity
V - Particle volume
v_t - Particle settling velocity on a infinite media
Re - Reynolds Number (Eq. 9)
Re_{gen} - Generalized Reynolds Number (Eq. 21)

Greek Letters:

α - Function of sphericity (Eq. 4)
ϕ - Particle sphericity
ρ - Fluid density
ρ_s - Particle density
θ - Function of sphericity
$\dot{\gamma}$ - Shear rate

τ - Shear stress
χ - Function of sphericity
μ - Newtonian fluid viscosity
Ω - Function of sphericity

DETAILED FLOW PATTERNS IN THE CYLINDRICAL CYCLONE DUST COLLECTOR

*AKIRA OGAWA,*TETZURO KATO,*AKIRA HIRONAKA,**HISAO NAGABAYASHI,
* Department of Mechanical Engineering,** Department of Civil Engineering,
College of Engineering,Nihon University,
T-963. Tokusada, Tamura-machi,Kooriyama-shi,Japan.

ABSCTRACT

Up to this time many papaers concerning the distributions of the tangential,axial and radial velocities, the Reynolds stress and also the equi-flow rate lines, the particle concentration, the fractional collection efficiency and the collection efficiency in the various types of the cyclone dust collectors were reported. In this paper, basing upon the results of the measurements of the distributions of the axial velocities in the cylindrical cyclone of the diameter D_1=139.5 mm and the total length H=385 mm with a cylindrical Pitot-tube for Reynolds number $Rec=Q_0/H_i\nu$ =2159(V_0=5.0 m/s),3023 (V_0=7.0 m/s) and 4318(V_0=10.0m/s), the detailed equi-flow rate lines were shown. From these results, the remarkable non-symmetrical flow patterns were shown and also from the results of the tangential velocities, the fluid dynamical unstable region due to the non-symmetricalrotational flow was shown. In addition to this, a result of theReynolds stress $\hat{\tau}_{\theta z}=-\rho\overline{v_\theta v_z}$ for V_0=10.0 m/s measured with a hot-wire with X-probe is shown. Here Q_0(m^3/s) is the flow rate into the cyclone, H_i(m) is the imaginary cylindrical length, ν (m^2/s) is the kinematic viscosity of air, V_0(m/s) is the mean inlet velocity in the inlet pipe, ρ(kg/m^3) is the density of air.

1.INTRODUCTION

Up to this time there are many papers concerning the turbulent rotational gas flows in the cylindrical cyclones[1] , in the cyclone combustion chambers [2], in the rotatry flow dust collectors or another types of the vortex chambers[3,4,5,6].However the detailed report of the flow patterns in the above stated vortex chambers is a few. While, in order to recognize the separation mechanisms of the fine solid particles in the turbulent rotational air flow in the various kinds of the cyclones or in the rotatry dust collectors and also the energy dissipation of the turbulent rotational flow, in addition to this, the transport of the angular moment of the rotational flow or the stability criterion of the rotational flows, it is neccessary to study the detailed flow patterns in the vortex chamber which is one of the most simple constructions. From the above stated point of view, the paper of Smith(1962) are very interesting, but the several problems are pointed out from the experimental results in the vortex chambers as follows;(1) the stream lines(equi-flow rate lines) are shown in the quasi-free vortex region and these stream lines are distributed nearly axis-symmetric configurations,(2)in the quasi-forced vortex region,it looks like to be stagnant state for the axial flow along the vortex core,(3) as a result of the stream lines, it may be assumed that the radial flow is homogeneous and inward direction along the center line. Therefore in this paper the authors study the detailed flow patterns of the turbulent rotational gas flow in the cylindrical vortex chamber.

2. EXPERIMENTAL APPARATUS AND EXPERIMENTAL CONDITIONS

Fig.1 shows the detailed sizes of the vortex chamber which made of the transparent acrylic resin, so the inner surface of this vortex chamber is very smooth. The diameter of the inlet pipe , of the vortex chamber and of the exit pipe are D_0=50.0 mm,D_1=140.0 mm and D_2=50.0 mm. The total length is

Advances in Fine Particles Processing
John Hanna and Yosry A. Attia, Editors

L=385 mm and the imaginary cylindrical length H_i which is the imaginary length from the edge of the exit pipe to the bottom surface is H_i=303 mm. And the distributions of the tangential velocity and the static and total pressures are schematically shown. In order to measure the time mean value of the tangential(V_θ) and axial(V_z) velocities of gas flow, a cylindrical Pitot-tube of the diameter d=3 mm is used. The axial positions beneath the edge of the exit pipe are Z=10,40,100,150,200,250, and 280 mm. The peripheral angles θ from the connection point of the inlet pipe to the vortex chamber are θ=0°, 45°,90°,180°,270° and 315°, respectively.The mean inlet velocity in the inlet pipe is V_0=5.0, 7.0, and 10.0 m/s. The Reynolds number Rec defined as

$$Rec = Q_0/H_i \cdot \nu \qquad (1)$$

corresponding to V_0=5.0,7.0 and 10.0 m/s is Rec=2159,3023 and 4318,respectively, where Q_0(m^3/s) is the flow rate of gas into the vortex chamber.

3. EXPERIMENTAL RESULTS

3.1 PRESSURE DROP OF THE VORTEX CHAMBER

When the static pressure in the inlet pipe and the atmospheric pressure are p_0(Pa) and p_a(Pa), so as a result of the pressure feed method, the pressure drop Δp_c(Pa) of the vortex chamber can be determined as

$$\Delta p_c(\text{Pa}) = (p_0 - p_a) + \frac{1}{2}\rho V_0^2 \qquad (2).$$

Then the coefficient of the pressure drop can be defined as

$$\zeta_c(1) = \Delta p_c / \frac{1}{2}\rho V_0^2 \qquad (3).$$

The experimental results between Δp_c and V_0 and also between ζ_c and R_{ec} are shown in Fig.2. Roughly speaking, the pressure drop is nearly proportional to the square of the inlet velocity V_0, therefore the coefficient of the pressure drop is nearly independent of Rec.

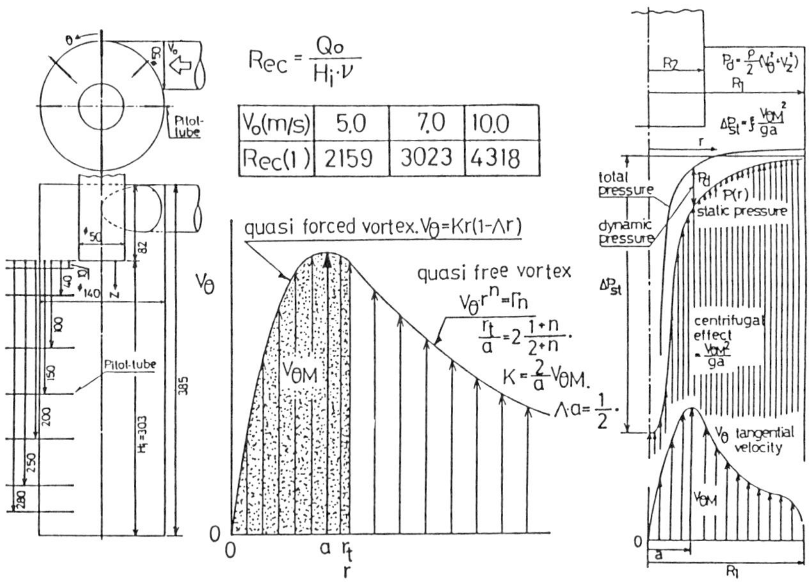

V_0(m/s)	5.0	7.0	10.0
Rec(1)	2159	3023	4318

Fig.1 Schematic illustration of the vortex chamber and the Ogawa combined vortex model and the distributions of the static and total pressures.

3.2 VELOCITY DISTRIBUTIBUTIONS

Fig.3 shows the distributions of the tangential velocities V_θ which are made dimensionless by the inlet velocity V_0. From these results, the velocity distribution in the quasi- free vortex can be written as

$$V_\theta \cdot r^n = \Gamma_n = \text{constant} . \qquad (4),$$

where the velocity exponent n(1) depends on Reynolds number Rec and the construction of the vortex chamber or of the cyclone types. The value of n is nearly $0.2 < n < 0.95$ for the ordinary types of the vortex chambers but as shown in the later figures, n depends on the measured positions and is not always smaller than n=1.0. The radii r=a corresponding to the maximum tangential velocity $V_{\theta M}$ are located nearly at about $a \doteqdot (2/3)R_2$. This radius a can be estimated by the assumption of the minimum energy flux[7]. One of the experimental results is shown in the later figure.

Fig.4 shows the distributions of the axial velocity V_Z at 0=0°,90°,180° and 270°, respectively. From Fig.4, it is interesting to note that the axial velocities along the center axis show the reverse flow. БАЛУЕВ and ТРОЯНКИН (1967) were derived the equation of the axial velocity which showed the reverese flow near the center of the vortex core by the Reynolds equation [8]. This type of the reverse flow near the center of the vortex core was observed with the visualization method with water into the vortex chamber by Fujita and Ogawa [9]. The radii r=b corresponding to $V_Z(r)=0$ which means the boundary radius between the downward flow and the upward flow are located nearly at $b \doteqdot (2/3)R_1$.

3.3 DISTRIBUTIONS OF $V_{\theta M}/V_0$, a/R_2 and n.

Fig.5 shows the maximum tangential velocity $V_{\theta M}/V_o$ at the angle θ=90° for the Z-axis. With increasing the mean inlet velocity V_0(Reynolds number Rec is also increased .) the value of $V_{\theta M}/V_o$ increases.

Fig.6 shows the radial position r=a corresponding to the maximum tangential velocity $V_{\theta M}$ at 0=90° for the Z-axis. The value of a/R_2 show the clear distribution from Z=0 to Z=150 mm and depends on V_0 or Rec. But a/R_2 is nearly independent of V_0 or Rec in the region from Z=150 mm to Z= 300 mm, respectively.

Fig.7 shows the velocity exponent n at θ=90° for the Z-axis. As already described above, n is not always smaller than n=1(potential flow) and n becomes to higher values with increasing the values V_0 or R_{ec}. As you known, n=1 is potential flow(free vortex) and basing upon upon the stability criteria by Rayleigh [10] and Prandtl [11], n=1 is a critical point

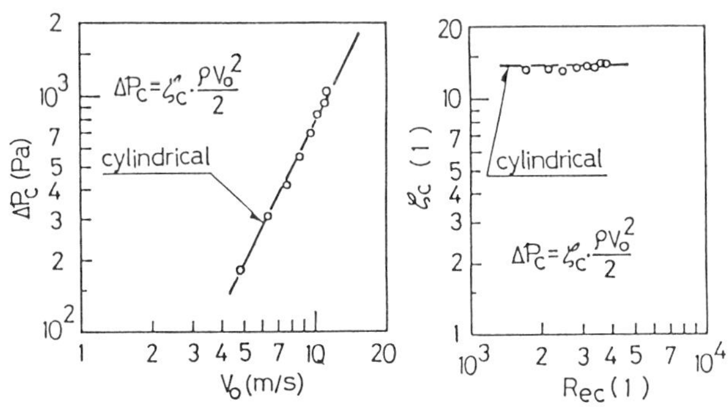

Fig.2 Correlation between the pressure drop and the mean inlet velocity and also correlation between the coefficient of the pressure drop and the Reynolds number

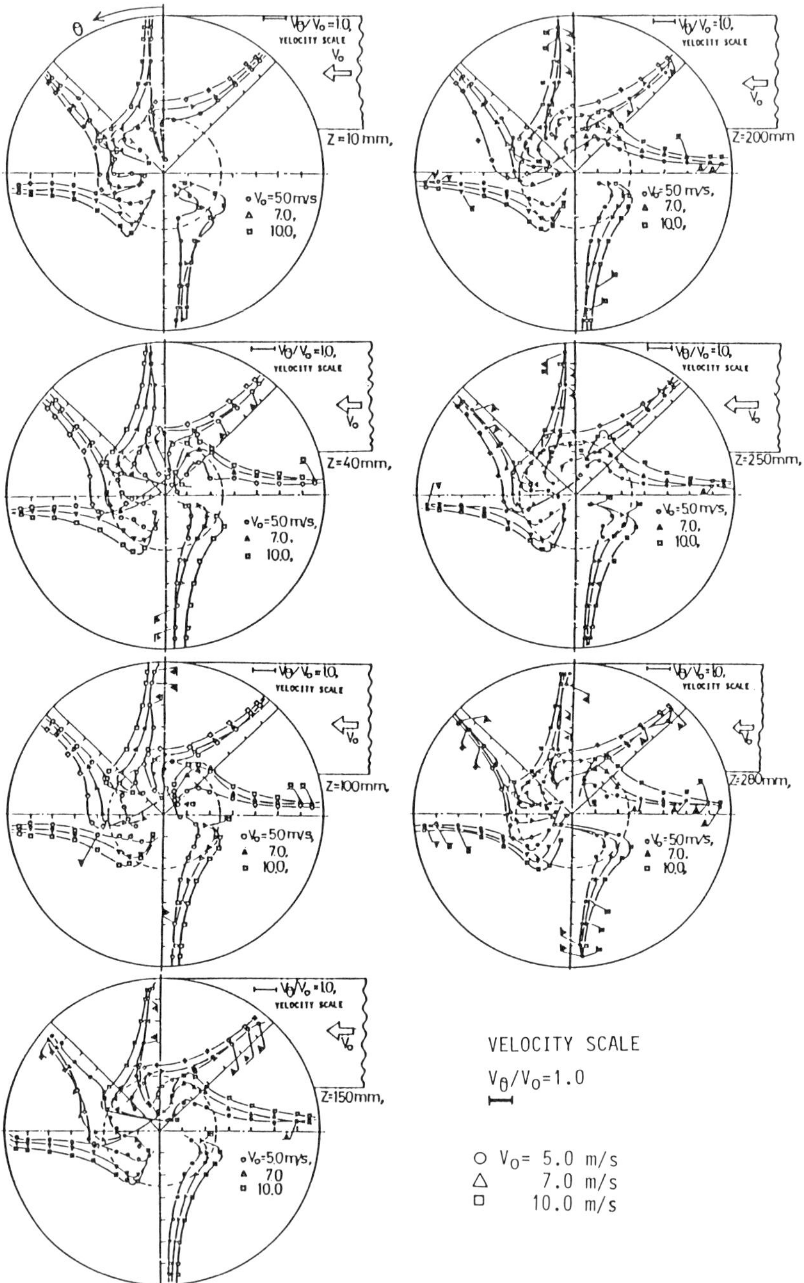

Fig.3 Distributions of the tangential velocities at the various Z-planes.

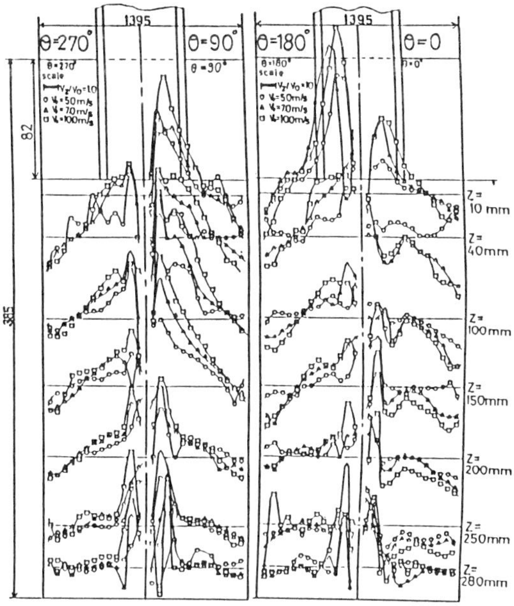

Fig.4 Distributions of the axial velocities at the various Z-planes.

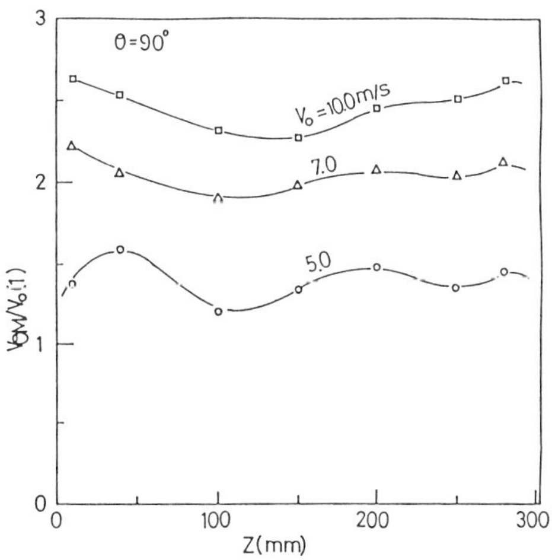

Fig.5 Maximum tangential velocity along the Z-axis

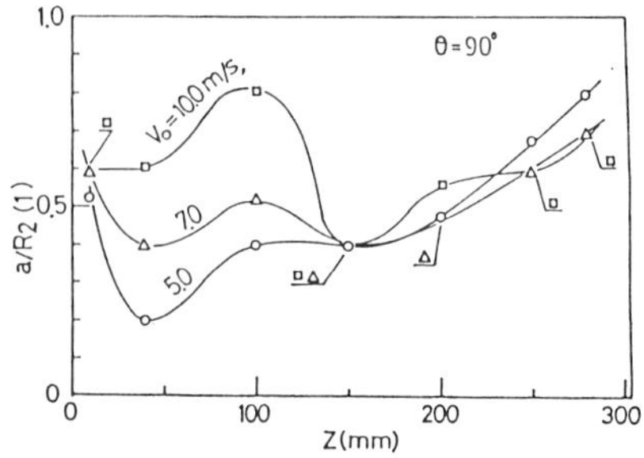

Fig.6 The radii r=a along the Z-axis at θ=90°.

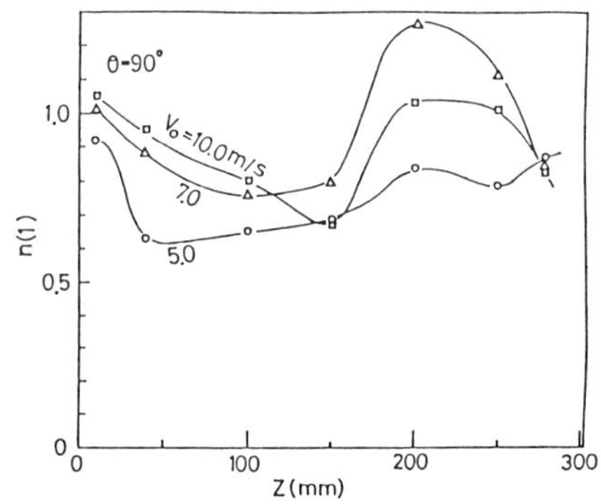

Fig.7 Velocity exponent along the Z-axis at θ=90°.

between the stable and unstable flows. When the value of n is higher than n= 1, so the rotational flow becomes to unstable condition. But when the value n is lower than n=1, so the rotational flow becomes to stable condition. However these criteria were considered under the assumption of these two dimensional flow in spite of the laminar or turbulent flows. But as shown in Fig.4, the rotational flow is not always two dimensional but three dimensional flows and also the radial gradient of the axial velocity is very high value. Therefore by the theory of Ludwieg [12], Scorer[13], described above, the stable condition is

$$\frac{dV_\theta}{dr}\frac{r}{V_\theta} - \frac{\left(\frac{dV_z}{dr}\right)^2\left(\frac{r}{V_\theta}\right)^2}{1 - \frac{dV_\theta}{dr}\frac{r}{V_\theta}} > -1, \qquad (5),$$

and the vorticity components are defined as

$$\zeta_z = \frac{dV_\theta}{dr} + \frac{V_\theta}{r}, \qquad \zeta_\theta = -\frac{dV_z}{dr},$$

even if the stable condition is satisfied by the theory of Rayleigh, the stability condition is not always satisfied by the correlation between dV_θ/dr and dV_z/dr in the quasi-free vortex region which depends on the velocity exponent n. One of the calculated results shows in Fig.8. From this figure it will be recognized that in the quasi-free vortex region the velocity exponent n is lower than n=1, according to the criterion of Ludwieg or

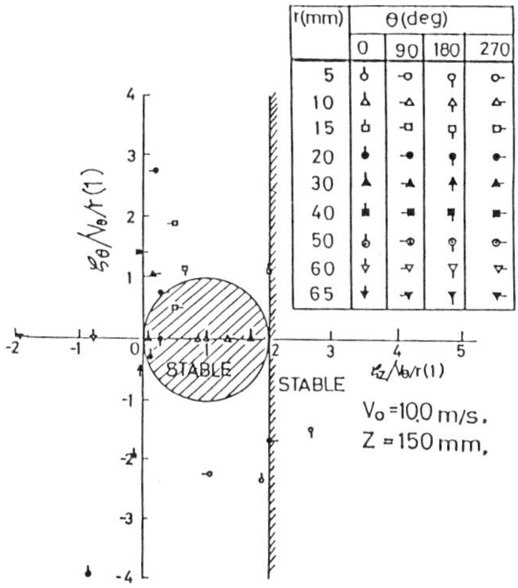

Fig.8 The stable and unstable region of the rotational flow by the criterin of Ludwieg and Scorer.

Scorer, the rotational flow becomes to unstable condition. So it can be possible to generate the torus of the vortex ring around the quasi-forced vortex like a Taylor vortex in the co-axial cylinders.

3.4 EQUI-FLOW RATE LINES.

The equi-flow rate lines(strean lines by Smith) are calculated by the following equation as

$$\Psi_i(r) = \frac{\sum \Delta S_i \cdot V_{zi}}{\sum \Delta Q_i} = \frac{1}{n(A/S)} \sum_{i=1}^{i} \frac{V_{zi}}{V_o} \qquad (6),$$

where $\Delta Q_i = (\sum_{i=1} \Delta S_i/S)Q_o$, $S(m^2)$ is the cross sectional area, $A(m^2)$ is the cross sectional area of the inlet pipe, $V_o(m/s)$ is the mean inlet velocity and $Q_o(m^3/s)$ is flow rate into the vortex chamber. The calculated results by eq. (6) are shown in Fig.9. From this figure it will be find that the generation of the torus of the circulation which is similar to the Taylor vortex around the rotating inner cylinder in the outer fixed cylinder is suggested. This kind of the circulation may be based upon the instability of the rotational flow. It will be reasonable to apply the Ludwieg or Scorer theory rather than Rayleigh stability theory for the criterion of this instability in the quasi-free vortex region. In comparison these results with the result of the equi-flow rate lines of Smith[14,15] , the following points are suggested that(1) in the quasi-forced vortex region the flow patterns are complicated, and also in the quasi-free vortex region the flow patterns are very complicated, and also in the quasi-free vortex region the flow patterns are unsymmetrical states to the center axis,(2) in the quasi-forced vortex region the upward and downward flows are complicated labyrinthinely and also the stagnant region in the vortex center may be generated,(3) as a result of these stream lines the radial flow is not homogeneous to the inward direction along the Z-axis, but the radial outward and inward flows are co-existed. Concerning the radially inward and outward flow problems, the following conditions are considered. When the Euler equation satisfies the following equation as

$$\frac{1}{\rho} \cdot \frac{dp}{dr} - \frac{V_\theta^2}{r} = \frac{d}{dr}\left(\frac{V_r^2}{2}\right) = 0 \qquad (7).$$

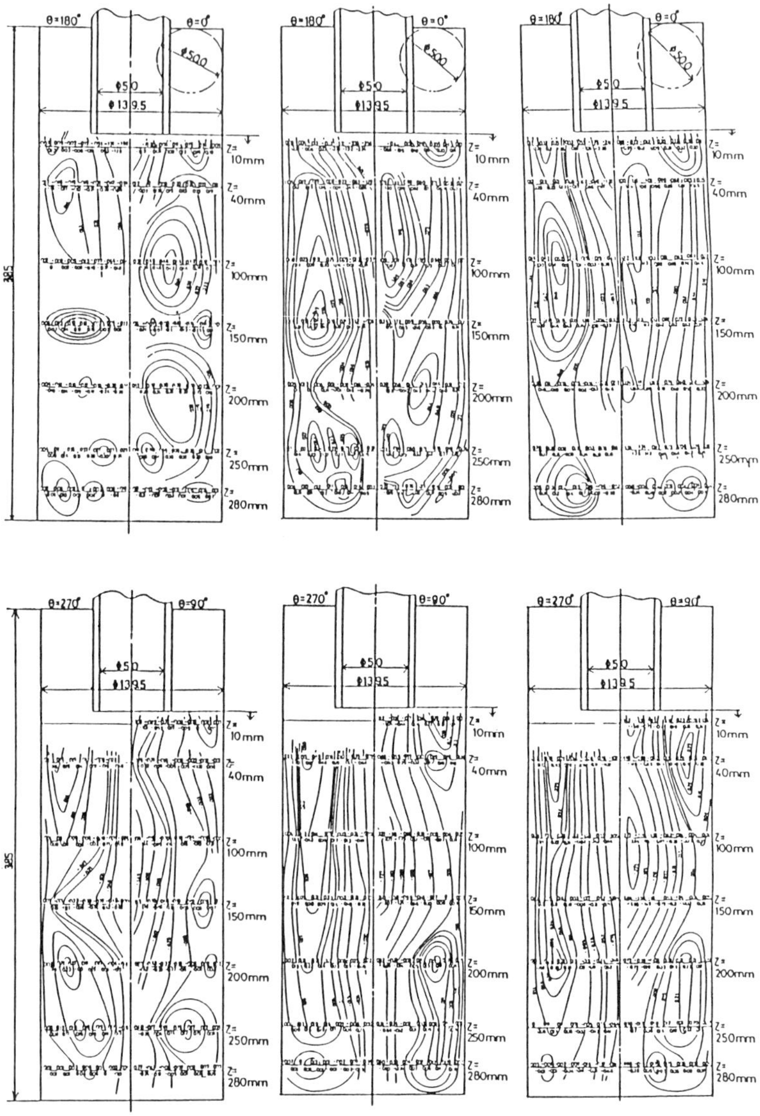

V_O=5.0 m/s V_O=7.0 m/s V_O=10.0 m/s

Fig.9 Equi-flow rate lines for V_O=5.0,7.0 and 10.0 m/s by eq.(6)

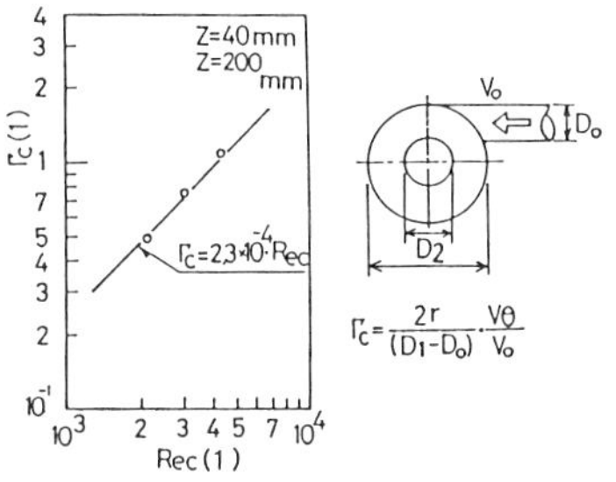

Fig.10 Correlation between circulation and Reynolds number

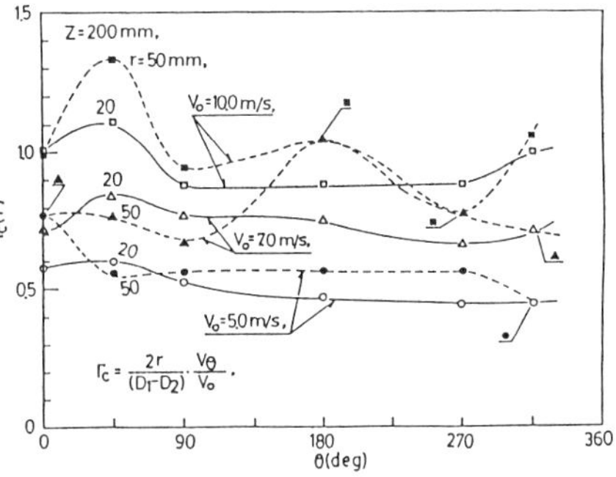

fig.11 Distributions of circulation along the peripheral angle at Z=200 mm

The radial pressure gradient is equal to the centrifugal force per unit mass, so the component of the radial velocity becomes to zero. Then there is no radial flow. Therefore the rotational flow is formed the circular line. But as shown in Fig.5, the value of $V_{\theta M}/Vo$ is not the constant value but is depended on the Z-axis position. Then the following inequality is kept in the vortex chamber as

$$\frac{1}{\rho}\frac{dp}{dr} - \frac{V\theta^2}{r} \gtrless \frac{d}{dr}\left(\frac{Vr^2}{2}\right) \qquad (8).$$

As a result of this inequality, the single or double re-circulating flows due to the secondary flow are generated in the vortex chambers [16] or in the cyclones [17]. Of course one of the other reasons is based upon the instability of the rotational flow based upon the theories by Rayleigh, Taylor or Ludwieg and Scorer. In addition to this, it is enough possibility to generate the unstable state of the rotational flow in the vortex chambers or in the cyclones due to the small deviation of the vortex core, since the peripheral gradient $dp/d\theta$ of the static pressure along a closed circular line becomes to the positive value $dp/d\theta > 0$ on the unstable condition and to the negative value $dp/d\theta < 0$ on the stable condition. This behaviours are suggested by the distributions of the tangential velocities of a solution of the Navier-Stokes equation [18]. Concerning the above problems, one of the experimental results in the eccentric cyclone dust collector by Seito &Ogawa [19] showed that the distributions of the tangential velocities distinctly showed the unstable states which were related to the complicated

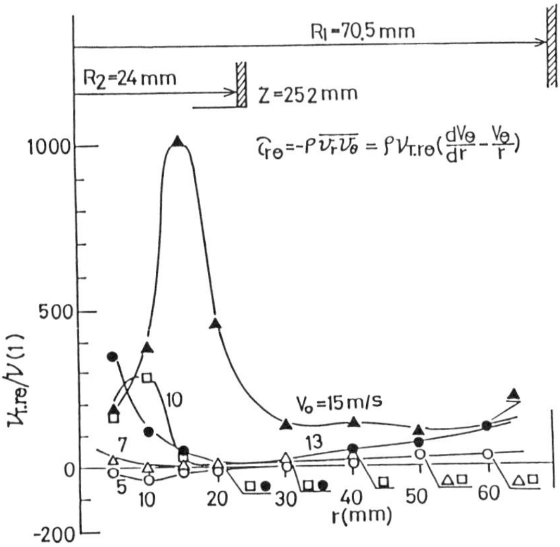

Fig .12 Distributions of the eddy kinematic viscosity

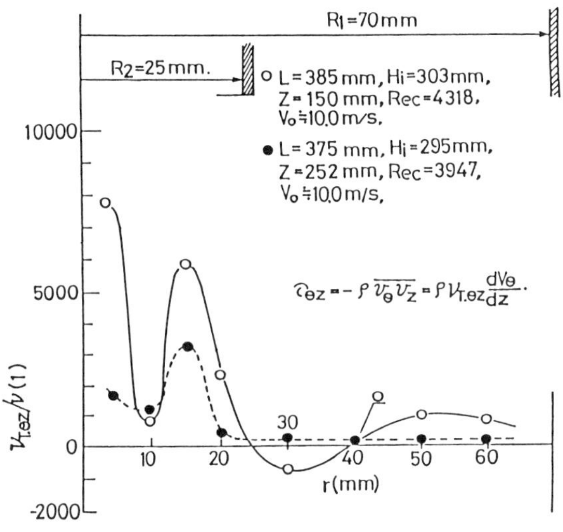

Fig.13 Distributions of the eddy kinematic viscosity

distributions of the axial velocities in comparison with those in the normal cyclones in the region of $dp/d\theta > 0$.

3.5 CIRCULATION IN THE QUASI-FREE VORTEX REGION

Fig.10 shows the dimensionless circulation Γ_c which is defined as

$$\Gamma_c = \frac{2r}{(D_1 - D_0)} \cdot \frac{V_\theta}{V_0} \quad (9),$$

where the experimental results at Z=40 mm and 200 mm are used. The empirical equation can be obtained as

$$\Gamma_c = 2.3\times10^{-4} \cdot Re_c. \quad (10).$$

Therefore Γ_c is proportional to Rec. Fig.11 shows one of the examples of the values of Γ_c along the peripheral angles θ at r=20 and 50 mm at Z= 200 mm. Roughly speaking, the value of Γ_c along the periphery is nearly constant value.

3.6 DISTRIBUTIONS OF THE EDDY KINEMATIC VISCOSITY

In order to measure the Reynolds stresses $\tau_{r\theta}$ and $\tau_{\theta z}$ of the turbulent rotational flow in the vortex chamber, X-type of the hot-wire probe was applied. As shown in the figures of the equi-flow rate lines, the rotational flow is very complicated, so it may be assumed that the stresses $\tau_{r\theta}$ and $\tau_{\theta z}$ show the different values, then the eddy kinematic viscosity $\nu_{T,r\theta}$ and $\nu_{T,\theta z}$ are calculated by the following equation as

$$\tau_{r\theta} = -\rho\overline{v_r v_\theta} = \rho\nu_{T,r\theta}\left(\frac{dV_\theta}{dr} - \frac{V_\theta}{r}\right) \quad (11)$$

$$\tau_{\theta z} = -\rho\overline{v_\theta v_z} = \rho\nu_{T,\theta z}\frac{dV_\theta}{dz}. \quad (12).$$

The distributions of $\nu_{T,r\theta}$ as shown in Fig.12 for the cylindrical vortex chamber has been calculated [20]. Therefore from the distribution of , the calculated results of $\tau_{\theta z}$ is shown in Fig.13, in where it is rather difficult to estimate exactly the value of dVθ/dz due to the short length of the vortex chamber. The ratio of $\nu_{T,\theta z}$ to $\nu_{T,r\theta}$ is from two to seven. So the value of $\nu_{T,\theta z}$ is higher than that of $\nu_{T,r\theta}$ in the quasi-forced vortex region. But in the quasi-free vortex region the value of $\nu_{T,\theta z}$ is nearly constant value and becomes to lower value in comparison with that in the quasi -free vortex region. And also the total configurations of $\nu_{T,\theta z}$ are very similar to those of $\nu_{T,r\theta}$

By the way the interesting experimental results by Shook&Sagar [21] for the helical ribbed pipe of the diameter D=138 mm at $Re=7\times10^4$ showed that (1) the mean tangential velocity could be expressed as

$$V_\theta/V_{zo}=0.17\cdot(r/R_1)^{1.9} \quad (13),$$

where Vzo is the center velocity, (2) the Reynolds stress $\tau_{r\theta}$ could be expressed as

$$-\overline{v_r v_\theta}/u_\tau^2 = 0.4(r/R_1). \quad (14),$$

where u_τ is the frictional velocity defined as $u_\tau=(D_e\cdot\Delta p_c/4\rho\cdot\Delta z)^{0.5}$, here D_e is the hydraulic equivalent diameter, (3) the Reynolds stress could be expressed as

$$-\overline{v_\theta v_z}/u_\tau^2 = (r/R_1)^{2.4} \quad (15),$$

(4) the ratio of $\tau_{\theta z}$ to $\tau_{r\theta}$ was written as

$$\tau_{\theta z}/\tau_{r\theta} = 2.5(r/R_1)^{1.4} \quad (16).$$

(5) the eddy kinematic viscoaity $\nu_{T,r\theta}$ could be expressed as

$$\frac{\nu_{T,r\theta}}{\nu} = \frac{2.61\cdot u_\tau}{V_{zo}}\cdot\frac{R_1\cdot u_\tau}{\nu}\left(\frac{r}{R_1}\right)^{0.1} \quad (17).$$

Therefore except near the center region of the vortex core, the distribution of $\nu_{T,r\theta}$ is nearly constant value,(6) for calculating $\nu_{T,\theta z}$, it is necessary to use the velocity gradient(dVθ/dz), but from this paper it was difficult to estimate the value of (dVθ/dz).

4. CONCLUSIONS

(1) The velocity exponent n depends on the periphery angle and are not always lower than n=1(potential flow) due to the deviation of the center of the vortex core.

(2) The radius r=a corresponding to the maximum tangential velocity $V_{\theta M}$ is located near the radius $r=a\fallingdotseq(2/3)R_2$, but depends on the axial position and also the peripheral angle.

(3) The axial velocity Vz in the vortex core is not always upward direction,

the downward flow can be recognized.
(4) Even if the stable condition of Rayleigh is satisfied in the quasi-free vortex region, this stable condition is not always satisfied bu the correlation between (dVθ/dr) and (dVz/dr) by the stability criterion of Ludwieg and Scorer.
(5) Equi-flow rate lines show the circulation zones in the quasi-free vortex region and also not the axis symmetrical configurations, Equi-flow rate lines in the vortex core are very complicated.
(6) The dimensionless circulation Γ_c in the quasi-free vortex is proportional to the Reynolds number Rec.
(7) The eddy kinematic viscosity $\nu_{T,\theta z}$ in the quasi-forced vortex region is higher than the eddy kinematic viscosity $\nu_{T,r\theta}$ in that.

REFERENCES

1. A.Ogawa, O.Seito,H.Nagabayashi; Particulate Science and Technology, Vol.6, No.1(1988),p.17-p.28.
2. J.Jacobs,R.Günther; Forscg.Ing.Wes.,Bd.41,Nr.4(1975),s.113-s.119.
3.E.Kriegel;Tech.Mitt.Krupp.Forsch.-Ber.,Bd.25,Heft1/2(1967),s.21-s.30 und s.31-s.36.
4.A.Rochino,Z.Lavan;J.Applied Mech.(T.A.S.M.E.)(June,1969),p.151-p.158.
5.K.Buick;Verfahrenstechnik,Bd.4,Nr.11(1970),s.511-s.513.
6.R.M.G.Boucher,G.Vert; Génie Chimique,Vol.74,No.1(1955),p.1-p.18 et Vol.74, No.2(1966),p.38-p.50.
7.A.Ogawa, J.College of Eng.,Nihon Univ.,Vol.A-24(March,1983)p.93-p.102.
8. Е. Д. Балуев, Ю. В. Троянкин ; Исследование аэродинамической структуры газового патока в циклонной камере, ТЕПЛОЭНЕРГЕТИКА, №1, (1967), С. 63–С. 65.

9. Y.Fujita,A.Ogawa; J.College of Eng.,Nihon Univ.,Vol.A-19(1978)p.185-p.197.
10.L.Rayleigh; Scientific Papers,No.413p.447-p.453,Dover,Vol.5&6(1964).
11.L.Prandtl;Gesammelte Abhandlungen,Zweiter Teil,Springrt-Ver.(1961).
12.H.Ludwieg;Z.Flugwiss.,Bd.8,Nr.5(1960),s.135-s.140.
13.R.S.Scorer; J.Inst.Math.Applics.,Vol.3(1967),p.250-p.265.
14.L.J.Smith;J.Basic Eng.,T.A.S.M.E.,(Dec.1962),p.602-p.608.
15.L.J.Smith;J.Basic Eng.,T.A.S.M.E.,(Dec.1962),p.609-p.618.
16.M.Escudier; Ann.Rev.Fluid Mech., Vol.19,(1987),p.27-p.52.
17.A.Ogawa;Particulate and Multiphase Processes,Vol.1,(1987),p.129-p.146.
18.A.Ogawa,T.Kato,H.Nagabayashi;First report(in japanese),preprint of JSME, No.880-6(1988),p.227-p.228.
19.O.Seito,A.Ogawa;Bulletin of JSME.,Vol.28,No.243(1985),p.1949-p.1954.
20.A.Ogawa; Separation of Particles from Air and Gases,Vol.II.CRC-Press (1984).
21.C.A.Shook,S.K.Sagar; Can.J,Chem.Eng.,Vol.54,(Dec.1976)p.489-p.496.

UNIVERSAL BLENDER
FOR COHESIVE AND FREE FLOWING POWDERS
(World wide patented)

presented by
Dr.Ir. I.A.S.Z. Peschl
Consultant of Powder Technology
Siloproject B.V.
Kennedylaan 8, 6029 RE Sterksel
The Netherlands
TEL 31 4907 2447 / FAX 31 4907 1628

INTRODUCTION

Many conventional blenders and mixers, are applicable only for free flowing materials or they can be built only for small batches. For most of such blenders it is not possible to calculate the necessary blending time or blending quality . Therefore decisions for practical application require experience or experiments. For new materials or large blenders, large errors resulting from extrapolation or inproper assumptions are possible.

The universal blender offers a new dimension in blending technique by controlling the flowpatterns of the content in the blending silo. It combines a high blending accuracy with NO HEATING of the powder and NO DEGRADATION of particles.This advantage makes the UBL especially applicable for the most critical applications in the pharmaceutical, chemical, food and ceramic industry.

GENERAL DESCRIPTION

The universal blender consists of a blending silo, blending bottom, collecting hopper, and recycling device.
A specially designed blending/discharge bottom allows the activation of a predetermined cross-sectional area of the silo. In this way the flow pattern of the silo can be changed to yield three different types as shown in Fig 1.

1. IDEAL FLOW - the whole silo content moves uniformly. In this period the vertical columns will be mixed.

2. FUNNEL FLOW - only a partial column of the content moves. In this period the vertical columns will be displaced to each other so that the horizontal layers will be mixed.

3. MASS FLOW - with predicted velocity over the horizontal cross section.

Using these flow patterns in an prescribed sequence a very intensive blending or mixing will be achieved. In the most practical cases, the required level of blending will be achived within two recyclings of the blending content.

Advances in Fine Particles Processing
John Hanna and Yosry A. Attia, Editors

The blending procedure can be run fully automatically by predicting the number of recyclings and the time necessary for discharge. In all our tests the blending quality was so accurate that measured differences have been within the deviation of the analytical. equipment. UNIVERSAL BLENDER is able to handle very COHESIVE to FREE FLOWING material and is applicable for small blenders of a few liters to 1000 m3. The BLENDING QUALITY is predictable because the BLENDING PROCEDURE can be described in MATHEMATICAL MODELS. Theoretical analysis and practical experiments on small and big blenders shows that, a DEVIATION OF THE QUALITY of 1% or less can be achived within 2 to 3 recyclings.

When using uniform discharge over the whole silo-cross-section (ideal flow) the equipment can be applied as an ANTI-SEGREGATION and an ANTI-DEGRADATION device.

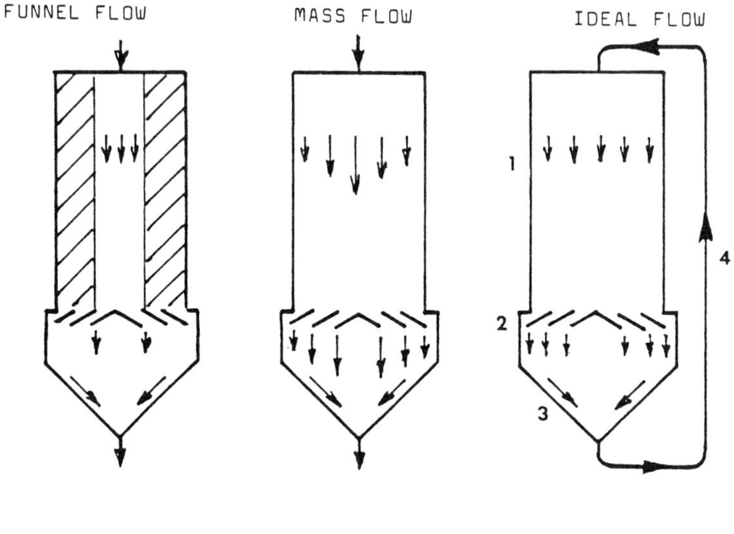

Fig. 1. SCHEMATIC DIAGRAM OF THE MAIN PARTS OF UNIVERSAL BLENDER AND THE MAIN FLOW PATTERNS OF THE SILO.
1.SILO 2.BLENDING BOTTOM 3.COLLECTING HOPPER 4.RECYCLING

THEORETICAL ASSUMPTIONS

Unlike many other systems for handling of powders the UNIVERSAL BLENDER utilizes the tendency of powders to build ratholing or a funnel flow in order to achieve a high blending quality in a short time.

A specially designed blending bottom allows the activation of a predetermined cros sectional area of the silo. In this way flow pattern of the silo can be changed to yield three different types as shown in Fig. 1.

The blending procedure can be specifically designed, using all three possibilities mentioned under 1 - 3, to achieve very short blending times or a minimal number of recyclings of the silo content.

The discharged material will be replaced in the blending silo. By repeating the replacement of the partial column and the movement of whole mass according to the predetermined blending procedure very intensive blending can be achieved. Theoretical models and computer simulation have been developed for many shapes of original deviation.

As an example,blending of a linear distribution of quality of a batch will be described. This distribution can be supposed due to different chemical or other reactions at the begin and at the end of the batch.

In Fig.2 is shown the distribution of deviation of the quality at the begin and at various significant stages of blending process.

ORIGINAL DEVIATION
Original distribution of deviation A from the average quality.

FIRST STEP
The middle section of the blending bottom is activated, and thus funnel flow is obtained.B1 and B2 represent two partial vertical columns having the same size of the horizontal cross section. By sinking and simultaneously recycling of partial column B2 over half silo height,a quality average distribution B from the average is obtained however, within a horisontal cros section is stil a deviation B1 and B2.

RECYCLING
Activating the whole silo bottom and thus achieving ideal flow, each horisontal cross section is mixed to a quality distribution B.

SECOND STEP
Now the flow pattern is changed again to funnel flow and the partial column C2 is sunk over a quarter of the height; thus a quality distribution C over the horisontal cros section is obtained.

DISCHARGE
If the deviation is small enough, the silo can be discharged by adjusting the blending bottom to IDEAL FLOW; otherwise the blending procedure should be continued.

The total recycling quantity for the shown blending procedure is 1.375, which is sufficient to obtain a reduction of the deviation of 1:4.

0,25	recyclings	for the funnel flow B
1,	"	for ideal flow - recycling
0,125	"	for funnel flow C

The mathematical model for blending of linear distribution of deviation can be found as follows:

$$D = D0 \times (0.5) \exp N$$

$$R = N + (1 - (0.5) \exp N) / 2$$

Where:
DO - initial deviation
D - deviation after blending / mixing
N - number of blending steps
R - number of recyclings
exp - exponent of the expression between the parentheses

In Figure 2. is shown the theoretic curve and the schematic few of the distribution of the deviation.

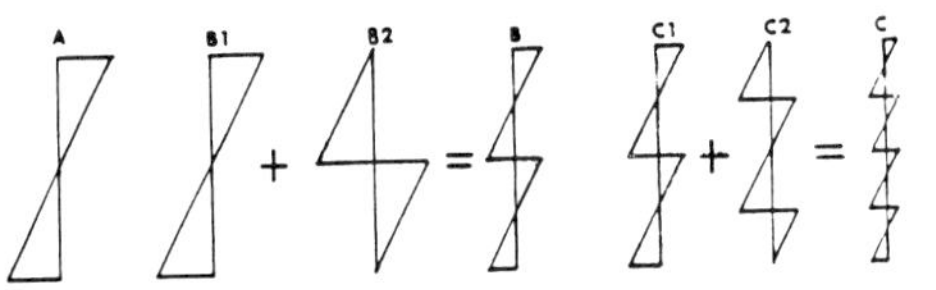

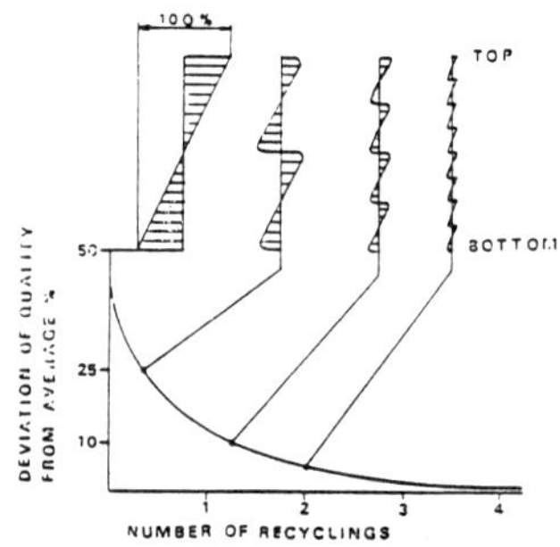

Fig. 2a. Distribution of the deviation: A-original deviation
B-after first blending step, C-after second blending step
B1 and C1- not moved columns, B2 and C2-replaced columns

Fig. 2b. Theoretic curve for the reduction of the deviation as a function of a number of recyclings

APPLICATIONS IN PRACTICE

The UNIVERSAL BLENDER can be built for a wide range of capacity and for a wide range of powders.The capacity of universal blender/mixer ranges from 2 liter up to 1000 m3. The small blender can be used as a stationary or as a transportable blender, in which case it has the function of a portable blender/container.

Because of the "life bottom" function of the blending bottom the universal blender can be used for extremely cohesive as well as for free flowing materials.The practical blending time depends only on the capacity of the recycling system.It is important to note that as THE RECYCLING CAPACITY is increased, THE BLENDING TIME will be decreased.Typical transport capacity of an elevator or a pneumatic system, is 100-200 m3/hr. This means that for 50 m3 approximately 30 minutes are needed for two recyclings in order to achieve a reduction of the deviation of 1% of original deviation.

In Figure 3. is shown a blending installation of a capacity of 20 m3 which is able to mix extremely cohesive powders,but can be used for free flowing materials also. The BLENDING TIME depends on the choice of the RECYCLING SYSTEM. For shown installation, where the capacity of the recycling is 30 m3/h, the required blending time is approximately 20 min.A laboratory blender is available in capacity of 1 to 10 litre as shown in Fig. 4.

Figure 5. shows a PORTABLE BLENDER with incorporated recycling installation. The portable blender shown has a capacity of 500 litre, and the recycling installation, incorporated vertical screw conveyor, has a transport capacity of 5000 l/h. The required blending time, corresponding with the installed recycling capacity, is approximately 10 min.

In Figure 6. is shown a PORTABLE BLENDER / CONTAINER with a detachable pneumatic recycling station. By this solution a recycling system with a high recycling capacity can be used so that a very short blending time can be obtained.The blending time for the portable blender is between 3 and 10 min, depending of the capacity and the required deviation of quality.

Fig.3 UNIVERSAL BLENDER (shown 20 m3)
1. Silo
2. Blending bottom
3. Collecting hopper
4. Recycling system

CAPACITY 1-1000 M3

Fig.4 LABORATORY BLENDER
CAP. 10 LITRE

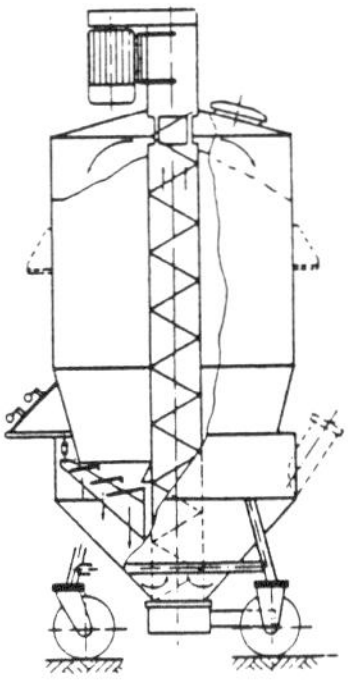

Fig.5 PORTABLE BLENDER
CAPACITY 100-2000 LITRE

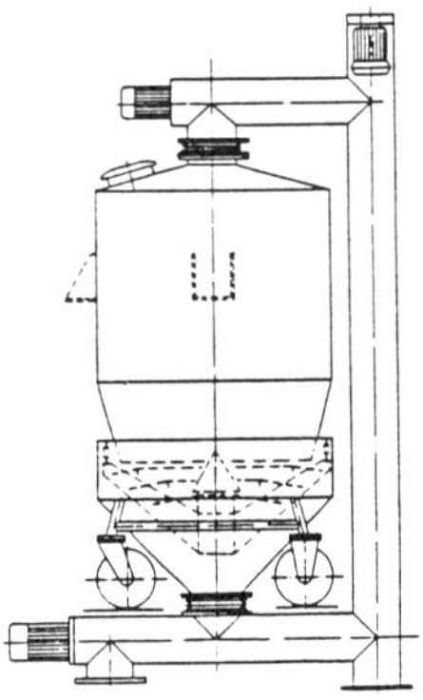

Fig.6 PORTABLE BLENDER/CONTAINER
CAPACITY 100 - 2000 LITRE

EXPERIMENTAL RESULTS

In cooperation with the variouse clients a number of full scale experiments have been done. All analyses have been done in the laboratories of the clients using test equipment which is used for normal routine tests of the production quality.The experiments show that a blending quality of 1% deviation or less can be achieved after 2 or 3 recyclings. All tests have been stopped when further blending did not result in a measurable improvement, because the limit of instruments for analysis has been reached.

BLENDING OF CORK POWDER WITH VARYING PARTICLE SIZES

Blending quantity: 10 m3

Cork powder is used as a basic material for the production of linoleum.Inconsistent particle size distribution may cause inacceptable color variances. To prevent this, a large quantity of cork powder should be mixed in order to obtain a uniform particle size distribution throughout the entire mass of the content to be blended.The analysis of blending results is shown in Fig. 7a. The accuracy of the particle size is approx. 2%. Further improvement of blending quality could not be established because 2% is already the limit of accuracy for sieve analysis.

CERAMIC MIXTURE OF VARIOUS COMPONENTS

Blending quantity: 0,65 m3

The blending of ceramic mixture, shown in Fig. 7b, has been done in a small transportable blender, shown in Fig. 5.
The melting process and the viscosity of glass mass depend on the correct proportion of all components.In Fig. 7b are shown the results of analysis of pigment Fe_2O_3, using the "Roentgen Fluorescence Method". After 2 recyclings no further improvement of blending quality could be established because the limit of 0.1 % of this method is reached.

MILK POWDER WITH VARYING FAT CONTENTS

Blending quantity: 40 m3

The composition of many natural products varies over a wide range.Milk powder is such a product. Variations of fat or protein, or both, determine the quality of the product and the selling price.
Blending samples were tested for fat content because this test was less time consuming.
In Fig. 7c the results of analysis show that after 2 recyclings the deviation of fat content is reduced from the original approx. 6% to approx. 0,1%. Blending was broken off because the limit of accuracy of the method of analysis was reached.An additional test for blending a very small quantity of a component was carried out by putting a thin layer of grass flour on the top of the silo content. This was done in order to follow the blending quality visually during the test run and to study the behavior of a small component in a large mass, which can be important for the mixing of medical products.

Fig. 7. TEST RESULTS OF BLENDING WITH UNIVERSAL BLENDER

CORK POWDER
Blender size 10^3
Mixing of various particle size:
15% of particle size of 80 micron
85% of particle size of 330 micron

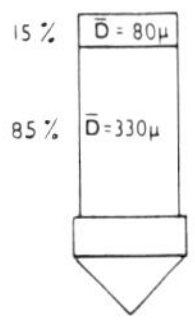

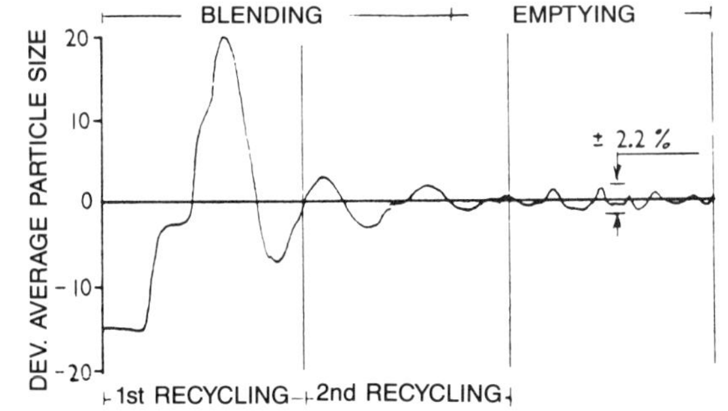

MILK POWDER
Reduction of variation of fat over the silo content from approx.
6% to approx. 0.1%.
Blender size 40 m^3

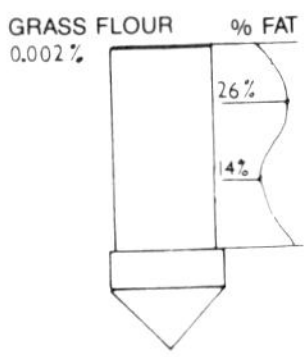

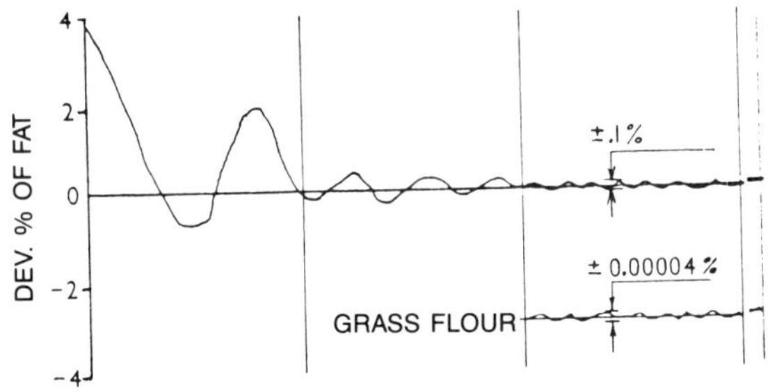

CERAMIC MIXTURE
Mixing of multicomponent batches placed in layers one over each other in the blender

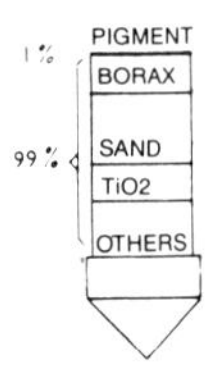

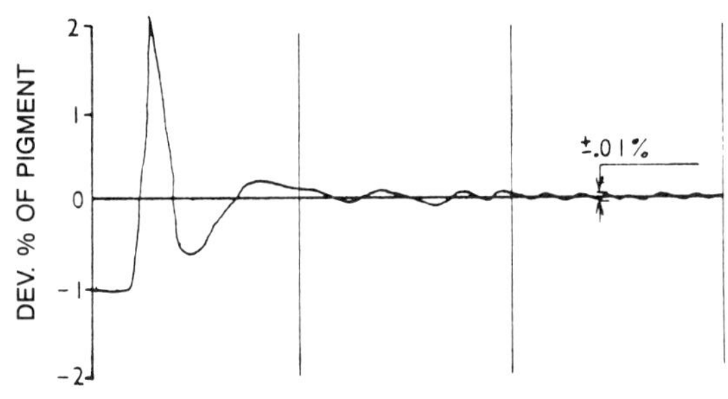

PHARMACEUTIC MIXTURE
Spray dried powder composed from soybean meal, fat etc. on which surface a pharmaceutic product.
Blender size 600 Liter.

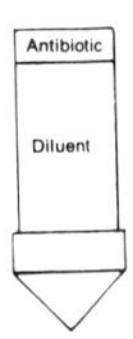

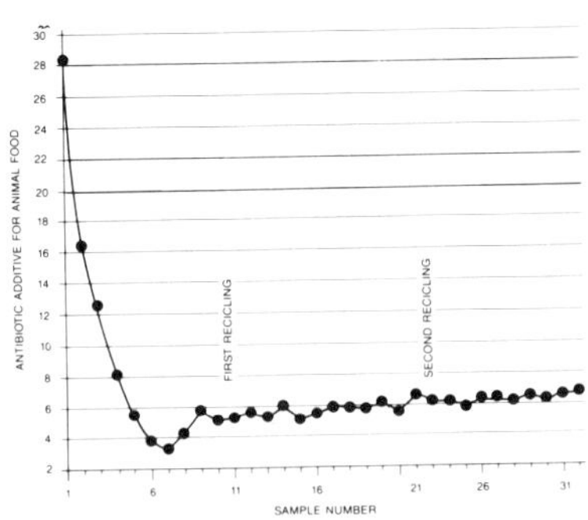

Because of a considerable difference in particle size we have been able to separate and measure the grass flour.
Surprisingly the grass particles were homogeneously distributed throughout the entire mass of the silo content of 40 m3.

PHARMACEUTIC MIXTURE

On the surface of a diluent, composed of soybean meal and fat, an antibiotic ADDITIVE is provide. This mixture can be supplied to the animal food. The curve in Fig.7d shows a typical intensive mixing in the first minute of the mixing. The necessary mixing time was 3 min.

CONCLUSIONS

Theoretical assumptions and experimental results shows that the Universal Blender meets the requirements of present-day industry.

The universal blending system has the additional advantage that a possible segregation of material - larger particles at the outside and the finer ones at the centre of the silo - will automatically be remixed in the last phase of the blending procedure when the silo operates with ideal flow.

A further advantage is a very small energy consumption,that results in a very little energy absorption by the material itself, so that increase of temperature, degradation of particles or other energy consuming effects will be reduced to a minimum.

***** THE MAIN CHARACTERISTIC OF THE UNIVERSAL BLENDER: *******

1. The resulting blending quality and required blending time can be calculated. That means that the blending installation can be designed near to a technical and economical optimum without experimental improvements.
2. The blender can be used for COHESIVE AND FREE FLOWING materials.
3. The blender can be used for EQUALIZING THE POWDER QUALITY OR MIXING DIFFERENT POWDERS TO A UNIFORM PRODUCT.
4. The blender can be used as a ANTI SEGREGATION SILO.
5. The blender can be used as a ANTI-DEGRADATION SILO.

REFERENCES

PRECHL - UNIVERSAL BLENDER - BULK SOLIDS HANDLING , 1986 , NO 3

PESCHL - IDEAL FLOW SILO - ADVANCES IN FEED TECHNOLOGY, 1989,NO 1

PESCHL/COLIJN - NON SYMETRICAL BIN FLOW PROBLEMS - BULK SOLIDS HANDLING 1981, NO 3

PESCHL - INTERACTION OF FLOW PROPPERTIES, SILO GEOMETRY AND SILO STATICS - INTERNATIONAL WINTER MEETING OF THE AMERICAN SOCIETY OF AGRICULTURAL ENGENERS, CHICAGO 1988

PART 3.

SURFACE AND COLLOIDAL PHENOMENA IN FINE PARTICLE PROCESSES

THE ROLE OF PARTICLE FORCES IN DETERMINING THE RHEOLOGICAL PROPERTIES OF CONCENTRATED DISPERSIONS. AN EXPERIMENTAL STUDY

PAUL F. LUCKHAM* and M. ALI ANSARIFAR†
Department of Chemical Engineering and Chemical Technology,
Imperial College, Prince Consort Road, London, SW7 2BY, U.K.; † Present address Cavendish Laboratories, University of Cambridge, Madingley Road, Cambridge, CB3 OHE, U.K..

* To whom all correspondence should be addressed.

ABSTRACT

The forces of interaction between two mica surfaces bearing an adsorbed layer of a poly–2–vinyl–pyridine/poly<u>tert</u>–butylstyrene (P2VP/PtBS) AB block copolymer of molecular weight 21 400 has been measured. The forces commence at a surface separation of some 50 nm and increase approximately exponentially with decreasing surface separation. An increase of three orders of magnitude in the force is measured. Subsequently polyacrylonitrile (PAN) particles, stabilised by the same P2VP/PtBS AB block copolymer, have been prepared and the rheological properties of these dispersions determined up to high volume fraction $\phi < 0.6$. Particular emphasis has been put on the storage modulus of these dispersions, which increases exponentially with increasing ϕ also by three orders of magnitude. The storage modulus should be proportional to the differential of the particle force–distance profile. Using a cellular model the two sets of experimental data may be related and correspondence between the data achieved.

INTRODUCTION

The forces acting between particles in a suspension control the bulk properties of a wide range of materials, from synthetic particulates such as paints and pharmaceuticals formulations to natural products such as blood and caesein micelles. Such forces include "field type" interactions such as electrical double layer repulsive forces and van der Waals attractive forces and molecular interactions such as the short range solvation forces and the long range steric forces which arise when polymers adsorb from solution onto the surface particles. It is the subtle interplay of these forces which gives rise to the physical properties of suspensions. It is difficult to modify solvation or van der Waals forces for any specific system, in addition in many instances electrical double layer forces are weak (in non polar solvents, for example or in high electrolyte solutions). Polymers, however, have dimensions of 5–100 nm, the range of importance for particulate systems and are commonly adsorbed to particles in order to modify particle interactions and hence <u>produce</u> materials of the desired consistency. Man has utilised polymers in this way for many thousands of years, for example the paints used by neolithic man in the caves at Lascaux and the early Egyptian and Chinese civilisations, consisted of particles of soot stabilised by a variety of naturally occurring polymers such as caesein (from milk). Today many formulations bear adsorbed polymers of the suspensions. Despite the recognition of the role played by interparticle forces on modifying the bulk rheological properties of a suspension, there have been few experimental studies to confirm this thesis. This has been due primarily to the difficulties in measuring interparticle forces. However, recently Goodwin and coworkers [1, 2] and Benzing and Russel [3, 4] have related the elastic or shear modulus of charge stabilised suspensions to theoretically predicted particle

Advances in Fine Particles Processing
John Hanna and Yosry A. Attia, Editors

interactions and found a close correspondence. In this paper we report data for experimentally determined forces [5, 6] and correlate these data to rheological measurements. [7]

The forces measured were those between mica surfaces bearing an adsorbed layer of an AB block copolymer of poly–2–vinylpyridine/poly<u>tert</u>butylstyrene (P2VP/PtBS) immersed in an aromatic hydrocarbon solvent. The mica force balance has been used to measure forces between bare sheets of mica in organic liquids [8], aqueous electrolyte solutions [9], and between mica sheets bearing adsorbed polymers [5,6,10] and surfactants [11]. The technique is capable of measuring small forces (~ 10 nN) between mica surfaces and of measuring the distances between the surfaces to an accuracy of some 0.2 nm. Subsequently polyacrylonitrile particles containing the same AB block copolymer adsorbed on the surface were prepared. The rheological properties of these suspensions were subsequently measured. By ensuring that the solvent for both the force experiments and the rheology experiments have similar solvency conditions for the AB block copolymer (in these experiments the solvent for the poly–2–vinylpyridine was poor and the solvent for poly<u>tert</u>butylstyrene was good) the configuration of the adsorbed polymer would be similar in the two experiments. Namely with the poly–2–vinylpyridine laying flat on the mica or polyacrylonitrile particles with the poly<u>tert</u>butylstyrene (which is non adsorbing to mica or polyacrylonitrile) extending away from the surface (see Figure 1).

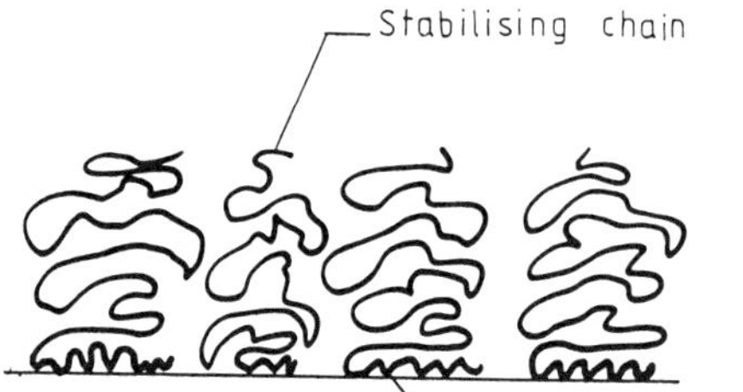

Figure 1 The configuration adopted by the poly—2—vinyl pyridine—polytertbutyl styrene block copolymer on both mica and polyacrylonitrile surfaces in these experiments.

In a rheological experiment energy is supplied to the system by the application of shear. In viscoelastic materials that energy is either lost through friction, the viscous component, or stored, the elastic component. Conceptually, in a suspension the viscous energy relates to the energy required to move one particle past another whilst the elastic component, relates to the energy required to "push" two particles together. If the suspension is concentrated, this corresponds to the energy required to "push" against the pair potential of the particle. Thus the elastic modulus G is related to the particle pair potential by the relation (V(D))

$$G \propto \frac{d^2V(D)}{dD^2} \qquad (1)$$

where D is the particle surface separation.

By performing direct measurements of the pair potential we shall test this relation.

EXPERIMENTAL

Materials

The polymer used in this experiment was a poly–2–vinylpyridine – polytertbutylstyrene (P2VP/PtBS) block copolymer of molecular weight (weight average) 21 400. The molecular weight of the polytertbutylstyrene (PtBS) portion was 15 000 and the poly–2–vinylpyridine (P2VP) was 6 400. (Mw/Mn = 2.2) The polymer was prepared by the sequential anionic polymerisation of tert butylstyrene and 2–vinyl pyridine and was terminated with methanol. (Further details of polymerisation procedure may be found elsewhere [12]).

Polyacrylonitrile (PAN) particles stabilised by P2VP/PtBS block copolymer were prepared using a free radical polymerisation technique similar to that described by Barrett [13]. Acrylonitrile (the monomer) and solvent (cyclohexane) were heated to 80–85 °C with stirring under a nitrogen atmosphere, the block copolymer was dissolved in a minimum of toluene (a better solvent than cyclohexane) and added, followed by the initiator (2, 2′ Azobisisobutylnitrile). Particle formation occurred soon after the commencement of initiation. Three further aliquots of monomer, stabiliser and initiator, were added at hourly intervals. Further details of this procedure are given elsewhere [14]. The PAN particles were transfered to Solvesso 200, (a high boiling point 234 °C, industrial aromatic solvent, viscosity 2.63 mPa, density 989 kg.m^{-3} refractive index 1.59) for the rheological experiments. This was achieved by adding Solvesso 200 to the PAN particles and removal of cyclohexane by evaporation in a vacuum oven (40 °C, 1 mm Hg for 24 hours). The particle size was measured using a transmission electron microscope and found to be 150 ± 20 nm (Z average).

The Surface Forces Apparatus

The apparatus used to measure the forces between surfaces bearing adsorbed P2VP/PtBS is similar to that described by Israelachvili and Adams [9] and is shown schematically in Figure 2. The forces measured are those between two polymer coated mica sheets (about 1 cm x 1 cm x 1–3 μm thick), which are partially silvered (i.e. allow about 10% of the light to be transmitted). The sheets are glued with molten glucose to two optically polished glass discs with cylindrically curved surfaces, which are positioned mutually perpendicular.

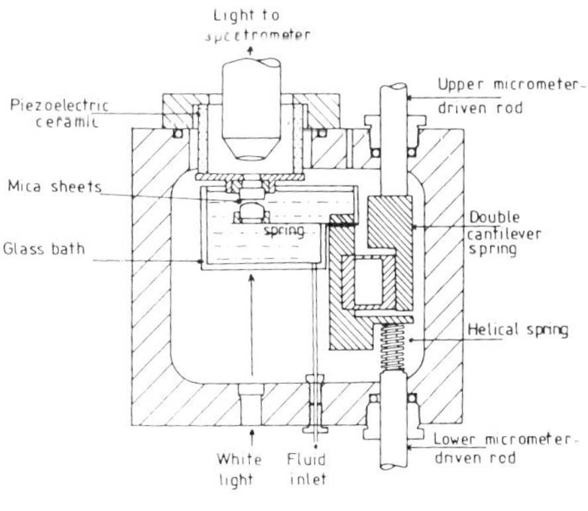

Figure 2 The force measuring apparatus

The separation between the surfaces is controlled via three stage mechanism. The upper micrometer driven rod may be moved up and down by a stepper motor, coupled directly to the rod. This movement constitutes a coarse control and allows the surfaces to be positioned to an accuracy of about one micrometer with a total range of about 2.0 cm. Springs are used to eliminate backlash, wobble and creep. The lower micrometer driven rod is moved by a similar motorised mechanism using a two way synchronised motor. This constitutes the medium control stage. The lower rod pushes against a helical spring which in turn pushes upon a stiff cantilever stainless steel spring which is about one thousand times stronger than the helical spring. Therefore, a one micrometer movement of the lower rod corresponds to ~ 1.0 nm movement of the mica surfaces. The lower rod is connected to a high precision, linear resistance potentiometer which enables the measurement of the applied displacement. The fine control of surface separation is achieved using a rigid piezoelectric tube, which expands by approximately 1.0 nm per volt. This non–mechanical fine control is used to position the two surfaces to better than 0.1 nm and has a total range of about 500 nm. These three mechanisms allow the separation between the surfaces to be easily varied during an experiment.

The separation between the mica surfaces may be measured, to within 0.05 nm by allowing white light to pass normally through the partially silvered mica and observing the interference fringes (fringes of equal chromatic order (FECO)) in a spectrometer [15]. The distance between the surfaces, D, is essentially measured by comparing the wavelengths of the interference fringe when the mica surfaces are in contact, (Figure 3a) and when the surfaces are separated (Figure 3b). As an illustration, in Figure 3b the surfaces are some 3 nm from molecular contact (note the shift of the wavelengths of the fringes with respect to the calibration mercury line).

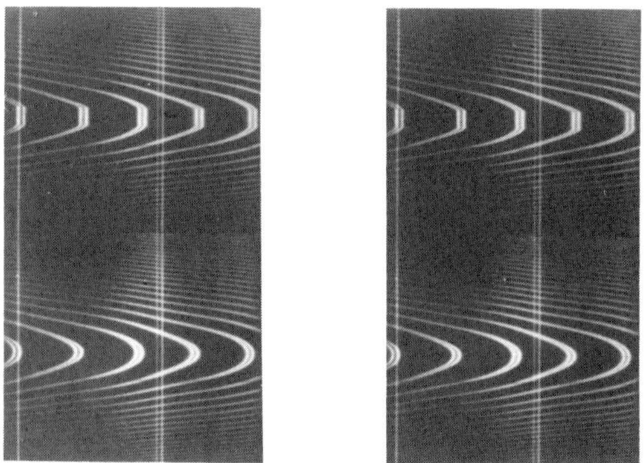

(a) (b)

Figure 3 A typical set of fringes of equal chromatic order (FECO). In (a) the mica surfaces are in molecular contact. The strong attractive force distorts the shape of the mica such that the contact is flat and the FECO have a region of constant wavelength. In (b) the surfaces are separated by some 5 nm. The shape of the fringes reproduces that of the cylindrical forms used to mount the mica in these experiments.

The force, F(D) between the two surfaces is measured by applying a known relative displacement, ΔD_0, to one of the surfaces by applying, say, a known voltage to piezoelectric tube, and simultaneously measuring the actual motion, ΔD of the surfaces relative to each other using the optical technique. If there is no force between the surfaces, then $\Delta D = \Delta D_0$; if they attract, $\Delta D > \Delta D_0$ and if they repel $\Delta D < \Delta D_0$; the difference in both cases being taken up by bending of the single cantilever spring. In general,

$$F(D) = K\,(\Delta D_0 - \Delta D) \qquad (2)$$

where K is the constant of the leaf spring, which is approximately 100 Nm^{-1}. Since $(\Delta D_0 - \Delta D)$ may be as small as 1.0 nm, forces as small as 10^{-7} N may be measured. By commencing measurement with surfaces far apart, where $F(D) = 0$, force–distance profiles may be measured.

Prior to all experiments, all parts of the apparatus coming into contact with the solution were thoroughly cleaned and dried. The apparatus was assembled in a clean air cabinet and placed on a vibration table. The mica surfaces were then brought into contact in air and the wavelengths of the FECO noted. The surfaces were separated, toluene, the solvent was added to ensure that both mica surfaces were covered by the solvent and after allowing time for the system to come to thermal equilibrium, a force distance profile was measured. The block copolymer solution (10–100 $\mu g.ml^{-1}$) was added to the required concentration, and the surfaces were allowed to incubate for 12–16 hours, to permit adsorption. Force profiles in the polymer solution with polymer adsorbed to the mica surfaces were determined. The time to measure each profile varied between 15 minutes to 1 hour. In some experiments the polymer solution was removed and replaced by pure solvent.

Rheological Measurements

A Bohlin rheometer (Bohlin Rheologie, Lund, Sweden) was used for all measurements. Concentric cylinder or cone and plate geometries were used depending on the volume fraction. The measurements were taken at 20 °C.

The instrument used in both continuous and oscillatory shear modes. In the continuous shear mode, a constant shear is applied and the shear stress measured. The applied shear was increased incrementally over a range $10^{-3} - 10^{3}\ s^{-1}$ from which the viscosity of the suspension as a function of the applied shear was measured. In the oscillatory mode a frequency range of 10^{-2} – 10Hz was used. The amplitude of the oscillation was varied between 0–20 m rads. In oscillatory experiments, the response in stress of a viscoelastic material to a sinusoidally varying strain of known amplitude γ_0 is monitored. The stress amplitude τ_0 is also a sinusoidally varying function, but for a viscoelastic material it is shifted out of phase from the strain by a phase angle δ. From the amplitudes of stress, τ_0, strain γ_0 and the shift in the phase angle, δ, the rheological parameters G*, the complex modulus, G′, the storage modulus, G″, the loss modulus, may be measured for various fraction, ϕ of the solid.

Such that

$$G^* = \tau_0/\gamma_0 \qquad (3)$$

$$G' = G^* \cos\delta \qquad (4)$$

$$G'' = G^* \sin\delta \qquad (5)$$

and hence

$$G^* = G' + iG'' \qquad (6)$$

where i is the square root of minus one.

RESULTS

1) Surface Forces Experiments

Before the addition of polymer in any experiment the force–distance profile between the bare mica surfaces immersed in toluene was determined. Figure 4 shows a typical force profile. The force axis is normalised by dividing by the radius of curvature of the mica surfaces in their mutual cross cylinder configuration. According to the Derjaguin approximation [13] which is nearly exact for $R >> D$, as in the current experiments, this gives the interaction profile per unit area, E(D) of flat parallel surfaces or distance, D apart obeying the same force law, i.e.,

$$F(D)/R = 2\pi E(D) \qquad (7)$$

This normalisation procedure is used in all the force profiles presented here. In pure solvent (toluene) (Figure 4) no forces were detected as the surfaces approached from large D (~ 300 nm) down to D $\lesssim$ 10 nm, when an attraction was observed, on further compression the surfaces jumped spontaneously from D ~ 5 ± 2 nm to an new position very close to the 'contact' position of the mica surfaces in air. Such jumps are due to the mechanical instability of the spring which one of the mica surfaces is attached, and are expected whenever dF(D)/dD $\geq$ K. The attraction between the bare mica surfaces in pure toluene resembles van der Waals type interaction with a suitable Hamaker constant and is similar to that obtained previously[10]. Despite extensive drying of the toluene no molecular structuring close to the mica surface was noted, as has been observed by Christenson and coworkers in many organic solvents[8,16]. The absence of any long range repulsive forces between the mica surfaces greatly simplifies the results obtained in the presence of adsorbed polymer.

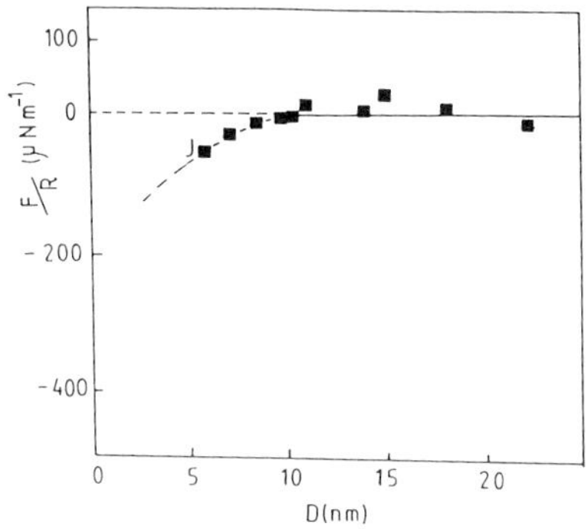

Figure 4 Force—distance profile between mica surfaces immersed in toluene. J represents the position of an inward jump due to the mechanical instability of the force measuring spring.

Figure 5 shows the force profiles determined after introducing the P2VP–PtBS block copolymer into the apparatus (at a concentration of 100 μg.ml^{-1}) and allowing the surfaces to incubate in the solution at a separation of 2mm for 16h. The results are

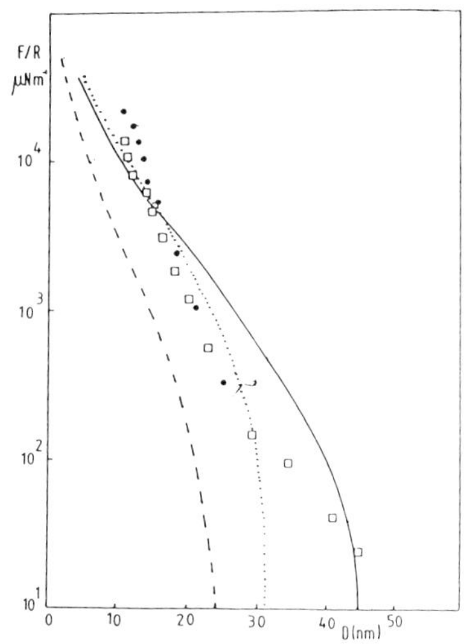

Figure 5 Force—distance (on a log force linear distance scale) following 16 $\pm$ 4 h incubation in a 100 mg.dm^{-3} solution of P2VP/PtBS copolymer (Mwt 21 400). Filled symbols compression of surfaces, open symbols decompression. The dotted line is the theoretical force profile predicted from scaling theory. The dashed line is the theoretical profile predicted by Milner *et al*, whilst the solid curve is the Milner prediction which accounts for the poly dispersity of the polymer.

presented on a semilogarithmic scale to enable the display of several orders of magnitude in F(D). On approach of the surfaces, following incubation in the polymer solution, no interaction was measuring from large surface separations down to D = 50 nm, when an approximately exponentially increasing repulsive force was observed. The surfaces could be compressed to separations of some 10 nm whereupon a steep repulsion was observed, and using the present apparatus, it was not possible to compress the surfaces any further together. On separation of the surfaces, the same force profile (within experimental error) was observed. No attractive or adhesive component in the interaction was noted.

Rheological Measurements

The rheological measurements were performed on the PAN latex particles stabilised by the same P2VP/PtBS block copolymer as was used in the study of the forces between the polymer coated mica surfaces. In Figure 6 are the results obtained in the continuous shear mode. The results are presented as log η v log γ where η is the viscosity and γ the shear rate. The results for different volume fraction of particles ϕ are given. In Figure 6 it is clear that η decreases as the shear rate, γ increases, a typical result for moderately concentrated particulate systems. In addition, this non–Newtonian flow behaviour is more pronounced at higher ϕ. The origin of this non–Newtonian flow is the breakdown in 'structure' in the stable colloidal dispersion.

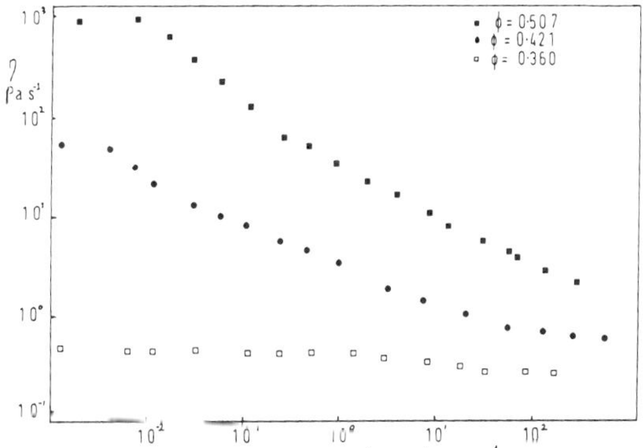

Figure 6 The shear modulus viscosity η as a function of shear rate for polyacrylonitrile particles stabilised by P2VP/PtBS block copolymer, for particle volume fractions in the range 0.1 > ϕ > 0.50

Figure 7 shows the variation of the complex, storage and loss modulus, G*, G′ and G″ respectively, with frequency ω at various volume fractions of PAN latex. At low ϕ, G″ > G′, in other words the dispersion in behaving as a viscous fluid. This is not surprising since at these low volume fraction there are to be expected to be only weak particle–particle interactions and so most of the applied shearing force will be lost as particles move past each other. The adsorbed polymer layers, will by and large be non–interactive and the dispersion will behave mainly as a viscous fluid. Increasing the volume fraction of the particles results in an increase of all the rheological parameters with the rate of increase of G′ being greater than G″ such that when $\phi \sim$ 0.45 G′ > G″. This increase of the storage modulus is a consequence of the compression and interpenetration of the adsorbed polymer layers as the particles come into a closer proximity with each other. As ϕ increases further G′ >> G″, this indicates that the interactions between the steric barriers have dramatically altered the rheological characteristics of the dispersion and at these high particle volume fractions the sterically stabilised dispersions behave as a nearly elastic body.

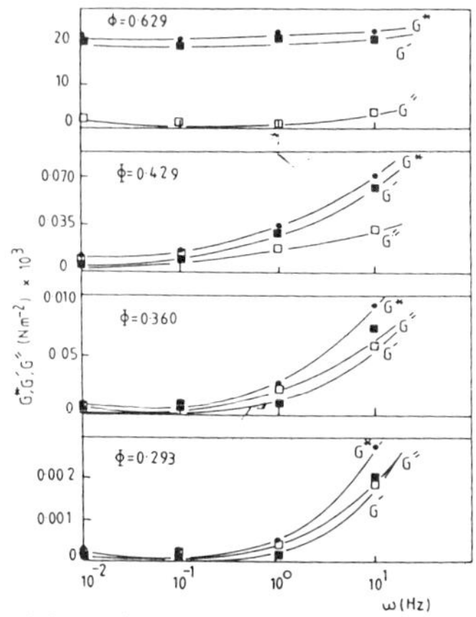

Figure 7 The complex modulus, G* ■ , the storage modulus G′ ; and the loss modulus G″ versus frequency ω for volume fractions in the range 0.293 > ϕ > 0.629.

DISCUSSION

Before commencing the discussion of the results for the adsorbed copolymer, it is necessary to comment on the results obtained for the two homopolymers, P2VP and PtBS. Toluene is a non–solvent for P2VP, in fact P2VP with a molecular weight of only 5000 proved to be insoluble. The results obtained with PtBS, have indicated that there was no adsorption of polymer on to mica, as had also been observed with polystyrene [10]. These observations lead us to conclude that when the AB block copolymer of P2VP/PtBS is adsorbed to mica the P2VP is laying flat on the mica surface (i.e. minimising its contact with toluene) whilst the PtBS is extending away from the particle surface as illustrated in Figure 1).

The repulsive forces between the polymer coated mica surfaces are due principally to osmotic interactions between segments from opposing adsorbing layers as they come into overlap. Note that the forces measured on separation of the surfaces are, within error, identical to the forces measured on approach. This is in contrast to the results observed for the interactions between adsorbed homopolymers, [10] , where the forces measured on separation of the surfaces, were the same only when the rate of separation was slow.

This study is sufficiently direct to enable comparison of the results with predictions from various theoretical models of interaction between two surfaces bearing terminally attached polymers. Two theories enable direct comparison namely the scaling theory of de Gennes and Alexander [17,18], and the mean field theory developed by Milner, Whitten and Cates [19,20].

The de Gennes model assumes that the concentrations of polymer in the adsorbed layer is constant (i.e. a step profile). The forces between the two polymer coated surfaces will then have two components: i) an osmotic repulsion and ii) an elastic restoring force which will be trying to "thin out" the adsorbed layer and is essentially an attractive term, thus

$$E(D) = \frac{KT}{S^3}\left\{\left[\frac{(L)^{9/4}}{1.25(D)^{5/4}} + \frac{D^{7/4}}{1.75(L)^{3/4}}\right] - \left[\frac{L}{1.25} + \frac{L}{1.75}\right]\right\} \qquad (8)$$

where k is Boltzmann's constant, T, the absolute temperature, S is the distance between the anchor points of the polymer, D is the distance between the surfaces and L is the distance of the onset of interaction.

The treatment of Milner, Whitten and Cates on the other hand does not assume the concentration of polymer in the adsorbed layer to be constant, parabolic segment density distribution is assumed instead, which is more consistent with experimental segment density profiles Cosgrove *et al* [21].

Using mean field theory for the adsorbed polymer layer the energy E(D) was found to be given by

$$E(D) = (9/10)\,(\pi^2/12)^{1/3}\, N\sigma^{5/3}\, \omega^{2/3}\, \nu^{1/3} \qquad (9)$$

and D is given by,

$$D = (12/\pi^2)^{\frac{1}{2}}\, N\sigma^{1/3}\, \omega^{1/3}\, \nu^{-1/3} \qquad (10)$$

where N is the number of monomers in the polymer chain ω is an excluded volume parameter and is given by

$$\omega = 2\Pi\phi^2 \qquad (11)$$

where Π is the osmotic pressure ϕ is the volume fraction of polymer in the adsorbed layer. σ is the surface coverage of polymer and ν is given by

$$\nu = \frac{N}{2\,Rg^2} \qquad (12)$$

More recently the effect of polydispersity in the forces has been considered such that equation 10 is modified as

$$D^* = D(1+\Delta((2N)) \qquad (13)$$

where D* is the calculated distance for the force including a polydispersity correction and Δ is related to the polydispersity P by

$$P = 1+\Delta^2(3N^2) \qquad (14)$$

The dotted line in Fig. 5 is the prediction of the force according to the de Gennes equation assumming the separation between each attached polymer to be 5 nm, which corresponds to a surface coverage of some 2 mg.m^{-2} and L = 20.

A reasonable fit is noted at large surface separations but a weaker force is predicted at shorter separations.

The treatment of Milner *et al* does not have so many unknown variables as the de Gennes prediction. The only poorly known variable is the surface coverage (which is of the order 3±1.5 mg. m^{-2}) other data are readily available from the literature.[20]

Ignoring the polydispersity factor gives poor agreement between theory and experiment (dashed line), at all surface separations the force is greater than that predicted. However, when the polydispersity of the adsorbed polymer is taken into account, close agreement between theory and experiment is observed at all surface separations. (It is important to emphasise here that there are no parameters in this theory which enables one to "adjust the distance scale to fit the data", as was the case with de Gennes treatment, the distance for the onset of interaction is dependent only on the molecular weight and polydispersity of the polymer and the surface coverage).

Thus the force profile for this polymer, adsorbed to mica surfaces may be explained in a fundamental way in terms of a modified osmotic pressure exerted by the adsorbed polymer. These forces will also be exerted by the polymer when it is in polyacrylonitrile latex particles, and will modify the rheological behaviour of these suspensions.

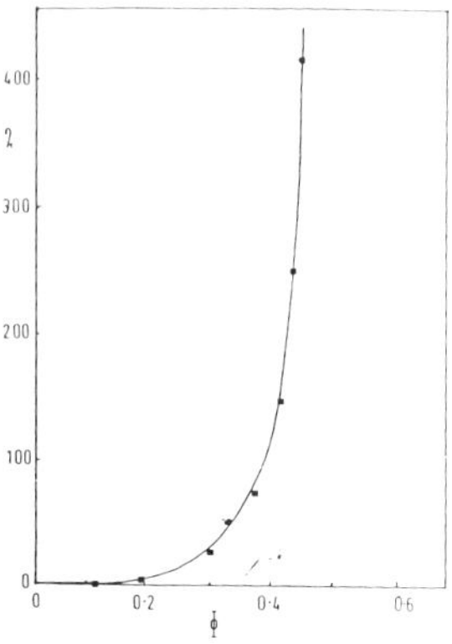

Figure 8 The high shear limiting viscosity (plastic viscosity) $\eta_{\rho l}$ versus particle volume fraction ϕ for P2VP/PtBS stabilised PAN particles.

In considering the rheological data we shall concentrate initially on the continuous shear measurements. From the data presented in Fig. 6 it is possible to calculate the relative high shear limiting viscosities (see Fig. 8). This high shear limiting viscosity increases with increasing volume fraction ϕ (Fig. 8) of the P2VP.PtBS stabilised polyacrylonitrile particles. These data obey the Dougherty–Krieger equation.[22]

$$\eta_r = [1 - \phi/_{\phi_{max}}]^{-2 \cdot 5 \phi_{max}} \qquad (15)$$

where η_r is the relative viscosity, ϕ_{max} the maximum packing fraction of particles (0.68 for body centred cubic) assuming that the volume fraction of particles includes the volume of the adsorbed PtBS layer (i.e. an effective volume fraction is used). As the true volume fraction of particles is directly measured in these experiments, the Dougherty–Krieger equation may be used to estimate the volume and hence the thickness of the PtBS layer. These data are presented in Table 1. Where it is clear that at relatively low volume fractions of particles ~ 0.3 the thickness of the PtBS layer is some 25 nms (this is very similar to the value estimated from the surfaces forces measurement where the onset of interaction D ~ 50 nm, remember that in this experiment two surfaces are involved). However, as the volume fraction of particles is increased, the PtBS layer is progressively decreased. We conclude that as particle concentration increases the adsorbed polymer layer becomes compressed and this is reflected in the plastic viscosity results.

TABLE I. Table of calculated effective volume fraction and adsorbed layer thickness δ from the effective viscosity.

ϕ	η_r	ϕ_{eff}	δ(nm)
0.101	2.92	0.302	27
0.174	5.78	0.420	21
0.293	28.9	0.573	15
0.321	51.7	0.602	14
0.360	75.7	0.617	12
0.398	155.9	0.638	10.4
0.421	253.0	0.648	9.4
0.429	420.5	0.655	9.2
0.469	1600	0.663	7.4

In the oscillatory shear measurements both the storage and loss moduli of the suspension are measured. Conceptually we may consider the storage modulus to be related to the energy required to 'push' two particles together, whilst the loss modulus is the amount of energy required to move one particle past another. Thus the storage modulus data as a function of particle volume fraction (or indirectly the particle–particle surface separation) is a rheological analogue of the direct surface forces experiment.

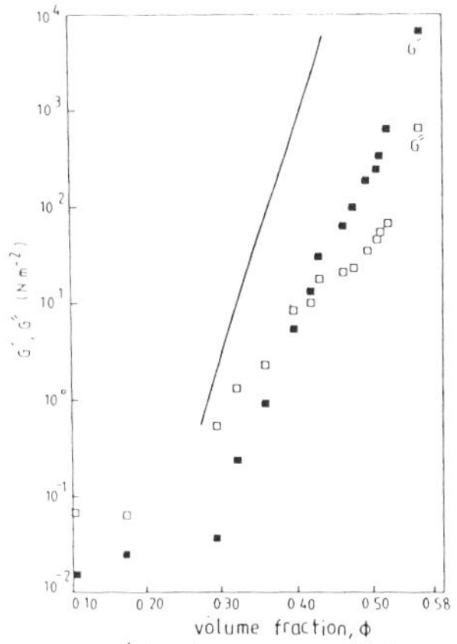

Figure 9 G′ the storage modulus versus ϕ (and average particle–particle surface separation D) for the P2VP/PtBS stabilised PAN particles at a frequency of 10 Hz. The solid curve is a theoretical prediction for G′ based on equation 17 and the results of Figure 5.

Inspection of Fig. 7 reveals that as ϕ is increased, both the storage and loss modulus increase, in addition, however we note that at low ϕ $G'' > G'$ this is not surprising as the particle–particle surface separation is greater than twice the adsorbed thickness and as most of the energy supplied is lost as viscous flow of the suspension. However as ϕ is increased to 0.45 we note that $G' \sim G''$, this increase of the storage modulus is a consequence of the compression and interpenetration of the polymer layers as the particles come into close proximity. Further increase in ϕ results in $G' >> G''$ and the fluid behaves as a nearly elastic body. Similar data have been obtained by Prestidge and Tadros in their study of the viscoelastic properties of polymer stabilised latexes [23]. In Fig. 9 we investigate the behaviour of G′ with ϕ in more detail and plot G′ versus ϕ at frequency of 10 Hz. At low ϕ we note that G′ is very low (indeed this is the sensitivity of the instrument) however as ϕ is generally increased we note that G′ begins to increase. The increase is approximately exponential with increasing ϕ and increases by some four orders of magnitude. We note that the increase G′ with ϕ is approximately the same as the force increases with decreasing surface separation. By assuming some form of packing for the particles in the dispersions, it is possible to calculate the average particle–particle surface separation D′ for a particular ϕ from the expression

$$D' = 2a\,[(\phi_{max}/\phi)^{1/3} - 1] \qquad (16)$$

where a is the particle radius and ϕ_{max} is the maximum packing fraction.

It is not clear how colloidal particles pack however assuming body centred cubic close packing (where $\phi_{max} = 0.68$) we calculate that the onset of interaction (as measured by an increase in G′) is ~ 50 nm and increases approximately exponentially as the particle–particle surface separation decreased to some 10 nm. These results are qualitatively very similar to the mica force results.

In order to compare the force profiles to the experimentally determined rheology results it is necessary to derive a relation between G′ and the force profile. Buscall *et al* [2] have used a simple cell method to achieve this and found that the high frequency limit of the elastic modulus (the shear modulus G_0) was related to the interparticle pair potential dV/dD by the expression

$$G_0 = \frac{3}{32 \phi_{max} n \quad 6a} \left[\frac{\phi}{0.68}\right]^{\frac{1}{2}} \frac{d^2V}{dD^2} \qquad (17)$$

where n is the number of nearest neighbours each particle has. The particle pair potential dV/dD is obtained directly from the force profiles thus enabling G_0 to be estimated. The theoretical G_0 vlaues thus obtained the continuous line of Figure 9. These results show that the theoretical results scale in the same way as the experimental results but are larger by some 1–2 orders of magnitude. This is largely due to the inadequacies of the cell model theory which would tend to overestimate the modulus.

In conclusion these experiments conclusively demonstrate a direct relation between the particle forces and the elastic moduli for polymer stabilised systems and show that measurement of the viscoelastic properties of concentrated suspensions spreads light on the nature and strength of the particle–particle interactions themselves.

ACKNOWLEDGEMENTS

We would like to thank Mr. B. A. Costello for some computational work and the SERC specially promoted programme in particle technology for their financial support of this work. (Grant number GR/D/24326).

REFERENCES

1. J. W. Goodwin amd A. M. Khidir in M. Kerker Ed., *Colloid and Surface Science*, vol 4, pp 529–539, Academic Press, New York 1976.
2. R. Buscall, J. W. Goodwin, M. W. Hawkins and R. H. Ottewill, *J. Chem. Soc. Faraday Trans. I*, vol 78, pp 2873–2899 (1982).
3. W. B. Russel and D. W. Benzing, *J. Colloid Interface Science*, vol 83, pp163–177, 1981.
4. D. W. Benzing and W. B. Russel, *J. Colloid and Interface Science*, vol 83, pp 178–190, 1981.
5. M. A. Ansarifar and P. F. Luckham, *Polymer*, vol 29, pp 329–35, 1988.
6. M. A. Ansarifar and P. F. Luckham, *Polymer Communications*, vol 2, pp 177–181, (1988).
7. M. A. Ansarifar and P. F. Luckham, *Colloid and Polymer Science*, in press.
8. H. K. Christenson, R. G. Horn and J. N. Israelachvili, *J. Colloid Interface Science*, vol 88, pp 79–88, (1982).
9. J. N. Israelachvili and G. E. Adams, *J. Chem. Soc. Faraday Trans. I.*, vol 74, pp 975–1001, (1978).
10. P. F. Luckham and J. Klein, *Macromolecules*, 18, 721, (1985).
11. P. F. Luckham and J. Klein, *J. Colloid and Interface Science*, vol 117, pp 149–159, (1987).
12. G. P. H. L. de Silva, P. F. Luckham and Th. F. Tadros, *Polymer Communications* 30, (1989).
13. K. E. J. Barrett, (1975) *"Dispersion Polymerisation in Organic Media"*, Wiley New York.
14. M. A. Ansarifar and P. F. Luckham, *Colloid and Polymer Science*, 266, 1020, (1988).
15. J. N. Israelachvili, *J. Colloid Interface Sci.*, vol 44, 259–272, (1973).
16. H. K. Christenson and R. G. Horn, *J. Colloid Interface Sci.*, vol 103, pp 50–55, (1985).
17. P. G. de Gennes, *Advances Colloid Interface Science*, vol 27, pp 189–209, (1987).
18. S. Alexander, *J. de Physique*, vol 38, pp 483–987, (1977)
19. S. T. Milner, T. A. Whitten and M. E. Cates, *Macromolecules*, vol 21, pp 2610–2619, (1988).
20. S. T. Milner, *Europhysics Letters*, vol 7, pp 695–700, (1988).
21. T. Cosgrove, N. Finch, B. Vincent and J. Webster, *Colloids and Surfaces*, vol 31, pp 33–47, (1988).
22. I. M. Krieger, *Adv. Colloid Interface Science*, vol 3, pp 111–136, (1972).
23. C. Prestidge and Th. F. Tadros, *J. Colloid and Interface Science*, vol 124, 660–665, (1988).
24. B. V. Derjaguin, *Kolloidnyiz*, vol 69, pp 155– (1934).

SELECTIVE SEPARATION OF FINE PARTICLES AT A CHARGED SOLID/ LIQUID INTERFACE

R.A. WILLIAMS and X. JIA
Department of Chemical Engineering, University of Manchester Institute of Science and Technology, P.O. Box 88, Manchester M60 1QD, United Kingdom.

ABSTRACT

This paper considers recent developments in theoretical models which may be used to simulate particle capture processes and the feasibility of separating colloidal species according to their surface charge, by selective and reversible adsorption onto a collector surface. Selectivity is achieved through control of the electrochemical potential of the collector. The principles of controlled–collector–potential chromatographic separation (CCPCS) methods are illustrated for the separation of alumina and titania particles (200 nm radius) at a macroscopic platinum (oxide) collector in aqueous electrolytes for a flow–through sandwich–cell separation module. For this ceramic system separation *via* the secondary minimum interaction energy level is not possible, however, selective separation based on differences in the DLVO energy barrier using potential control seems feasible. Practical limitations on the selection of appropriate approximations to describe particle–collector interaction and the role of heterogeneities in particle/charge properties are discussed.

INTRODUCTION

The adsorption of fine particles and biomaterials (<<10 μm) onto a solid surface immersed in an aqueous medium is frequently encountered as part of controlled–separation or coatings processes, and in an uncontrolled fashion through adventitious fouling of membranes and heat exchanger surfaces. For either mode of particle deposition, prediction and control of such physicochemical phenomena is rarely achieved due to the need for simultaneous solution of the equations describing surface forces, hydrodynamics and multi–body particle interactions.

The difficulties associated with this type of problem were identified in some of the earliest reputable measurements reported by Marshall and Kitchener [1], in which particle adsorption under the well defined hydrodynamics of the rotating disc electrode differed from behaviour predicted by the Levich equation and the DLVO theory. This, together with additional seemingly anomalous behaviour, has since been reported by many other experimentalists. Qualitative predictions and simulation models have been reported for a number of ideal dispersions [2–4]. However, the possibilities for exploiting particle adsorption at a macroscopic (collector) surface for the purpose of controlled–separation of particles has received comparatively little attention. This paper discusses the feasibility of a novel separation method *Controlled Collector Potential Chromatographic Separation (CCPCS)* for selective (and sometimes reversible) capture of colloidal species at large collector electrode whose electrochemical potential, and hence double layer properties, is controlled independently. Practical problems that hinder a full understanding of process in which colloidal species are captured in the electrical double layer of a secondary collector surface are also discussed with respect to the practicalities of measuring and predicting particle adsorption and desorption.

Advances in Fine Particles Processing
John Hanna and Yosry A. Attia, Editors

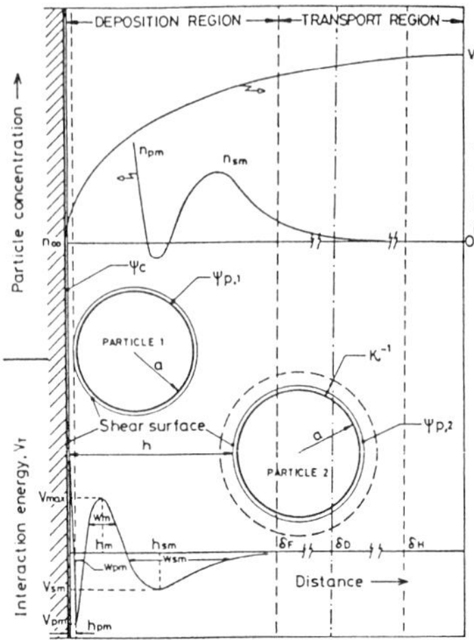

FIG. 1. Schematic diagram of particle adsorption at a macroscopic collector under controlled potential conditions. Typical orders of magnitude for which this representation is valid are: a: <<5 μm, h_{pm}: 0.4–1.0 nm, h_m: 1–10 nm, h_{sm}: 10–10^2 nm, δ_F: 10^2 nm, δ_D: 10^3–10^5 nm, δ_H: 10^5–10^7 nm, ψ_P : -10^2–10^2 mV, ψ_C : -10^3–10^3 mV, n_∞: 10^{-8} cm^{-3}, V_∞: 0.1–10.0 cm s^{-1}.

Fig. 1 shows the principles of particle adsorption in the electrical double layer of a macroscopic collector surface. The adsorption process may be considered as two–stages involving mass transport to the interphase region (controlled by hydrodynamics) followed by adhesion (controlled by van der Waals attraction, electrical double layer interaction and Born repulsion). Since the second stage occurs very close to the collector surface over distances much shorter than the thickness of the diffusion boundary layer, the effects of fluid convection within this layer can often be neglected. The efficiency of the adhesion or capture process depends upon the form of the total interaction energy between the particle and the collector, as a function of their distance of separation. For high ionic strengths (typically, above 0.1 mol dm^{-3}) the interaction profile may exhibit one mimimum and no maximum, resulting in net attraction and adsorption. For low ionic strengths (below 10^{-4} mol dm^{-3}) a single maximum is observed, which presents a large repulsive barrier inhibiting adsorption. For intermediate ionic strengths two minima (the so–called primary and secondary minima) and one maximum may exist, and weak adsorption may occur in the (outer) secondary minimum, or in the (inner) primary minimum if the particle can surmount the intermediate repulsive energy barrier. Hence once a particle has been transported to the vicinity of the collector surface the adsorption (and desorption) processes are governed by the balance of the short range forces, which depend most critically on the particle size, electrolyte composition, the relative surface potentials of the collector and the particle, and the effective Hamaker constant for the particle/electrolyte/collector system.

The possibility of particle desorption is of considerable interest since predictions of the kinetics of detachment can be made on the basis of the expected valve of the velocity of a particle at the peak of the energy barrier (V_{max}) subject to a number of simplfying assumptions [5], including the presence of an infinite energy well at zero–separation between a particle and the wall of the macroscopic collector. Subsequently Ruckenstein and Prieve [6] made a major improvement in quantitative analysis of the phenomenon by taking Born–repulsion forces into account explicity. Rates of adsorption/desorption were calculated by considering Born, electrical double layer (EDL) and van der Waals forces whose effects were lumped into a first order, reversible and "reaction–like" boundary condition on the usual convective– diffusion equation (described later). On the basis of these results, a separation method termed "Potential–Barrier Chromatography" was mooted. Detachment kinetics for sphere/ plate geometries have also been estimated in terms of solution of Kramer's equation, by combining Ruckenstein and Prieve's methodology [6] and the earlier model proposed by Barouch *et. al.* [8].

All of the work cited above was largely concerned with separation facilitated by the means of the primary minimum (PM) in the EDL energy profile (Fig.1), whereas, the presence of the secondary minimum (SM) is of considerable importance. SM effects have been studied quite widely in connection with the reversible aggregation of particles. Hence, the possible exploitation of both SM and PM may offer opportunities for selective capture and/or rejection of particles at solid/liquid interface.

CONTROLLED–COLLECTOR POTENTIAL CHROMATOGRAPHIC SEPARATION (CCPCS)

Principle of CCPCS

The aim of the present paper is to investigate the reversible and selective separation of particle species i with surface potential $\psi_{p,i}$ in a system where the potential at a collector surface ψ_c is controlled independently. Fig.1 shows a schematic diagram of such an arrangement where two smooth spherical particles $\psi_{p,1}$ and $\psi_{p,2}$ are retained in the primary and secondary minima regions of the double layer of a planar macroscopic electrode.

Experimentally, this can be achieved by controlling the surface potential ψ_c of a macroscopic collector (such as a doped–oxide or a metal oxide) by means of a conventional potentiostatic circuit with respect to a standard reference electrode and secondary counter electrode. The surface potentials of the collector are modest (quite unlike high–gradient electric field separators) and must lie well within the potential range between the electrochemical reduction/oxidation of water.

It is assumed that the surface potential $\psi_{p,1}$ of the colloidal species can be approximated from electrophoretic mobility measurements performed in an indifferent electrolyte, and adjusted by modifying the ionic composition of the liquid medium. This may be justified on the basis that the potential at the shear plane of the double layer (*i.e.* the ζ–potential) is the effective potential of interaction between the particle and the collector surface.

The key question to be addressed concerns the degree of control over particle deposition that may be achieved by modifying the EDL energetics profile, and the possibility of bulk or selective particle ejection from the SM interface region back into transport region. These cases will be examined in Section 4.

Experimental separation module

The time–scale of fine particle separation methods involving various electrophoretic field–flow fractionation devices tends to limit their application to low–throughout specialist chemical/bio–separations. For our studies we chose to employ a sandwich–type cell, for which the hydrodynamics are obviously less well–defined than for rotating disc or wire collectors, but if successful may be configured as lamaella collectors capable of handling larger quantities of material. The characteristics of such sandwich systems have been elucidated by Bowen and Epstein for deposition of silica spheres on to passive (plastics) collector surfaces [9].

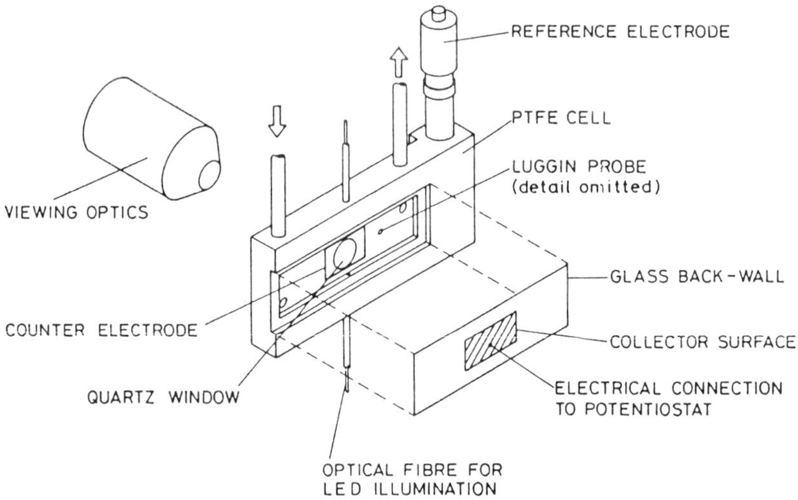

FIG. 2. Exploded view of a separation module for CCPCS

Fig.2 gives an exploded view of a separation module that may be used to perform experimental CCPCS measurments. It consists of a PTFE cell with a removable rear glass (or plastic) wall, to which the macroscopic collector electrode is attached. The colloidal mixture is admitted to the cell at a controlled rate, and flows over the collector, whose potential is controlled with respect to a micro–luggin probe (located near the upper edge of the collector surface) and reference electrode, *via* a symmetric counter–electrode embedded in the opposite wall. The potentiostat used (Thompson Autostat 401) is controlled by an Archimedes 301 microcomputer, enabling operation under static or dynamic potential conditions. For the module illustrated, the channel depth is $2x10^{-3}m$ and $50x10^{-3}m$ in length for investigation of flow rates up to 1 $cm^3\ s^{-1}$.

Particle deposition is observed through a quartz window using focused optics mounted on a micrometer driven stage. Illumination is provided *via* two optical fibres. Deposition rates may be assessed using either still or video photography, and analysis of the resultant images with time.

PREDICTION OF PARTICLE CAPTURE

Simulation models

Conventional theoretical approaches to predict particle adsorption on a collector demand an intimate knowledge of the properties of the system. In general this includes the properties of the particle/liquid and collector/liquid interfaces which give rise to electrical forces between the solid surface according to the nature of the interphase region and the geometry of the surfaces. Providing that information on the hydrodynamic characteristics of the suspended particulate phase(s) are known, it is then possible to model the deposition process by computer simulation, subject to a number of limiting approximations. The mode of the simulation may be Monte Carlo, molecular dynamic or based on Brownian dynamic or other perturbation methods depending upon the size of the colloidal species and the information that is sought (or available).

Two basic theoretical approaches are commonly used, namely, Lagrangian methods, in which the trajectories of individual particles are followed, and Eulierian methods, which treat the problem in terms of the spatial concentration and orientation of particles. Since an excellent and detailed account of these two methods and their various sub–divisions has been given by Adamczyk *et. al.* [3], here it is only pertinent to highlight the factors associated with choosing an appropriate method, and any inherent limitations. However it should be mentioned that even the most sophisticated of models for simple particle/collector geometries can only be used with moderate success, since although the Navier–Stokes equations defining the fluid flow can be solved for many configurations [2] uncertainties regarding the form of electrical double layer and short range forces and inhomogeneties in the physical properties of the solid phases abound. Difficulties may also arise in the selection of appropriate geometric boundary conditions and computationally–imposed limitations on the number of colloidal species that can be considered.

Lagrangian methods are based on Newton's laws of motion and were first employed for cases in which Brownian effects (*e.g.* thermal motion) could be ignored. Hence it is possible to define the limiting trajectory of a particle approaching a collector surface within which capture is assured, for example, as in the classical approaches used to define particle capture onto an air–bubble in flotation proceses. It is necessary to identify whether inertia and diffusion effects can be ignored, since the method has been used most widely for relatively large particle (> 1 μm). For smaller colloidal particles Brownian effects become more dominant and the trajectories are no longer deterministic, so a dynamic simulation must be sought. Such methods are well established for liquids [10] and are now being applied on a macroscopic scale to dispersions. In cases where the time–scales of the respective forms of motion vary greatly (*e.g.* Brownian motion *cf.* gravitational motion) the Monte Carlo and molecular dynamic methods can be problematic due to the large number of small time–steps required to simulate the process, whereas the Brownian dynamics approach which employs a Langevin–type equation that can be integrated in a matter that is not so prohibitive [11].

Eulerian methods are less constrained when it is necessary to take Brownian effects into account, since analytical equations are used, which are more readily and rapidly solved than the force–balance type of equations demanded by Lagrangian methods. This method expresses particle motion by means of equations for the probability density distribution *i.e.* Fokker–Plank equation [12] which is also capable of taking interactions *between* particles into account. In the limiting case, approaching infinite dilution, they may be reduced to the convective diffusion (continuity) equation, that can be solved to yield the particle concentration distribution:

$$\frac{\partial n}{\partial t} + \nabla . j = Q \tag{1}$$

$$j = -\mathscr{D} . \nabla n + U n \tag{2}$$

in which n is the particle number density, j is the particle flux, Q is a source term, $\mathscr{D}$ is the diffusion tensor and U is the particle velocity vector. The success of this method lies in the boundary conditions stipulated with regard to the specific interaction between the particle and collector surfaces, and the relative magnitudes of the interaction force and mass transfer terms.

The simplest assumption may be referred to as the *perfect sink model* in which it is assumed that once a particle has reached a distance δ from the surface of the collector it will be irreversibly captured, and thereafter disappears from the colloidal system. Many different geometries and flow regimes have been considered in the literature, of which the most interest to the present work is obtained by assuming that the interaction forces act over much smaller distances than the thickness of the diffusion boundary layer (*i.e.* $\kappa a_c Pe^{-1/3} >> 1$, where a_c is the radius of the spherical or cylindrical collector). Hence the boundary condition takes on the form of an irreversible first–order chemical reaction on the collector surface:

$$j_\perp \Big|_{h=\delta_F} = D \frac{\partial n}{\partial h} = K n \tag{3}$$

$$K^{-1} = \int_0^{\delta_F} \frac{\exp(\Phi_S)}{D(h)}\, dh$$

where δ_F is the interaction force boundary layer, Φ_S is the scaled interaction potential (V_T/kT), and the constant K is the collector surface *via* eqn.2. In this way eqn.3 can be used as the boundary condition for the usual convective–diffusion equation and developed to describe the rates of adsorption in both the PM and SM regions of the interaction energy profile [13].

The second assumption is less idealised, and is known as the *non–penetration model.* In this situation the particles do not disappear on adsorption, but are located at some defined position near the collector wall, thus more closely representing the real adsorption process. The most fruitful approximations are achieved by effectively lumping all the forces that cannot be specified explicitly into the source term (Q) of eqn.1, and the results from analytical expressions of this form (for various particle–collector geometries) have been compared and validated against numerical solutions for Peclet numbers up to one [14].

Estimation of surface interaction potentials

The form of the interaction potential used in predictive simulations are, of necessity, a simplified representation of the actual potential–distance functions and originate from studies of liquid–state physics. The simplest being a discontinuous potential step (hard sphere model; for which $V_T(h \leq \sigma) = \infty$, $V_T(h > \sigma) = 0$) and the more realistic soft–sphere models with continuous potentials. The most common example of the latter category is the Lennard–Jones "12–6" potential (eqn.4) and the "exp–6" potential (eqn.5):

$$V_T(h) = 4E[(\sigma/h)^{12} - (\sigma/h)^6] \tag{4}$$

$$V_T(h) = [E/(1 - 6/\beta)][\exp\{\beta(1 - h/\sigma)\}] - (h/\sigma)^6] \tag{5}$$

where E is the depth of the energy well, σ is the collision diameter and β is a parameter which determines the steepness of the repulsive part of the interaction. Some other potential functions of this type can be tuned to reflect the salient features of colloidal dispersions.

Unfortunately there is no universal equation or approximation that can describe adequately the surface interaction potential between, say, a smooth spherical particle and a smooth macroscopic collector for all distances of surface–to–surface separation. The individual components and limitations in descriptions of Born repulsion, van der Waals attraction and EDL forces have been discussed in detail elsewhere [2], so for the purposes of the analysis which follows our principal interest (Section 1) is in the energy barrier and SM of the interaction energy profile, both of which are relatively far away from the collector surface compared to the PM. Hence it will suffice to consider only van der Waals and EDL interactions, but, nevertheless, a number of trade–offs have to be accepted in deriving suitable approximations for the interaction terms.

The energy barrier and the SM usually occur at distances where retardation effects are significant [2], consequently the best approximation for van der Waals attraction is that given by Czarneki [15]

$$V_A = A\left[\frac{2.45\lambda}{60\pi}\left[\frac{h-a}{h^2}-\frac{h+3a}{(h+2a)^2}\right]-\frac{2.17\lambda^2}{720\pi^2}\left[\frac{h-2a}{h^3}-\frac{h+4a}{(h+2a)^3}\right]\right.$$

$$\left.+\frac{0.59\lambda^3}{5040\pi^3}\left[\frac{h-3a}{h^4}-\frac{h+5a}{(h+2a)^4}\right]\right] \tag{6}$$

where λ is the characteristic wavelength in the medium (which usually takes the value of 100 nm), h is the surface–to–surface distance between the two interacting surface and A is the effective Hamaker constant. This approximation has been shown [16] to be in good agreement other exact calculations at all separation distances except below about 5 nm in aqueous solutions. However, at very high ionic strengths the energy profile is likely to become so compact that the SM/PM extrema may well be confined to distances below 5 nm. In this case, an alternative equation derived by Gregory [16] may be adopted for calculation purposes:

$$V_A = -\frac{Aa}{6h}\left[1-\frac{bh}{\lambda}\ln(1+\frac{\lambda}{bh})\right] \tag{7}$$

where b=5.32. This equation is a good approximation for retarded van der Waals attraction up to about h=a/10.

A large number of expressions have been reported for the EDL interaction using either the constant–potential, constant–change or some intermediate assumption as the respective EDLs come sufficiently close to interact with each other, and various methods for solution of the linear or the non–linear Poisson–Boltzmann equation [2]. Clearly the major requirement in selecting an approximation for the EDL interaction for CCPCS assessment is the ability to allow for a potential–controlled system and wider applicability to the potentials of the interacting surfaces. Therefore, we chose to use the formula derived and used by Ruckenstein and Prieve [6]:

$$V_R = 16\epsilon\,(\frac{kT}{e})^2\, a\gamma_1\gamma_2 \exp(-\kappa h) \tag{8}$$

where ϵ is the dielectric constant for the system ($\epsilon=\epsilon_0\epsilon_r$), κ is the reciprocal Debye–Huckel length, $\gamma_i = \tanh(e\psi_i/4kT)$, and k is the Boltzmann constant. Although eqn.8 is reported to be appropriate for 1:1 electrolyte solutions when $\kappa h>2$ and $\kappa a>>1$, partial breach of these conditions may not incur serious error, indeed Ruckenstein and Prieve [6] applied this equation to a range of separation distances well below the imposed limit.

Combining the two relevant equations for EDL interaction (eqn.8) and van der Waals attraction (eqn.6 or eqn.7), yields the total interaction potential energy (V_T). These provide the basis for an initial assessment of the influence of the key variables in CCPCS. For example, from eqns.6 and 8 it is evident that V_T will decrease as the Hamaker constant increases, and as a result the SM will become deeper and the height of the energy barrier (if present) will become smaller. With regard to the influence of the potentials of the interacting surface ψ_1, ψ_2 the total interaction (eqns.6 and 8) can take the form:

$$V_T = V_A + \gamma_1\gamma_2 f_R(h,a,T) \tag{9}$$

thus

$$\frac{\partial V_T}{\partial \psi_1} = \frac{ef_R}{4kT}\,\gamma_2\,\mathrm{sech}^2(\Phi_1/4)\begin{cases} > 0, \text{ when } \psi_2 > 0 \\ < 0, \text{ when } \psi_2 < 0\end{cases} \tag{10}$$

$$\frac{\partial^2 V_T}{\partial {\psi_1}^2} = -(\frac{e}{4kT})^2 f_R\,\gamma_1\gamma_2\begin{cases} > 0, \text{ when } \psi_1\psi_2 < 0 \\ < 0, \text{ when } \psi_1\psi_2 > 0\end{cases} \tag{11}$$

where $\Phi_1 = (ze\psi_1/kT)$. (Note that it is not necessary to distinguish or assign ψ_1 or ψ_2 explicitly to either ψ_c or $\psi_{p,i}$.) Hence, when $\psi_2>0$, increasing ψ_1 leads to an increase in the total interaction energy, *i.e.* shifting the profile (Fig.1) upwards. When $\psi_2<0$, the converse occurs.

It is also evident that when the two charged surfaces are of the same sign, increasing ψ_1 will reduce the sensitivity of V_T to ψ_1. However, if the two surfaces are oppositely charged, the reverse will be true. In fact, very high value of ψ_1 causes γ_1 to approach unity (*e.g.* when $\psi_1 = 260$ mV at 25°C, $\gamma_1 = 0.9874$). Thus, V_T seems likely to depend solely on ψ_2 when ψ_1 is sufficiently large.

Similarly, considering the likely effect of electrolyte concentration on the interaction energy, a rise in the electrolyte concentration will result in the profile shifting downwards, *i.e.* reducing the height of the maximum and deepening the SM. If $\psi_1\psi_2<0$, the opposite situation will apply. For $\psi_1\psi_2>0$, the total interaction energy becomes increasingly sensitive to electrolyte concentration.

FEASIBILITY OF CCPCS OF COLLOIDAL Al_2O_3/TiO_2 DISPERSIONS

Characteristics of the ceramic dispersion

The feasibility study is based on alumina (Al_2O_3) and titania (TiO_2) which represent two common materials that are used widely in the pigment and electronics industries. It may be postulated that CCPCS could be used to separate a given colloidal species on the basis of a difference in their surface potentials. For the purpose of this assessment it is assumed that a given particle is spherical, smooth, exhibits negligible heterogeneity in surface charge, and interact only with the macroscopic collector surface. Experimental measurements of electrophoretic mobilities of mixtures of Al_2O_3 and TiO_2 ceramics (a = 200 nm) in KCl electrolytes are reported elsewhere [17], but here we will use data for the zeta–potentials of the pure components (Table 1) given by Wiese and Healy [18]. Other important physical parameters of the system are the Hamaker constant for the collector (Pt–oxide) and the colloidal ceramics. The literature reveals that significant variations exist in "typical values" of A for each material, with values for the overall Hamaker constant term A varying between 5×10^{-20} to 13×10^{-20} J. However, only two values of A were selected, 5×10^{-20} J and 10×10^{-20} J, for calculation of the ceramic/collector interactions for Al_2O_3/Pt(oxide) and TiO_2/Pt(oxide), respectively.

Effect of process variables

The effect of variations in electrolyte concentration (M, mol KCl dm^{-3}), electrolyte pH, Hamaker constant (A,J), particle radius (a) and surface potentials (ψ_i,ψ_c) on the viability of a selective separation process will each be considered. In general the range of values considered was $10^{-6} \leq M \leq 0.1$, $5\times10^{-20} \leq A \leq 10^{-18}$, $10 \leq A(nm) \leq 10^4$, $0 \leq \psi(mV) \leq |500|$, pH values between 5–10, and for separation distances (h) between 5–1000 nm at 298 K.

Increasing the Hamaker constant makes the interaction energy profile more compact in the direction perpendicular to the collector surface. The depth of the SM is more sensitive to changes in A than the energy barrier,a nd both extrema show reduced sensitivity if the electrolyte concentration is further reduced, say, from 10^{-3} to 10^{-4} mol KCl dm^{-3}.

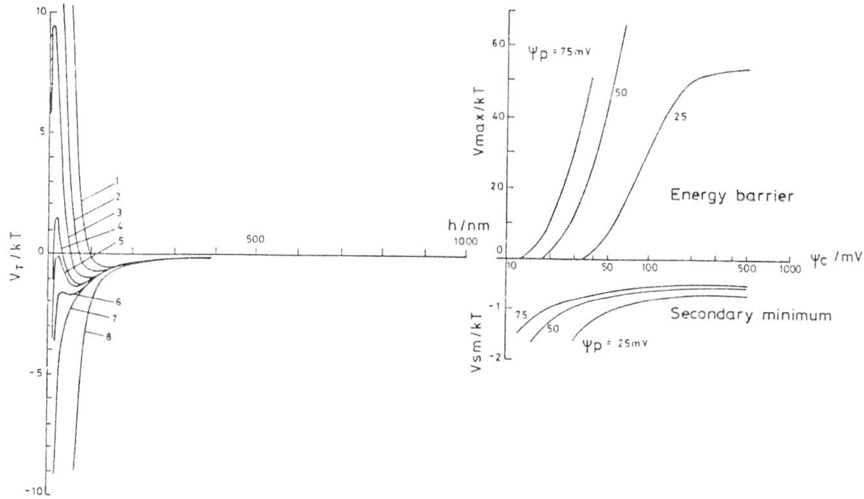

(a) (b)

FIG. 3. (a) The total interaction energy profile for different values of collector potential at T=298 K, a=500 nm, A=3×10^{-20} J, M=10^{-3} mol dm^{-3}, ψ_p=50 mV, $\psi_{c,1}$=500, $\psi_{c,2}$=50, $\psi_{c,3}$=28, $\psi_{c,4}$=20, $\psi_{c,5}$=18, $\psi_{c,6}$=15, $\psi_{c,7}$=10, $\psi_{c,8}$=−100 mV, and (b) variation of peak values of the energy barrier and secondary minimum with collector potential at different particle ζ-potentials.

For given particle ζ-potential, increasing the potential of the collector has the effect of "lifting" the energy profile upwards. Consequently, the energy barrier becomes larger but the SM is shallower. If the collector potential ψ_c is fixed, higher particle ζ-potentials ψ_p yield greater changes in the height of the energy barrier, as shown in Fig.3(a). At given values of the electrolyte concentration, particles possessing a small ζ-potential require a higher collector potential to give the same height of the energy barrier. However, Fig.3(b) demonstrates that both the potential of particles and the collector potential have little effect on the depth of the SM.

As might be expected from simple DLVO theory there is no possibility of selective separation and high electrolyte concentrations (> 10^{-2} mol dm^{-3}) due to the absence of any extrema in the total interaction energy curves. For lower electrolyte concentrations (*e.g.* 5×10^{-3} mol dm^{-3}), although a significant SM can be obtained the energy barrier tends to be small and this situation is not significantly improved by adjusting ψ_c. Therefore, use of the SM alone does not seem feasible for the purposes of separating TiO_2/Al_2O_3 materials.

In order to assess other separation strategies using the energy barrier or PM it is necessary to consider the variation of the energy barrier and SM with particle potential. Figure 4 shows the results at different collector potentials for A = 5×10^{-20} and 10×10^{-20} J. Fig.5 compares the interaction energy profiles for TiO_2 and Al_2O_3 for pairs of ζ-potentials corresponding to several possible values of pH (Table 1).

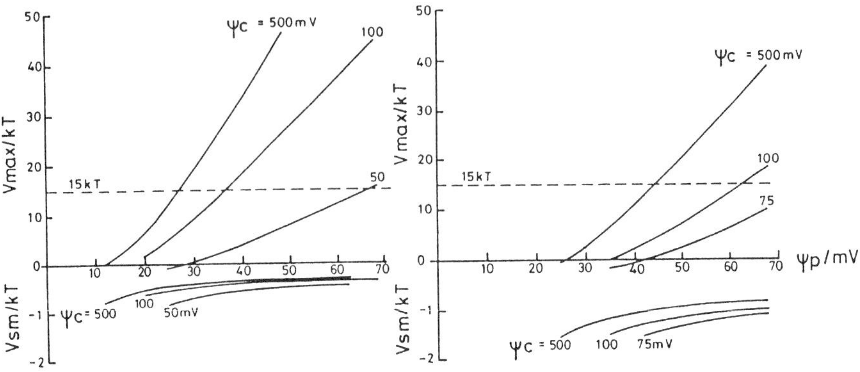

FIG. 4. Variation of the peak values of the energy barrier and secondary minimum with particle ζ-potential at different collector potentials. T=298 K, M=10^{-3} mol dm^{-3}, a=200 nm. (a) A=5×10^{-20} J, (b) A=10×10^{-20} J.

TABLE I. Zeta potentials (mV) of aluminia and titania as a function of pH at different electrolyte concentrations

M /mol KCl dm^{-3}	Particle	pH										
		5.0	5.5	6.0	6.5	7.0	7.5	8.0	8.5	9.0	9.5	10
10^{-2}	TiO_2	10	5	0	−17	−28	−35	−38	−40	−40	−40	−40
	Al_2O_3	37	35	33	31	29	27	21	10	0	−12	−20
10^{-3}	TiO_2	25	12	0	−28	−46	−51	−57	−60	−62	−64	−68
	Al_2O_3	53	50	49	45	43	42	35	20	0	−20	−35
10^{-4}	TiO_2	31	15	0	−36	−52	−66	−70	−72	−74	−77	−78
	Al_2O_3	72	70	68	65	60	55	45	25	0	−28	−50

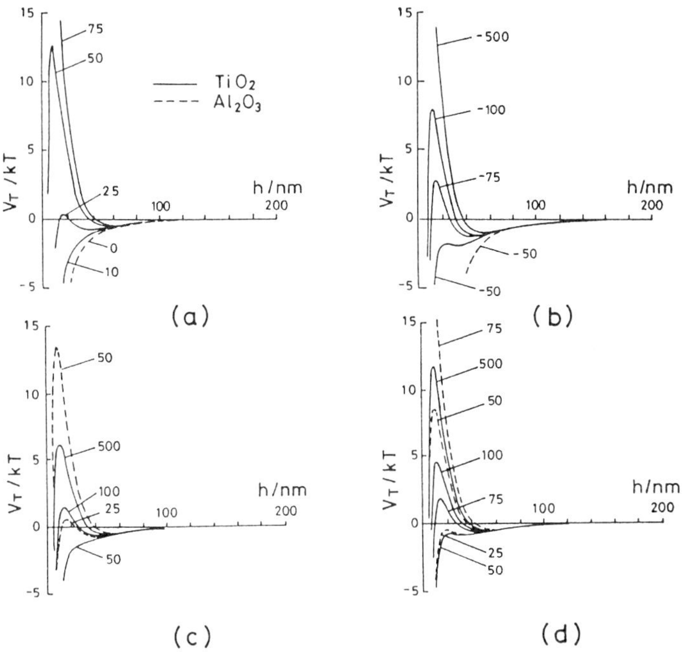

FIG. 5. Comparison between the energy profiles of TiO_2 (solid lines) and Al_2O_3 (broken lines) at given pH values for different collector potentials, T=298 K, $M=10^{-3}$ mol dm^{-3}, a=200 nm. (a) pH=9.0, $A=5x10^{-20}$ J; (b) pH=7.5, $A=10x10^{-20}$ J; (c) pH=9.5, $A=5x10^{-20}$ J; (d) pH=5.0, $A=5x10^{-20}$ J.

At the isoelectric point of one species, it is possible for the other to develop an insurmountable enrergy barrier (Fig.5(a)). Therefore, in principle, the two species can be separated readily. In Fig.5(b) where the two species are oppositely charged, the role of the energy barrier is more evident. In this case, the collector potential can be adjusted in favour of one species against the other although this is not a practical regime under which to operate since particle coagulation is likely to occur. Even when the two species are of the same sign of charge, providing the difference in the ζ–potentials between the two species is not too small, the collector potential can be used to enhance the energy barrier of one species while depressing that of the other. Figs.5(c) and (d) show some examples of the degree of selectivity that can be achieved by controlling the collector potential. Under these experimental conditions it is generally assumed that the energy barrier may be regarded as being insurmountable if it exceeds 15 kT, although in some cases values above 10 kT may be sufficient. For the conditions examined in 10^{-3} mol KCl dm^{-3} the most probable scenario for selectivity on the basis of an insurmountable energy barrier occurs at pH 10 with ψ_c = 500 mV ($A = 5x10^{-20}$ J). Similar trends in the changes of the energy barrier occur when in 10^{-4} mol KCl dm^{-3}. However, at this electrolyte concentration, even the smallest ζ–potential could result in a large energy barrier of sufficient height to be regarded as insurmountable and thus offer possibility of separation. As the electrolyte solution becomes so dilute, no significant SM appears even assuming the combination of the highest plausible particle and collector potentials. The implication of this is that it is not possible to make use of the SM for separation processes at such low electrolyte concentrations. Even for materials with a high Hamaker constant the height of the energy barrier may be reduced but the SM remains undeveloped in 10^{-4} mol dm^{-3}

solutions. Nevertheless, there are numerous possibilities for separations based on differences in the height of energy barriers at modest collector potentials ($\psi_c \geq 50$ mV) in 10^{-4} mol KCl dm^{-3}.

CONCLUSIONS

The selective separation of titania and alumina species on the basis of the SM alone is questionable since the necessary prerequisite of the presence of an insurmountable energy barrier is not met for the range of variables examined. Given the ubiquity of the PM in the energy profile it seems more feasible to make use of the energy barrier and the PM to achieve selective separation of these species by controlling the potential of the collector. The value of the effective Hamaker constant is critically important in determining the form of the interaction curves, hence optimization of the Hamaker constant by judicious choice of the collector material may offer significant improvement in the flexibility of this system, and further exploitation of the SM.

It must be recognized that in the present Al_2O_3/TiO_2 system the possibility of heterocoagulation and the effects of chemical dissolution/adsorption of ionic species has been omitted from our analysis. At present we are verifying the practicalities of such separations using the CCPCS module shown in Fig.1 using both model latex and (non–equilibrium) oxide dispersions and by computer simulation [19]. This is particularly important since a number of non–ideal phenomena can affect the process, namely: surface heterogeneities of charge and roughness; ionic equilibrium/dissociation and other time–dependent effects; secondary interactions between particles in the dispersion; particle blocking and loading of the interphase region.

REFERENCES

1. J.K. Marshall and J.A. Kitchener, The Deposition of Colloidal Particles on Smooth Solids, *J. Colloid Interface Sci.*, 22, 342–351, (1966).
2. X. Jia and R.A. Williams, Particle Deposition at a Charged Solid/Liquid Interface: Review Article, *Chem. Eng. Commun.*, (in press).
3. Z. Adamczyk, T. Drabos, J. Czarnecki and T.G.M. van der Ven, Particle Transfer to Solid Surfaces, *Adv. Colloid Interface Sci*, 19, 183–252, (1983).
4. J.P. Hsu and S.S. Sun, A Probabilistic Analysis of the Adsorption of Particles on Solid Surfaces, *J. Colloid Interface Sci.*, 122, 73–77, (1988).
5. B. Dahneke, Kinetic Theory of the Escape of Particles From Surfaces, *J. Colloid Interface Sci.*, 50, 89–107, (1975).
6. E. Ruckenstein and D.C. Prieve, Adsorption and Desorption of Particles and Their Chromatographic Separations, *A.I.Ch.E.J.*, 22, 276–283, (1976).
7. E. Barouch, T.H. Wright, and E. Matijevic, Kinetics of Particle Detachment, *J. Colloid Interface Sci.*, 118, 473–481, (1976).
8. E. Barouch, E. Matijevic, T.A. Ring and J.M. Finlan, Heterocoagulation, *J. Colloid Interface Sci.*, 67, 1–9, (1978).
9. B.D. Bowen and J.N. Epstein, Fine Particle Deposition in Smooth Parallel– Plate Channels, *J. Colloid Interface Sci.*, 72, 81–97, (1979).
10. M.P. Allen and D.J. Tildesley, *Computer Simulation of Liquids*, (Clarendon Press, Oxford, 1987).
11. D. Gupta and M.H. Peters, On the Angular Dependence of Aerosol Diffusional Deposition onto Spheres, *J. Colloid Interface Sci.*, 110, 286–291, (1986).
12. T.J. Murphy and J.L. Aguirre, Brownian Motion of N Interacting Particles, *J. Chem. Phys.*, 57, 2098–2104, (1972).
13. D.C. Prieve and M.M.J. Lin, Adsorption of Brownian Hydrosols onto a Rotating Disc Aided by a Uniform Applied Force, *J. Colloid Interface Sci.*, 76, 32–47, (1980).
14. Z. Adamczyk and T.G.M. van der Ven, Kinetics of Particle Accumulations at Collector Surfaces, *J. Colloid Interface Sci.*, 97, 68–90, (1984).
15. J. Czarneki and van der Waals, Attraction Energy Between Sphere and Half–Space, *J. Colloid Interface Sci.*, 72, 361–362, (1979).

16. J. Gregory, Approximate Expressions for Retarded van der Waals Interaction, *J. Colloid Interface Sci.*, 83, 138–145, (1981).
17. X. Jia, M.Sc. Dissertation, University of Manchester, (1988).
18. G.R. Wiese and T.W. Healy, Heterocoagulation in Mixed TiO_2–Al_2O_3 Dispersions, *J. Colloid Interface Sci.*, 51, 458–467, (1975).
19. R.A. Williams and X. Jia, Simulation of Particle Deposition at a Charged Solid/Liquid Interface, (to be published).

ADSORPTION OF COLLECTORS ON MINERALS - EFFECTS OF LATERAL INTERACTION AND MOLECULAR SIZE

AHMED YEHIA*, BADR G. ATEYA** AND A.A. YOUSSEF*
*Central Metallurgical Research and Development Institute, CMRDI, Cairo, Egypt; ** Department of Chemistry, Faculty of Science, Cairo Univ., Egypt

ABSTRACT

The Adsorption isotherm of CTAB on synthetic carbonate apatite minerals, with different CO_3/PO_4 molar ratios, was determined at controlled ionic strength, pH and temperature. Flotation recovery of apatite minerals were also determined for the same conditions. The results were analyzed according to the Langmiur, Flory-Huggins and Frumkin isotherms.

In this study the Frumkin isotherm was modified in order to account for the large size of the CTAB molecule. While the Langmiur and Flory-Huggins isotherms give rise to energies of adsorption $\Delta G°$ independent of coverage, the Frumkin isotherms give rise to values dependent on coverage due to lateral interaction effects. The effect of the size of the CTAB molecule on the free energy of adsorption is evaluated. The results of flotation recovery were discussed in view of this analysis.

INTRODUCTION

The importance of the adsorption of collectors on the surfaces of minerals cannot be overestimated. A clear understanding of the thermodynamics of the process is quite essential for the interpretation of their flotation and electrokinetic behavior. The adsorption data are frequently analyzed using simple isotherms, e.g. the Langmiur isotherm [1-3]. Application of the Langmiur isotherm pre-supposes that the heat of adsorption is independent of coverage, which is only rarely true [4]. Furthermore, such simple isotherms are inherently inadequate for the description of the adsorption of bulky long chain molecules. These isotherms neglect one or both of the following factors:

1. Lateral interaction between the adsorbed molecules on the free energy change of the process. This is particularly important for the adsorption of collectors with long hydrocarbon chains protruding from the mineral surface into the solution.
2. The displacement of a number of water molecules off the mineral surface for the adsorption of each collecttor molecule. This is an important consideration in the adsorption of molecules with large cross-sectional area e.g. CTAB which has a cross-sectional area of 50 $A°^2$ [5] (comparable to the areas of 4 water molecules). Proper account of the above factors in the thermodynamic treatment of the adsorption process leads to isotherms which are considerably different from those commonly used. The resulting thermodynamic quantities will also be different.

The purpose of this paper is to account for the above factors in the analysis of the adsorption behavior of CTAB on apatite. Furthermore, the effects of varying the CO_3^{-2}/PO_4^{-3} ratio in the mineral on the extent of adsorption are evaluated.

Advances in Fine Particles Processing
John Hanna and Yosry A. Attia, Editors

THEORETICAL BACKGROUND

The surfaces of hydrophilic minerals are hydrated. The adsorbed water molecules are attached to the surface by partial electrostatic or covalent bonds. Some of these molecules have to be desorbed off the surface before the surfactant molecules are adsorbed, i.e. the process is essentially a substitutional adsorption reaction:

$$\text{SRFCT (sol.)} + n\ H_2O\ \text{(min.)} \rightleftharpoons \text{SRFCT (min.)} + n\ H_2O\ \text{(sol.)} \qquad (1)$$

where SRFCT (min.) and SRFCT (sol.) refer to the surfactant on the mineral surface and in the solution, respectively.

The above process attains equilibrium when the sum of the chemical potentials of reactants and products are equal [1], i.e.:

$$\mu_1\ \text{(sol.)} + n\ \mu_2\ \text{(min.)} = \mu_1\ \text{(min.)} + n\ \mu_2\ \text{(sol.)} \qquad (2)$$

where μ_1 and μ_2 refer to the chemical potentials of collector and water molecules, respectively under the conditions indicated in parentheses.

The chemical potential of a certain species is related to its activity by [1,6]:

$$\mu_i = \mu_i^\circ + RT \ln a_i \qquad (3)$$

where μ_i° is the standard chemical potential of the ith species and a_i is its activity. Equation (3) is the simplest form of the chemical potential of a certain component in solution [6]. Depending on the prevailing conditions, other terms (one or more) may be added to the right side of Eq. (3). For example for a charged species of charge Z under a potential of Ø volts, a quantity of energy ZFØ KJ/mole must be added to the r.h.s. of Eq. (3). For a species at the liquid-gas or liquid-liquid interface, a quantity γA must be added, where γ is the free energy of the surface (per unit area) and A its area.

For long chain collector molecules, adsorbed on the mineral surface and protruding out in the solution, the chemical potential must be influenced by lateral interaction forces between the adsorbed molecules. These may be due to short or long range forces [4]. While the former are due to the interaction of the electron clouds of adjacent adsorbed molecules, the latter are due to the interaction of the dipoles of these molecules. The mathematical expressions of the free energy terms due to lateral interaction can be quite complex [7,8]. Furthermore, they often involve variables with either unknown or unmeasurable values. Therefore, the lateral interaction term is frequently represented by $\alpha\ \theta$, where θ is the degree of surface coverage of the mineral with the collector molecules and α is an empirical constant independent of coverage, θ [4].

In order to apply Eq. (3) to the species involved in Eq. (2), we will assume unit activity coefficients and express the concentrations in terms of mole fractions. The mole fractions of the surfactant are $C/(55.5+C) \simeq C/55.5$ and θ, in solution and on the mineral surface, respectively. The corresponding values for water are $55.5/(55.5+C) \simeq 1$ and $(1-\theta)$. Thus:

$$\mu_{H_2O}\ \text{(sol.)} = \mu_{H_2O}^\circ\ \text{(sol.)} \qquad (4a)$$

$$\mu_{H_2O}\ \text{(min.)} = \mu_{H_2O}^\circ\ \text{(min.)} + RT \ln (1-\theta) \qquad (4b)$$

$$\mu_{SRFCT}\,(sol.) = \mu^{\circ}_{SRFCT}\,(sol.) + RT\ln(C/55.5) \tag{4c}$$

$$\mu_{SRFCT}\,(min.) = \mu^{\circ}_{SRFCT}\,(min.) + RT\ln\theta + \alpha\theta \tag{4d}$$

substituting with the above Eqs. in Eq. (2), and rearranging gives rise to:

$$n\,\Delta\mu^{\circ}_{H_2O} + \Delta\mu^{\circ}_{SRFCT} + \alpha\theta = RT\ln\frac{C(1-\theta)^n}{(55.5)\theta} \tag{5}$$

where $\Delta\mu^{\circ}_{(SRFCT)}$ is the free energy change for the adsorption of one mole of surfactant molecules from the solution onto the mineral surface, while $\Delta\mu^{\circ}_{(H_2O)}$ is the same for the desorption of one mole of water off the mineral surface.

The two values are given by:

$$\Delta\mu^{\circ}_{SRFCT} = \mu^{\circ}_{SRFCT}\,(min.) - \mu^{\circ}_{SRFCT}\,(sol.) \tag{6a}$$

$$\Delta\mu^{\circ}_{H_2O} = \mu^{\circ}_{H_2O}\,(sol.) - \mu^{\circ}_{H_2O}\,(min.) \tag{6b}$$

The sum of the quantities on the l.h.s. of Eq. (5) is the free energy change of the substitutional adsorption process at coverage θ, i.e. $\Delta G^{\circ}_{(\theta)}$: It is composed of coverage-dependent and coverage-independent terms. It may be represented by:

$$\Delta G^{\circ}_{(\theta)} = \Delta G^{\circ} + \alpha\theta \tag{7}$$

where ΔG° is the free energy change of the process (Eq. 2) per mole of adsorbed surfactant at zero (infinitely low) coverage, i.e. in absence of lateral interaction. Equation (5) can be rearranged to give:

$$\frac{\theta}{(1-\theta)^n}\exp.(a\theta) = KC \tag{8}$$

where $K = (1/55.5)\exp.(-\Delta G^{\circ}/RT)$
is the adsorbability of the surfactant at infinitely low coverage, and a is the lateral interaction coefficient given by $a = \alpha/RT$

If a = 0 i.e. no lateral interaction, Eq. (8) reduces to:

$$\theta/(1-\theta)^n = KC \tag{9}$$

which is quite similar to the Flory-Huggins isotherm [9]. Furthermore, if n=1, Eq. (9) reduces to the Langmiur isotherm. On the other hand, if $a \neq 0$ and n=1, one obtains the Frumkin isotherm [10]:

$$\frac{\theta}{1-\theta}\exp.(a\theta) = KC \tag{10}$$

The above isotherms have been applied to the experimental results of adsorption of CTAB on apatite.

MATERIALS AND EXPERIMENTAL PROCEDURES

1. Materials

Synthetic carbonate apatite minerals were precipitated at 95°C [11,12]. After precipitation, they were filtered, washed with distilled water and dried in vacuum. Three different samples (S_1, S_2 and S_3) were prepared with CO_3^{-2}/PO_4^{-3} mole ratios of 0.5, 10 and 25%, with nitrogen surface area of 28.56, 25.45 and 19.85 m^2/gm, respectively. Typical analyses of the apatite samples showed a Ca/P ratio of 1.6 which is very close to the theoritical value (1.66). X-ray diffraction and I.R. analyses proved that the majority of the samples corresponds to carbonate apatite.

Cetyltrimethylammonium bromide (CTAB) was supplied by B.D.H. It was purified by extraction with petroleum ether (b.p. 60-80°C) then crystallized from hot acetone containing about 2% water [13].

2. Experimental Procedures

The adsorption isotherms of CTAB were determined at 30°C. A known weight of apatite (1 gm) was stirred in 50 ml solution of CTAB dissolved in $2x10^{-2}$ M sodium chloride (as indifferent electrolyte). The mixture was shaken for 1 hr. The amount adsorbed was calculated from the total surfactant concentration and the equilibrium concentration of the free surfactant. The determination of surfactant was based on the two-phase dye-transfere method [14]. The quaternary ammonium salt was determined by titrating against purified Aerosol OT using methylene blue as indicator.

RESULTS AND DISCUSSIONS

Figure (1) illustrates the adsorption isotherms for different carbonate apatites as a function of CTAB equilibrium concentration. The adsorption density varies with concentration in a manner suggestive of monolayer adsorption with plateau values of 1.02, 1.62 and 1.61 μ mole/m² for S_1, S_2 and S_3, respectively. The adsorption densities in the plateau region correspond to a loose-packed monolayer of 162.28, 102.37 and 103.15 A°² per CTAB molecule with percent coverage of 30.81, 48.84 and 48.47% for S_1, S_2 and S_3, respectively. This indicates that the surface of the mineral is not completely covered with the collector molecules.

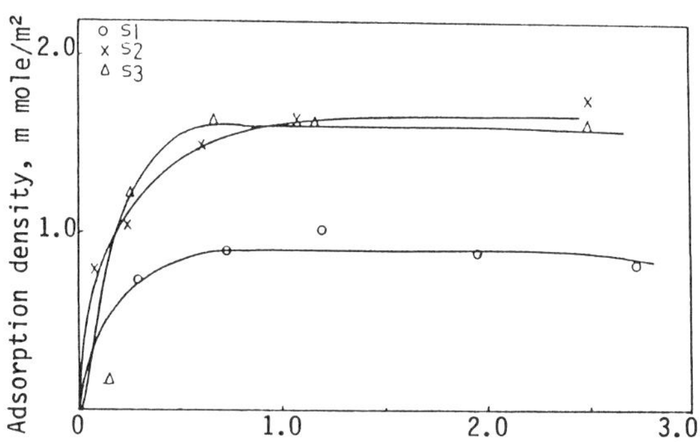

Figure 1. Adsorption isotherms for CTAB on carbonate appatite.

It is generally believed that the adsorption of collectors, and their orientation on mineral surfaces are dependent on the surface properties of the mineral including charge characteristics, chemical composition, and crystal structure. Thus, the adsorption behaviour of the three tested minerals can be correlated to the percentage of carbonate in them. Table (I) lists the mole fraction of carbonate in the bulk of the mineral X along with the saturation adsorption density, and the percent coverage of the surface, at the plateau. The table also includes the values of the a-axis of the crystal. Significantly, the replacement of a phosphate by a carbonate ion is known to produce two related effects, relevant to the present discussion:

i) As the mole fraction of CO_3^{-2} in the mineral increases, the unit cell dimension decreases [12]. i.e. there will be more unit cells per cm² of the mineral surface, and

ii) The carbonate ion, which carries two negative charges, has a planar structure with an ionic radius of about 1.85 A° while the phosphate ion is tetrahedral with an ionic radius of 2.38 A°, and carries three negative charges distributed over the tetrahedron. Thus, the replacement of a PO_4^{-3} by CO_3^{-2} ion causes structural defects as well as electrostatic imbalance. This is partially corrected by substitution of F^- or OH^- in the vacant oxygen sites within the crystal. It follows, then, that the combined effects of the above two factors produce an increases in the density of the negative charges on the surface of the mineral with increase of the mole fraction of carbonate in the mineral.

TABLE I. Effect of the mole fraction of carbonate in the mineral on the saturation adsorption density, % coverage at saturation and unit cell a-axis of the crystal.

Mineral	**CO_3/PO_4 mole ratio**	**X**	**$\Gamma_{max.}$ (μ mole/m²)**	**% Coverage at plateau**	**a-axis, (A°)**
S_1	0.5	0.33	0.90	30.8	9.4016
S_2	10.0	0.91	1.63	48.8	9.3365
S_3	25.0	0.96	1.62	48.5	9.2831

Table (I) shows that both the adsorption density and the percentage coverage at the plateau $\Gamma_{max.}$ are directly proportional to the mole fraction of carbonate X and hence to the density of negative charges on the mineral surface.

Test of the Isotherms

A certain isotherm is usually applied to a set of data to determine the parameters of the isotherm n, a and K (and hence $\Delta G°$). For the case of CTAB, n is known from independent sources. The cross-sectional area of CTAB = 49 A^{o2} per molecule [5] and that of water is 12.5 A^{o2}, i.e n = 4. Thus we need only determine a and K. This can be done graphically. Upon rearranging Eq. (8) and taking logarithms, one obtains:

$$\ln \frac{\theta}{C(1-\theta)^n} = -a\theta - \Delta G°/RT - \ln 55.5 \qquad (11)$$

Therefore, a plot of ln $\theta/C(1-\theta)^n$ versus θ (for n=4) gives a straight line with a slope of a, and an intercept of ($-\Delta G°/RT$ - ln 55.5). The test of the Frumkin isotherm can be achieved by plotting the results according to Eq. (1) with n=1. On the other hand, the Flory-Huggins isotherm can be tested by plotting the results according to Eq. (12) [which results from rearranging and taking logarithms of Eq. (9)] taking in consideration that n=4. Thus:

$$\ln \frac{\theta}{(1-\theta)^n} = \ln C - \Delta G°/RT - \ln 55.5 \qquad (12)$$

The Langmiur isotherm can be tested by plotting the results according to Eq. (12), with n=1. The applicability of Eq. (12) to a certain set of data is proved only when a plot of $\ln \theta/(1-\theta)^n$ versus ln C gives a straight line with unit slope. A non-unit slope of such a plot indicates that some contributions to the free energy of adsorption have been ignored.

The degree of coverage θ was calculated at various concentrations from $\theta = \Gamma/\Gamma_{max}$.

Figure (2) shows plots of the iata according to the Frumkin (28) [Eq. (11) with n=1] and the modified Frumkin [(2A) Eq. (11) with n=4] isotherms. In the case of Frumkin isotherm (2B) the results are divided into two straight line regions.

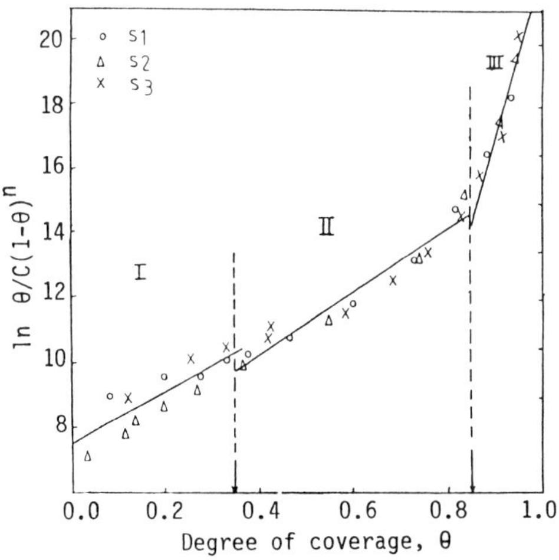

Figure 2 A. Plot of the data according to the modified Frumkin isotherm, Eq. (11), using n=4.

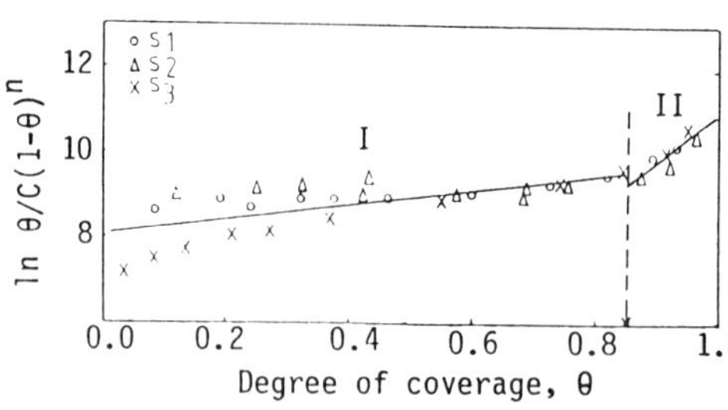

Figure 2 B. Plot of data according to the modified Frumkin isotherm, Eq. (11), using n=1.

Thus: region I $0 < \theta < 0.85$ $a = -1.78$
II $0.85 < \theta < 1$ $a = -10.845$

On the other hand, the results of the modified Frumking isotherm (2A) are clearly divided into three straight line regions with different slopes, i.e. different lateral interaction coefficients.

Thus: region I $0 < \theta < 0.35$ $a = -7.980$
region II $0.35 < \theta < 0.85$ $a = -9.797$
region III $0.85 < \theta > 1$ $a = -51.818$

It is noteworthy that the Frumkin isotherm (i.e. the isotherm which neglects the effect of molecular size) predicts lateral interaction coefficients sufficiently different from the modified Frumkin isotherm.

It is noticed that the region of $0.85 < \theta < 1$ is distinctly different from the other regions. The slope of the straight line in this region is considerably greater than in the other regions. A very large a value at high coverage is due to:

i) Increased lateral interaction effects due to the close proximity of the vertically adsorbed long chain collector malecules due to short range forces which increase quickly at very high coverage. The magnitude of these forces is proportional to X^{-3} where X is the distance between nearest neighbours [4], and
ii) The high sensitivity of the configurational factor $\theta/(1-\theta)^n$ to the increase in θ at high θ and n values. At high θ values, the adsorption of CTAB molecules is less probable than at lower θ since a much fewer number of vacant adsorption sites is available for adsorption of the collector molecule.

It is clear, then, that the adsorption results on all the surfaces are represented satisfactorily by a single straight line in each region. Thus the lateral interaction coefficient and the standard free energy change of the adsorption process, as well as the number of water molecules exchanged per molecule of CTAB are essentially the same on all surfaces. The observed differences between the results obtained on the different surfaces are thought to be of no statistical significance. Therefore, it follows that the differences in the maximum adsorption density between the various surfaces must be related to the structure of the surfaces and/or the density of negative charge on them.

The fact that the lateral interaction coefficient has a negative value indicates that there is attractive interaction between the adsorbed species. The extent of this lateral attraction increases with the degree of coverage. This attraction is believed to be between the long hydrocarbon chains of the CTAB molecules. This point is related to the flotation recovery which is discussed later on.

Adsorption Energy

The analysis of adsorption data is often performed to obtain the free energy of the adsorption process. It is clear from the above analysis that proper choice of the adsorption isotherm is necessary for obtaining a correct value of the free energy of the process. The adsorption of CTAB on synthetic apatites, with different C/P ratios, was analyzed according to the following isotherm: Langmiur, Flory-Huggins, Frumkin isotherm, and the present isotherm, [Eq. (11)]. The latter is considered as a modification of the original Frumkin isotherm in order to account for the large size of the CTAB molecule. The value of free energy change at some degrees of coverage and the lateral interaction coefficient calculated from the various isotherms are listed in Table (II). Table (II) reveals the magnitude of the differences observed in the values of $\Delta G°$ obtained from application of the various isotherms. Thus, while the Langmiur and Flory-Huggins isotherms give rise to $\Delta G°$ values independent of coverage, the Frumkin isotherms give rise to $\Delta G°$ values dependent on coverage, due to lateral interaction effects [see Eq. (7)]. Linear regression of the adsorption data according to the Langmiur or the Flory-Huggins isotherms [Eq. (12) with n=1 and n=4, respectively] gave satisfactory straight lines but with slopes invariably greater than unity (about 1.5 and 5, respectively). This is taken as an evidence for improper account of some free energy terms in the development of the isotherm and hence in the free energy changes calculated therefrom. On the other hand, the observed differences between the $\Delta G°$ and a values calculated from the Frumkin and modified Frumkin isotherms are due to the presence of the size of the adsorbing collector molecule in Eq. (11) and its absence in the original Frumkin isotherm.

TABLE II. Free energy changes of adsorption of CTAB on synthetic apatite minerals and lateral interaction coefficient as calculated from the various isotherms.

Isotherm	**Free energy of adsorption $\Delta G^{\circ}_{\theta}$ KJ/mole**				**lateral interaction coefficient, a**
	θ = 0.0	**θ = 0.1**	**θ = 0.5**	**θ = 0.9**	
- Langmiur [Ea.(12),n=1]	-44.59	same	same	same	0.0
- Flory-Huggins [Eq.(12),n=4]	-118.90	same	same	same	0.0
- Frumkin [Eq.(11),n=1)	-30.482 (region I)	-30.931	-32.725	-	-1.781
	-10.38 (region II)	-	-	-34.967	-10.845
- Modified Frumkin [Eq.(11),n=4]	-28.98 (region I)	-30.98	-	-	-7.98
	-26.02 (region II)	-	-38.36	-	-9.80
	+64.91 (region III)	-	-	-52.55	-51.82

Relation to Flotation Recovery

Figure (3) illustrates the flotation recovery of the tested minerals at various concentrations of CTAB. The flotation recovery increases with CTAB concentration up to a certain maximum beyond which it starts to decrease. This maximum occurs at concentrations of 0.29, 0.27, 0.68 m mole/L for S_1, S_2 and S_3 respectively. The values of θ corresponding to these concentrations are 0.78, 0.72, 0.99 for S_1, S_2 and S_3 respectivley. Inspection of Fig. (2) reveals that this observed decrease in flotation recovery occurs in the region III, i.e. in the region where there is a strong lateral attraction forces between the hydrocarbon chains of the adsorbed CTAB molecules. The attractive forces become particularly profound at high coverages where the hydrocarbon chains are so close together that short range attractive forces come into play. Furthermore the hydrocarbon chains may gather the hydrocarbon radicals forming a second layer of CTAB with the positively charged quaternary ammonium group protruding out into the solution causing hydrophilicity and the flotation recovery decreases.

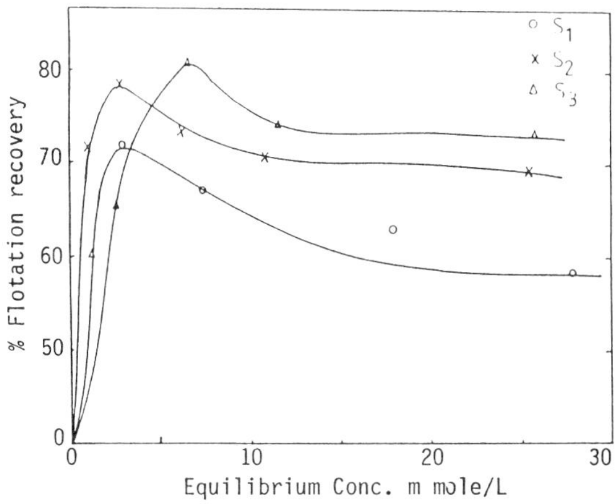

Figure 3. Flotation of carbonate apatite using CTAB as a collector.

CONCLUSIONS

The adsorption of CTAB by synthetic carbonate apatite minerals, with different CO_3/PO_4 molar ratios, has been analyzed using the isotherms of Langmiur, Flory-Huggins and Frumkin. In this investigation, the Frumkin isotherm was modified in order to account for the large size of the CTAB molecule. When the Langmiur and Flory-Huggins isotherms were tested, a plot was obtained with a non-unit slope which indicates that some contributions to the free energy of adsorption have been ignored. Also, the obtained $\Delta G°$ values were independent of coverage. On the other hand, when the data were analyzed using the Frumkin isotherms the $\Delta G°$ values were dependent on coverage due to lateral interaction effects. Besides, using the modified Frumkin isotherm, it was found that the lateral interaction coefficient, the standard free energy change of the adsorption process and the number of water molecules exchanged per molecule of CTAB are essentially the same on all surfaces. The observed difference between the adsorption results obtained on the different surfaces must be related to the structure of the surfaces and/or the density of negative charge on them.

In developing the above model, the process was viewed as substitutional adsorption. The free energy change of the process was expressed as the sum of two terms. The first is coverage-independent $\Delta G°$ which includes the sum of chemical and electrical free energy changes in the desorption of water molecules and adsorption of the collector molecule. This is essentially a composite quantity given by [15]:

$$\Delta G° = \Delta G°_{elec} + \Delta G°_{chem} + \Delta G°_{CH_2} + \Delta G°_{solv} + \ldots etc.$$

The value of $\Delta G°$ is given in the first column in Table II. No trial has been made in the present study to separate $\Delta G°$ to its components.

The second is a coverage dependent term which accounts for lateral interaction between the hydrophobic long chain alkyl groups. The results show clearly that the coefficient of lateral interaction depends on the degree of coverage, having low or moderate values at low and intermediate coverage and very high values at high coverage. In correlating the results in Fig. 2A, they were divided into three regions. Alternatively the results in regions I and II were also analyzed with a view to represent them with a single line. The correlation was somewhat less satisfactory than that adopted here. Furthermore, conceptually the lateral interaction coefficient at a coverage of 0.2 must be different from that at a coverage of 0.7.

The results show that the plateau coverage is proportional to the mole fraction of phosphate in the mineral. This is the only observed difference between the three surfaces which were studied. Otherwise, their adsorption behaviour is rather similar, in terms of the form of adsorption isotherm, the free energy change of the process and the value of the lateral interaction coefficient. Furthermore, the plateau coverage is less than 50% in all cases. This observation has to be combined with a seemingly contradictory one i.e. strong lateral interaction coefficient at high coverage, i.e. near the plateau region. Both observations can be reconciled if we accept that once the adsorbed ions reach a certain critical concentration at the mineral-water interface, they begin to associate into essentially two dimensional patches forming hemi-micelles.

REFERENCES

1. M.C. Fuerstenau, Editor, "Flotation, A.M. Gaudin Memorial volume, (AIME, New York, 1976) pp. 45.
2. J.O'M. Bockris, M.A.V. Devanathan, and K. Müller, Proc. Roy. Soc., A 274, pp. 55, (1963).
3. J.O'M. Bockris, E. Gileadi, and K. Müller, Electrochim. Acta, 12, pp. 1301 (1967).
4. W. Adamson, Arthur, "Physical Chemistry of Surfaces" 3rd edition, (John Wiley, New York, 1976) pp. 644.
5. F.Z. Saleeb and J.A. Kitchener, J. Chem. Soc., pp. 911 (1965).
6. G.N. Lewis and M. Randall, "Thermodynamics, Revised by K.S. Pitzer and L. Brewer, (McGraw-Hill, New York 1961).
7. R. Guidelli, J. Electroanalyt. Chem., 110, pp. 205 (1980).
8. P. Nikitas, Electrochim. Acta, 32, pp. 205 (1987).
9. H.P. Dhar, B.E. Conway, and K.M. Joshi, Electrochim. Acta 18, pp. 789 (1973).
10. A. Frumkin, Z. Physik Chem. 116, pp. 466 (1925).
11. R.Z. LeGeros, Ph.D. Thesis, (New York Univ., 1967).
12. G.B. Soleimani, Ph.D. Thesis, (Univ. of Leeds, London, 1978).
13. T. Nash, J. Appl. Chem., 8, pp. 440 (1958).
14. J.J. Oliver Barr and W.V.J. Stubbings, Soc. Chem. Ind., 67, pp. 45 (1948).
15. D.W. Fuerstenau, J. Pure and Appl. Chem., 24, pp.135-164 (1970).

ADSORPTION AND WETTING CHARACTERISTICS OF PURE NON-METALLIC MINERALS IN CONTACT WITH CATIONIC SURFACTANTS.

JOHN HANNA
Mineral Resources Institute, The University of Alabama
Box 870204, Tuscaloosa, AL 35487-0204

ABSTRACT

The adsorption, zeta potential and wetting characteristics of pure calcium carbonate (synthetic), calcite, quartz and kaolinite minerals in contact with cetyltrimethylammonium bromide (CTAB) and dodecylammonium chloride (DDAC) were studied at constant temperature, pH and ionic strength. Further studies were made on the effects of pH and inorganic electrolytes on the adsorption of CTAB on the synthetic calcium carbonate.

Adsorption of the weakly ionizable surfactant DDAC showed multilayer formation, but it was restricted to a monomolecular layer for systems with the strongly ionizable CTAB surfactant. In case of monolayer adsorption, the CTAB molecules were oriented with their polar head groups directed towards the solid surface and the hydrocarbon chains exposed to the aqueous phase. A direct correlation of the adsorption, zeta potential, contact angle and flotation was established and an adsorption mechanism was proposed. It involves mainly electrostatic attraction between the cationic head groups and the oppositely charged sites on the mineral and possibly some ion exchange, chemical and hydrophobic interactions, depending on the activity of the surface groups.

INTRODUCTION

When a solid is brought in contact with an aqueous solution of a surface-active agent, the surfactant molecules will be adsorbed at the solid-liquid interface. For nonpolar solids such as graphite, coal and sulfur, the surfactants adsorbed at the interface will be oriented in such a way that, on the whole, they turn a majority of their hydrophilic groups towards the aqueous phase so that wetting occurs. The adsorption is then non-specific. On the other hand, adsorption on polar solids may occur due to chemical or electrostatic interaction between the polar head groups of the surfactant and the oppositely charged surface sites, leaving the hydrocarbon chains directed towards the aqueous phase. As a result, the surface is rendered non-wettable or hydrophobic. The contact angle becomes finite and flotation is achievable. A second layer of the surfactant may be built up at higher equilibrium concentrations on top of the first layer through chain-chain cohesion [1,2]. Reverse orientation of surfactant molecules results in rewetting of the surface, a decrease in contact angle and poor flotation. Such adsorption phenomena are of fundamental importance in technical applications such as detergency, flotation, water proofing and paper manufacture.

The present work is devoted to obtain information on the adsorption and wetting characteristics of three major non-metallic minerals--calcite, quartz and kaolinite. Contact angles, flotation recoveries, zeta

Advances in Fine Particles Processing
John Hanna and Yosry A. Attia, Editors

potentials and adsorption densities were determined as functions of the concentration of certain cationic surfactants to obtain more evidence about the mechanism of adsorption. Whenever it was possible, all of the above parameters were measured on one and the same equilibrated solid sample.

EXPERIMENTAL

Materials

Synthetic calcium carbonate (A.R.) was a B.D.H. product. Adsorption and flotation tests were made on the as-received material with nitrogen surface area of $5.5 m^2/g$ and particle size of 20-40 microns. For contact angle measurements the compacted dry material was polished and stored under water after careful washing with distilled water to avoid surface contamination. Highly pure natural calcite crystals were provided by the National Research Center, Cairo, Egypt. Adsorption and flotation measurements were conducted on wet-ground calcite particles of 50-70 microns in diameter with a nitrogen surface area of $0.2 m^2/g$. Fresh calcite surfaces were prepared for contact angle measurements by careful cleavage of the mineral under water.

High purity quartz mineral was obtained from Aswan, Egypt. Polished surfaces of $4\text{-}6 cm^2$ were carefully washed with hot 2N HCl and water and then stored under distilled water until used for contact angle measurements. Crushed and ground quartz particles of 105-150 microns in size were prepared and cleaned as described above for use in flotation and adsorption experiments.

A well crystallized sample of Georgia kaolinite was obtained from the clay repository at the University of Missouri. The sample averaged 20 microns in particle size and gave a nitrogen surface area of $9.8 m^2/g$. For adsorption and zeta potential measurements the sample was treated with NaCl to prepare the homoionic "Na-kaolinite" according to the procedure described elsewhere [3].

Cetyltrimethylammonium bromide (CTAB) was a B.D.H. product and dodecylammonium chloride (DDAC) was a Light product. The CTAB was purified by the method of Nash [4] while DDAC was used as received. Both surfactants gave no minimum in the surface tension-concentration curve, indicating the absence of surface-active impurities. The concentration of the surfactant solutions was determined by the dye-transfer method [5] which was accurate to $\pm$ 1% above 0.1 mM. Lower concentrations of CTAB were determined colorimetrically by measuring the optical density at 490 mu of the complex formed with the Orange II extracted in chloroform [6].

Analytical grade reagents such as NaCl, Na_2SO_4, NaOH and HCl were used to control the pH and ionic strength of the system. Double distilled water was employed in all experiments.

Apparatus and Technique

The adsorption of CTAB was determined at $35 + 0.5^oC$ on 1-10g samples. Equilibrium adsorption was reached in less than 2 hours, but analysis was carried out after 4 hours of shaking. The zeta potential was calculated from electrophoretic mobility of the fine mineral particles. Mobilities were measured using a microelectrophoresis cell of the bent cylindrical capillary tube type [7] at a constant ionic strength of 10 mM NaCl to keep the thickness of the double layer almost constant. The zeta potential was calculated from the mobility measurement, using the well known Smoluchowski equation [8].

Contact angles were measured directly on free bubbles by a travelling microscope. Details of the method are given elsewhere [9,10]. Flotation tests were carried out in a froth column using the pulp remaining after adsorption measurements. A constant flow of purified air set at 1.2 l/min. was used to collect the froth products for a period of 2 minutes. In most cases, no external frother was needed because of the inherent frothing properties of the residual surfactants in the pulp.

RESULTS AND DISCUSSION

Systems with Quaternary Amine Salts

Calcium Carbonate-Calcite: Adsorption isotherms of CTAB on calcium carbonate from water at pH 8.3 and 10.5 and various salt solutions are presented in figure 1. At natural pH of 8.3 the isotherm of CTAB from water is S-shaped (curves characterized by small affinity at lower equilibrium concentrations) with a shallow adsorption maximum occurring around the critical micellar concentration (CMC). At pH 10.5 adsorption of CTAB from water is almost doubled (figure 1) and the isotherm becomes a normal Langmuir type isotherm. The disappearance of the low affinity part of the isotherm (S-shaped) under these conditions may be attributed to specific adsorption of the negatively charged OH^- ions on the $CaCO_3$ surface so that it becomes more favorable for CTAB adsorption. Also, addition of 5mM Na_2SO_4 (at natural pH) to the system is shown (figure 1) to double CTAB adsorption and to change the shape of the adsorption isotherm to Langmuir-type. Thus, both OH^- ions and the negatively charged SO_4^{--} anions appear to activate CTAB adsorption through their co-adsorption on the $CaCO_3$ surface.

In contrast, addition of 10 mM NaCl or $MgCl_2$ to the system is shown (figure 1) to decrease CTAB adsorption as compared to that from water at constant pH of 8.3 (figure 1), but the isotherms remain S-shaped. This suggests that both Na^+ and Mg^{++} cations are competing with the positively charged $CTAB^+$ for adsorption on the available negative sites of the $CaCO_3$ surface. In addition, partial solubility of $CaCO_3$ in water may release significant amounts of Ca^{++} cations into the system (during equilibration). These cations will compete with surfactant adsorption as depicted by the low affinity part of the isotherms of CTAB from water at natural pH of 8.3.

The maximum adsorption capacity at pH 8.3 amounts to 4 M mole/g, which on comparison with the nitrogen surface area of $5.5 m^2/g$ gives an area of 228 A^2/CTAB molecule. This area is large compared with 40-51 A^2 found for a close-packed monolayer of vertically oriented molecules on carbon [11] and calcium phosphate [12] surfaces, respectively. Thus, the surface is covered by a loosely packed monolayer of the surfactant. Surface roughness cannot be the main cause for the low adsorption capacity for CTAB since the total nitrogen area of calcium carbonate is available for anionic surfactants [2], and since CTAB adsorption can be greatly increased in the presence of sulfate and hydroxyl ions in solution (figure 1).

Cetyltrimethylammonium bromide can be adsorbed in two ways: (a) physical adsorption (due to surface activity) with the polar head group oriented towards the aqueous phase and adsorption would increase by increasing the ionic strength of the solution or (b) electrostatic adsorption with the polar head group directed towards oppositely charged sites on the solid surface. For the positively charged $CTAB^+$ ion to be electrostatically adsorbed, a negative site on the calcium carbonate surface has to be available. Adsorption would be expected to increase by increasing negative sites and vice versa. This is clearly illustrated in figure 1, where sulfate and hydroxyl anions increase adsorption while magnesium and calcium cations decrease it. If this orientation is true, the calcium carbonate surface would show an optimum in hydrophobic character around saturated adsorption and the net surface charge should be about 30-40 mV.

Figure 2 shows the results of zeta potential measurements at a constant pH of 8.3 and ionic strength of 10mM NaCl. The data indicate that with increase of surfactant concentration the negatively charged $CaCO_3$ particles become neutral then positively charged above $2\text{-}3x10^{-5}$ M CTAB (zero point of charge). The positive potential increases with CTAB concentration to reach a limiting value of +27 mV above the CMC. This potential is much less than the +80 to +90 mV reported for carbon blacks, where a CTAB monolayer was physically adsorbed with the head groups oriented outermost (i.e., towards the aqueous phase) [13].

Contact angle measurements (figure 2) show that at the low surface coverage (i.e., CTAB concentrations) a measurable contact angle is established. Thereafter, the contact angle increases with increased surface coverage to reach a limiting value of 42° above the CMC and then decreases slightly at higher surfactant concentrations. The hydrophobic behavior of the $CaCO_3$ particles was confirmed by the flotation results shown in figure 2. In fact the shape of the flotation recovery curve is very similar to the corresponding adsorption isotherm, indicating the direct correlation of adsorption and wetting properties of the surface.

The results of measurements of zeta potential, contact angle flotation and adsorption of CTAB on calcite from 10 mM NaCl solutions are shown in figure 3. The adsorption isotherm obtained shows that calcite, unlike precipitated $CaCO_3$, has a greater affinity to CTAB as represented by the steep initial part of the isotherm. Saturation adsorption is reached at an equilibrium CTAB concentration of less than 0.05 m mole/1, which is far less than the CMC. The adsorption capacity in the plateau region amounts to 0.44 M mole/g which corresponds to a loose packed monolayer with molecular parking area of 151 A^2 per CTAB molecule. This parking area is much lower than the 228 A^2 found for the synthetic calcite, but it is still very large compared to the 51 A^2 found for calcium phosphate [12] and 40 A^2 for carbon black-CTAB systems [11].

The difference in the shape of the adsorption isotherms and the higher affinity of CTAB for calcite over calcium carbonate may be due to the difference in their crystallinity and dissolution properties. Goujon, et al., in their studies on a calcite/N-dodecylammonium chloride (DDAC) system, distinguished two main types of adsorption sites for the well crystallized calcite surface, while only one was identified for the ultra finely ground calcite with a poorly defined crystalline surface [15]. This may explain the higher affinity of the surfactant observed for calcite. Moreover, solubility measurements indicate that the well crystallized calcite particles are less soluble (15 mg/l) than the precipitated calcium concentrate (35 mg/l). As discussed earlier, the dissolved Ca^{++} ions in the system may compete with the $CTAB^+$ cations for the negatively charged sites on the surface and may lead to the appearance of the low affinity part of the S-shaped adsorption isotherm calcium carbonate (figure 2).

Results of the zeta potential (figure 3) indicate that the original negative charge of calcite is reversed to positive values at very low surfactant concentrations to reach limiting values above the CMC of CTAB. The increase in the positive charge while the adsorption capacity is unchanged may be due to variation in charge distribution in the electrical double layer caused, for example, by changes in the number of specifically adsorbed ions in the Stern layer at high CTAB concentrations. Also, a marked increase in flotation and contact angle results are observed above the CMC, indicating the importance of zeta potential in determining the wetting behavior of calcite despite the constancy in CTAB adsorption.

The above results for precipitated calcium carbonate and calcite-CTAB systems show that surface phenomena which reflect conditions at the solid/liquid interface (adsorption and zeta potential) can be correlated directly with surface phenomena which reflect conditions at the solid/liquid/gas interface, namely, contact angle and flotation.

Quartz: The results of zeta potential, contact angle, flotation and adsorption of CTAB on quartz from 10 mM NaCl solutions at pH 6.8 are shown in figure 4. The adsorption of CTAB on quartz gives a two-step adsorption isotherm. The adsorption capacity at the first step amounts to about 1.5×10^{-8} mole/g, and 7.5×10^{-8} mole/g at the second step which is reached at about 0.3 mM. Based on an estimated surface area of 175 cm^2/g, the adsorption density at the second plateau amounts to 4.3 m $mole/m^2$. This means that the surface is covered by a close-packed monolayer of CTAB with a parking area of 39 A^2 per molecule. Two-step adsorption isotherms with comparable adsorption densities of 3.8 and 5.3 m $mole/m^2$ have been reported for adsorption of dodecyltrimethylammonium bromide (DTAB) and CTAB, respectively, on precipitated silica [16] and for the tetradecyl-(TTAB) and hexadecyl-(HTAB) homologs on vitreous silica at various pH's [17].

To understand the reasons for the two-step adsorption behavior of CTAB on the pure quartz mineral, two adsorption mechanisms are considered. The one proposed by Bijsterbosch [16] suggests that at low surfactant concentrations a monolayer is formed due to interaction of negatively charged sites on the silica surface and the oppositely charged surfactant ions. At higher concentrations a different mechanism (probably hydrophobic bonding) causes a bilayer to be built up. According to this model the hydrophobic character of the system would be expected to reach maximum in the region of the first plateau, where the monolayer is completed, then decline with the build up of the second layer in which the surfactant molecules are oriented with their polar head groups directed towards solution. Also a zeta potential in the range of +80 mV would be expected for the silica particle with the completion of the second layer [13]. However, this is found not to be the case for the quartz/CTAB system. The contact angle and flotation data given in figure 4 show that the hydrophobicity of quartz continues to increase with surfactant concentration to reach a maximum above the CMC (i.e. in the region of the second plateau). Also, the zeta potential values obtained do not exceed +38 mV.

The other model proposed by Ter-Minassian-Sarga [17] provides a better explanation for our results. According to this model the surfactant molecules adsorbed at low concentrations (first step) are most likely to be oriented with the apolar hydrocarbon chain parallel to the solid surface. At higher concentrations, the surfactant molecules would gradually be standing up, thus exposing more sites and room on the surface for adsorption to proceed until a close-packed monolayer is reached (second plateau). The new adsorption sites are shown to be different from the primary dissociated silanol groups. Association of the CTAB long hydrocarbon chains exposed to the aqueous phase would also con-

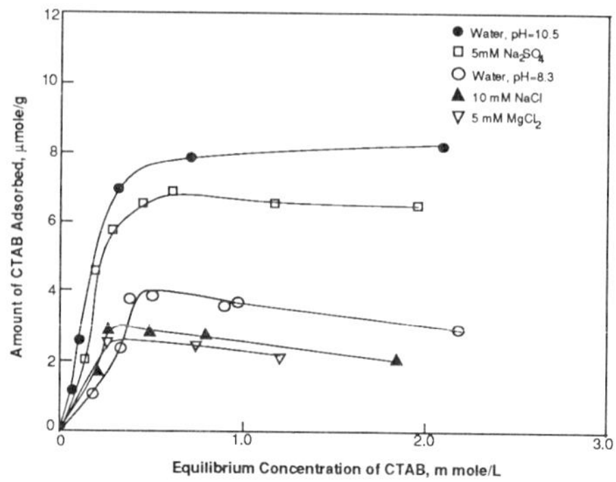

Figure 1. Effect of Electrolytes on Adsorption Isotherms of CTAB on Synthetic Calcium Carbonate

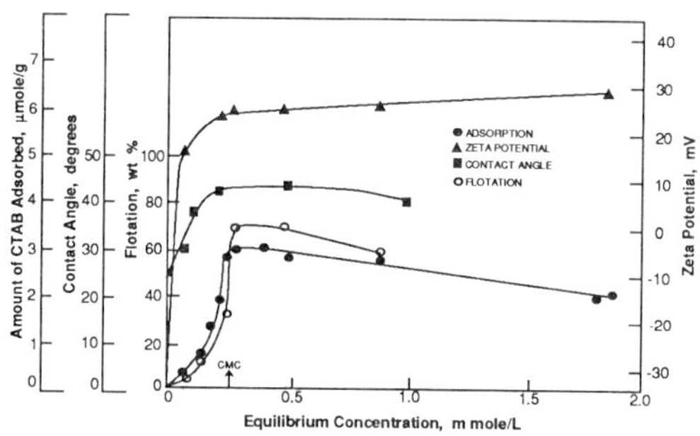

Figure 2. Adsorption, Zeta Potential, Contact Angle, and Flotation of Synthetic Calcium Carbonate in CTAB Solutions with 10 mM NaCl

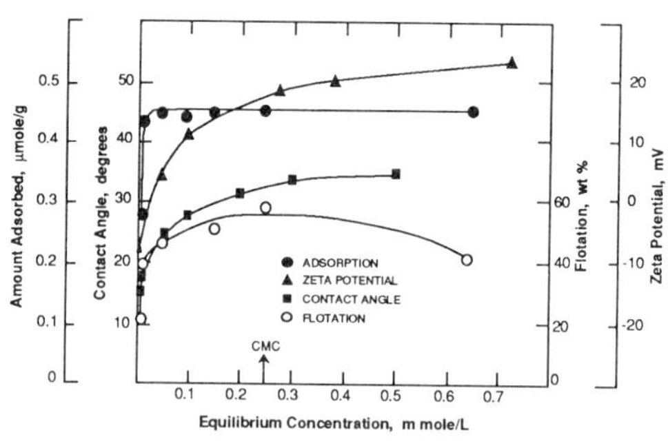

Figure 3. Adsorption, Zeta Potential, Contact Angle, and Flotation of Calcite (50-70 μm) in CTAB Solutions with 10mM NaCl

tribute to the formation of a close-packed monolayer at high surfactant concentrations [18].

The discussion of quartz/CTAB system shows the importance of zeta potential, contact angle and flotation measurements in understanding the adsorption process. Accordingly, a two-step adsorption mechanism is proposed. At low CTAB concentration adsorption takes place on the dissociated silanol groups (primary sites) through electrostatic and possibly chemical forces. At higher surfactant concentrations adsorption takes place on the adsorption sites probably via siloxane bonds. Hydrophobic bonding may also occur under these conditions.

Kaolinite: Adsorption, zeta potential and flotation results of kaolinite as a function of CTAB equilibrium concentration are shown in figure 5. The adsorption isotherms show that initially, adsorption increases linearly with surfactant concentration to reach limiting values in the region above the CMC. At pH 5.3 the limiting adsorption capacity of 19 M mole/g increases, to about 33.8 M mole/g at pH 9.5. Based on the kaolinite nitrogen surface area of 9.8 m^2/g the adsorption density at pH 9.5 corresponds to a closely packed monolayer with a parking area of 49 A^2/CTAB molecule. This parking area agrees well with the 51 A^2 per molecule found for CTAB calcium phosphate [12]. At pH 5.3 (natural pH) the kaolinite surface is expected to be covered by less packed monolayer with 89 A^2/CTAB molecule. Similar results and conclusions were reported for the kaolinite/DTAB system at pH's of 5-10 [20] and for silica/DTAB at pH's of 8-10 [16].

Information about the mode of adsorption and orientation of the CTAB molecules on the kaolinite surface may be obtained from the zeta potential and wetting properties (flotation) measurements. The data shown in figure 5 (at natural pH of 5.3 $\pm$ 0.2) indicate that increased adsorption of CTAB produced a rapid increase in the positive surface charge (zeta potential curve) and the hydrophobicity of the surface (flotation curve), to reach limiting values around the CMC of the surfactant.

The above results suggest that adsorption takes place via electrostatic interaction between the positively charged CTAB head groups and the negatively charged basal surfaces of kaolinite, as indicated by the charge reversal above 5 x 10^{-5} M CTAB. The marked increase in adsorption at high pH's may be explained in terms of increased adsorption of OH^- on the surface as well as increased adsorption of CTAB on the aluminum centers of the edge sides or those contaminating the basal sides of the kaolinite crystal lattice [21,22,23]. Ion exchange between surface cations and H^+ ions on the kaolinite surface and the positively charged surfactant may, or may not, be ruled out as indicated by the shift of the initial pH of kaolinite/CTAB suspensions towards lower pH values [20]. In both cases the surfactant molecules should be oriented with their polar head groups toward the solid surface leaving the hydrocarbon chains exposed to the aqueous phase.

Systems with Primary Amine Salts: Results of the adsorption and wetting (flotation) tests conducted on calcite, precipitated $CaCO_3$/-, quartz/-, and kaolinite/dodecylammonium chloride (DDAC) systems at the natural pH of each mineral suspension are shown in figure 6 and 7, respectively. The published data reported by Goujon et al., on calcite [15] and deBruyn on quartz [24] were used for comparison with our results on kaolinite. The data shown in figure 6 indicate that at low equilibrium concentration of DDAC, where electrostatic mode of adsorption is expected, the affinity of the minerals towards $DDAC^+$ ions appears to decrease in the order of quartz>calcite>kaolinite (i.e., the order of in-

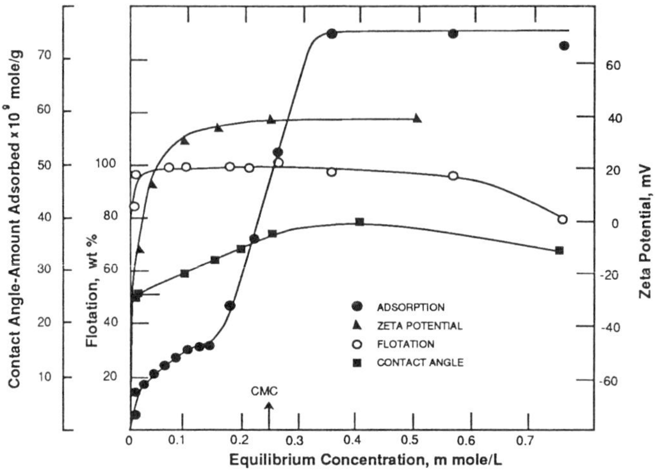

Figure 4. Adsorption, Zeta Potential, Contact Angle, and Flotation of Quartz (105-150 m) in CTAB Solutions with 10mM NaCl

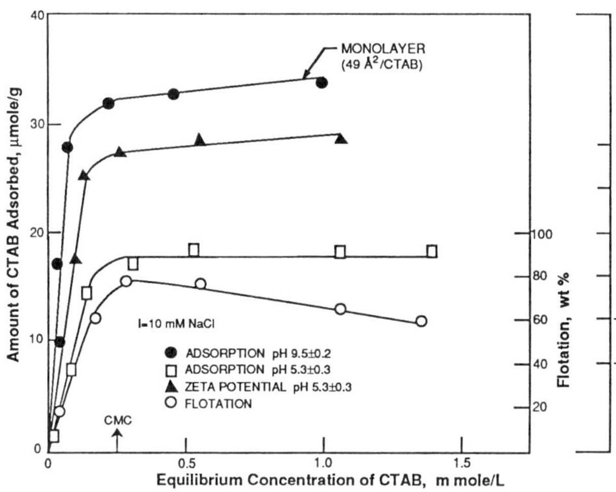

Figure 5. Adsorption, Zeta Potential, and Flotation of Kaolinite

creasing competition of $DDAC^+$ ions with the $Na^+ < Ca^{++} < Al^{+++}$ ions present on the mineral surfaces). Also, the slope of the adsorption isotherms, which is almost the same for all three solids, shows a sharp increase above 0.1 m mole/l for calcite and at 0.3 m mole/l for quartz indicating multilayer adsorption. For kaolinite, however, the slope of the adsorption isotherm remains more or less constant, up to the CMC of DDAC (7 m mole/l), then declines at higher concentrations. In this case the break in the curve occurs after the formation of close-parked monolayer with molecular parking area of about 23 A^2/DDAC molecule. In all cases multilayer adsorption appears at high surfactant concentrations. This occurs at surfactant concentrations below the CMC for calcite and quartz, but above the CMC for kaolinite.

Multilayer adsorption may result from strong chemical interaction between DDAC and certain ionic species on the mineral surface such as CO_3^{--} of calcite or the silanol groups of quartz, leading to the formation of less soluble amine salts, i.e. chemisorption [15,17]. Other reasons may include adsorption of mixed layers of amine salts and their neutral hydrolyzable products, association of the apolar groups as "hemicells" and bulk precipitation on the mineral surface [18,24,25]. The fact that kaolinite did not show any sign of multilayer formation below the CMC may be attributed to the well known surface acidity of kaolinite [21,22,23]. Surface acidity is expected to enhance the ionization of the DDAC molecules at the kaolinite/water interface, leading to more electrostatic adsorption over a wide concentration range, as in the case of CTAB (figure 5). With the completion of the first layer normal hydrolysis of the surfactant will resume, leading to the multi-layer formation on the kaolinite surface.

Figure 7 shows the flotation results of quartz, calcite, calcium carbonate, and kaolinite using DDAC solutions with various concentrations. The data clearly show that quartz floats almost completely at all surfactant concentrations used. The flotation recovery of the other materials increases with surfactant concentration to reach limiting values at about 1.0 mM DDAC. At low amine concentrations, which are normally used in flotation practice, the floatability of the solids increases in the order kaolinite < calcite < $CaCO_3$, i.e., in the order of increasing adsorption affinity, as depicted in figure 6.

CONCLUSIONS

Adsorption of the cationic surfactants CTAB and DDAC on pure calcium carbonate (synthetic), calcite, quartz and kaolinite minerals was shown to follow different types of isotherms. The shape of the isotherms and the adsorption capacities of the solids are largely dependent, among other things, on pH, polarity of the surface, ionic strength, type of dissolved or added ionic species and the ionization of the surfactant. Hydrolyzable surfactants such as DDAC in most cases gave multilayer adsorption. For the strongly ionizable, non-hydrolyzable CTAB, adsorption normally does not exceed a close-packed monolayer of the surfactant oriented with their apolar hydrocarbon chains directed towards the aqueous phase. Thus, the surface becomes hydrophobic, as confirmed by flotation, contact angle and zeta potential measurements. An adsorption mechanism was proposed, which involves an initial electrostatic attraction between the positively charged head groups of the surfactant and the negative sites on the solid surface, followed possibly by ion exchange, chemical or hydrogen interactions at high surfactant concentrations. A direct relation between surface zeta potential and wetting properties of the solids (contact angle and/or flotation) was established for systems with quarternary amine surfactants.

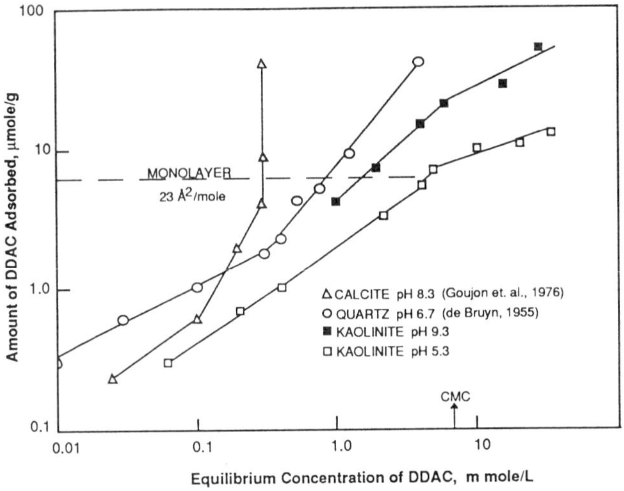

Figure 6. Adsorption Isotherms of DDAC on Calcite, Quartz, and Kaolinite

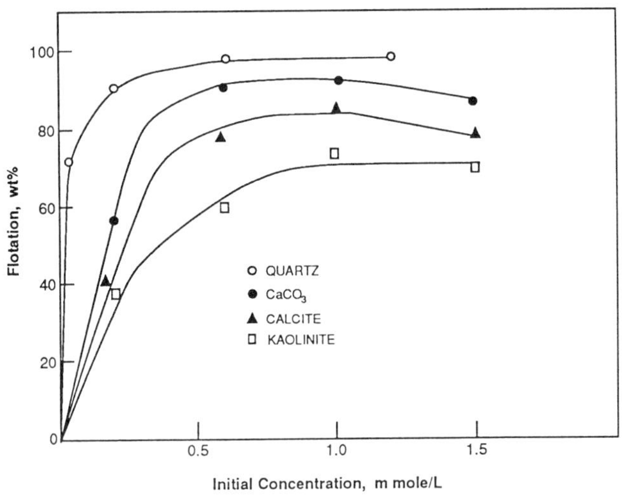

Figure 7. Flotation of Quartz, Synthetic Calcium Carbonate, Calcite, and Kaolinite with DDAC

ACKNOWLEDGEMENTS

The author is indebted to Dr. F.Z. Saleeb, General Foods Co., for many enlightening discussions and helpful criticism of the manuscript.

REFERENCES

1. B.D. Cuming and J.H. Schulman, Austral. J. Chem 12, 413 (1959)
2. H.S. Hanna and Saleeb, F.Z., "Adsorption and Wetting Properties of Calcite and Calcium Carbonate with Anionic Surfactants" 14th ACS Midwest Regional Meeting, Fayetteville, Arkansas (1978).
3. H.S. Hanna and Somasundaran, P.J., Colloid, Interf. Sci. 70, 1, 181 (1979).
4. T. Nash, F. Appl. Chem. 8, 440 (1958).
5. E.F. Hillenbrand, Jr., Sutherland, W.E., and Hogsett, J.N., Anal. Chem. 23, 606 (1951).
6. A.V. Few, and Ottewill, R.H., J. Colloid, Sci. 11, 34 (1956).
7. J.H. Schenkel and Kitchener, J.A., Experientia 14, 425 (1958); Trans. Faraday Soc. 56, 161, (1960).
8. M. Smoluchowski, Z. Phys. Chem. 92, 129 (1918).
9. M.L. Corrin, E.L. Lind, A. Roginsky, and W.D. Harkins, J. Colloid Sci. 4, 485 (1949).
10. R.I. Razouk, Saleeb, F.Z., and Hanna, H.S., Proc. 5th Int. Congr. Surf. Activity, Barcelona, 1968, Vol. 2, pp. 695-702 (1969).
11. F.Z. Saleeb and Kitchener, J.A., J.Chem. Soc. 911 (1965).
12. H.S. Hanna and Saleeb, F.Z.; Colloids and Surfaces, 1, 295 (1980)
13. F.Z. Saleeb, Ph.D. Thesis, Univ. of London (1963).
14. J.T.G. Overbeek, in Colloid Science, Kruyt, (ed. Elsevier, 1952) Vol. I, p. 194.
15. G. Goujon, Cases, J.M., and Mutaftschiev, B.J. Colloid, Interf. Sci. 56, No. 3, 587 (1976).
16. B.H. Bijsterbosch, J. Coll. Interf. Sci. 47, No. 1, 186 (1974).
17. L. Ter-Minassian-Sarga, AIChE Symposium Series No. 150, 71, 68 (1975).
18. A.M. Gaudin and Fuerstenau, D.W., Trans. AIME 202, 958 (1955).
19. H. Rupprecht, Kolloid, Z.Z. Polym. 249, 1127 (1971).
20. R.W. Smith and Wen, W.W., 27th Am. Min. Symp. Vol. Univ. of Minn. (1966) 193.
21. R.F. Conley and Althoff, A.C., J. Coll. Interf. Sci. 37, 1, 186 (1971)
22. A.S. Buchanan and Oppenheim, R.C., Aust. J. Chem. 21, 2367 (1968) and 25, 1857 (1972).
23. B. Rand and Melton, I.E., J. Colloid Interface Sci., 60, 2, 308-326 (1977)
24. P.L. DeBruyn, Trans. AIME 202, 291 (1955).
25. B. Tamamush and K. Tamaki, Proc. 2nd Int. Congr. Surface Activity, London, p. 449, (1957).

SURFACE CHARACTERIZATION OF SURFACTANT-MODIFIED COLLOIDAL ALUMINA

C.A. MALBREL,* P. SOMASUNDARAN,*
M. FRANCOIS,** J.E. POIRIER,** AND J.M. CASES**
* *Langmuir Center for Colloids and Interfaces, Columbia University, New York, NY 10027*
** *Centre de Recherche sur la Valorisation des Minerais, B.P. 40, 54501 Vandoeuvre Cedex, France*

ABSTRACT:

Surface modification by surfactant and polymer adsorption is being increasingly used to adjust the surface properties of mineral fines for use in technological applications such as formulation of fillers for elastomers or dispersion of pigments in paints and other coatings. ESR spectroscopy is used here to monitor *in-situ* adsorption of Aerosol OT on colloidal alumina prior to adsorption of water vapor. Analysis of the adsorption isotherm of water on the surfactant-modified alumina particles shows that water adsorbs only on the fraction of the alumina surface not coated with the surfactant. It is found also that no significant interaction exists between water and surfactant molecules during the hydration of the surface.

INTRODUCTION

Chemical modification of fine particles by surfactant and polymers is becoming increasingly useful as a means for altering their surface properties for specific technological applications such as fillers formulation [1], pigments dispersion [2], magnetic recording [3], and lubrication [4]. In order to characterize the surface of these solids, techniques involving contact angle measurements [5], gas and surfactant adsorption [6], and calorimetry [7] are used. These techniques are helpful for developing an insight into the mechanisms controlling the adsorption of the additives. However, because they are based on the measurement of average macroscopic properties, these techniques give only a partial picture of the phenomena occuring at the interface. For instance, they do not provide direct information on the orientation of the molecules adsorbed at the interface although this information is critical in many interfacial phenomena (dispersion, flocculation, wetting). Furthermore, these techniques often require specific sample preparation which may affects the properties to be measured.

With the recent applications of spectroscopic techniques to the study of interfaces, it has become possible to probe directly the solid/liquid interface and to follow *in-situ* changes occuring at the interface as the surface is modified [8]. In addition, spectroscopic techniques such as fluorescence or electron spin resonance spectroscopy (ESR) offer the advantage of not requiring undesirable sample preparations. In this paper, we report results obtained with ESR spectroscopy on the adsorption of a surfactant, Aerosol OT, (sodium bis-(2 ethyl hexyl sulfosuccinate)) from a cyclohexane solution on alumina. These data are correlated with results on the adsorption of water vapor on the treated alumina.

ESR SPECTROSCOPY OF ADSORBED SURFACTANT LAYERS

Electron spin resonance spectroscopy is based on the fact that an electron placed in a magnetic field shows a typical energy absorption spectrum, that is highly sensitive to the environment of the electron . Free electrons are provided by paramagnetic species such as transition metal ions or free radicals [9]. ESR can be used to obtain information on the microenvironment of these species. However, this technique would

Advances in Fine Particles Processing
John Hanna and Yosry A. Attia, Editors

be of limited use without the development of stable free radicals that can be used as probes. Developed primarily for microenvironmental studies of biological membranes and membrane-mimetic systems such as micelles, vesicles and emulsions,[9] the spin probing technique has been extended to the study of adsorbed layers at the solid/liquid interface [10]. The experiments reported here have been carried out using 12-doxyl stearic acid which is a stearic acid molecule on which a stable radical, nitroxide, is tagged to the 12th carbon atom of the alkyl chain of the molecule. This probe was selected because structural similarities with the surfactant used, Aerosol OT, allow it to be integrated into surfactant aggregates in solution as well as at an interface [11].

Several pieces of information are contained in the ESR spectrum of this type of probe (Figure 1):
- Viscosity/structural ordering of the probe environment: when the probe is rotating freely in a medium of low viscosity, the spectrum obtained is characterized by three sharp lines of equal height as shown on Figure 1a. Hindrance in the probe motion induces line broadening of the spectrum as in Figure 1b. This property can be used to assess the fluidity of the probe environment.
- Polarity of the probe environment: when the probe has a relatively high mobility, the distance separating the low field and central line of the spectrum is called hyperfine splitting constant and it can be used to get an indication of the polarity of the probe environment. This information is important since it can reveal the location of the probe, i.e. whether it is situated in aqueous or hydrocarbon media for example.
- State of aggregation of the probe molecules: when two probe molecules interact, the spectrum obtained is characterized by a phenomenon called spin-spin relaxation that leads to a unique broad line (Figure 1c). This feature can be used to control whether probe molecules are aggregated or not.

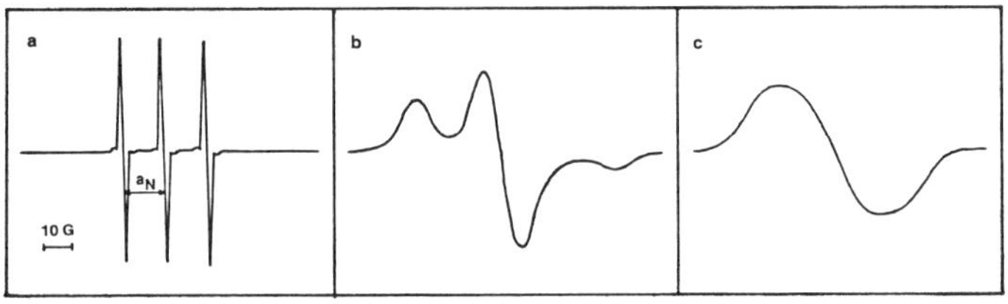

Figure 1: ESR spectra of nitroxide spin probe under various conditions of probe mobility: (a) isotropic spectrum for nitroxide tumbling in a medium of low viscosity; (b) powder spectrum for nitroxide randomly distributed but rigidly oriented in a frozen solution; (c) line boadening due to spin-spin relaxation indicative of probe-probe interactions.

In the experiments reported here, samples were prepared by adding 15 cm^3 of cyclohexane solution of AOT (Fisher Sc.) containing approximatively 10^{-5} mole/l of 12-doxyl stearic acid (Aldrich) to 1 g of alumina (Linde A grade, Union Carbide Corp.; non-porous, 14 m^2/g by nitrogen adsorption) previously desiccated at 200°C for 5 hours and allowed to cool in a desiccator connected to a vacuum pump using P_2O_5 as desiccant. The suspension was conditioned for 24 hours with a wrist action shaker. The ESR spectra were obtained with an IBM-Bruker Model 100D X-band spectrometer using both the supernatant and the sediment. Aerosol OT was analyzed by a two-phase titration technique where the surfactant is titrated against hexadecyltrimethylammonium bromide in chloroform with dimidium disulfine blue as end-point indicator.

Figure 2 shows the adsorption isotherm of AOT on alumina together with the corresponding ESR spectra. It can be seen from the shape of the adsorption isotherm that AOT shows a high affinity for the surface since maximum adsorption is almost reached before any residual surfactant concentration can be detected in the supernatant. Similarly, the ESR spectra of the supernatant showed that the probe also has a high affinity for the surface since no residual probe concentration could be detected in the supernatant.

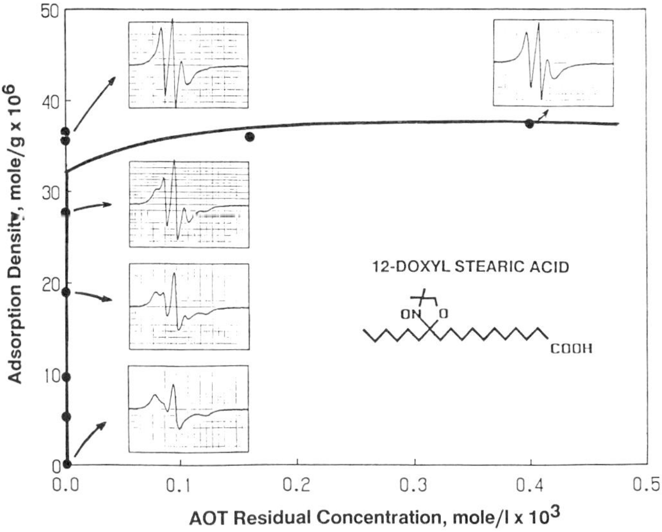

Figure 2: Adsorption isotherm of Aerosol OT on alumina from cyclohexane solution and corresponding ESR spectra obtained using 12-doxyl stearic acid as probe (inset).

The changes observed in the ESR lineshape with surfactant adsorption density are consistent with changes in the probe conformation as more and more surfactant molecules co-adsorb with the probe (Figure 3):

1 - In the absence of co-adsorbed surfactant, the spectrum obtained is very broad and corresponds to mobility of a probe placed in a highly constricted environment. The only thing that can hinder the probe motion in the absence of any surfactant is the alumina surface itself. We postulated recently that, in the absence of surfactant, the probe adsorbs in a flat configuration with both carboxylic and nitroxide groups anchored to the mineral surface [11].

2 - The maximum adsorption density obtained at the plateau of the adsorption isotherm corresponds to a complete monolayer coverage of the surface by the surfactant [12]. Under these conditions, interactions between the adsorbed probe and the surfactant molecules adsorbed around it lead to a very mobile spectrum consistent with a model where the carboxylic group of the probe carrier molecule (stearic acid) is adsorbed on the polar surface of alumina while the nitroxide is rotating in the solution above the adsorbed surfactant layer.

3 - At intermediate surfactant adsorption densities, the complex spectra obtained can be interpreted as a combination of the two previously described spectra: a fraction of the probe population adsorbed on the surface is in a flat configuration while the other fraction is pushed up by the surfactant molecules adsorbed around the probes. As the surfactant adsorption density increases, the contribution of the latter population increases. The hypothesis that two populations of probe molecules are responsible for the complex spectra obtained was confirmed by the observation that any experimental spectrum could be simulated using a linear combination of the two extreme ESR lineshapes.

One of the objectives of this work was to investigate the adsorption of water vapor on an oxide surface modified by a adsorbed surfactant layer. In order to perform water adsorption from the vapor phase, a thorough desiccation of the samples is required prior to water adsorption. This was done by subjecting the samples to a mild heat treatment at 110-120°C under vacuum (10^{-2} Torr) for 5 hours. ESR spectroscopy was used to control the effect of desiccation procedure on the adsorbed surfactant layer.

Figure 4 shows the spectra obtained with different drying procedures. In this experiment, the sample was resuspended in cyclohexane after heat treatment to allow an exact comparison of the spectra before and after the heat treatment. It can be seen that the freeze drying procedure does not alter the adsorbed surfactant layer whereas the desiccation required for the water adsorption modifies the structure of the adsorbed layer significantly. When this spectrum is compared with those shown in Figure 2, it becomes clear that this sample preparation leads either to desorption of the surfactant or reorganization of the adsorbed layer which leaves behind bare alumina surface on which the ESR probe can readsorb in its flat configuration.

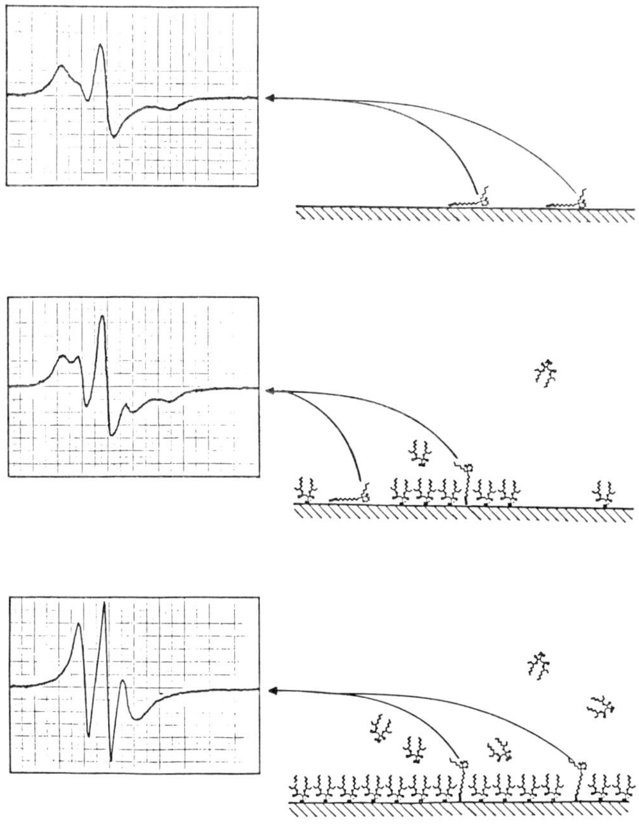

Figure3: Typical ESR spectra of 12-doxyl stearic acid at the alumina/cyclohexane interface obtained in the presence of co-adsorbed Aerosol OT. Schematic representation of the change in probe conformation at the interface due to co-adsorption of Aerosol OT molecules.

We have presented here a technique allowing in-situ measurement of surfactant adsorption density. This technique has been used to detect surfactant adsorption density changes owing to different sample treatments. Potentially, this method has several applications: it can be useful when changes in the adsorption density occur only by reorganization of adsorbed species at the interface; it also facilitates measurement of surfactant adsorption when monitoring of the surfactant concentration in the supernatant is difficult or impossible.

ADSORPTION OF WATER ON SURFACTANT-MODIFIED ALUMINA

Adsorption of water on a substrate treated with a surfactant is a subject of considerable importance for both fundamental and practical reasons: such investigation is necessary for understanding the nature of the interactions between water molecules and adsorbed surfactant species. It can also provide much needed information on the wetting behavior of a surfactant-modified powder.

The samples used for this experiment were prepared following the procedure described above and using surfactant concentrations at which complete coverage was obtained. In addition to the treatment with Aerosol OT, another sample was prepared using lauric acid as the surface modifying agent. After conditioning the samples for 24 hours in surfactant solution, the samples were washed twice in pure cyclohexane to prevent surface precipitation of the excess of surfactant that could occur during subsequent desiccation of the samples.

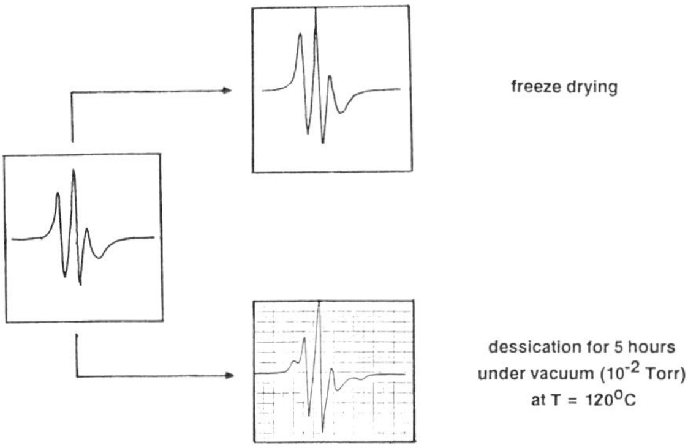

Figure 4: Effect of two desiccation procedures on the ESR response of 12-doxyl stearic acid co-adsorbed on alumina with a monolayer of Aerosol OT.

Figure 5 shows data for adsorption of water on alumina after surfactant pretreatments over the entire range of water vapor pressure at 30°C. These results were obtained by adsorption gravimetry using an instrument described elsewhere [13]. The water vapor is supplied from a liquid source kept at 41°C at a slow, controlled flow rate to ensure quasi-equilibrium conditions at all time. The increase in sample weight and the pressure in the chamber are recorded continuously on an X-Y recorder. It is clear from the data presented in figure 5 that the amount of water adsorbed on the powder precovered by the surfactant is much lower than that when the alumina is not pretreated. Calculation of the volume of the adsorbed monolayer of water using the B.E.T. method for the three isotherms shown on Figure 5 gives 174, 98, and 62 μM/g of adsorbed water respectively for the untreated alumina, the one precovered with Aerosol OT, and the one treated with lauric acid (Table 1).

Using the B.E.T. method, it is possible to calculate the C value of the water adsorption isotherm. This constant gives an indication of the energetics involved in the water adsorption. Most interestingly, the value calculated for the three adsorption isotherms are remarkably constant, suggesting that the adsorption on the three substrates is of very similar nature (Table 1). These results are consistent with the observations made with ESR that the drying by a heat treatment of the surfactant coated powder prior to water adsorption leads to surfactant removal or reorganization of the adsorbed layer and to the creation of bare areas on the particle surface.

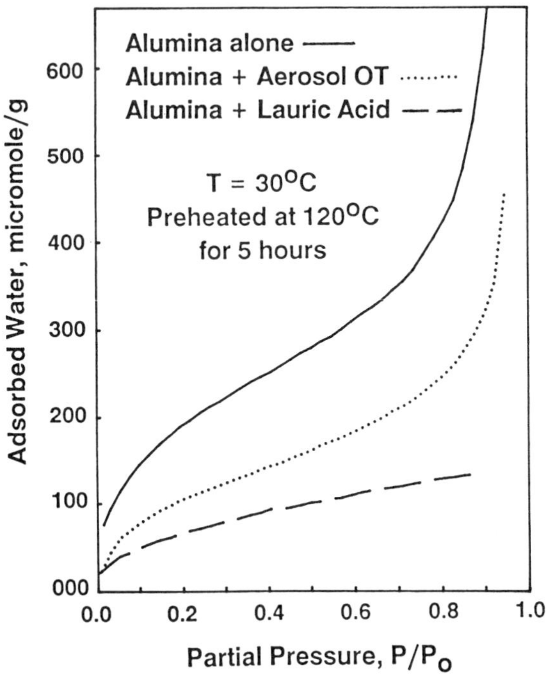

Figure 5: Adsorption isotherms of water vapor on alumina and on surfactant-modified alumina.

An examination of the adsorption data plotted as a function of θ versus P/P_0 shows how the water adsorption takes place as the partial pressure of water increases (θ corresponds to fractions of a monolayer using the monolayer volume as calculated by the B.E.T method). Such plots given in Figure 6 show that the three adsorption isotherms are superimposed on each other. When these data are compared with the adsorption isotherm of water on another alumina (Degussa alumina: 100 m^2/g and $V_{\theta=1.0}$ = 1097 μM/g) [14], it becomes clear that, even though water adsorbs on alumina in all cases, water adsorption plotted as θ versus P/P_0 shows significant differences, suggesting that this plot is very sensitive to changes in the nature of the substrate [15]. As a consequence, it can be concluded that the superimposition of the three adsorption isotherms obtained with Linde alumina suggests that the water adsorption has taken place in all cases on the same alumina surface and that, under these conditions, no significant interaction exists between water and the surfactant molecules.

Table 1

	untreated alumina	alumina + Aerosol OT	alumina + lauric acid
$V_{\theta=1.0}$ (μM/g)	173	98	62
Constant C	28	26	27

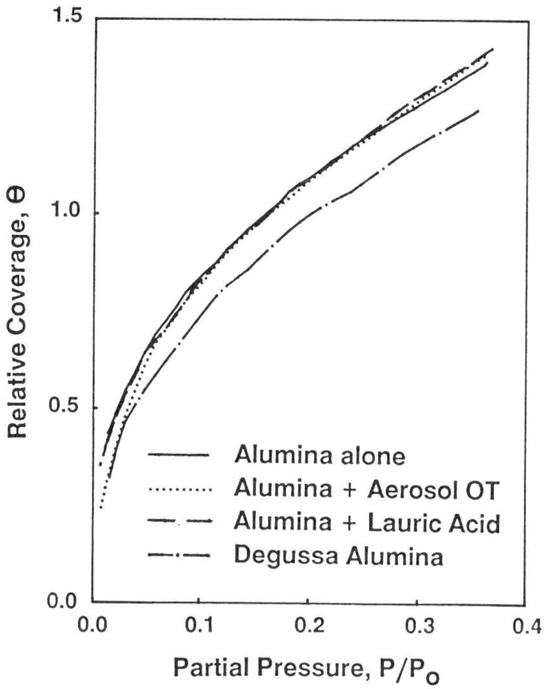

Figure 6: Adsorption data of Figure 5 plotted in terms of relative coverage. The adsorption of water vapor on Degussa alumina is also given for comparison.

CONCLUSIONS

ESR spectroscopy can be used to perform in-situ measurements of surfactant adsorption. Using this technique, it was shown that heat treatment (110-120°C under vacuum) of alumina treated with surfactant leads to alterations of the adsorbed layer owing to a desorption of some surfactant and/or reorganization of the adsorbed layer. This modification of the adsorbed layer results in the exposure of bare areas on the treated mineral surface.

Water vapor adsorption on surfactant-treated alumina is markedly lower that that on the untreated alumina. An analysis of the adsorption isotherms reveals that water adsorption takes place only on the alumina surface that is free of any adsorbed surfactant. Such result would suggest that there is no significant interaction between water and surfactant molecules on the particle surface.

ACKNOWLEDGEMENT

The financial support of the National Science Foundation is acknowledged (MSM-86-17183 and INT-87-14518).

REFERENCES

1 - (a) D.H. Salomon and D.G. Hawthorne, Chemistry of Pigments and Fillers, A. Wilay & Sons: New York, 1983; (b) F. Thomas, J.Y. Bottero, J.M. Cases, and Y. Grillet, *J. Chimie Physique*, **85**, p. 807 (1988).
2 - R.B. McKay in Interfacial Phenomena in Apolar Media, H-F. Eicke and G.D. Parfitt eds., Marcel Dekker: New York, 1987, p. 361.
3 - V. Novotny, Colloids & Surfaces, **24**, p. 361 (1987).
4 - B. Briscoe and D. Tabor in Interfacial Phenomena in Apolar Media, H-F. Eicke and G.D. Parfitt eds., Marcel Dekker: New York, 1987, p. 327.
5 - A.W. Neumann, Adv. in Colloid & Interf. Sci., 4, p. 105 (1974).
6 - J.M. Cases, P. Levitz, J.E. Poirier, and H. van Damme in Advances in Mineral Processing, P. Somasundaran ed., AIME: New York, 1986, p. 171.
7 - S. Partyka, M. Lindheimer, S. Zaini, and B. Brun in Solid-Liquid Interactions in Porous Media, J.M. Cases ed., Technip: Paris, 1985, p. 509.
8 - (a) P. Levitz, H. van Damme, and D. Keravis, *J. Phys. Chem.*, 88, p. 2228 (1985); (b) K.C. Waterman, N.J. Turro, P. Chandar, and P. Somasundaran, *J. Phys. Chem.*, **90**, p. 6828 (1986).
9 - L.J. Berliner, Spin Labelling I: Theory and Applications, Academic Press: New York, 1979.
10 - (a) H. Hommel, A. P. Legrand, H. Balard, and E. Papirer, *Polymer*, 24, p. 959 (1983); (b) P. Chandar, P. Somasundaran, K.C. Waterman, and N.J. Turro, *J. Phys. Chem.*, **91**, p. 150 (1987).
11 - C.A. Malbrel. P. Somasundaran, and N.J. Turro, *Langmuir*, **5**, p. 490 (1989).
12 - C.A. Malbrel and P. Somasundaran, *J. Coll. Interf. Sci.*, in press.
13 - Y. Grillet, J.M. Cases, M. Francois, J. Rouquerol, and J.E. Poirier, *Clay and Clay Minerals*, **36**, p. 233 (1988).
14 - F. Thomas, J.Y. Bottero, and J.M. Cases, *Colloids & Surfaces*, **37**, p. 269 (1989).
15 - (a) J.M. Cases, *Bull. Mineralogie*, **102**, p. 683 (1979); (b) J.M. Cases, J.E. Poirier, and D. Canet in Solid-Liquid Interactions in Porous Media, J.M. Cases ed., Technip: Paris, 1985, p. 335.

FOURIER BESSEL CHARACTERIZATION OF POLISHED METAL SURFACES

NAGJOON CHOI, YOUNGTAIK LIM*, KEITH PRISBREY, GENE BOBECK
Department of Metallurgy, *Department of Mathematics and Applied Statistics, University of Idaho, Moscow, Idaho 83843

ABSTRACT

The amount of information available from metallographic cross sections about material properties remains only subjectively accessible via trained metallurgical experts. Image analysis has not fulfilled the promise to quantify, standardize, or transmit this subjective information. Our solution is to describe different alloys with Luerkens-Beddow style Bessel Fourier descriptors. When AISI-SAE 1035, 1045, and 1060 steels undergo spheroidizing or normalizing heat treatments, the relative amounts and morphology of pearlite, bainite, and other phases, which determine material properties, change even more. The Bessel Fourier descriptors enable accurate classification (up to 93% accuracy) according to heat treatment history and alloy composition, indicating their superior ability to describe metallographic surfaces.

INTRODUCTION

Properties of materials can be explained and predicted from examination of their polished and etched metallographic cross sections. The distribution and abundance of constituents such as pearlite, bainite, and ferrite in steel, for example, explain hardness, toughness, yield strength and other material properties. The amounts of these constituents depend on the alloy content and heat treatment history. AISI-SAE 1035, 1045, and 1060 steels contain 0.35, 0.45, and 0.60 percent by weight of carbon respectively, which determines the quantity of the pearlite and ferrite constituents. Further heat treating causes formation, disappearance, and morphological modification of these and other constituents, and causes corresponding changes in material properties. If the surface textures and morphologies can be accurately characterized, quantitative interpretation and prediction of material properties should be possible. Furthermore, unsuspected heat treatment history could be inferred. The problem calls for a better measurement of surface morphology, and analysis with the use of Luerkens' Bessel Fourier coefficients (1). Such analysis must describe broad categories of alloy composition and heat treatment history with more accuracy than conventional techniques. Advantages of such a system would be more standardized interpretation, better quality control, and improved process control.

Current image analysis techniques such as those based on gray level distributions of digitized images are not adequate. Even though they extract both quantities and distributions of the component phases, they don't have enough information about grain shape, size, and other morphology. Any pattern recognition features extracted from this conventional data are either too specific for generalizations beyond the immediate sample, or too general for anything but qualitative use.

Luerkens (1) investigated the functional forms for a general quantitative representation of surfaces. From a variational principle he derived the following Bessel Fourier series which gives the gray level of each pixel in a digitized image with:

Published 1990 by Elsevier Science Publishing Co., Inc.
Advances in Fine Particles Processing
John Hanna and Yosry A. Attia, Editors

$$G(r,\theta) = [\Sigma\ (a_{m,n} \cos n\theta + b_{m,n} \sin n\theta)\ J_n(Y_{m,n})]^{m,n}$$

$G(r,\theta)$ is the gray level at each pixel coordinate,

$a_{m,n}$, $b_{m,n}$ are the set of Bessel Fourier coefficients describing the surface features, and

$Y_{m,n}$ are the m zeros of the nth Bessel Function, J_n.

The application under consideration was heat treatment and alloy classification. For our experimental method we used the Bessel Fourier coefficients as pattern recognition features. Since by increasing the number of coefficients an image could be reproduced to any degree of accuracy, the coefficients provided a means for identifying the few features of greatest importance to an application.

Seventy metal samples of three different alloy compositions were heat treated, polished, and etched to expose the grain structure and surface morphology. The samples underwent three different heat treatments. In addition, three different locations on each surface were photographed through a high quality metallurgical microscope giving 210 images total. Bessel Fourier coefficients were extracted from the digitized images and tested for accuracy as pattern recognition features. We used nested effects models of multivariable regression analysis to establish correlations (PROC GLM in SAS), stepwise discriminant analysis to determine the most efficient subset of classification features (PROC STEPDISC in SAS), and discrimination analysis to show classification accuracy (PROC DISCRIM in PC SAS, version 6) (2).

The Bessel Fourier coefficients proved to be a good solution. Using only two or three coefficients, heat treatment could be accurately detected within an alloy category by use of linear discriminant functions. Conversely, alloy composition could be accurately detected within a heat treatment category. An even better result was the independent detection of both heat treatment and alloy category, which suggests that these morphological features provide a link between quantitative image analysis and property prediction.

METHOD AND MATERIALS

The materials were cut from half inch bar stock of plain carbon steels from three different alloy classes (AISI-SAE 1035, 1045, and 1060) into approximately half inch lengths. Ten pieces each of 1035 were normalized, quenched, and spheroidized (30 samples). In addition, ten pieces each of 1045 and 1060 were normalized and quenched (40 samples). The pieces were normalized by heating to 900 C and cooling in still air. For quenching the pieces were heated to 900 C and quenched in water. For spheroidizing the pieces were held at 700 C for four hours and air cooled.

After polishing and etching, each sample was photographed at three locations through a metallurgical microscope at 200X using 35mm color slide film. Slide images were used because they could be conveniently transferred to half-inch video tape, from which they were digitized through hardware connected to an IBM PC. The images were digitized at a resolution of 64 gray levels, from white to black, at each pixel point. A matrix of 244 X 256 pixel points was used for each alloy image.

Bessel Fourier coefficients were extracted using a "slow" Fourier transform modified by repeated Bessel coefficient evaluation (3). Twenty pairs of Bessel Fourier coefficients were extracted from each digitized image (ten values of m at n = 0 and 1 each for $a_{m,n}$, $b_{m,n}$ in Equation 1). Finally, we produced 20 values of combined coefficients for each image, $C = (a^2+b^2)^{1/2}$, and examined them for differences due to alloy and heat treatment. Computer code for a "fast" Fourier Bessel coefficient transform was needed because twenty coefficients required almost two hours on an AT level computer, and is under development by us. We transformed the 244 X 256 pixels into radial coordinates for convenience and correspondence to Equation 1. Lagrangian interpolation to equally spaced radial coordinates was not done. The slight accuracy loss probably wasn't significant for the comparisons in this study.

Data analysis consisted of linear regression, and stepwise and normal discriminant analysis.

RESULTS

We extracted Bessel Fourier coefficients from each metal sample in order to characterize heat treatment and alloy composition, the two factors which determine material properties.

Due to the hierarchical nature of the data (i.e., three images from each metal piece, ten pieces for each heat treatment, approximately three heat treatments for each alloy, and three alloy classes), we used a nested effects linear regression model to test the null hypothesis that there were no differences in the images. The null hypothesis was resoundingly rejected at a significance level of 0.0001 The software (PROC GLM) compared observed and "random" effects of heat treatment and alloy. Since 0.1 is considered good and 0.01 is much better, this significance level is excellent.

Stepwise discriminant analysis (PROC STEPDISC) selects only those best coefficients necessary for pattern recognition. We found that the coefficients C2 and C6 were the most important, with C16 and C20 marginally important. All others were relatively unimportant. Using these coefficients only, we found that we could distinguish heat treatment within an alloy type more accurately than the converse problem--distinguishing alloy within a heat treatment category. For example, Figure 1 shows the almost clean division between normalizing (1's) and quenching (2's) for 1045 steel. Each 1 or 2 is an image plotted according to its value for C2 and C6. Figure 2 shows a less clean division for 1060 steel. However, the overall classification accuracy using a linear discrimination function for 1060 remains good, as discussed further.

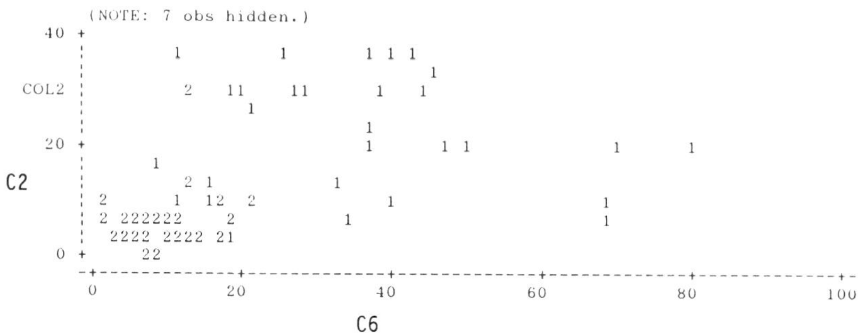

Figure 1 Bessel Fourier Coefficients for Classifying Alloy 1045's Heat Treatment

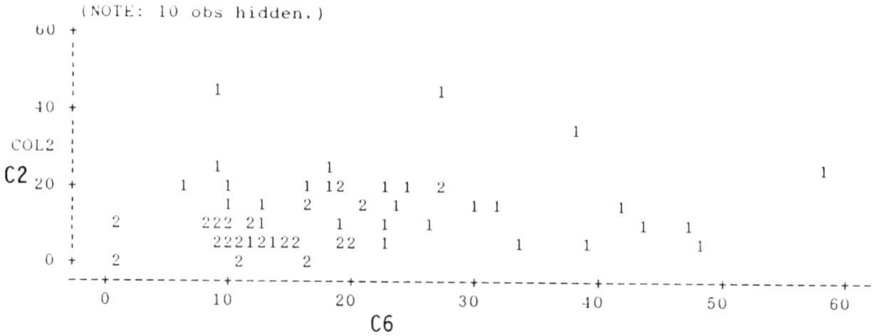

Figure 2 Bessel Fourier Coefficients for Classifying Alloy 1060's Heat Treatment

Linear discrimination functions for detecting 1045 and 1060's heat treatments are derived using the complete data base as a "training set" (PROC DISCRIM):

Alloy 1045: treatment = .094*C2 + .032*C6 + .0010*C16

Alloy 1060: treatment = .083*C2 + .074*C6

Using these functions gives 93% and 80% overall classification accuracy for 60 images each of 1045 and 1060 (Table 2,3). The overall classification accuracy is the average of the individual accuracies in the diagonal positions.

Table 1. Classification Accuracy for Alloy 1045 Using A Linear Discriminating Function

From Heat Treatment Category	Number of Observations and Percent Classified into Heat Treatment Category	
	Normalizing	Quenching
Normalizing	27 90%	3 10%
Quenching	1 3%	29 97%

Table 2. Classification Accuracy for Alloy 1060 Using A Linear Discriminating Function

From Heat Treatment Category	Number of Observations and Percent Classified into Heat Treatment Category	
	Normalizing	Quenching
Normalizing	21 70%	9 30%
Quenching	3 10%	27 90%

The same Bessel Fourier coefficients detected normalizing heat treatment in 1035 with 85% accuracy, but had trouble with quenching and spheroidizing (only 48%).

Linear discrimination functions for the converse problem of detecting an alloy within a heat treatment category were also derived:

Normalizing: alloy = .065*C2 + .033*C6 + .0010*C16

Quenching: alloy = .00017*C3 + 0.0015*C16 + .00082*C20

Using these functions gives 55% and 62% overall accuracy for 90 images each of normalized and quenched structures. Since there are three alloy types one would expect 33% accuracy by random chance alone. Thus these higher accuracies show a strong ability of the three Bessel Fourier coefficients. Nevertheless, it is harder to distinguish 1035, 1045, and 1060 after normalizing or quenching than it is to distinguish the heat treatment, normalizing or quenching, within an alloy.

In order to improve the classification results it is possible to use more parameters. However, in another attempt we visually screened the 210 image data set for those that were obviously blurred, quenched poorly, or apparently heat treated improperly. This screening used our subjective metallurgical experience, and we discarded 69 images to leave 141 images as a training set.

After this screening, each of the above classification accuracies increased by only 5 to 15 percent, indicating that our subjective screening was not a significant improvement.

However, in a final test we used PROC STEPDISCRIM to get a linear discrimination function with the nine best coefficients to distinguish seven categories: 1) normalized 1035, 2) quenched 1035, 3) spheroidized 1035, 4) normalized 1045, 5) quenched 1045, 6) normalized 1060, and 7) quenched 1060. There were originally 30 images in each of the seven categories before screening, and about 20 images in each category after screening. The overall discrimination accuracy was 57%, which is considerably better than a random 1/7 = 14% expected by chance alone.

DISCUSSION

The relationships between Bessel Fourier coefficients and heat treatment/alloy categories were strong enough for quality control and standardization (93% in our best case). The best accuracy was obtained in distinguishing heat treatment for a specific alloy type.

Even though this exploratory study was based on a rather bland set of mild steels, the principle was important to establish. A quantitative connection was made between morphological features and alloy composition and heat treatment history. The significance is that this makes it possible to extend the connection and predict material properties from metallographic images. The most important applications are expected to be in high performance materials such as superalloys. The most common problem in addition to composition control, is heat treatment and prior history. This study showed the most promise in heat treatment identification.

In an industrial quality control setting, discrimination functions or other empirical relations would be derived from a training set and extended to new or independent data, something that was not done in this study. However, the results are strong enough to suggest that extrapolating the regression functions beyond the training set would be no problem.

Traditional or conventional morphological features could be used to predict material properties, yet there is little evidence in the literature that this has happened. Extensive qualitative interpretations and correlations abound, but no quantitative standards exist such as could be possible with Bessel Fourier coefficients. In order to explore this lack further, we used the same search for features on gray level histograms, the basis for most conventional morphological features, with no success (4). Our searches included all kinds of means, moments, maximums, and histogram fractions. The best discrimination accuracies were only slightly better than random. This underlined the importance to us of the quantitative abilities of the Bessel Fourier coefficients.

Much work remains to be done in order to capitalize on these quantitative abilities. Probably of primary importance is bridging the gap between metallurgical skill or qualitative interpretation and abstract "scores" from discrimination functions. Luerkens has explained one of the most promising approaches to this problem (5). Coordinate systems are defined by industrial applications, such as phase transformation, parent phase shadows, angularity key shapes, and material properties, and then are developed in Bessel Fourier coefficient space--a linear algebra approach. We are excited about these possibilities.

REFERENCES

1. Luerkens, D.W., "Surface Representations Derived from a Variational Principle 1. The Gray Level Function", Particulate Science and Technology 4, 361-369, 1986.

2. SAS, "SAS/STAT User's Guide: Release 6.03 Edition", SAS Institute, Cary , North Carolina.

3. Press, W.H. et. al. Numerical Recipes Cambridge Univ. Press, 1986 ISBN 0 521 30811 9.

4. Prisbrey, Keith, "Comparison of Bessel Fourier Coefficients to Gray Level Histograms as Pattern Recognition Features", presented at Fine Particle Society Meeting, San Jose, CA, August, 1988.

5. Luerkens, D. W. and Beddow, Keith, "Morphological Analysis", Short Course presented at the Fine Particle Society Meeting, San Jose, CA, August, 1988.

PART 4.

SURFACE AND COLLOIDAL CHEMISTRY IN THE PROCESSING OF CLAYS

INTERPARTICLE FORCES OF CLAYS

PHILIP F. LOW
Department of Agronomy, Purdue University, West Lafayette, Indiana 47907

Clays modify the physical behavior of any porous medium in which they are incorporated. If they are present in small proportions, they lend plasticity and strength to the medium when it is wet, especially if the other components of the medium are relatively large in size. But, if they are present in large proportions, their swelling or disjoining pressure can destabilize the medium. Then, when drying occurs, their shrinkage can produce planes of weakness and fissures that affect its structure and integrity. All of these effects depend on forces that operate between the clay particles. Despite their importance, these forces are not well understood. However, enough is known about them, especially as a result of recent research, to warrant the present discussion. It should be emphasized that this discussion is not intended to be an exhaustive review. It is intended to give the author's perception of the subject and to justify this perception with a few examples of the kinds of supporting evidence that are available.

Both short-range and long-range forces exist between clay particles. The short-range forces are the Born repulsive force and the Coulomb attractive force. The Born repulsive force develops when the electron shells of atoms in adjacent surfaces interpenetrate; whereas, the Coulomb attractive force develops when the particles (or layers within particles) are sufficiently close together for their surfaces and the intermediate cations to act as oppositely charged condenser plates. The long-range forces are the van der Waals' force, the double-layer force and the hydration force. The van der Waals' force is attractive and is due primarily to the in-phase oscillation of the electron shells of atoms in adjacent particles. The double-layer force can be either attractive or repulsive depending on whether the particle surfaces bear unlike or like charges, respectively. It arises when the ionic atmospheres of adjacent particles overlap. The hydration force is repulsive and is due to the surface-induced modification of the interparticle water. It arises when the hydration shells (zones of modified water) of neighboring particles overlap. Although the hydration force has often been assumed to be a short-range force, this assumption is contrary to the available evidence, some of which will be presented hereafter.

The interplay of the different forces can produce any of the three fundamental arrangements illustrated in Parts (a), (b) and (c) of Fig. 1. These arrangements will be referred to hereafter as the edge-to-edge (EE), edge-to-face (EF) and face-to-face (FF) arrangements, respectively. Let us now consider the conditions under which the different arrangements exist.

The data in Fig. 2 were replotted from the results of Norrish [1] and Foster et al. [2] who determined the effect of NaCl concentration on the c-axis spacings of Upton and Belle Fourche montmorillonite at atmospheric pressure by X-ray diffraction. It should be noted that X-ray diffraction peaks do not appear unless the montmorillonite particles are oriented in a face-to-face (i.e., parallel) arrangement and so the data in Fig. 2 apply to this arrangement. From the figure, it is evident that the parallel layers of the montmorillonite approached each other monotonically as the concentration of NaCl increased until a

Published 1990 by Elsevier Science Publishing Co., Inc.
Advances in Fine Particles Processing
John Hanna and Yosry A. Attia, Editors

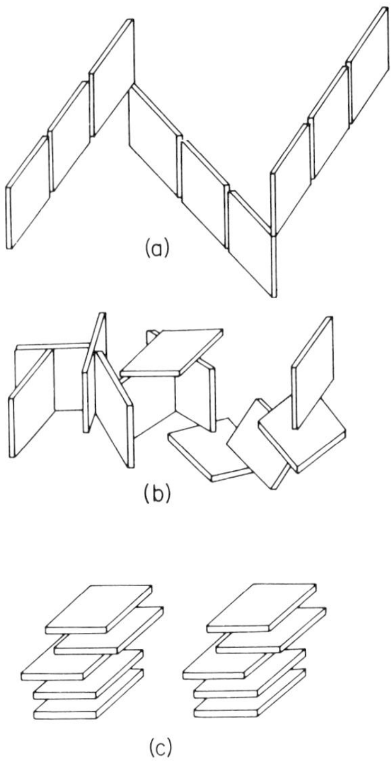

FIG. 1. Possible arrangements of plate-shaped clay particles.

concentration of ~ 0.25 M was reached. Then they jumped into a partially expanded state. This jumping together can be regarded as flocculation and the partially expanded state can be regarded as the flocculated state. Hence, it appears that flocculation of Na-montmorillonite in the face-to-face arrangement does not occur until the electrolyte concentration reaches ~ 0.25 M.

There is rheological evidence [3] and evidence from light scattering [4] which indicates that, in dilute suspensions of Na-montmorillonite, flocculation occurs when the electrolyte concentration reaches ~ 5×10^{-3} M. From the evidence presented in Fig. 2, it appears that the flocculation that occurs at this concentration cannot be in the face-to-face arrangement. It must be in the edge-to-edge or edge-to-face arrangement.

When the edges of clay particles are amphoteric, as they are in kaolinite [5], they have an isoelectric point below which they are positively charged and above which they are negatively charged. Consequently, at pH values below the isoelectric point, double-layer attraction exists between the positively charged edges and the negatively charged faces. This attraction reinforces the existing van der Waals' attraction. As a result, the edge-to-face arrangement prevails and the suspension behaves rheologically as shown in Fig. 3, which is based on the work of Rand and Melton [6].

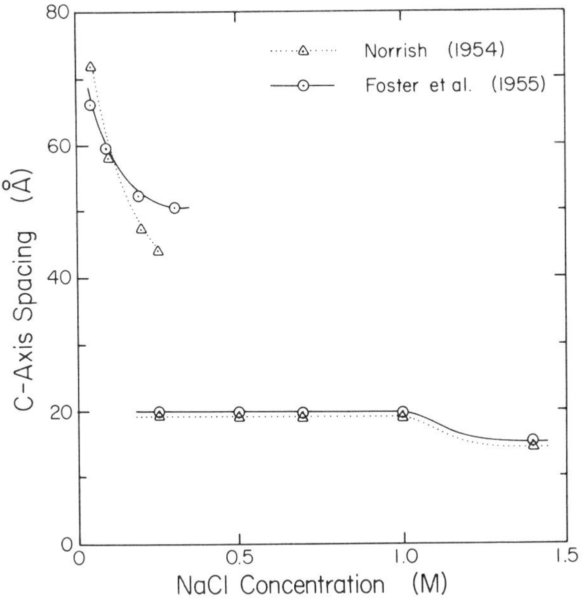

FIG. 2. Relation between the c-axis spacing and the NaCl concentration for Na-saturated Upton and Belle Fourche montmorillonite.

Before considering Fig. 3, it will be convenient to define the extrapolated shear stress, θ. When rheological data are used to plot a curve of the shear stress, s, against the shear rate, σ, for a thixotropic clay suspension, s is found to be an exponential function of σ at low values of σ and a linear function of σ at high values of σ (e.g., [7, 8]). The equation for the linear portion of the curve may be written.

$$s = \eta_{pl}\ \sigma + \theta \qquad (1)$$

in which η_{pl} is the plastic viscosity and θ is the extrapolated shear stress or Bingham yield point, i.e., the intercept of the extrapolated linear portion of the curve on the s-axis. According to Goodeve [9, 10], the first term in Eq. (1) is ascribable to hydrodynamic effects and the second term is ascribable to particle interactions. Following Goodeve [9, 10], Gillespie [11] developed equations that can be combined to give

$$\theta = E_A\ K_1\ N^2/2 \qquad (2)$$

in which E_A is the energy of the interparticle bonds that are constantly breaking and reforming during viscous flow, K_1 is a rate constant descriptive of the rate at which bonds are formed by shear and N is the number of particles per cubic centimeter having bonds that are being broken and reformed. It should be noted that N may be a function of shear rate. Thus, θ is supposed to reflect the number and energy of the interparticle bonds.

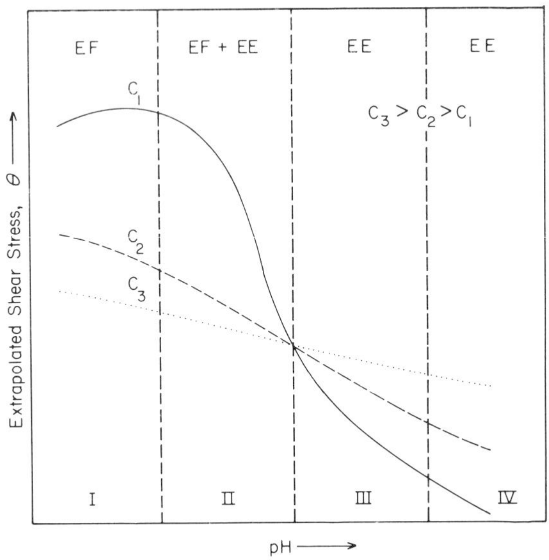

FIG. 3. Qualitative relation between the extrapolated shear stress, θ, and the pH of a clay suspension at different values of the electrolyte concentration, c, when positive edge-negative face interaction exists.

Now let us return to Fig. 3. The pH is low enough in Region I for the positive charge on the particle edges to be fully expressed. Hence, edge-to-face attraction is at a maximum and θ is high. In Region II, the pH has increased to a value that is high enough for edge-to-face attraction to be weakened and some of the particles have reorganized into the edge-to-edge arrangement. As a result, θ is lower. In Region III, the pH has passed the isoelectric point and the edge-to-edge repulsion is only slightly exceeded by the van der Waals' attraction. Hence, θ is smaller still. By the time the pH is in the regime of Region IV, the edge-to-edge repulsion exceeds the van der Waals' attraction and θ falls to zero, at least when c, the electrolyte concentration, is small. As c increases, the electric double layers are repressed so that edge-to-face attraction and edge-to-edge repulsion are decreased. The result is a decrease in θ below the isoelectric point and an increase in θ above the isoelectric point. Hence, the isoelectric point can be identified as the point at which the curves of θ vs pH intersect.

The rheological behavior of kaolinite conforms to that depicted in Fig. 3 [6]. However the rheological behavior of montmorillonite does not [3, 12, 4]. It is reasonable to assume, therefore, that the edges of montmorillonite particles are not amphoteric in character and that the edge-to-edge arrangement is preferred at all pH values. This assumption is consistent with data obtained by light scattering [13, 14] and by electron microscopy [15]. It is also consistent with the theory of Vold [16] who showed that van der Waals' forces are strongest in the edge-to-edge arrangement.

Although no theory has been developed that will allow the interparticle bond energy in the edge-to-edge arrangement to be calculated, a method has been developed recently by Akae and Low [8] that will allow this energy to be measured. The apparatus used for the measurement is diagrammed in Fig. 4. It consists of a Stormer

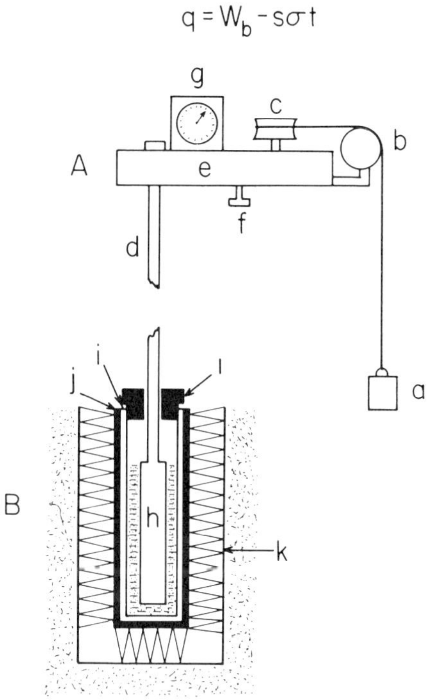

FIG. 4. Apparatus for measuring interparticle bond energy.

viscometer, A, and a Calvet calorimeter, B. A weight, a, is attached to a string that passes over the pulley, b, and winds around the drum, c. The drum is connected to the metal shaft, d, through gears in box, e. When the brake, f, is released, the weight falls and the shaft rotates. A meter, g, counts the rotations of the shaft and the duration of these rotations is measured by a stopwatch. A rotor, h, is connected to the shaft and rotates with it. The stainless-steel cup, i, is concentric with the rotor and fits into the silver socket, j, which is surrounded radially by 500 thermocouples, k. The silver socket is encased in a mica sheath so that it is electrically insulated from the thermocouples. A clay suspension occupies the annulus between the rotor, h, and the cup, i. Evaporation of water from the suspension is prevented by the cap, l. When the rotor turns, energy is used to: (1) break the interparticle bonds that give the suspension its consistency, and (2) to maintain the flow thereafter. The energy is dissipated as heat, q, which is measured by the calorimeter. Hence,

$$q = w_b - s\sigma t \tag{3}$$

where w_b is the energy required to break the interparticle bonds in a cubic centimeter of suspension and t is the time during which shear occurs. The experiment yields appropriate values of q, σ and t. The value of s at the measured value of σ is obtained from a previously prepared curve of s vs σ for the given suspension. Thus, w_b can be calculated. Now,

$$w_b = -E_A\, N/2 = -E_A\, m_c/2gV \tag{4}$$

where N is the total number of bonds that have been broken, E_A is the energy of these bonds, m_c is the mass of clay in the volume, V, and g is the mass of the clay particle. It is evident, therefore, that $E_A/2g$, the interparticle bond energy per unit mass, can also be calculated.

Figures 5 and 6 illustrate the kinds of results that can be obtained with the apparatus depicted in Fig. 4. These figures were taken from the paper of Akae and Low [4]. The sol-gel transition is responsible for the breaks in the curves of both figures. Now, consider Fig. 5 in the light of Eqs. (2) and (4). In Eq. (2), N refers to the number of particles that are constantly breaking and reforming bonds with their neighbors at relatively high values of σ, where Eq. (1) applies, and E_A is the energy of the bonds being broken and reformed. Some of the stronger bonds that have to be broken to initiate shear do not have time to reform during the shearing process. N in Eq. (4) refers to the particles with such bonds and E_A is the average energy of these bonds. Nevertheless, comparison of Eqs. (2) and (4) indicates that w_b should be related to θ and this indication is borne out in Fig. 5.

It is interesting to note from Fig. 6 that the interparticle bond energy per gram of clay increases with clay concentration above the sol-gel transition. This means that the number of interparticle bonds increases as the particles come closer together. If we knew the molecular weight of the clay, i.e., the weight of Avogadro's number of particles, we could use Fig. 6 to obtain the interparticle bond energy per mole of clay. Data published by Low [17] on the osmotic pressure of very dilute sols of Na-saturated Upton montmorillonite suggest a molecular weight of ~ 24,000 g/mole. If this suggestion is valid, we see from Fig. 6 that, when $m_c/V = 0.08$, the interparticle bond energy per mole of particles would be ~ 240 kJ/mole (57 k cal/mole), a value that is comparable with the energy of a chemical bond.

In condensed systems, the clay particles may still be linked edge-to-edge but limitations of space require that they also assume the face-to-face arrangement. This is the arrangement which is conducive to swelling. Swelling is the separation of the particles (or layers within particles) with the admission of interparticle or interlayer water. For many years, swelling has been attributed to repulsion arising from the

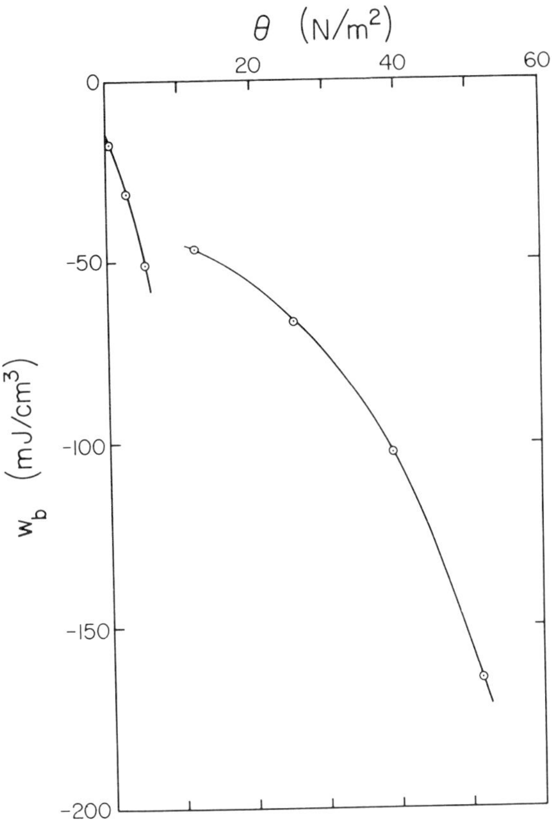

FIG. 5 Relation between w_b, the interparticle bond energy per cubic centimeter of suspension, and θ, the extrapolated shear stress, for suspensions of crude Wyoming bentonite in water at 26.9°C.

overlap of the electric double layers of adjacent particles. A model of the electric double layers of such particles is shown in Fig. 7 where:

ψ_o = the electrostatic potential at the clay-water interface
ψ_δ = the electrostatic potential at the Outer Helmholtz Plane, i.e., the outer limit of the Stern layer.
ζ = the electrostatic potential at the plane of shear
ψ_h = the electrostatic potential midway between the parallel clay particles
σ_o = the charge density in the plane of the clay-water interface
σ_δ = the charge density in the Outer Helmholtz Plane
δ = the thickness of the Stern layer
τ = the distance from the Outer Helmholtz Plane to the plane of shear.

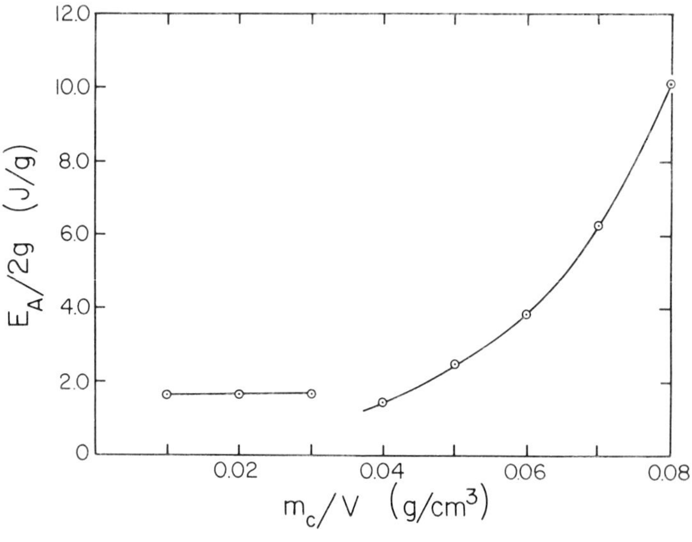

FIG. 6. The relation between $E_A/2g$, the interparticle bond energy per gram of clay, and m_c/V, the clay concentration, for suspensions of crude Wyoming bentonite in water at 26.9°C.

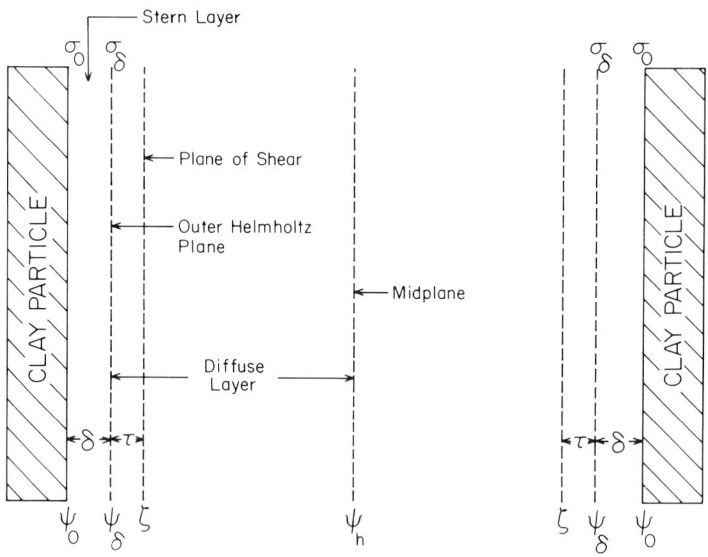

FIG. 7. Model of the overlapping double layers of clay particles.

According to the electric double-layer theory of swelling (e.g., [18, 19]), the following equations relate the variables defined above when a single symmetrical electrolyte is present

$$p = 2n_o kT(\cosh u - 1) \tag{5}$$

$$\int_z^u dy/\sqrt{2(\cosh y - \cosh u)} = \kappa h \tag{6}$$

$$\sigma_\delta = \sqrt{\frac{n_o DkT}{2\pi}}\, \sqrt{2(\cosh z - \cosh u)} \tag{7}$$

$$\kappa = \sqrt{8\pi\nu^2 e^2 n_o/DkT} \tag{8}$$

where:

p = the repulsive pressure between the particles or layers within particles
n_o = the electrolyte concentration in the external solution where the electrostatic potential is zero
k = the Boltzmann constant
T = the absolute temperature
h = the distance between the Outer Helmholtz Plane and the midplane
D = the dielectric constant
ν = the common valence of the ions of the electrolyte
e = the electronic charge
ψ = the electrostatic potential at any distance from the particle surface

and the reduced variables, y, u and z are defined by $y = \nu e\psi/kT$, $u = \nu e\psi_h/kT$ and $z = \nu e\psi_\delta/kT$. Moreover, the swelling pressure, Π, of the clay is assumed to equal the difference between p and f, the van der Waals' attractive force. Thus,

$$\Pi = p - f. \tag{9}$$

The relation between f and λ, the distance between the parallel surfaces of adjacent particles, is given by

$$f = A/6\pi\lambda^3 \tag{10}$$

in which A is the Hamaker constant.

In their studies of swelling, the author and his colleagues [20, 21, 22] determined the relation between Π and λ for nine different Na-saturated smectites and a Li-saturated vermiculite by using an X-ray diffractometer in conjunction with the environmental chamber diagrammed in Fig. 8. In this method, an oriented gel, g, on the membrane filter, h, is placed on a porous ceramic plate, i, set into a stainless-steel support containing parallel grooves, j, that are connected to the outside atmosphere by the drain tube, k. The assembly, B, is properly aligned within the chamber, A, on the axis of an X-ray goniometer and the edge aperture, e, is lowered into place. A thin mica deposit, f, serves as an internal standard by which the alignment can be checked. The beryllium window, a, is transparent to X-rays but is strong enough to support pressures < 10 bars. Water-saturated nitrogen gas at a fixed pressure is allowed to fill the chamber through the port, b. The nitrogen gas acts as a piston and expresses water from the clay gel until equilibrium is attained. Then the X-ray diffraction pattern of the clay is recorded and, from it, the value of λ is determined. Since the swelling pressure of the clay equals the pressure of the gas at equilibrium, the value of Π corresponding to this value of λ is obtained by reference to a pressure gauge between the nitrogen tank and the environmental chamber.

Low [23] used data obtained by the method just described to construct Fig. 9. The dotted lines in the figure indicate the limits of the Π-λ relations for all of the Na-smectites and the Li-vermiculite studied thus far. On the basis of this figure, it is reasonable to conclude that all smectites swell alike and that variables such σ_o have little effect. This conclusion is reinforced by Fig. 10, which was taken from Viani et al. [20]. The values of σ in Fig. 10 were determined by dividing the cation exchange capacity of each smectite by its surface area and so they are actually values of σ_o.

Figures 9 and 10 show that all the Na-smectites had essentially a common value of λ at any specific value of Π, regardless of the value of σ_o. Ionic substitutions that distinguish the various smectites should have very little effect on A. Therefore, in keeping with Eqs. (10), (9) and (5), a common value of λ indicates common values of f, p and u. If $\delta = 0$, i.e., if there is no Stern layer, $h = \lambda/2$ and h would also have to have a common value. Then, it follows from Eqs. (6) and (7) that the same would be true of z and σ_δ. But if there is no Stern layer, $\sigma_\delta = \sigma_o$ and we know that σ_o is not constant. The necessary conclusion is that a Stern layer must be present.

The foregoing conclusion was reached earlier by Low [24] who showed that almost all of the counterions of a Na-smectite are in the Stern layer. This conclusion will receive additional substantiation in a paper which has been submitted for publication in Langmuir. In the latter paper, it will be shown that $\psi_\delta = \zeta \simeq -60$ mv. By assuming that ψ_δ remains constant at -60 mv as λ changes and using the above equations, the theoretical curve in Fig. 9 was obtained. Obviously, the contribution of double-layer repulsion to Π is relatively small.

Any discrepancy between the observed curve of Π vs. λ and that based on double-layer theory is usually ascribed to the force produced by hydration of the particle surfaces. This force has been called the hydration force or the structural component of the disjoining pressure (e.g., [25, 26]). However, it is possible that the discrepancy arises from another force of unknown origin. To establish the role of surface hydration in clay swelling, it is necessary to demonstrate that clay surfaces modify the properties of the adjacent water and that Π depends on the degree of modification. For this purpose, the following discussion is included.

The author has studied the nature and properties of water near clay surfaces for many years. His results prior to 1979 have been summarized and integrated by Low [27]. Later results are presented in papers by Oliphant and Low [28], Mulla and Low [29] and Sun et al. [30]. These results show that

$$J_i = J_i^o \exp [\beta_i (m_c/m_w)] \qquad (11)$$

and that

$$\beta_i = k_i S \qquad (12)$$

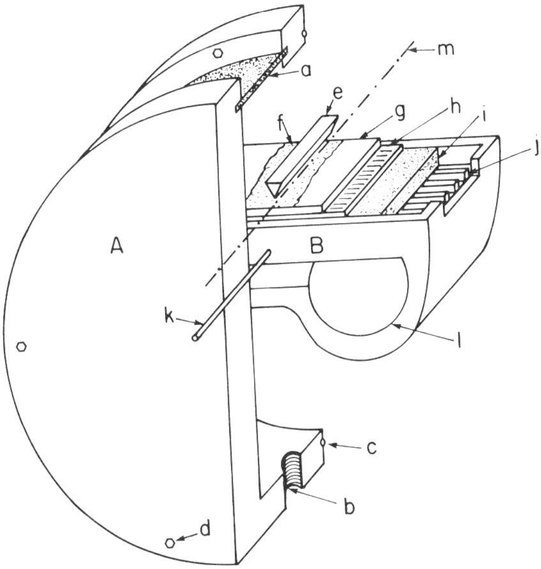

FIG. 8. Schematic drawing of the environmental chamber for X-ray diffraction.

where J is the value of any property, i, in the clay-water system, J^o is the value of the same property in pure bulk water, β is a constant that is characteristic of the clay and the property, m_c/m_w is the mass ratio of clay to water, k_i is a constant that is characteristic of the property only and S is the specific surface area of the clay. Now we know that t, the average thickness of the water films on the clay surfaces, is given by

$$t = (m_w/m_c)/\rho S. \qquad (13)$$

where ρ is the density of the water. Combination of Eqs. (11), (12) and (13) yields

$$J_i = J_i^o \exp (k_i/\rho t). \qquad (14)$$

As a close approximation, $\rho = 1.0$ g/cm^3. Hence, J is exponentially related to t. To illustrate this relation, Fig. 11 is included. In Fig. 11, which was taken from Low [31], ϵ is the molar absorptivity of the water ($H_2O + D_2O$) at the frequency of O-D stretching and $(\partial\phi_v/\partial T)_P$ is the apparent specific expansibility of the water (H_2O). The apparent

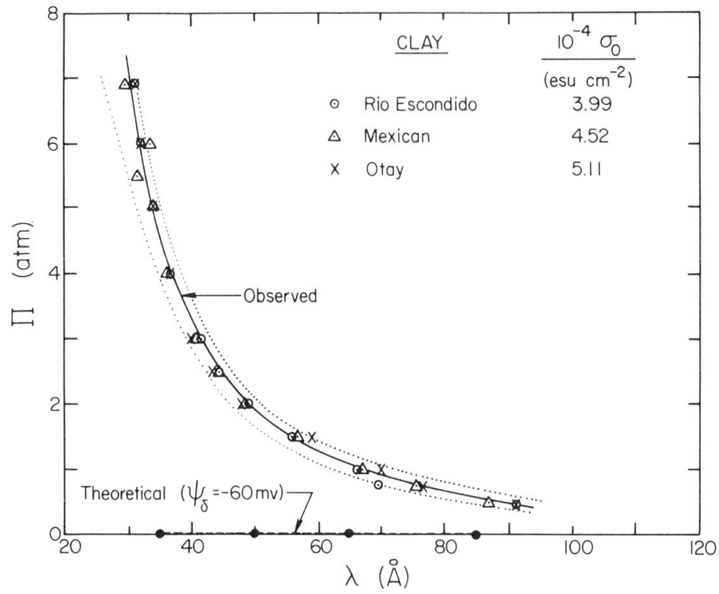

FIG. 9. Observed and theoretical curves of Π, the swelling pressure, versus λ, the interlayer distance, for Na-smectites (Dotted lines indicate extremes for all clays studied thus far).

specific expansibility can be regarded as the thermal expansibility per gram. Both ϵ and $(\partial\phi_V/\partial T)_P$ are sensitive to structural modification. Note that, for each property, the data for all of the several clays fall on a common line. This line is described accurately by Eq. (14) with appropriate values for J^o and k. Obviously, the clay has a marked influence on the properties of the water associated with it.

By using Eq. (14) with constants appropriate to ϵ and $(\partial\phi_V/\partial T)_P$, the solid lines in Fig. 12 were obtained. The dashed lines are extrapolations of the linear portions of the dotted lines. Now, suppose that there are two kinds of water in the clay-water system, namely, water that is modified by the particle surfaces and bulk water. Let the symbols representing the modified water be identified by the * superscript and the symbols representing the bulk water be identified by the zero superscript. Then we can write

$$J_i = f^* J_i^* + f^o J_i^o \qquad (15)$$

where f^* and f^o are the gram fractions of the modified and bulk water, respectively. But $f^* + f^o = 1$ and $f^* = m_w^*/m_w$ and so

$$J_i = (J_i^* - J_i^o) \frac{m_w^*}{m_w} + J_i^o. \quad (16)$$

In keeping with Eq. (13), we know that

$$m_w^*/m_w = \rho^* t^* S/\rho t S \simeq t^*/t. \quad (17)$$

Combination of Eqs. (16) and (17) gives

$$J_i = (J_i^* - J_i^o) \frac{t^*}{t} + J_i^o. \quad (18)$$

The values of both J^* and t^* will remain constant until the films of modified water begin to overlap and then they will change.

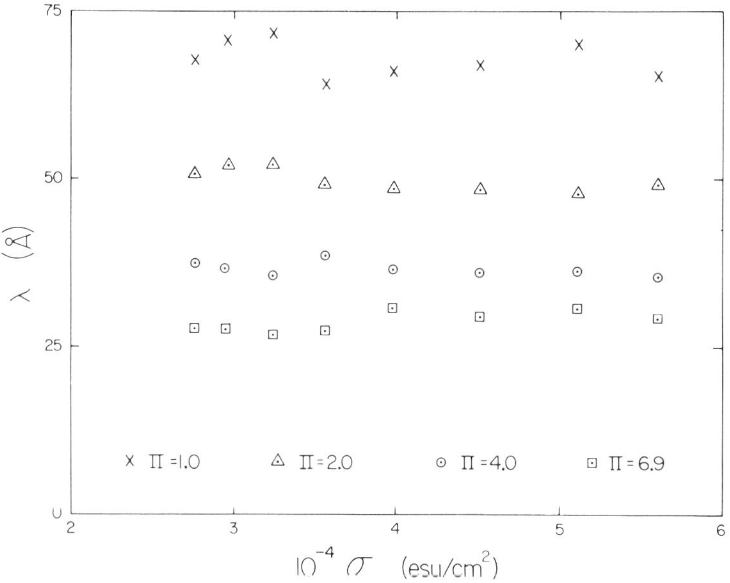

FIG. 10. Relation between λ, the interlayer distance, and $\sigma = \sigma_o$, the surface charge density, for eight Na-smectites at different values of Π, the swelling pressure.

Consequently, J_i will be a linear function of 1/t until $t = t^*$ and the zones of modified water begin to overlap. Then the relation between J_i and 1/t will depart from linearity. Observe from Fig. 12 that, for both water properties, this departure occurs at $1/t \simeq 0.0275 \times 10^8$ cm^{-1} or $t = t^* \simeq 36.4 \times 10^{-8}$ cm. In other words, the films of modified water were about 36 Å thick. Other water properties with different sensitivities to structural modification will have different values of t^*. However, from Fig. 12 we can be assured that the surfaces of the montmorillonite particles affect the water to a depth of at least 36 Å.

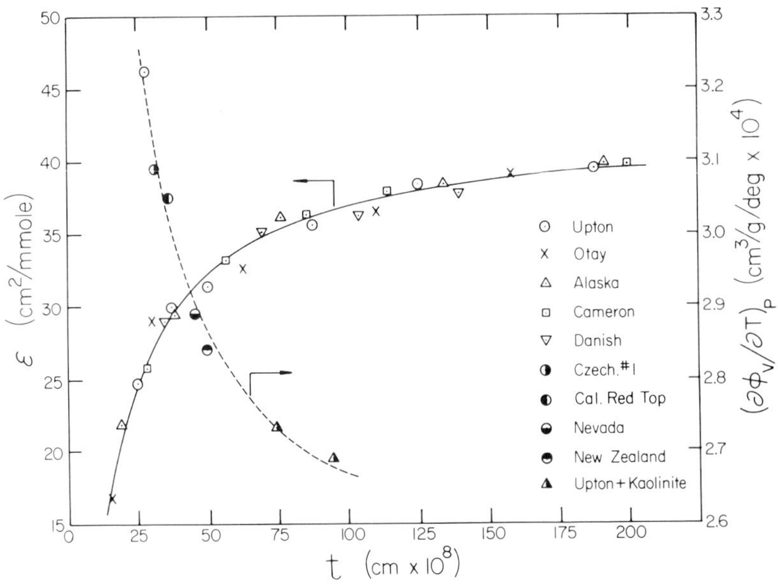

FIG. 11. The dependence of ϵ, the molar absorptivity, and $(\partial\phi_V/\partial T)_p$, the apparent specific expansibility, of the water on t, the average thickness of the water films on the surfaces of Na-smectites.

The molar absorptivity, ϵ, is related to the change in dipole moment of the water molecule during the course of a vibration and, hence, reflects the strength of intermolecular bonds. For example, it is known that ϵ for O-H stretching decreases as the strength of intermolecular hydrogen bonds decreases [32]. The strength of the intermolecular bonds in water cannot change without changing its escaping tendency and, hence, its relative partial molar free energy, $\overline{G}_w - G_w^o$. Thermodynamics shows that

$$\overline{G}_w - G_w^o = -\overline{v}_w \Pi \tag{19}$$

in which $\overline{v}_w$ is the partial molar volume of the water. We see, therefore, that ϵ and Π should be related. Similar arguments could be advanced to support the concept that all properties of the water should be related to Π. Verification of this concept is given in the following paragraph.

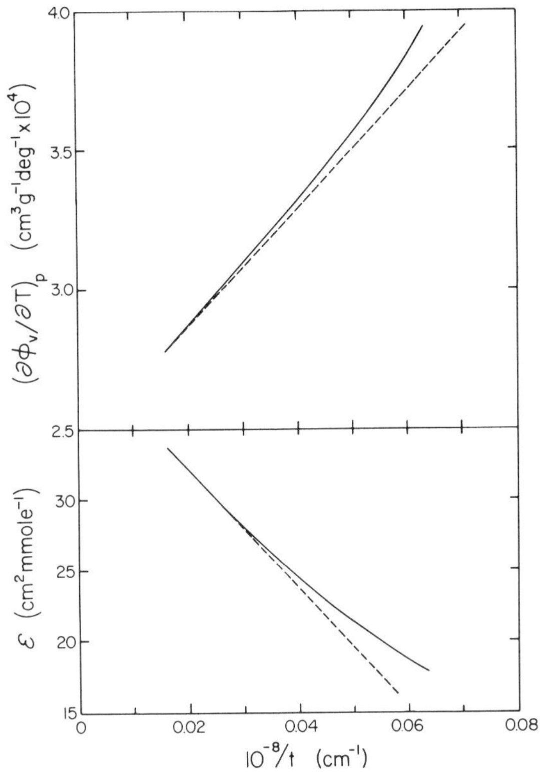

FIG. 12. The relation between ϵ, the molar absorptivity, and $(\partial\phi_v/\partial T)_p$, the apparent specific expansibility, of the water and 1/t, the reciprocal of the average film thickness of the water films on the surfaces of Na-smectite particles.

The observed curve in Fig. 9 is described by the expression

$$(\Pi+1) = b \exp(k/\lambda) \tag{20}$$

in which b and k are constants. Since $t \propto \lambda$, Eq. (19) can be transformed to

$$(\Pi + 1) = b' \exp (k'/t) \tag{21}$$

where b' and k' are also constants. This equation describes the curve in Fig. 13, which was published initially by Low (1987). Evidently, Π is also exponentially related to t.

Since both Eqs. (14) and (21) have a common variable, t, they can be combined by the elimination of this variable. The result is

$$(\Pi + 1) = b' (J_i/J_i^o)^{k'/k_i} \tag{22}$$

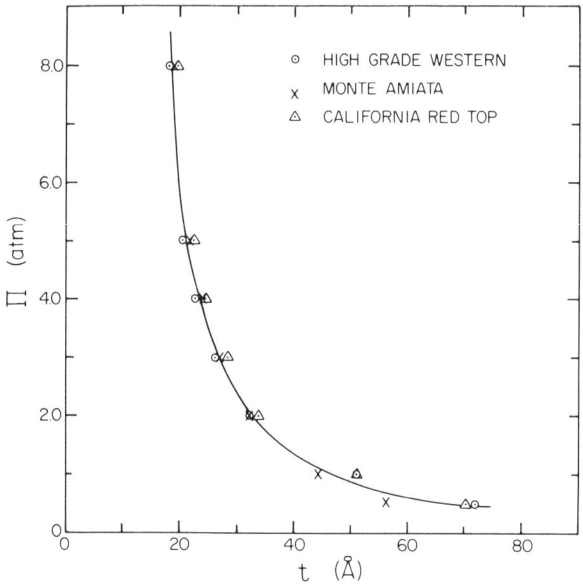

FIG. 13. Relation between Π, the swelling pressure, and t, the average thickness of the films of water on the surfaces of three Na-smectites.

This relation between Π and J/J^o is demonstrated in Fig. 14 for three different water properties. They are the molar absorptivity, the isothermal compressibility, and the thermal expansibility. Data used to construct the figure were obtained from Mulla and Low (1983) and Sun _et al_. (1986).

We have demonstrated that clay surfaces modify the properties of the adjacent water to an appreciable depth and that Π depends on the degree of this modification. Moreover, we have shown that double-layer repulsion makes only a relatively small contribution to Π. It is reasonable to assume, therefore, that the swelling of smectites is primarily due to the hydration of their surfaces.

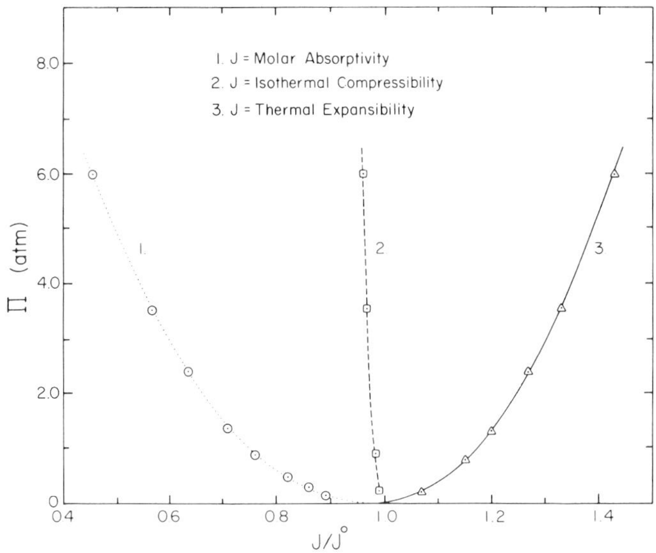

FIG. 14. Relation between Π, the swelling pressure of the clay-water system, and J/J^o, the ratio of the value of the water property in the system to that in bulk water, for three different water properties.

REFERENCES

1. Norrish, K. (1954) Disc. Faraday Soc. 18, 120-134.
2. Foster, W. R., Savins, J. G. and Waite, J. M. (1955) Clays and Clay Minerals 3, 296-316.
3. Rand, B., Pekenc, E., Goodwin, J. W. and Smith, R. W. (1980). J. Chem. Soc. Faraday Trans. 76, 225-235.
4. Chen, J. S., Cushman, J. H. and Low, P. F. (1989) Clays and Clay Minerals. In press.
5. Schofield, R. K. and Samson, H. R. (1953) Clay Minerals Bull. 2, 45-51.
6. Rand, B. and Melton, I. E. (1977) J. Colloid Interface Sci. 60, 308-320.
7. Davey, B. G., and Low, P. F. (1971) Soil Sci. Soc. Amer. Proc. 35, 230-236.
8. Akae, T. and Low, P. F. (1988) J. Colloid Interface Sci. 124, 624-631.
9. Goodeve, C. F. (1939) Trans. Faraday Soc. 35, 342-358.
10. Goodeve, C. F. (1949) In Proc. Internatl. Rheo. Congr. (Schaveningen, Holland, 1948) Part 2. North Holland Pub. Co., Amsterdam, p. 5-11.
11. Gillespie, T. (1960) J. Colloid Interface Sci. 15, 219-231.
12. Heath, D. and Tadros, Th.F. (1983) J. Colloid Interface Sci. 93, 307-319.

13. M'Ewan, M. B. and Pratt, M. I. (1957) Trans. Faraday Soc. 53, 535-547.
14. M'Ewan, M. B. and Mould, D. L. (1957) Trans. Faraday Soc. 53, 548-564.
15. Tessier, D. and Pedro, G. (1982) In Proc. 7th Internatl. Clay Conf. (Bologna and Pavia, 1981. H. van Olphen and F. Viniale, eds.) Elsevier, Amsterdam, pp. 145-176.
16. Vold, M. J. (1957) Proc. Indian Acad. Sci. 46A, 152-156.
17. Low, P. F. (1974) In Transactions of the 10th Internatl. Congr. Soil Sci. (Moscow, 1974) Vol. 1, pp. 42-46.
18. Verwey, E.J.W. and Overbeek, J.Th.G. (1948) Elsevier, New York, 205 pp.
19. Van Olphen, H. (1963) Interscience, London, 301 pp.
20. Viani, B. E., Low, P. F. and Roth, C. B. (1983) J. Colloid and Interface Sci. 96, 229-244.
21. Viani, B. E., Roth, C. B. and Low, P. F. (1985) Clays and Clay Minerals 33, 244-250.
22. Wu, J., Low, P. F. and Roth, C. B. (1989) Clays and Clay Minerals. 37, 211-218.
23. Low, P. F. (1987) Langmuir 3, 18-25.
24. Low, P. F. (1981) Soil Sci. Soc. Amer. J. 45, 1074-1078.
25. Israelachvili, J. N. and Adams, G. E. (1978) J. Chem. Soc. Faraday Trans. I. 74, 975-1001.
26. Derjaguin, B. V. and Churaev, N. V. (1974) J. Colloid Interface Sci. 49, 249-255.
27. Low, P. F. (1979) Soil Sci. Soc. Amer. J. 43, 651-658.
28. Oliphant, J. L. and Low, P. F. (1983) J. Colloid Interface Sci. 95, 45-50.
29. Mulla, D. J. and Low, P. F. (1983) J. Colloid Interface Sci. 95, 51-60.
30. Sun, Y., Lin, H. and Low, P. F. (1986) J. Colloid Interface Sci. 112, 556-564.
31. Low, P. F. (1987) In Proc. Internatl. Clay Conf., Denver, 1985 (L. G. Schultz, H. van Olphen and F. A. Mumpton, eds.) pp. 247-256.
32. Pimentel, G. C. and McClellan, A. L. (1960) W. H. Freeman and Co. San Francisco, 475 pp.

APPLICATION OF SIMS TO THE STUDY OF POLYCATION ADSORPTION ON KAOLIN

J.K. LAMPERT,* L.J. MORGAN,* AND B.L. BENTZ**
*Engelhard Corporation, P.O. Box 2900, Edison, NJ 08818;
**David Sarnoff Research Center, Princeton, NJ 08543

ABSTRACT

Secondary Ion Mass Spectrometry (SIMS) is used to study polycation adsorption behavior on hydrophilic, negatively charged aluminosilicate surfaces. In the SIMS experiment, bombardment of a sample surface by energetic particles gives rise to a population of charged atoms, molecules, and molecular fragments. These so-called secondary ions are extracted and focused into a mass spectrometer where they are separated and identified according to their mass to charge ratio. Because the secondary ions are sputter desorbed from the uppermost surface layers of the sample, recording their mass spectra affords a sensitive probe of surface composition.

We used SIMS to study kaolin surfaces after polycation adsorptions from 0.0 and 0.1 M NaCl aqueous solutions. When the secondary ions characteristic of the kaolin matrix and the polycation are monitored as a function of sputtering time, we observe behavior that is dependent on the adsorption conditions. For equal adsorption densities, secondary ions from matrix aluminium are delayed relative to those from the polycation for the zero ionic strength adsorption case, whereas the two signals appear simultaneously when the polycation is adsorbed from 0.1 M NaCl. This suggests the polymer occupies less surface area per molecule when adsorbed from the NaCl solution, consistent with models predicting a coiled polymer configuration under these conditions.

INTRODUCTION

Polyelectrolytes are often used as flocculants in clay and cellulose colloid systems for water clarification and filter aid applications[1]. Recently, a unique application has been described for the polycation poly(diallyldimethylammonium chloride) (p(DADMAC)) as a bulking agent with kaolin clay to prepare structured paper coatings[2,3,4]. This end-use necessitates pigment dewatering while maintaining a bulked structure with the desired paper coating properties. Aqueous solution applications of polycations have generated numerous studies describing solution polycation-kaolin interactions, but there has been little work investigating the polycation conformation on the kaolin surface after pigment dewatering.

Published 1990 by Elsevier Science Publishing Co., Inc.
Advances in Fine Particles Processing
John Hanna and Yosry A. Attia, Editors

Adsorption isotherm, ion exchange, and theoretical modeling studies show p(DADMAC) adsorbs by an ion exchange mechanism, but not in a flat configuration with every charge site occupying adjacent cation exchange sites on the colloid surface. Rather, the molecule adsorbs with some flat sections ("trains"), and other portions of the polymer molecule extending into the solution ("loops" and "tails"). As the ionic strength of the polymer solution increases, higher polymer adsorption densities indicate the degree of looping also increases [5,6,7].

The polycation looping phenomenon has been investigated by Electron Paramagnetic Resonance (EPR) studies using nitroxide spin labels on copolymers of p(DADMAC) and acrylic acid[8,9,10]. EPR line broadening, indicative of restricted spin label motion, was observed when the labelled copolymer was adsorbed onto colloid surfaces from low ionic strength solution. When the solution ionic strength was increased, the EPR bands sharpened due to an increase in the number of freely rotating spin labels. The broadened bands are attributed to spin labels in trains at the colloid surface, while the sharp band component is due to freely rotating spin labels in polymer loops extending into solution and not constricted by the colloid surface. The model suggested by the spin label data was again corroborated by an increase in the colloid's adsorption capacity with solution salt concentration, indicating the polymer was adsorbed in a coiled configuration occupying fewer charge sites per molecule in high ionic strength environments.

We have used Secondary Ion Mass Spectrometry (SIMS) to probe the dewatered kaolin surface following p(DADMAC) adsorption from low and high ionic strength solutions. In the SIMS experiment, bombardment of a sample surface by energetic particles gives rise to a population of charged atoms, molecules, and molecular fragments. These so-called secondary ions are extracted and focused into a mass spectrometer where they are separated and identified according to their mass to charge ratio. Because the secondary ions are sputter desorbed from the sample's uppermost surface layers, recording their mass spectra affords a sensitive probe of the surface composition. SIMS has been used to study the inorganic surface chemistry of oxides and oxide films[11] and for polymer and polymer film characterization[12,13]. The study described in the present report is novel in its application of SIMS to explore polymer adsorption in a flocculated colloid system. Adsorption isotherms from deionized water and 0.1 M NaCl solutions are measured, and the surface kaolin composition is probed using SIMS.

EXPERIMENTAL

Colloid suspensions were prepared from sodium silicate/sodium carbonate treated kaolin from Washington County, Georgia, with a BET surface area of 12 m^2/g. The kaolin was used to prepare 10% aqueous slurries in deionized water and 0.1 M NaCl. The slurries had an equilibrium pH between 8.1 and 8.4.

The cationic polymer, p(DADMAC), obtained from Calgon Corporation, has a number and weight-average molecular weight of 14,000 and 68,900 daltons, respectively. It was added to the 10% kaolin slurries at final slurry concentrations from 1×10^{-4} to 1×10^{-3} g polymer/g slurry, and allowed to equilibrate overnight.

Adsorption isotherms were obtained by varying the amount of polymer added to the suspension, and measuring the residual solution polymer concentration after centrifugation. The residual polymer in the supernatant was determined by nitrogen analysis using an oxidative pyrolysis-chemiluminescence technique.

The kaolin samples for SIMS analysis were filtered through 0.2 μm polycarbonate filters, dried in a vacuum oven at 110°C, and lightly ground in a mortar and pestle. Each ground kaolin sample examined in the SIMS instrument was a 13 mm diameter, 0.5 mm thick pellet formed by compacting the powder in a press. Once formed, the pellet surface was roughened with a razor blade and mounted on a stainless steel sample holder using double-sided tape. Neat polymer samples were prepared by drying the 40% aqueous p(DADMAC) solution on copper foil.

Positive ion SIMS spectra were acquired with a Finnigan MAT triple stage quadrupole mass spectrometer (TSQ-15) extensively modified at the David Sarnoff Research Center for solid surface analysis[14]. A non-rastered beam of 3 keV xenon neutral atoms bombarded the sample at a 70° angle of incidence. Use of a neutral primary beam helps reduce sample charge-up when insulating samples are examined. The beam size at the sample was approximately 2 mm X 5 mm, and current densities were approximately 100 nA/cm^2 for the sputter profile experiments. Mass spectra were acquired by repetitively scanning the quadrupole between 10-140 daltons at 0.6 seconds per scan and summing the spectra recorded by the data system.

RESULTS AND DISCUSSION

The p(DADMAC) adsorption isotherms in Fig. 1 exhibit behavior similar to those reported in the literature. Polymer adsorption is quantitative from both deionized water and 0.1 $\underline{M}$ NaCl until a critical adsorption density is attained. For these kaolins, an adsorption rate decrease is observed in the deionized water system at approximately 0.3% nominal polymer dose. Adsorption continues at higher doses, but at the reduced rate. Quantitative polymer adsorption is observed from the 0.1 $\underline{M}$ NaCl solution at polymer doses above 0.1%.

SIMS spectra of homopolymeric materials generally contain peaks related to the monomeric unit and structurally diagnostic peaks at lower mass arising from fragmentation of the polymer backbone or side-chain substituents. The SIMS spectrum of neat, dehydrated p(DADMAC) (Fig. 2) has as its most intense peak m/z 58, identified as a highly stable iminium species formed from bond cleavages near the quaternary ammonium site in the repeat unit. The spectrum above m/z 140 is much lower in relative intensity, and is not considered further in this study.

An untreated kaolin secondary ion spectrum is shown in Fig. 3. The most intense peak, m/z 27, is due to Al^+. Lower intensity peaks are observed for Si^+ (m/z 28, 29, and 30), K^+ (m/z 39 and 41), and Na^+ (m/z 23). The peak cluster at m/z 43 to 45 is probably due to AlO^+ (m/z 43), SiO^+ and $AlOH^+$ (m/z 44), and $SiOH^+$ (m/z 45), from the kaolin structure. Surface organic contamination is evidenced by a weak signal at m/z 15 due to CH_3^+.

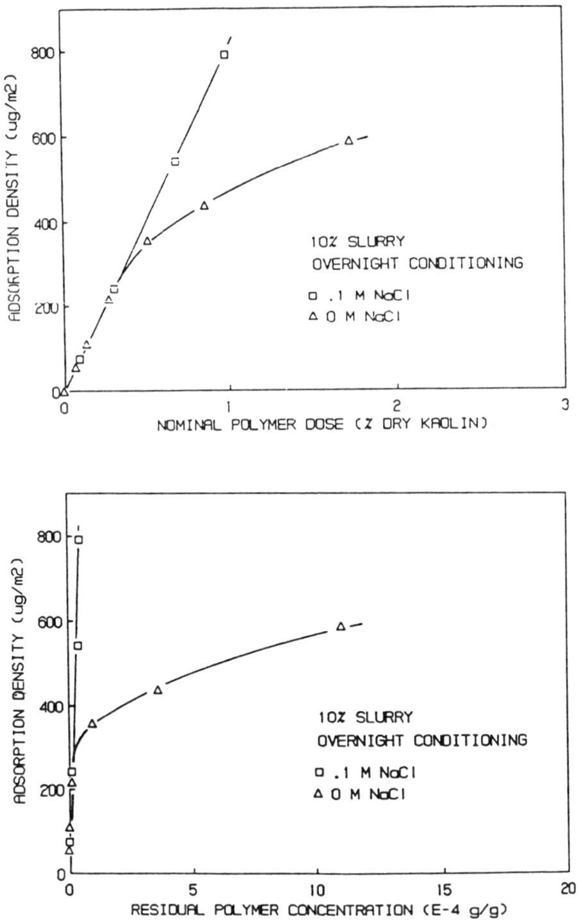

Fig. 1 Effect of strong electrolyte on polycation adsorption.

SIMS analysis of the polycation flocculated kaolin surface provides insight to its adsorption configuration. If p(DADMAC) effectively covers the kaolin surface after polymer adsorption, the atom beam interacts predominantly with the polycation on the surface, and the SIMS spectrum is initially dominated by p(DADMAC) secondary ions. As sputtering proceeds, the polycation is sputter desorbed from the kaolin, exposing the inorganic surface to the atom beam, and the secondary ion mass spectrum becomes increasingly characteristic of the untreated kaolin. Incomplete surface coverage will result in initial simultaneous polycation and kaolin secondary ion emission, with sputter desorption reducing the polycation surface concentration as the process proceeds.

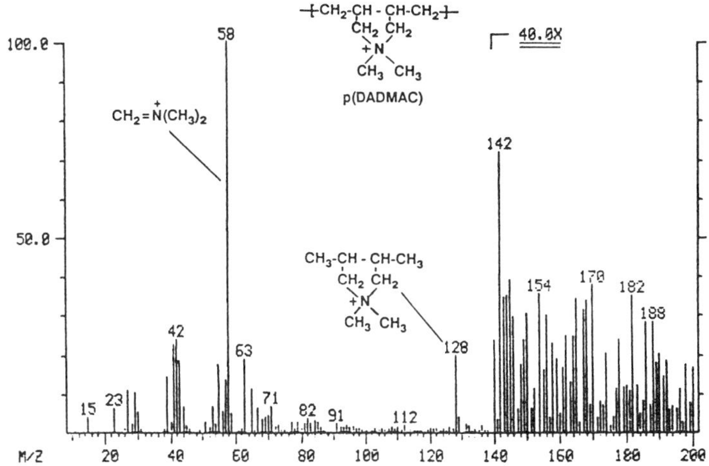

Fig. 2 Positive ion SIMS spectrum of dried p(DADMAC) film.

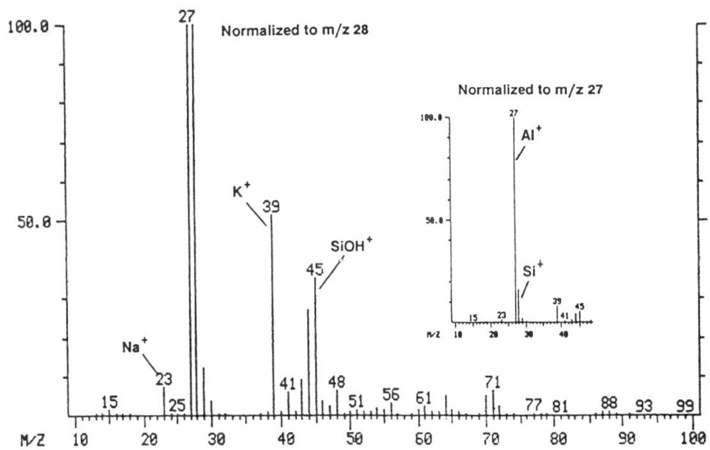

Fig. 3 Positive ion SIMS spectrum of untreated kaolin.

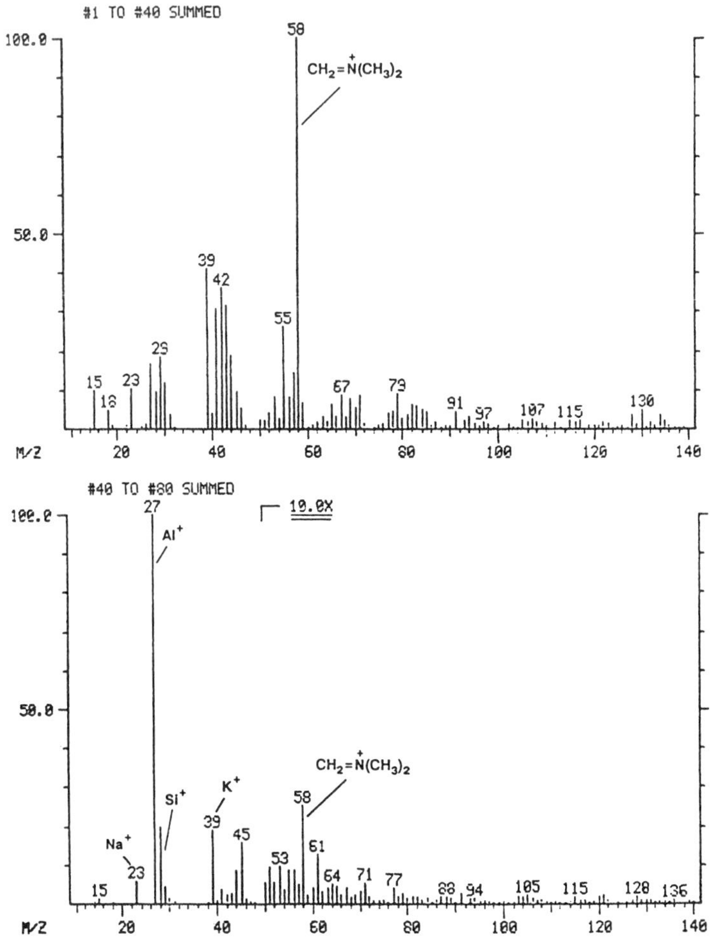

Fig. 4 Positive ion SIMS spectrum of 0.3% p(DADMAC) treated kaolin; adsorbed from deionized water.

Sputter-induced transformation from an organic to inorganic surface is shown for the 0.3% p(DADMAC) treated kaolin where the polycation was adsorbed from deionized water (Fig. 4). Kaolin surface coverage by the polycation lowers primary beam sputtering of the inorganic phase significantly. A mass spectrum generated by summing the first forty spectral scans shows an intense peak at m/z 58 from p(DADMAC) and clusters at m/z 29 and 42, observed also in the spectrum of neat p(DADMAC). Continued sputtering reduces the polycation surface concentration, and a spectrum characteristic of the inorganic surface is obtained as shown in Fig. 4 from summing the subsequent forty mass spectral scans. There, Al^+, m/z 27, dominates the spectrum with a low intensity signal from p(DADMAC) still evident.

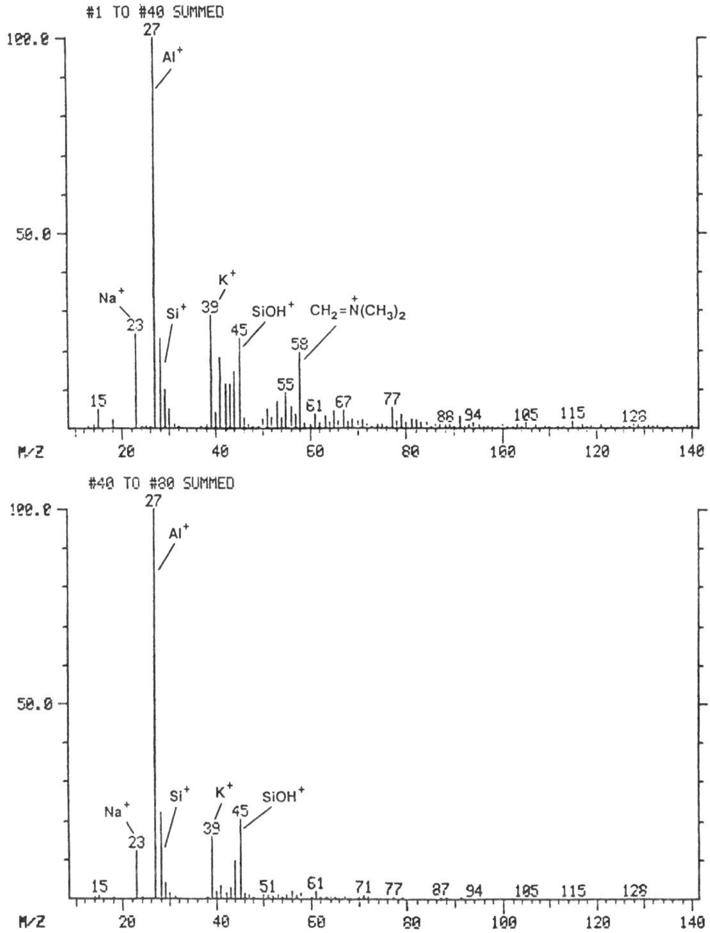

Fig. 5 Positive ion SIMS spectrum of 0.3% p(DADMAC) treated kaolin; adsorbed from 0.1 M NaCl.

Both p(DADMAC) and kaolin ion signatures are observed from the initial SIMS spectra of the 0.3% p(DADMAC), 0.1 M NaCl treated kaolin sample (Fig. 5), obtained under experimental conditions identical to those for the system described above. Kaolin secondary ion species are the most intense spectral peaks and a lower intensity signal from the polycation at m/z 58 is observed. This can only occur if the polymer does not completely cover the kaolin surface; the primary beam thus interacts with the polymer and the kaolin surface simultaneously.

Temporal behaviors of diagnostic peaks in the secondary ion spectra are shown in Fig. 6 and 7, so-called sputter profile plots. Displayed are the detected signals for Na^+ (m/z 23), Al^+ (m/z 27), the p(DADMAC) fragment (m/z 58), and the total secondary ion current. For the high ionic strength adsorbed system (Fig. 6), both the kaolin substrate Al^+ and Na^+ peaks are observed when the atom beam commences surface sputtering. Both gradually increase in intensity as sputtering proceeds. Very different behavior is observed for the p(DADMAC) fragment. Its intensity increases, reaches a maximum, and then decreases. This temporal profile may be due to shadowing and subsequent exposure of adjacent polymer adsorption sites to the atom beam as sputter desorption reduces the polycation surface concentration, or it may result from beam-induced polycation rearrangement on the kaolin surface. The simultaneous appearance of the p(DADMAC) and kaolin secondary ion signals again indicates the polymer does not completely cover the inorganic substrate. Sputter desorption eventually decreases the surface polycation concentration, exposing more of the kaolin matrix.

Very different temporal profiles are observed for the deionized water adsorbed system (Fig. 7). Aluminum matrix signal is delayed by twenty scans relative to the p(DADMAC) signal, indicating the polymer effectively prevents the atom beam from sputter desorbing the kaolin matrix surface. Plateaus are preceded by sharp Al^+ intensity increases, complemented by decreases in p(DADMAC) fragment intensity. This discontinuous behavior is unexpected, and may be due to atom beam-sample interactions. The atom beam may fragment the polycation and induce sudden polymer rearrangement on the kaolin surface.

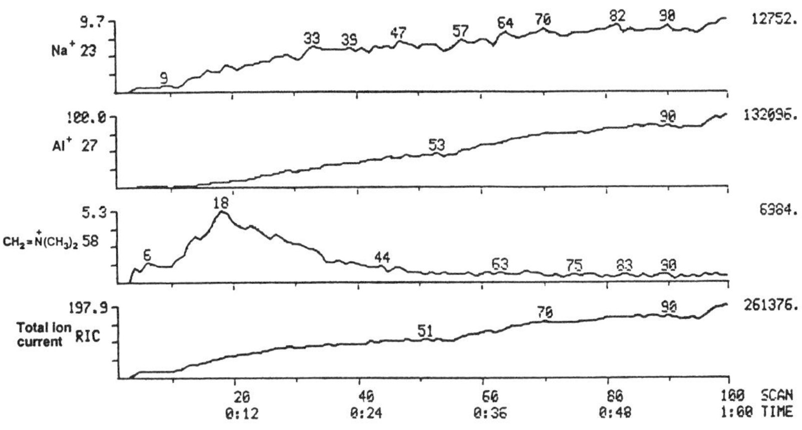

Fig. 6 Temporal behavior of secondary ions from 0.3% p(DADMAC) treated kaolin; adsorbed from 0.1 M NaCl.

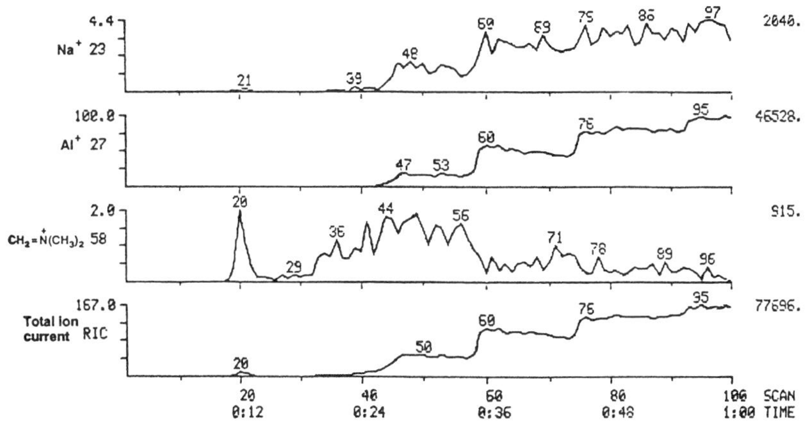

Fig. 7 Temporal behavior of secondary ions from 0.3% p(DADMAC) treated kaolin; adsorbed from deionized water.

CONCLUSION

SIMS data for the dewatered polycation - kaolin system are consistent with polycation adsorption models suggested by aqueous suspension data. SIMS data show the polymer occupies more surface area per molecule when adsorbed on the kaolin surface from low ionic strength solution than from 0.1 M NaCl. Consequently, for equal adsorption densities, surface coverage is higher, and kaolin substrate secondary ion emission occurs only when the polycation has been partially desorbed from the kaolin surface. Matrix and polycation sputtering occur simultaneously when the polymer is adsorbed from high ionic strength solution where the polymer is coiled and occupies less surface area per molecule.

REFERENCES

1. Hesselink, F.Th. (1983), "Adsorption of Polyelectrolytes from Dilute solution", in *Adsorption from Solution at the Solid/Liquid Interface*, G.D. Parfitt and C.H. Rochester, eds., Academic Press, NY.

2. Pratt, R.J., Slepetys, R.A., Nemeh, S., and Willis, M.J., US Pat. 4,738,726 (1988).

3. Johns, R.E., Berube, R.R., and Slepetys, R.A. (1988), "Chemically structured Kaolin - A New Coating Pigment", Fourth International Seminar for Paper-Making Technology, Seoul, S. Korea.

4. Slepetys, R.A. and Morgan, L.J., "Optics and Surface Chemistry of a Chemically Structured Coating Pigment", submitted to TAPPI, 1989.

5. Ueda, T. and Harada, S., "Adsorption of Cationic Polysulfone on Bentonite", J. Appl. Polym Sci. **12** (1968) 2395-2401.

6. Kokufuta, E. and Takahashi, K., Poly(diallyldimethylammonium chloride) on Colloid Silica from Water and Salt Solution", Macromolecules **19** (1986) 351-354.

7. Morgan, L.J. and Levine, S.M., "Simulating the Observed Behavior of Polyelectrolyte Adsorption on Kaolin", submitted to J. Colloid Interface Sci.

8. Fox, K.K., Robb, I.D., and Smith, R., "Electron Paramagnetic Resonance Study of the Conformation of Macromolecules Adsorbed at the Solid/Liquid Interface", J. Chem. Soc., Faraday Trans. 1 **70** (1975) 1186-1190.

9. Williams, P.A., Harrop, R., and Robb, I.D., "Adsorption of an Amphoteric Polymer on Barium Sulphate and its Effect on Colloid Stability", ibid. **80** (1984) 403-411.

10. Williams P.A., Harrop, R., and Robb, I.D., "Adsorption of an Amphoteric Polymer on Silica and its Effect on Dispersion Stability", J. Colloid Interface Sci. **102** (1984) 548-556.

11. Mitchell, D.F and Graham, M.J., "Quantitative SIMS Analysis of Hydroxyl Ion content in Thin Oxide Films", J. Electrochem. Soc.: Electrochem. Sci. Tech. **133** (1986) 936-938.

12. van Ooij, W.J. and Brinkhuis, R.H.G., "Interpretation of the Fragmentation Patterns in Static SIMS Analysis of Polymers. Part I. Simple Aliphatic Hydrocarbons", Surf. Interface Anal. **11** (1987) 430-440.

13. Briggs, D. and Ratner, B.d., "A Semi-Quantitative SIMS Analysis of the Surfaces of Random Ethyl Methacrylate: Hydroxyethyl Methacrylate Copolymer Films" Polym. Commun. **29** (1986) 6-8.

14. Bentz, B.L. and Gale, P.J., "Application of a SIMS/TSQ Spectrometer for Organic Surface Analysis in the Electronics Industry", presented at the 36th ASMS Conference on Mass Spectrometry and Allied Topics, June 5-10, 1988, San Francisco, CA, pp 347-348.

THE BEHAVIOR OF POLYELECTROLYTE ADSORPTION ON KAOLIN

L.J. Morgan, S.M. Levine and J.S. Thompson
Engelhard Corporation, Edison, NJ

ABSTRACT

Polyelectrolyte structuring of fine kaolin particles leads to enhanced optical performance in paper coating applications. In this investigation, the interaction between such a polymer and kaolin has been probed by applying traditional surface chemical concepts supplemented by computer modeling to simulate adsorption behavior. We measured polymer adsorption isotherms and monitored surface chemistry changes during polymer adsorption. In conjunction with the surface chemistry studies, a variety of molecular modeling techniques were used to investigate charge distribution, polymer flexibility and adsorption behavior. By combining these results, we have a more detailed picture of the surface interaction.

INTRODUCTION

Understanding polyelectrolyte adsorption behavior is important for a number of industrial processes. Applications range from waste water treatment and oil recovery to paper-making (1). We have taken advantage of the interaction between a quaternary ammonium polymer and kaolinite to structure fine particles into an enhanced optical performance pigment used in paper coating applications (2). This study reports part of a larger investigation into fundamental aspects of the polymer-kaolin interaction by comparing adsorption and surface chemical titration results with computer model predictions (3).

General characteristics of polycation adsorption have been described by Hesselink (1) and Theng (4). Various studies using polyammonium-type materials have examined interactions with silica (5), fibers (6) and clays (7). One concern in these investigations is the molecular conformation at the adsorption interface. In general, results suggest solution conformation is preserved. Polymer adsorbed from dilute or low salinity conditions tends to be much flatter than from concentrated solutions in which case loops form. ESR of spin-labeled co-polymer (5b), electrokinetics (6), and ion exchange capacities (8) have been used to examine adsorbed polymer conformation.

Theoretical studies have attempted to predict the conformation by deriving the partition function for free and adsorbed molecules, taking into account segment-segment and segment-solvent interactions (1,9). Others have correlated statistical predictions of segment density distribution with results from surface sensitive techniques such as small angle neutron scattering (10), or used mass action relationships (11), but it is difficult to compare these results with experiments.

In this study, static and dynamic energy minimization techniques are used to predict the energetics of the polymer molecule and to simulate adsorption on a model kaolin surface. The simulation predicts adsorbed polymer conformation. Surface chemistry changes during polymer adsorption are monitored using state of the art instrumentation and are compared to measured adsorption. We find a correlation between the simulation (computer prediction) on a microscopic scale and measured macroscopic behavior. From the computer model, we can visualize the polymer conformation on the kaolin surface.

Advances in Fine Particles Processing
John Hanna and Yosry A. Attia, Editors

EXPERIMENTAL

We used a sodium silicate dispersed kaolin from Washington County, Georgia whose surface area was 12 m^2/g by BET nitrogen analysis. Calgon Corp. supplied the polyelectrolyte, a poly(diallyldimethyl ammonium chloride), "DADMAC". NMR results have shown the nitrogen to be part of a five-membered ring (12). The molecular weights of the two polymers shown below were determined by American Polymer Standards Corp. using a Viscotek Differential Viscometer[1].

Table I: Polymer Molecular Weights

Nominal	Mn	Mw	Mw/Mn
4000	470	7,500	16.0
100,000	14,000	68,900	4.6

Flocculation was monitored in 1% slurries, otherwise the suspension was too concentrated to be able to follow the descending interface. Vigorous agitation was found to break up the flocs at this solids level, so after the polymer was added, mixing with an impeller at 500 rpm, the samples sat overnight without additional agitation. Following gentle resuspension, the slurry was transferred to flat bottomed 50 ml graduated cylinders to monitor settling. Transmittance was determined using a Varian DMS UV-Visible Spectrophotometer at 350 nm wavelength. Data is reported after 5 hours of sedimentation.

Adsorption was determined by the depletion technique. A known amount of polymer is added to 10% kaolin slurry which is stirred at 500 rpm with a one inch propeller. The residual concentration is determined after overnight conditioning on a wrist action shaker. The polymer depleted from solution is assumed to be adsorbed on the kaolin. Polymer analysis depends on pyrolysis and nitrogen detection using an Antek Chemiluminescent Analyzer with a detection limit of 3 ppm polymer.

Titrations were done using the Pen Kem System 7000, Acoustophoretic Analyzer. This instrument has been described elsewhere (13). In summary, it is an automatic titrator which simultaneously monitors particle mobility, pH, conductivity, and temperature. The mobility measurement is based on acoustic phenomena which permits the use of concentrated slurries.

COMPUTATIONAL PROCEDURE

Three types of calculations were performed in this study. Molecular Orbital (MO) calculations were used to determine orbital populations and atomic charges. Extended Huckel (EHT) type calculations (14) were used through the Chem-X molecular modeling software (15). Conformational analyses involved molecular mechanics energy calculations based on a description of atom-atom interactions and geometric terms (bond angles, etc.) using Alinger's MM2 force field (16). Molecular dynamics calculations were also performed to investigate the polymer's dynamic behavior. This technique considers simplified pairwise interactions (17) but allows more extensive flexibility of the molecules as well as "dynamic" motion.

The model monomeric unit, i.e. one repeating unit of the polymer terminated with methyl groups, was derived by minimizing van der Waals (VDW)

---

[1] Due to the numbers of calculations involved, the computer simulation is limited to smaller molecules. To be able to check its predictions, the smaller molecular weight analog was included.

energy using molecular mechanics. To determine the molecule's electrostatic field, the positively charged monomer (i.e. no chloride counterion) was subjected to Extended Huckel calculations. This produced net charges associated with each atom and the relative populations of atomic orbitals. The charges calculated for the monomer were used for the larger molecules as well. Although this should not drastically affect the overall conclusions of the study, absolute magnitudes may be shifted slightly.

Two connected monomeric units formed the model dimer. This structure was minimized for VDW interactions to a local minimum, and, unless otherwise stated, was the starting structure for conformational analyses (Figure 1).

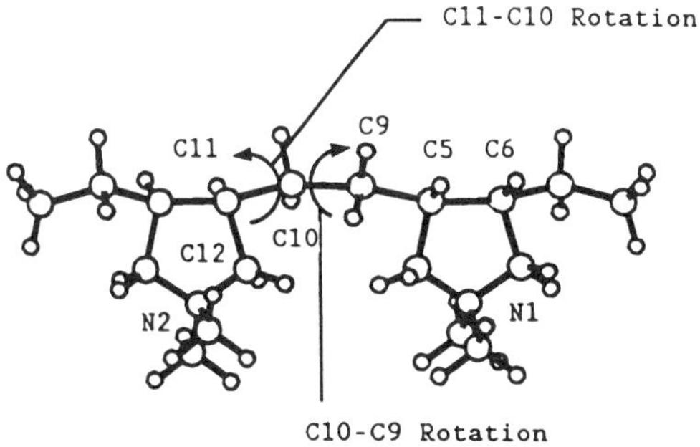

Figure 1. Ball and stick model of dimer molecule indicating rotation axes.

Conformational analyses (using molecular mechanics energy calculations) were performed by systematically varying bond angles and calculating the resulting energy of the system. In most instances, the structures resulting from each rotation were subjected to an additional molecular mechanics optimization.

Molecular dynamics calculations used 200 successive iterations at a simulated time interval of 10 picoseconds ($1x10^{-11}$ sec). The starting conformation was the same as that used for the conformational analysis. Only rotations about bonds were allowed and no subsequent relaxations of the generated structures were made. These simplifications were necessary to minimize calculation time and are not expected to alter the overall trends of the calculations or the conclusions.

Larger model polymer chains were constructed by successively joining model monomeric units. Conformations were minimized energetically prior to molecular dynamics calculations.

An idealized kaolin surface was built from X-ray crystal data to mimic typical alternating silica tetrahedral and alumina octahedral sheets. Two 60x100 Å sheets provided a model surface with a pseudo-random distribution of negative sites on the surface.

Simulated adsorption used molecular dynamics calculations with a 10 picosecond timestep at a temperature of 500 K. The internal geometry of the polymer's five-membered rings was fixed. All backbone bonds were allowed to vary and the surface atoms were treated as harmonic oscillators. This is a reasonable assumption at these time and temperature scales. Polymer-kaolin interactions were calculated via a combination of modified Born-Mayer-Huggins, Lennard-Jones (17a) and Rahman-Stillinger-Lemberg (17b) potential functions.

EXPERIMENTAL RESULTS

Polymer Adsorption

Polyelectrolyte flocculation results are shown in Figure 2 using the higher molecular weight polymer. The useful dosage range to achieve fine particle structuring is from 0.1% to approximately 0.4% on a dry kaolin basis.

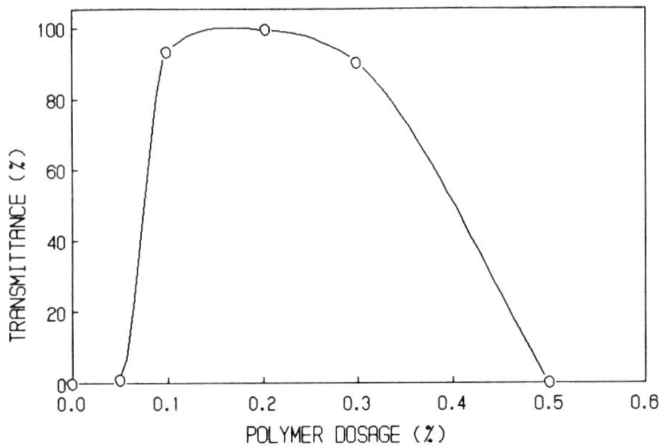

Figure 2. Poly(DADMAC) flocculation of kaolin.

Polymer adsorption density is given in Figure 3a. The isotherm has the shape of a classical high affinity isotherm. The initial vertical rise indicates strong adsorption and is followed by a levelling towards a plateau value.

Isotherm results may also be plotted against dosage, Figure 3b. Comparing the two plots, the linear rise in adsorption density at low dosage corresponds to the vertical rise of the isotherm and reflects the fact that, for doses up to at least 0.3%, all of the available polymer is sorbed by the kaolin. Beyond that dosage level, an equilibrium is established with residual polymer left in solution. Further adsorption is non-linear with dosage, less adsorbs per incremental addition until finally adsorption reaches a plateau.

Adsorption trends for the lower molecular weight material are similar but the plateau adsorption is lower. The effect of chain length is clearly indicated. The longer the polymer, the more attached per adsorption event. The effect of molecular weight on the adsorption plateau level is shown in Table II along with the calculated area per monomeric unit. The fact that the larger molecule seems to occupy less area suggests that it adsorbs forming loops at the solid/liquid interface.

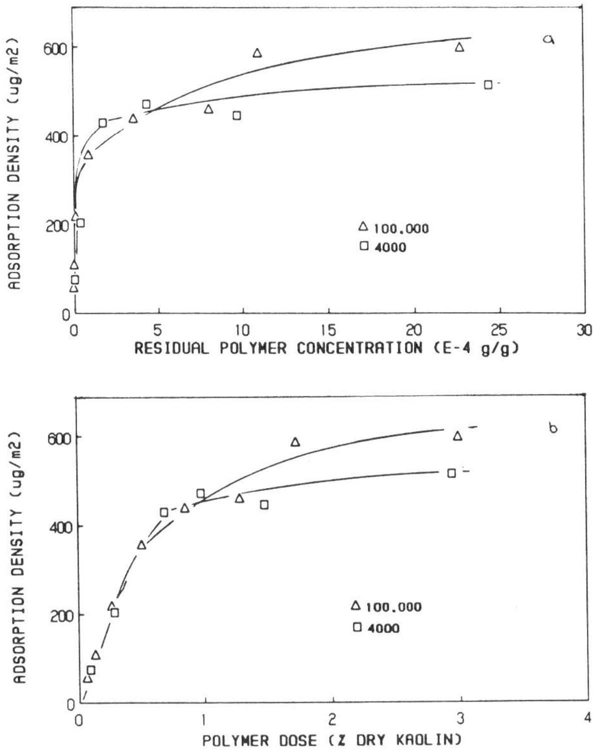

Figure 3. Poly(DADMAC) adsorption on kaolin;
a) isotherm, b) dosage basis.

Table II: Adsorption site area depends on polymer MW.

Molecular Weight (amu)	Adsorption Plateau (g/m^2)	Adsorption Area (A^2/monomer)
4,000	515	41
100,000	650	32

Comparing polymer adsorption using the higher molecular weight material with its with titration data gives insight to the kaolin surface. The titrator monitors four properties of the kaolin suspension: relative acoustophoretic mobility (RAM, related to the zeta potential), pH, conductivity and temperature. Titration results are shown in Figure 4 along with adsorption results. The cationic polymer is able to balance the negative surface charge of kaolin, attaining an isoelectric point (iep) at a given polymer dosage. As adsorption continues past this point, charge reversal is effected, and the kaolin becomes positively charged.

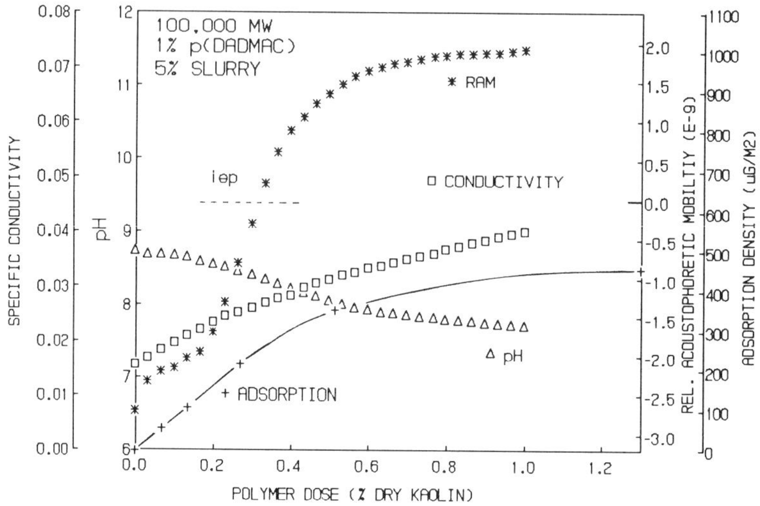

Figure 4. Acoustophoretic titration of kaolin with 1% poly(DADMAC).

We have identified several adsorption regimes and propose underlying phenomena. Initially, adsorption reduces particle negative mobility only slightly by replacing counter ions, e.g. sodium, in the particle double layer. With continued adsorption, surface hydrogen ions are displaced by adsorbing polymer, pH decreases and surface charge is neutralized or compensated. Polymer continues adsorbing above the dosage required to attain an isoelectric point (iep) and the particle becomes positively charged. The break in conductivity slope at this point is thought to reflect polymer looping. There is excess polymer in the interfacial region, not interacting with the kaolin, such that some chloride ions remain associated with the polymer and thus contribute less to the conductivity increase. As the particle becomes more positively charged, incremental adsorption decreases with dosage, tapering off towards a plateau level. The mobility and pH also reach plateau values.

Computer Simulation

Functional Group Charge Density. From the molecular orbital calculations, the overall +1 charge (due to the removal of the Cl^-) is not uniquely associated with the nitrogen center. The positive charge is delocalized with approximately 20% associated with each of the four carbons bonded to the nitrogen, 12% on the nitrogen itself, and the remainder associated with hydrogen atoms. These charges are the source of electrostatic attraction between the polymer and a negatively charged clay surface. With the charge delocalization, adsorption does not necessarily occur head down, backbone up.

Dimer Flexibility. Our conformational studies involved systematic searches through 360 degree rotations about the three backbone carbon-carbon bonds. The starting structure is indicated in Figure 5a. Five other calculated dimer structures are shown in Figure 5 with their relative potential energies. The structure labeled [f] has the lowest potential energy.

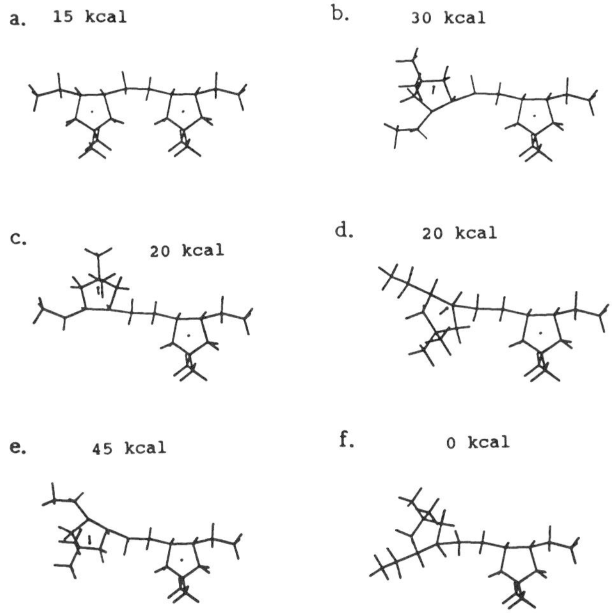

Figure 5. Stick drawings of the dimer molecule at various conformations generated by rotations about the back bone carbons and twisting of the rings. Relative potential energy of each conformation shows [f], with 180° between ring groups, to have the lowest energy.

The N1-N2 distance has been plotted vs. potential energy in Figure 6. The most stable conformation appears to have a distance of approximately 8 A. Calculating the torsional angle between the two 5-membered rings results in the minima at roughly 180 degrees, with considerable barriers near 0 (or ± 360) degrees.

Molecular dynamics calculations on the same molecule correlate well with the molecular mechanics energetics model. The agreement of minimized conformation calculated by the two techniques validates the use of the dynamics calculations which are necessary to calculate polymer adsorption behavior.

Expanding on these results for small molecules, a model polymer of approximately 3000 amu was built and energetically minimized. Adding some compensating counterions (N:Cl=3), the polymer coils. The relative positions of the rings create a spiral. Polymer looping puts a statistical number of ammonium functional groups in favorable and unfavorable positions for surface adsorption.

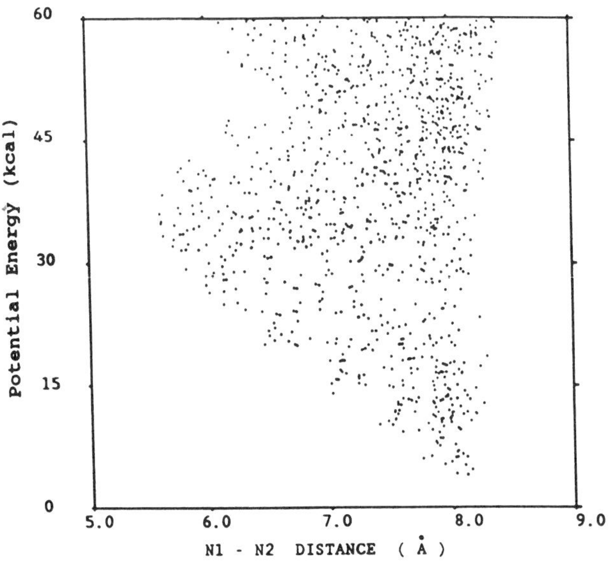

Figure 6. Potential energy vs. N1-N2 distance for dimer molecules subjected to systematic rotations about all adjoining bonds. The most stable conformation is with the N1-N2 distance at 8.13 A.

Modeled Adsorption - Molecular Dynamics

Monomer. The computer simulated adsorption of the DADMAC monomeric unit was done by introducing monomers in the vicinity of the modeled kaolin surface (25 A from surface). A snapshot of the surface at maximum adsorption is shown in Figure 7. Note that even for a small molecule, not all adsorption sites are accessible. The plateau level corresponds to 70 A^2 per monomeric unit which is in agreement with the measured value for the monomer (3).

Polymer. Adsorption was simulated using low molecular weight polymer (approximately 3350 amu) and higher molecular weight (approx 15,000 amu). The simulation used the same methodology as for the monomer. Molecules were placed sequentially in the vicinity of the surface up to a dosage level of 0.8% (wt. basis of dry kaolin of 12 m^2/g surface area).

Intermolecular repulsions dictated much of the flexibility when DADMAC contacted (or was nearby) the charged surface. The actual surface conformation of the molecule depended upon the strength and density of negative surface sites (i.e. simulated pH variations). At high density of adsorption sites, (i.e. high pH), the polymer adsorbs in a looped orientation as the probability of aligning an adsorption site with an available amine is increased. Much of the polymer is above the surface in loops, leaving unsatisfied adsorption sites on the surface itself. At lower polymer dosage, the electrostatic repulsion between an adsorbed molecule and another approaching the surface is sufficiently small to access sites buried beneath loops, Figure 8.

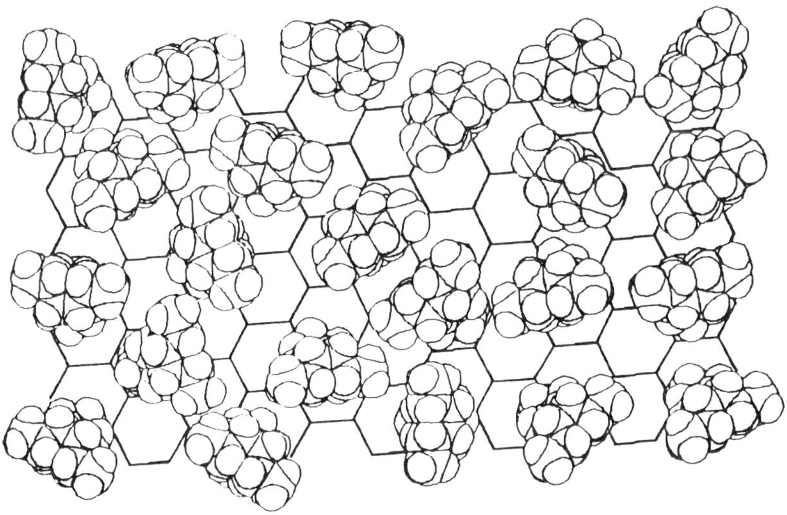

Figure 7. Computer generated kaolin surface at monomer saturation adsorption.

DISCUSSION

By measuring polymer adsorption and comparing it to surface chemical titration and flocculation results, we find that the polymer adsorbs electrostatically on the kaolin surface, probably via an ion exchange mechanism, and that low surface coverage is required to flocculate the kaolin. Taking the plateau adsorption value as monolayer coverage, flocculation occurs from about 10% to 34% of the possible surface coverage. The latter concurs with zero mobility. While further adsorption takes place, it leads to resuspension.

It should be noted that plateau adsorption is not in the true sense monolayer coverage. The polymer forms loops at the solid/liquid interface such that while no further adsorption can take place due to both steric and electrostatic effects, not all surface sites are occupied.

Flocculation begins while the particles are still negatively charged, which leads to the conclusion that flocculation occurs by polymeric bridging among particles. The extended polyelectrolyte molecule captures several particles, structuring them into an optically efficient pigment.

Evidence for polymer loops at the solid/liquid interface is found by comparing the area occupied per adsorbed monomeric unit at the plateau level for the two polymers. Table III compares areas/adsorbed monomeric unit found experimentally with predicted values. For the lower molecular weight material, the agreement is remarkably good, confirming the computer model prediction, 41 A^2 and 40 A^2. The experimental value for the larger polymer is only 31 A^2 while the predicted value for 15,000 amu material is 55-60 A^2.

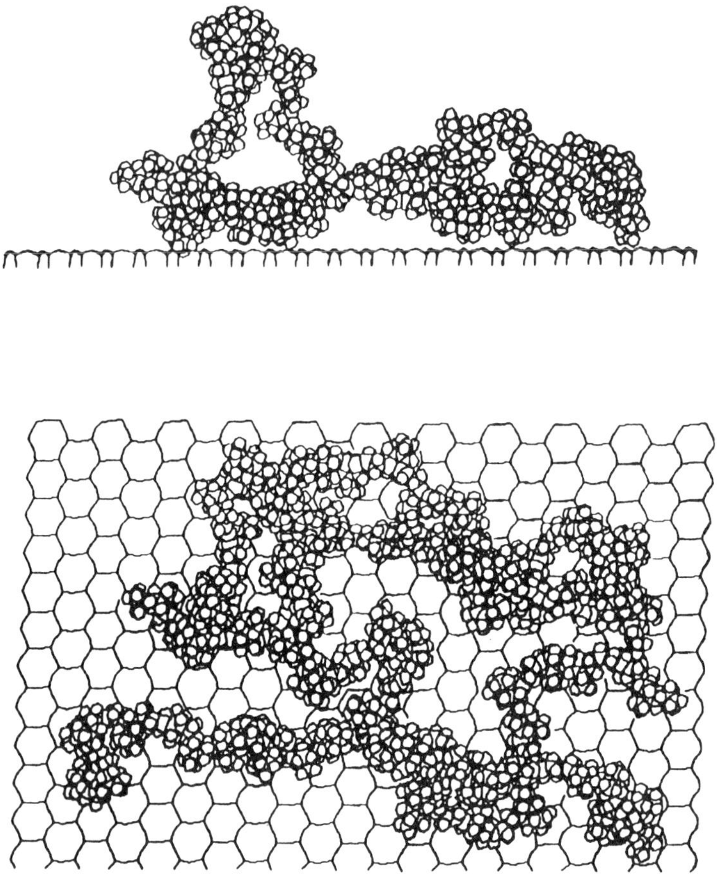

Figure 8. Computer generated top and side view of polymer adsorbed onto kaolin surface at basic pH. Some looping can be seen while sites nearby the loops can be accessed by an additional molecule.

Table III: Comparison of Experimental and Predicted Adsorption Areas

MW	Experiment (A^2/monomer)	Simulation (A^2/monomer)
4,000	41	40
15,000	-	55-60
100,000	32	..

*Calculated for approximately 15,000 amu.

The higher value predicted by the model confirms the presence of looping for the higher molecular weight polymer. The predicted area only includes polymer within 15 A of the surface and therefore does not include looped material above this region. On the other hand, all of the adsorbed polymer is included in the experimental value where the area decreases significantly.

Good correlation for the low molecular weight polymer justifies use of the simulation to semi-quantitatively estimate the fraction of loops, the size of loops and the fraction of surface covered at higher molecular weight. From the 15,000 amu adsorption simulation, we estimate that 30-60% of the adsorbed material is in loops (at full coverage) leaving 10-25% of the surface uncovered and inaccessible for further adsorption.

References

1. Hesselink, F. Th. (1983), in Adsorption from Solution at the Solid/Liquid Interface, G.D. Parfitt and C.H. Rochester, eds., Academic Press, NY.

2. (a) Pratt, Richard J., Slepetys, Richard A., Nemeh, Saad and Willis, Mitchell J., US Pat. 4,738,726 (1988); (b) Johns, R.E., Berube, R.R. and Slepetys, R.A. (1988), Fourth International Seminar for Paper-Making Technology, Seoul, S. Korea; (c) Slepetys, R.A. and Morgan, L.J., submitted to TAPPI, 1989.

3. Morgan, Leslie J., Levine, Steven M. and Thompson, Jacqueline S., submitted to the JCIS.

4. Theng, B.K.G. (1979), Formation and Properties of Clay-Polymer Complexes, Elsevier Pub. Co., NY.

5. (a) Kokufuta, Etsuo and Takahashi, Katsufumi (1986), Macromolecules, **19**, 351-354; (b) Williams, Peter A., Harrop, Raymond and Robb, I.D. (1984), JCIS, **10**, 2, 548-556.

6. (a) Onabe, Fumihiko (1984) Mokuzai Gakkaishi, **30**, 7, 553-559; (b) Onabe, Fumihiko, J.Appl'd Polymer Sci; **22**, 3495-3510, 1978; (c) **23**, 2909-2922, 1979; (d) **23**, 2999-3016, 1979.

7. (a) Gill, R.I.S. and Herrington, T.M. (1986), Colloids and Surfaces, 22, 51-76; (b) Gill, R.I.S. and Herrington, T.M. (1987), Colloids and Surfaces, **28**, 41-52; (c) Yorke, Monica A. (1973), Polymer Sci. and Tech., **2**, 93-104; (d) Kim, H.S, Lamarche, C. and Verdier, A. (1983), Colloid & Polymer Sci., **261**, 64-69 (French).

8. Ueda, Toshio and Harada, Susumu (1968), J. of Applied Polymer Science, **12**, 2395-2401.

9. (a) Silberberg, A. (1968), J. Chem. Phys. **48**(7), 2835-2851; (b) Scheutjens, J.M.H.M. and Fleer, G.J., J. Phys. Chem., (1979) **83**(12), 1619-1635; (c) (1980), **84**(2), 178-190; (d) Hesselink, F.Th. (1977), JCIS, **60**, 3, 448-466; (e) Silberberg, A. (1978), in Ions in Macromolecular and Biological Systems, D.H. Everett and B. Vincent, eds., Scientichnica, Bristol.

10. Cosgrove, T., Vincent, B., Crowley, T.L., Cohen-Stuart, M.A. (1984) in Polymer Adsorption and Dispersion Stability, E.D. Goddard and B. Vincent, eds., ACS.

11. Hogg, R. and Mirville, R.J. (1982), "Adsorption of Macromolecules at Solid-Liquid Interfaces," presented at 56th Colloid and Surface Science Symposium, Blacksburg, Va.; Hogg, R. (1984), "Evaluation of a Macroscopic Model for Polymer Adsorption," E.D. Goddard and B. Vincent, eds., ACS, NY.

12. Lancaster, J.E., Baccei, L. and Panzer, H.P. (1976), Polymer Letters Edition, Vol. 14, 549-554.

13. Marlow, B.J., Fairhurst, D. and Pendse, H.P (1988), Langmuir, **4**, 611-626.

14. (a) Lowe, J.P. (1978), Quantum Chemistry, Academic Press; (b) Hoffman, R. Quantum Chemical Program Exchange (QCPE), 1977, **11**, 344.

15. Chem-X, developed and distributed by Chemical Design Ltd, Oxford, England.

16. Burkert, U., Allinger, N.L. (1982) ACS Monograph 177.

17. (a) Woodcock, L.V. Advances in Molten Salt Chemistry, Plenum, N.Y. 1975, Vol.3 p.1; (b) Rahman, A., Phys Rev. Sect. A, 1964, **136**, 405; (b) Stillinger, Rahman A. (1968), J. Chem. Phys. 68(2), 666.

ULTRASONIC GELLING OF CHANNELIZED 2:1 CLAY IN IONIC MEDIA

Jim L. Elrod and Oscar E. Moore
Tennessee Valley Authority, National Fertilizer Development Center,
P.O. Box 1010, Muscle Shoals, Alabama 35660-1010

ABSTRACT

Ultrasonic energy was used to disperse and gel dry suspending clays in ionic fertilizer media. The method used involved bringing a mixture of fluid fertilizer and clay into contact with ultrasonic energy generated by either a piezoelectric or magnetostrictive transducer. The channelized 2:1 clays (attapulgite and sepiolite) were effectively gelled at a frequency of 20,000 Hz and a power density of about 2 watts per milliliter. The ultrasonic energy reduced the crystal agglomerates that make up the clay particles into their individual needlelike crystals. The dispersed clay crystals then formed a latticework capable of suspending solid particulates of up to about 20 mesh (850 micrometers) in size. A comparative study of mechanical shear (provided by a Waring blender) and ultrasonic methods of dispersing and gelling clays to produce ionic suspensions showed that ultrasonically induced gelation was more efficient, resulting in suspensions of equal or superior quality and with less energy consumption.

INTRODUCTION

Suspension fertilizers are fluids which have induced non-Newtonian flow properties. They usually contain suspended solid particulates (crystals) of up to 20 Tyler mesh (about 850 micrometers) in size. The suspended solids usually contain higher nutrient levels than the solution in which they are suspended. These solids also can contain impurities that would be unacceptable in clear solution fertilizers. The advantages of higher concentrations of plant nutrients and use of lower purity (lower cost) raw materials have led to the widespread use of suspension fertilizers.

The non-Newtonian flow properties necessary in a satisfactory suspension are induced by treating the fluid fertilizer with a suspending agent, such as attapulgite gelling clay [1], to achieve a rheological state (gel) in which flow does not occur until a critical shear stress, the yield point, is exceeded. In satisfactory suspensions, the yield point is greater than the gravitational force on the suspended solid particles. In today's suspension industry, mechanical agitators and recirculation pumps with high-shear impellers are used to disperse and gel clay in the production of fertilizer suspensions.

About 90 percent of all suspension fertilizers are made from six principal ingredients. These materials are urea-ammonium nitrate (UAN) solution (32 percent nitrogen), commercial 10-34-0 ammonium polyphosphate (APP) solution made from superphosphoric acid, solid monoammonium

Advances in Fine Particles Processing
John Hanna and Yosry A. Attia, Editors

phosphate (MAP), wet-process phosphoric acid (54 percent P_2O_5), anhydrous ammonia, and potassium chloride. Other common, but less widely used fertilizer ingredients [such as diammonium phosphate (DAP), micronutrient salts (for example, zinc oxide), etc.] also may be included in suspension formulations.

The principal criteria by which a suspension is judged as to its adequacy in preventing settling of fertilizer salt crystals are its measured viscosity and gel strength. Gel strengths of 2 to 10 gram-centimeters are generally considered acceptable for most suspensions. When no solids (other than the dispersed clay) are present in a suspension, as in 31-0-0 grade (2 percent clay) UAN base suspension, the gel strength is proportional to the viscosity. A satisfactory 31-0-0 grade UAN suspension with a viscosity of 250 centipoises would have a gel strength of about 10 gram-centimeters. The presence of fertilizer salt crystals increases the measured viscosity and decreases the correlation between viscosity and gel strength.

To gel attapulgite clay, small clay bundles that are typically smaller than 200 Tyler mesh (about 75 micrometers) must be separated into individual clay particles that are about 1 micrometer in length. A principal characteristic of suspension fertilizer nutrient materials is that they are practically all (except urea) ionic salts. Dry attapulgite clay strongly resists gelation in the presence of all but very dilute concentrations of ionic salts. This resistance to gelation, however, can be overcome by applying sufficient shear energy. The presence of relatively large amounts of solids and/or phosphates in the fluid also greatly assists the action of an agitator or pump in shearing the clay. The solids impinge on the clay particles and cause the particles to be broken down from their agglomerated state. Unless large amounts of undissolved solids and/or phosphates are present in a fluid fertilizer, dry clay normally is not used in making a suspension. The alternative is to first disperse the clay in water. [2] This method also requires the use of a chemical dispersant and high-shear mixing, but the resulting clay-water dispersion will give satisfactorily strong gels when added to ionic fertilizer solutions without assistance from large amounts of undissolved solids and/or phosphates.

Because of the problems with producing suspension fertilizers using conventional methods, tests were made in which ultrasonic energy was used in dispersing and breaking down attapulgite and sepiolite clay particles to form gels in ionic fertilizer media. Treatment of clay-water dispersions with ultrasonic energy has been studied in drilling mud applications [3], but not in suspension fertilizer applications.

Procedure

Tests were made to compare ultrasonic gelling with mechanical gelling provided by a Waring blender. Test samples were prepared in 250-gram batches containing 242.5 grams of fluid fertilizer and 7.5 grams of attapulgite clay (Min-U-Gel 200).

Samples for the ultrasonic gelling tests were prepared by sifting the dry clay into the fluid fertilizer over a 5-second period while the mixture was stirred with a three-bladed, 2-inch-diameter, propeller turning at a tip speed of 7 feet per second. Then the fertilizer-clay mixture was sheared with a piezoelectric horn-type transducer driven by a 400-watt generator with a variable power control. The transducer operated at a fixed frequency of 20 kilohertz. The sample volume of 187 milliliters had a diameter of 5.4 centimeters and a depth of 8.9 centimeters.

The end of the cylindrical transducer horn (2.8 cm^2) was submerged to depth of 0.6 centimeter during the tests.

Samples for mechanical gelling tests were prepared by sifting the dry clay into the fluid fertilizer over a 5-second period and then operating the Waring blender at a tip speed of 40 feet per second.

Suspensions were evaluated on the basis of apparent viscosity (centipoises), gel strength (gram-centimeters), energy requirements (kilowatt-hours per ton), and production time. Apparent viscosity was measured with a Brookfield viscometer, Model RVT digital-type, operating at 100 revolutions per minute. Gel strength was measured with a gelometer developed by TVA. A wattmeter was used to measure power requirements.

Results

The tests were set up for quantitative comparison of ultrasonic and mechanical shear gelling of dry suspending clays in ionic fluid fertilizers. Typical results with various fluid fertilizers are given in Table I. The test results with the UAN and APP fluids, which contained no solids, show that ultrasonic gelation is superior to mechanical gelation in many respects, including less power consumption and reduced time in producing suspensions of comparable viscosity. This is especially true in the UAN (32 percent nitrogen) solution in which no solids are present. Ultrasonic gelation of the clay in the UAN solution required less than 2 percent of the production time while consuming less than five percent of the power required by mechanical shear.

The 12-36-0 grade fluid contained phosphate ions and solids (diammonium phosphate crystals). In tests with this fluid, the mechanical shear gelation required less power than the ultrasonic gelation because the solids assisted with the mechanical gelation of the clay. The ultrasonic gelation method, however, still required less time.

In the tests with the 18-0-18, the solids present were potassium chloride (no phosphate ions). In tests with this fluid, the ultrasonic method again required considerably less energy and time. Thus the results with the 3-10-30 and 18-0-18 grade fluids indicate that ultrasonic gelation is independent of solids and phosphate ions and that both phosphate ions and solids must be present for mechanical gelation to be as effective as ultrasonic gelation.

The ultrasonic gelling method was found to be effective with sepiolite as well as with attapulgite clay. However, sodium bentonite (sodium montmorillonite), which is also used as a suspending agent in the fertilizer industry, did not form a gel when exposed to ultrasonic energy. This result was expected because sodium bentonite gels by a swelling process when mixed with water rather than by breaking apart of the clay particles as do attapulgite and sepiolite.

Effect of Power Density

The effect of power density (watts per milliliter) on the time required to obtain a specific level of gelling of the clay also was investigated. In these tests, the piezoelectric apparatus was used to gel attapulgite clay in 32-0-0 UAN solution containing 3 percent by weight of clay and in 10-34-0 APP solution containing 2 percent by weight of clay. In each test the viscosity was measured periodically until a reading of 250 centipoises was obtained. Gelation time was measured as the total time of exposure to ultrasonic energy. The power measurements were made on the 120-volt, 60 H supply to the converter and thus include losses in the converter and in the transducer. The results for tests with the UAN solution are shown in Figure 1. At less than about 1 watt per milliliter, retention times up to 12 minutes were required. Between 1 and 2 watts per milliliter, the required time decreased rapidly to about 1 minute. Further increasing the power density to 4 watts per milliliter only reduced the time needed to obtain 250 centipoises to about 30 seconds. The results for tests with the 10-34-0 APP solution (Figure 2) were very similar.

TABLE I. Stabilization of fluid fertilizers with dry attapulgite clay[a] by ultrasonic versus mechanical shear

Gelling method	Gelling time required, minutes	Power, watts	Total energy, kWh/ton of product	Gel strength, g-cm	Viscosity, centipoises
		UAN 32 Solidless Liquid			
Sonic	1.25	313	23.7	11.9	296
Mechanical	70.00	125	529.3	11.4	312
		10-34-0 Solidless, Phosphate Liquid			
Sonic	0.33	420	8.4	6.2	300
Mechanical	1.75	125	13.2	6.9	314
		12-36-0 Fine Solids (<20%), Phosphate Fluid			
Sonic	1.00	287	17.4	5.2	334
Mechanical	1.75	130	13.8	5.3	324
		18-0-18 Coarse KCl Solids (~25%), Fluid			
Sonic	1.5	430	39	-	530
Mechanical	20.0	130	157	-	442
		3-10-30 Coarse KCl Solids (~45%), Phosphate Fluid			
Sonic	0.25	430	6.5	-	480
Mechanical	1.00	125	7.6	-	460

a) Min-U-Gel 200, "as is" dry clay basis; UAN 32, 10-34-0, 12-36-0, and 18-0-18 were 3 percent by weight and 3-10-30 was 2 percent by weight.

Using the power density and retention time results shown in Figure 1, it is possible to calculate the energy consumption of ultrasonic gelling per unit of suspension produced. The results of this calculation are shown graphically as kilowatt-hours per ton versus retention time in Figure 3. With the clay used in the tests, the energy needed to reach 250 centipoises was constant at about 24 kilowatt-hours per ton for retention times from 0.5 to 1.0 minute and began increasing at longer retention times (and lower power densities). Together, these results indicate that the optimum conditions for ultrasonic gelling of this clay are 30 to 60 seconds exposure at 2.4 to 4.6 watts per milliliter. The optimum conditions for other clays would be expected to be somewhat different depending on the characteristics of the clay.

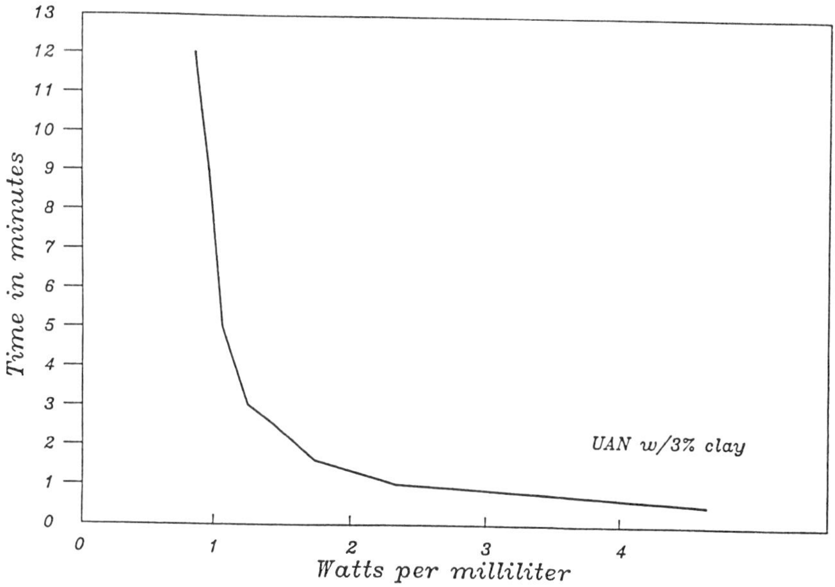

FIG 1. Power density requirements for ultrasonic gelation of 32-0-0 UAN solution with 3% clay.

While the most commonly used ultrasonic transducers are of the piezoelectric type, a magnetostrictive-type ultrasonic transducer also is commercially available. A limited number of tests have been made with a magnetostrictive transducer and the results were very promising.

Conclusions

In most fertilizer suspension systems, the use of ultrasonic energy may be a more efficient means of gelling the clay because little or no energy is used for bulk movement, or pumping, of the fluid which is unavoidable when using mechanical shearing equipment. A characteristic of pumps and agitators is that a large proportion of their energy goes into bulk movement of the fluid, with shear occurring only within the very immediate vicinity of the impeller or turbine. Although some fluid circulation is needed if fertilizer salts must be added and partially dissolved, the bulk movement of the liquid is conducive to shear only if large amounts of undissolved solids are present for collisions with the clay particles.

In some cases, the efficiency of ultrasonic gelation is much greater than that of mechanical shear gelation. For example, in the tests with the 32-0-0 UAN fluid, mechanical gelation required almost 20 times more energy than ultrasonic gelation. Also, the retention time required in these tests was much lower (1.25 minutes versus 70 minutes for mechanical gelation). The lower retention time is a result of the more concentrated ultrasonic energy (2 watts per milliliter versus 0.003 watts per milliliter for mechanical gelation). Thus, another advantage of ultrasonic gelation resulting from the lower retention times is that sizes of gelling vessels required are correspondingly smaller.

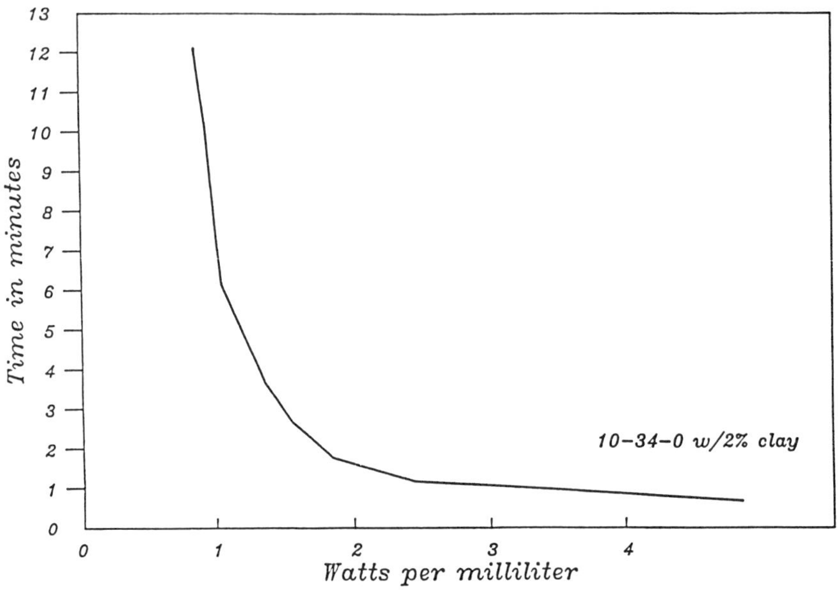

FIG 2. Power density requirements for ultrasonic gelation of 10-34-0 APP solution with 2% clay.

The potential advantages of ultrasonic gelling of suspending clay are (1) reduced power requirements, (2) substantially increased efficiency by which ultrasonic vibrations more fully utilize the full gel forming power of clay, (3) reduced production time, (4) reduced need for bulky and costly motors, pumps, recirculation lines, etc., and (5) less downtime since some ultrasonic transducers can operate full time for several years without malfunctions. Although ultrasonic gelling appears to have good economic potential, there is presently no commercial ultrasonic equipment available of sufficient size and power to use in a full-scale fertilizer suspension plant.

References

1. Tennessee Valley Authority, "Fluid Fertilizers," (NFDC Bulletin Y-185), September 1984, Muscle Shoals, Alabama, pp.86-102.
2. Jacobs, et al., U.S. Patent No. 3,509,066, April 28, 1970.
3. Kruglitskii, et al., Ukr. Khim. Zh. (Russ. Ed.), 40 (2), 141-145, 1974.

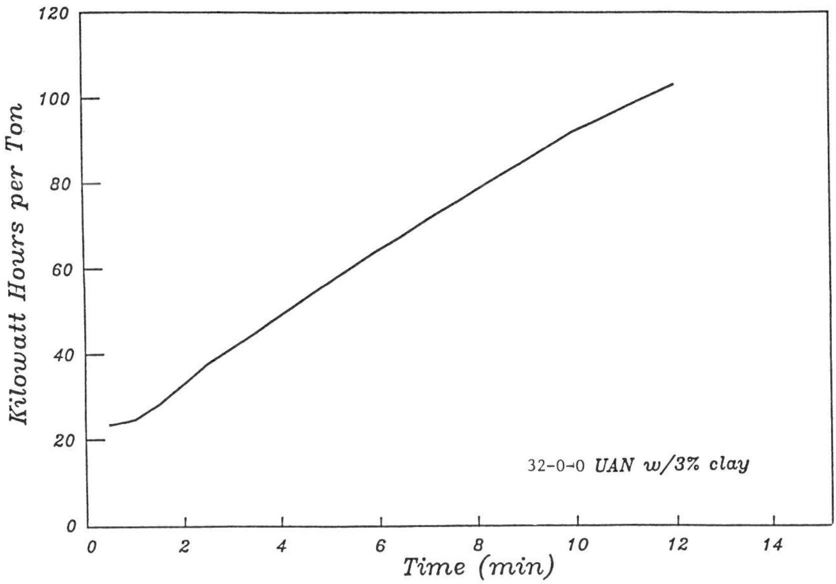

FIG 3. Power comsumption for ultrasonic gelation of 32-0-0 UAN solution with 3% clay.

PART 5.

PROCESSING OF FINE PARTICLES BY FLOCCULATION AND DISPERSION

EFFECTS OF POLYACRYLIC ACID CONCENTRATION ON ITS CONFORMATION AND ON THE STABILITY OF ALUMINA SUSPENSIONS

KUIRI F. TJIPANGANDJARA AND P. SOMASUNDARAN
Langmuir Center for Colloids and Interfaces, Henry Krumb School of Mines, Columbia University, New York, NY 10027.

ABSTRACT

Effect of polyacrylic acid concentration on the stability of alumina suspension is studied here and interpreted on the basis of the conformation of the adsorbed polyacrylic species. The conformation of polyacrylic acid species was monitored using fluorescence technique with pyrene as probe. By allowing polyacrylic acid first to adsorb on alumina in coiled form at low pH and then raising the pH, a drastic increase in flocculation was seen at low polyacrylic concentrations, while dispersion of the slurry was observed at higher concentrations. Analysis of the slurry showed that the adsorbed polyacrylic acid underwent concentration-dependent conformational changes which affected the stability of the slurry. At low concentrations conformation of the adsorbed polyacrylic acid transformed from coiled to stretched, resulted in better bridging of the extended polymer chain between the particles, and enhanced flocculation. At higher polymer concentrations such conformational transition was absent because of crowding of the polymer chains on the particles; as a result, the suspension remained dispersed. This study shows for the first time how correlation of adsorbed polymer configuration, determined by fluorescence, with stabilization can be utilized to obtain better dispersion or flocculation of colloidal suspensions.

INTRODUCTION

Colloidal stability is a key interfacial parameter determining the efficiency of many industrial processes: printing, drug delivery, detergency, cosmetics, microelectronics, high performance ceramics, mineral processing, effluent treatment, food processing, etc. [1-5]. It is equally important in biological processes involving, for example, blood, kidney stone as well as artificial organs. In most of these processes, dispersion or flocculation of particles is determined by macromolecular adsorption both in terms of the amount and the configuration of the adsorbed species [6-15]. Polymers can exist in different conformations depending on the solvent, pH, and ionic strength; its adsorption on the solid and the resultant suspension stability are influenced by these factors. Due to the non-existence of reliable *in-situ* techniques to determine the conformation and orientation of adsorbed polymer species, very little work has been done in the past to elucidate the mechanism by which polymer conformation control suspension stability. As a result, colloidal stability has been interpreted mostly on the basis of polymer adsorption density and the electrokinetics.

Recently, we have developed a multi-pronged approach involving simultaneous measurement of flocculation/dispersion responses, electrokinetic and configuration of adsorbed polymer species using fluorescence technique and pyrene-labelled reagents [16,17]. The rational behind the use of this technique is the observation that the extent of excimer formation, (i.e., an association of a ground state pyrene with an excited state pyrene), has a direct bearing on the polymer conformation. When a polymer is in coiled conformation, there is better probability for the intermolecular excimer formation between pyrene groups. Similarly when a

Advances in Fine Particles Processing
John Hanna and Yosry A. Attia, Editors

polymer is in stretched conformation, there is a low probability for the excimer formation. The difference between the two conformations is discernible in the intensity of their fluorescence spectra at certain wavelengths. Here, the ratio of the intensity of the excimer (Ie) to the monomer peaks (Im) is referred to as the "coiling index" of the polymer. In the absence of significant intermolecular interactions, a high Ie/Im ratio is obtained for a coiled conformation, while a low ratio is associated with a stretched one.

Objective of this study was to delineate the role played by polymer conformation in the stability of colloidal suspension. Configurational characteristics of polyacrylic acid at the alumina-solution interface on a molecular level were investigated for the same samples along with the stability and the zeta potential of particles. The effects of polymer concentration, pH and pH-perturbations on both the conformation of the adsorbed polymer chains and the stability responses were studied.

EXPERIMENTAL

Materials: Linde Alumina of 0.3 micron size purchased from Union Carbide was used for all the studies. The point of zero charge (pzc) of this sample was pH 8.3. A pyrene-labelled polyacrylic acid sample of molecular weight 88,000 was used in this investigation [18], along with an unlabeled polyacrylic acid of molecular weight 90,000 purchased from Polysciences, Inc. To prevent the intermolecular excimer formation at higher polymer concentrations, a mixture of pyrene-labelled PAA with pyrene-free PAA was used. Concentration of the pyrene-labelled polymer was maintained at 20 ppm while that of the unlabeled polymer was varied. All polymer solutions were prepared in 0.03 M NaCl solutions. Fisher-certified NaOH and HCl were used for pH adjustment, while constant ionic strength was maintained by the use of reagent grade NaCl.

Equipment: Emission spectra were made with a SPEX FLUOROLOG fluorescence spectrophotometer. The pH was measured with an Orion Research Digital Ionalyzer 501. Supernatant clarity, as percent transmittance, was measured using a Brinkman PC 600 Colorimeter at a wavelength of 670 nm. To determine the electrokinetic properties of the suspension, a Zeta-Meter was used. Residual concentration in the supernatant solutions was determined using the Dohrmann DC 90 Total Organic Carbon Analyzer ,TOC, [19].

Procedures: A ten gram sample of alumina was equilibrated with 194 ml of 0.03 M NaCl solutions in a 250-ml beaker for 45 minutes by stirring with a magnetic bar. After pH adjustment, the suspension was further equilibrated for 45 minutes. The magnetic bar was removed, a baffle with four 0.63 cm wide plates was inserted, and the suspension was stirred for 3 minutes using a 1"-diameter propeller, with three blades at 45° inclination at 600 rpm. Using a Sage syringe pump, 6 ml of polymer solution was added to the suspension dropwise at the rate of 6ml/min. The polymer-containing suspension was further stirred for 5 minutes, before transfer into a 250-ml flat-bottom graduated cylinder for flocculation response measurements. For the system in which pH adjustment was made after the polymer was added, an additional 5 minutes of stirring was made. Aliquots for emission spectra and zeta potential determinations for a given experiment were collected from the above sample cylinder.

Flocculation Tests: Percent solid settled was determined by suction technique. After allowing the suspension in the 250-ml beaker to settle for 45 seconds, the upper-half suspension was removed by suction and the percent solid settled estimated from the measurement of the solid content in the lower-half portion. Percent transmittance was measured by dipping a Brinkman probe to just below the surface of the liquid 10 minutes after the suspension was allowed to settle. It is to be noted that flocculation tests have been performed for both the system containing pyrene-labelled polyacrylic acid and that containing unlabeled polyacrylic acid. The results showed no significant flocculation difference between them, suggesting that the pyrene labelling had no effect on the flocculation behavior itself.

Adsorption Tests: Supernatant solutions were removed from these samples, centrifuged and analyzed for residual concentrations using the total organic carbon analyzer. Under all conditions polyacrylic acid was completely adsorbed and no residual concentration could be detected in the supernatant.

RESULTS AND DISCUSSION

Figure 1 shows the excimer-to-monomer(Ie/Im) ratio of adsorbed polyacrylic acid at the alumina-liquid interface as a function of polyacrylic acid concentration at pH 4.4 . Since PAA is only slightly ionized at pH 4 (pKa= 4.5, [20]) the adsorbed polymer chains have a coiled conformation (high Ie/Im ratio =0.6) on the alumina-liquid interface. For the concentration studied here, the conformation of polyacrylic acid *in-situ* remains unchanged in the complete range. Flocculation responses of alumina in the same samples do not exhibit the same constancy, fig. 2. While solids settled show a slight increase with the polyacrylic concentration, supernatant clarity goes through a maximum around 20 ppm. Interestingly, a significant reduction in zeta potential of alumina due to adsorbed polyacrylic acid is observed (Fig. 3), eventhough the stability of alumina did not show any marked change. Although the adsorbed polyacrylic in the coiled conformation was apparently not able to bridge onto many particles and to affect solids settled, the adsorbed polymer reduced the interparticle repulsion, making it possible for the fines to agglomerate and settle leading to an increase in the supernatant. By considering the fact this clarity is only a property of the supernatant regime, this trend cannot be considered to represent the overall flocculation behavior of the system, it only indicates that fewer fines remain in the supernatant at pH 4.4 . From this study it is clear that polyacrylic acid in coiled conformation does not have any significant effect on the stability of alumina under the tested conditions.

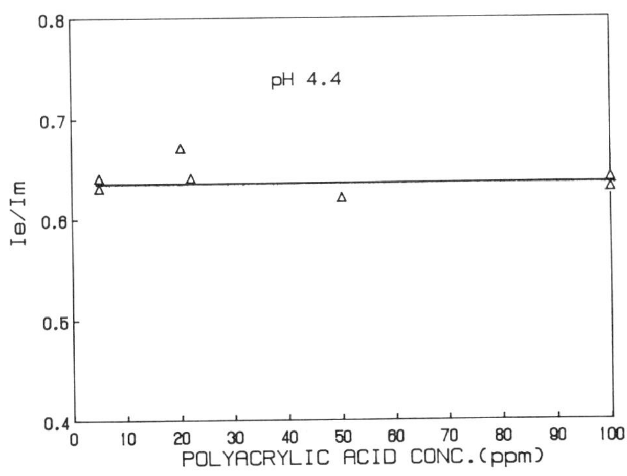

Figure 1. Excimer-to-monomer ratio, Ie/Im, of Polyacrylic acid at the Alumina-Liquid Interface as a function of Polyacrylic Acid Concentration at pH 4. (I.S.= 0.03M NaCl, S/L = 10g/200mL).

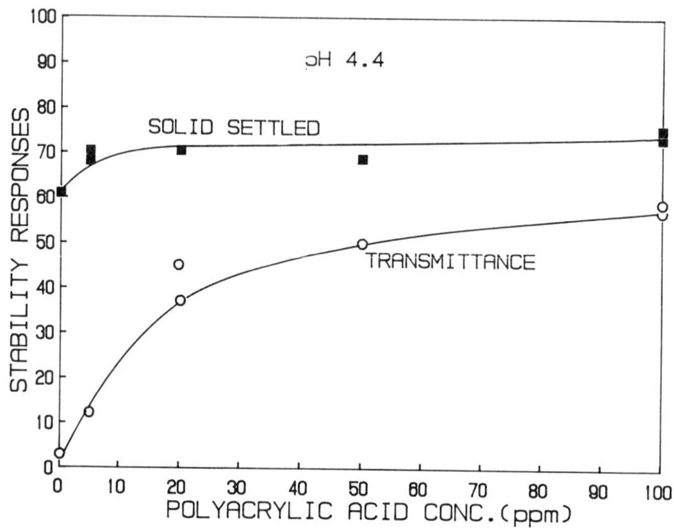

Figure 2. Stability Responses of Alumina Suspension as a function of Adsorbed Polyacrylic Acid at pH 4. (I.S. = 0.03M NaCl, S/L = 10g/200mL)

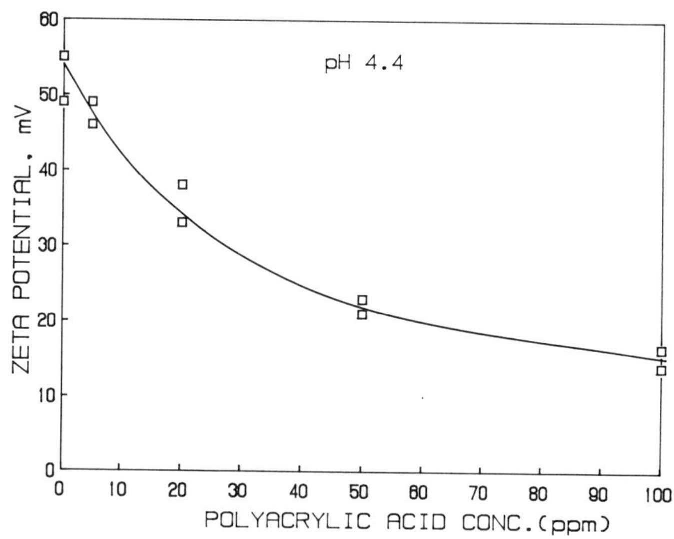

Figure 3. Zeta Potential of Alumina Suspension as a function of Polyacrylic Acid Concentration at pH 4. (I.S. = 0.03M NaCl, S/L = 10g/200mL)

In the case when polyacrylic aicd is adsorbed on alumina at pH 10.5, due to the repulsion between similarly charged polymer and alumina, PAA assumes a stretched conformation (lower Ie/Im ratio) on the alumina, figure 4. In contrast to pH 4.4, at pH 10.5 the conformation of polyacrylic acid depends on the concentration of the polymer. At lower polymer concentrations, the adsorbed chains seem to be stretched (Ie/Im = 0.25) on the particles. As the concentration of the polymer is increased, the polymer assumes as a lesser stretched conformation (Ie/Im = 0.35), apparently due to crowding of polymer chains on the particles.

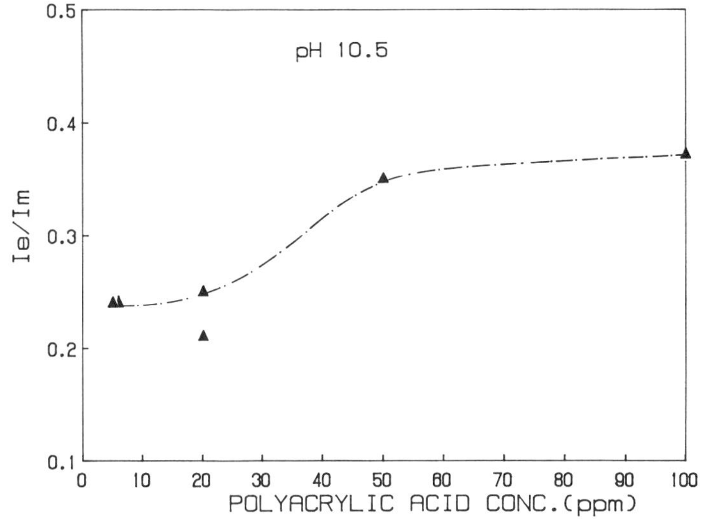

Figure 4. Excimer-to-monomer ratio, Ie/Im, of Polyacrylic acid at the Alumina-Liquid Interface as a function of Polyacrylic Acid Concentration at pH 10. (I.S. = 0.03M NaCl, S/L = 10g/200mL).

Stability tests, on the other hand, do not exhibit similar concentration-dependent behavior. As can be seen from figure 5, solids settled goes through a maximum at 20 ppm and is slightly higher than at pH 4.4, while the supernatant clarity is markedly reduced by an increase in polyacrylic acid concentration. Here, the effect of conformation on the alumina stability is clearly seen upon examining the solid settled data. Polymer chains in stretched conformation are able to better bridge between particles, than the coiled chains, bringing about higher flocculation and sedimentation; thus solids settled obtained at pH 10.5 is higher than that at pH 4.4. However, when the concentration of the polymer is increased, such bridging is prevented by the crowding of adsorbed polymer chains. As a result, the suspensions become more stable. The drastic decrease in supernatant clarity at pH 10 (i.e., many fines in the supernatant) can be explained on the basis of the changes in zeta potential. Zeta potential of alumina (in fig. 6) is seen to increase with polyacrylic acid concentration. Due to the increase in the electrostatic repulsion between the negatively charged alumina particles more fines remain suspended leading to a decrease in its transmittance.

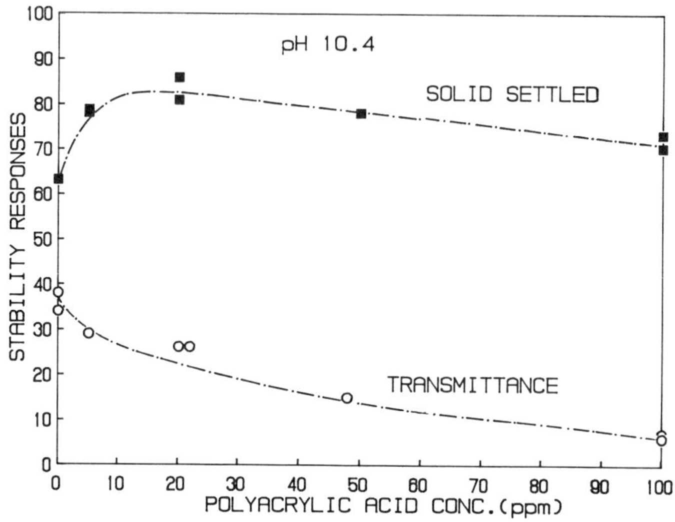

Figure 5. Stability Responses of Alumina Suspension as a function of Adsorbed Polyacrylic Acid at pH 10. (I.S. = 0.03M NaCl, S/L = 10g/200mL)

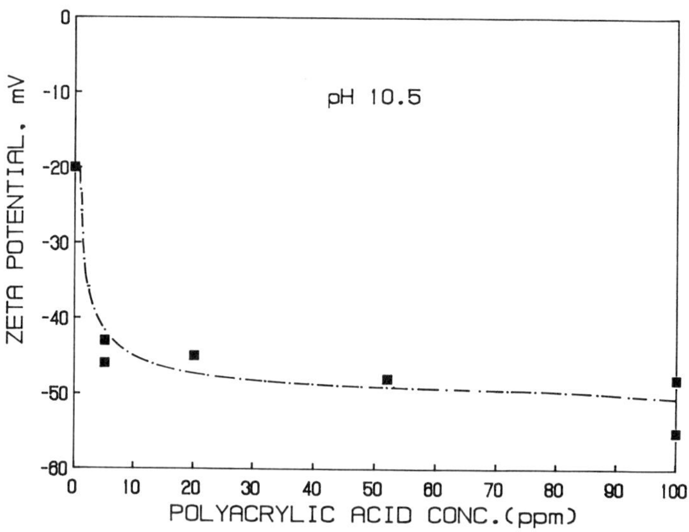

Figure 6. Zeta Potential of Alumina Suspension as a function of Polyacrylic Acid Concentration at pH 4. (I.S. = 0.03M NaCl, S/L = 10g/200mL)

The above study clearly shows that the conformation of polyacrylic acid has major effect on the stability of alumina suspension. Solid settled was higher with the stretched polymer chains due to better polymer bridging forming larger particles eventhough a number of fine particles remained dispersed. While better aggregation of finer particles was achieved with polyacrylic acid in coiled conformation, due to poor bridging of the unextended polymer chains the aggregate size was possibly smaller resulting in no improvement in the solid settled. In an attempt to control the stability of alumina with polyacrylic acid, a scheme involving pH changes after polymer adsorption was investigated. In this case, polyacrylic acid was first adsorbed on alumina at pH 4 and then the pH was raised to 10. As shown in fig.7, a marked increase in flocculation (solid settled and supernatant transmittance) is seen at lower polymer concentrations. The system was better flocculated under pH-perturbation conditions than either at pH 4 or at 10 where the stability tests were made with polyacrylic in coiled and stretched conformation respectively (compare fig. 7 with figures 2 and 5). At higher concentrations, the alumina suspensions were severely dispersed.

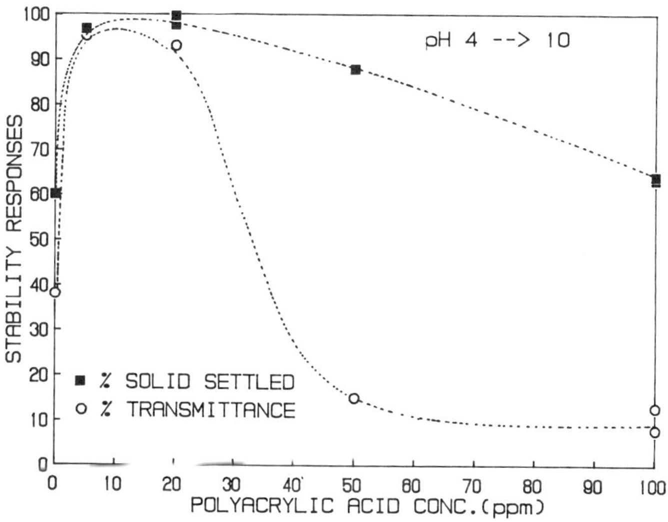

Figure 7. Stability Responses of Alumina Suspension as a function of Adsorbed Polyacrylic Acid under pH Changing Conditions , pH 4 to 10. (I.S.= 0.03M NaCl, S/L = 10g/200mL).

Zeta potential behavior of alumina particle as a function of polyacrylic acid (see fig. 8) alone cannot explain this flocculation/dispersion behavior. It must be noted that no residual polymer could be detected in the supernatant solutions under these conditions, even when the pH was raised confirming absence of desorption. Changes in the conformation of adsorbed polyacrylic acid due to pH increase are illusrated in figure 9. Adsorbing polyacrylic acid first in coiled form at pH 4 (high Ie/Im), and then raising the pH to 10, the conformation of polyacrylic acid changes from coiled to stretched (lower Ie/Im). However, adsorbed polyacrylic acid is more stretched at low concentrations at high pH than at higher concentrations. The marked increase in flocculation at low concentrations (see figure 7), is interpreted on the basis of changes in polymer conformation from coiled to stretched upon pH increase. When polyacrylic acid adsorbs initially on alumina suspension at pH 4, fines are aggregated into small flocs (*microfloccs*); with increase in pH the adsorbed polymer chains are extended due to ionization possibly causing aggregation of the microfloccs into *superfloccs* leading to enhanced flocculation and sedimentation [16]. At higher concentrations, when the pH of the alumina/polyacrylic acid system is raised from 4 to 10, adsorbed polyacrylic acid apparently is unable to stretch to the same extent as at lower concentrations due to crowding of these chains at the solid-liquid interface. Because of interpenetration and repulsion among the adsorbed polymer chains, the suspensions stabilize.

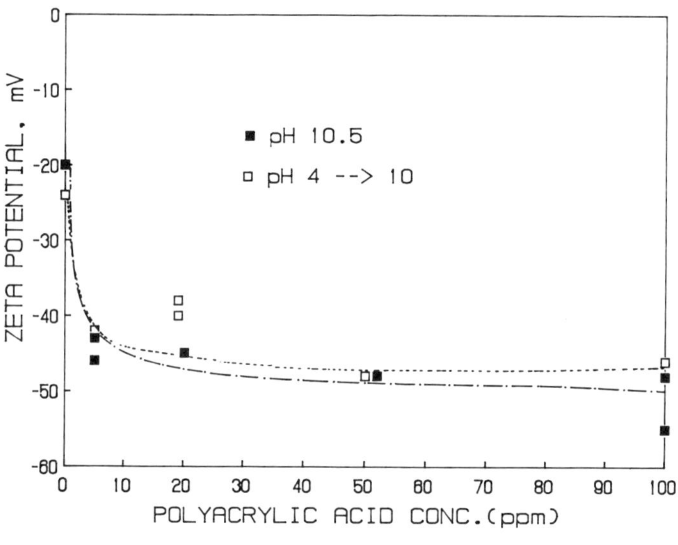

Figure 8. Zeta Potential of Alumina Suspension as a function of Polyacrylic Acid Concentration at pH 10. (I.S. = 0.03M NaCl, S/L = 10g/200mL).

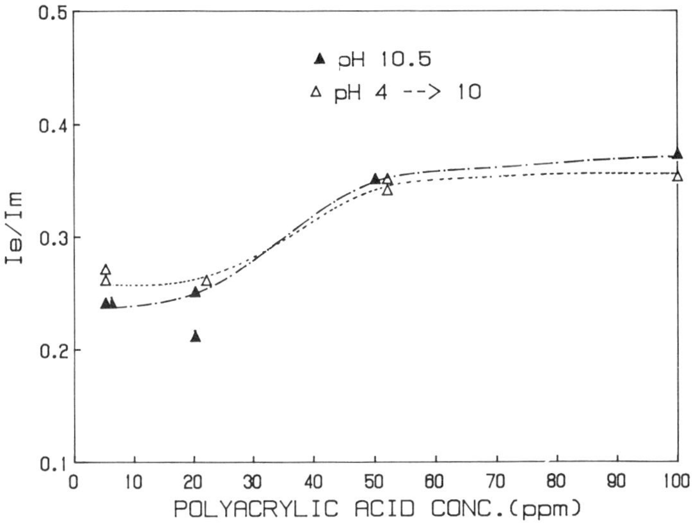

Figure 9. Excimer-to-monomer ratio, Ie/Im, of Polyacrylic acid at the Alumina-Liquid Interface as a function of Polyacrylic Acid Concentration under pH Changing Conditions, pH 4 to 10. (I.S.= 0.03M NaCl, S/L = 10g/200mL)

SUMMARY

It is evident that the conformation of polymer is a major controlling factor in determining the stability of mineral suspensions. pH-perturbation offers a new means to manipulate flocculation/dispersion of suspensions. At low dosages, polyacrylic acid conformational changes from coiled to stretch upon pH increase leads to flocculation, while stability is obtained at higher polymer dosage.

ACKNOWLEDGEMENTS

The authors acknowledge the financial suuport from the International Fine Particles Research Institute and National Science Foundation (CPE-83-18163). KFT, also thanks the United Nations Commissioner's Office for Namibia for financial support.

REFERENCES

1. D.H. Napper, Polymeric Stabilization of Colloidal Dispersion, Academic Press, New York (1983).
2. J.Th.G. Overbeek in: Surfaces and Coatings Related to Paper and Wood, Syracuse University Press, New York, (1967).
3. B.A. Stewart, Soil Condition, Soil Condition Sci. Soc. Of America, Madison, (1973).

4. E. Luck, et al. in: Ullmann's Encyclopedia of Industrial Chemistry, A11, Gerhartz Wolfgang ed., VCH Publishers, New York, (1988).
5. Brij.M. Moudgiland P. Somasundaran, Flocculation, Sedimentation and Consolidation, Proceeding of Engineering Foundation Conference, Sea Island, Georgia, USA, (1985).
6. J. Gregory in: Scientific Basis of Flocculation, K.J. Ives, ed., Sijthoft & Noorghoff, Rockville, MD, USA, (1978).
7. W.F. Linke and R.B. Booth, Trans. AIME, 217, 364-367 (1958).
8. Joseph 3rd Cesarano, I.A. Aksay and A. Bleier, Journal of American Ceramic Society, 71, p250 (1988).
9. (a) P. Somasundaran, Y.H.C. Wang and S. Acar in: Future Trends in Polymer Science and Technology, ITPR, CNR, Naples, (1984).
10. P. Somasundaran in: 13th Int. Mineral Proc. Congress, Part A, J. Laskowski ed., Elsevier, Amsterdam, (1981).
11. A.F. Hollander, P. Somasundaran and C.C. Gryte in: Adsorption from Aqueous Solutions, P.H. Tewari ed, Plenum, New York, (1981).
12. T.W. Healy and V.K. La Mer, Journal of Physical Chem., 66, p1935, (1962).
13. T. Sato, and R. Ruch in: Stabilization of Colloidal Dispersions by Polymer Adsorption, Marcel Dekker. Inc. New York, (1980).
14. P.A. Williams et al., in: The effect of Polymer on Dispersion Properties, Th.F. Tadros ed., Academic Press, (1983).
15. Th.F. Tadros in: The effect of Polymer on Dispersion Properties, Th.F. Tadros ed., Academic Press, (1981).
16. Kuiri F. Tjipangandjara, P. Somasundaran, Y-B. Huang and N.J. Turro, accepted for publications in Colloids and Surfaces, (1989)
17. P. Chandar, P. Somasundaran, N.J. Turro, and K.C. Waterman, Langmuir, 3, p298, (1987).
18. (a) N.J. Turro, and K.S. Arora, Polymer, 27, p783, (1986).; (b) N.J. Turro and O. Tusuneo, J. Phys. Chem., 86, p277, (1982)
19. Dohrmann, DC 90 Total Organic Carbon System Manual, by Rosemont Analytical, Santa Clara, California, (1988).
20. J.E. Gebhardt and D.W. Fuerstenau, Colloids & Surf., 7, p231, (1983).

SHEAR FLOCCULATION AND FLOTATION OF GALENA AND SYNTHETIC PbS

T. V. Subrahmanyam*, Zhongxi Sun**, K. S. Eric Forssberg*** and Willis Forsling****

*, ***Division of Mineral Processing, Luleå University of Technology, S-951 87 Luleå (Sweden).

,**Division of Inorganic Chemistry, Luleå University of Technology, S-951 87 Luleå (Sweden).

ABSTRACT

The present work deals with the shear flocculation and flotation of galena and synthetic PbS and emphasis is laid on the surface charge and the degree of hydrophobicity of the particles. The results are discussed based on the zeta potential measurements and the turbidity tests and supporting evidence is provided by the scanning electron micrographs.

INTRODUCTION

Among the methods to improve the floatability of fines, shear flocculation-flotation seems to be gaining ground and the work reported in literature is comparatively meagre on the aspects concerned.

The degree of flocculation is controlled by the probabilities of collision between particles (fines <5 μm and coarse <50 μm) and their adhesion after collision. The collision rates of particles suspended in a pulp increase at higher agitations and the adhesion occurs via hydrophobic association even though the particles are similarly charged. This energy barrier is overcome at higher energies than normal i.e. by intense agitation. A critical requirement for effective adhesion is the hydrophobicity of both the coarse and fine particles. The shear flocculated aggregates are considered to have better floatabilities.

The investigations reported in literature with respect to shear flocculation are: on scheelite-sodium oleate system [1,2]; and on shear flocculation and carrier flotation of fine hematite [3]. Other related papers deal with the separation of anatase from ultrafine kaolin by calcite as a coarse carrier mineral [4,5]; carrier flotation of wolframite and low grade oxidized ores [6-8]; and floc flotation of hematite-quartz mixture in tall oil emulsion [9]. A recent review discusses more details on shear flocculation and carrier flotation [10].

The present work investigates the shear flocculation behaviour of galena i.e. aggregation of fines ∿5 μm with coarse particles. In another series of tests synthetic PbS sample prepared by potentiometric titration was used in place of ∿5 μm galena particles in order to study their aggregation with coarse galena particles (-38+20 μm). The floatability of the aggregates of galena formed by shear flocculation is compared with the reco-

Advances in Fine Particles Processing
John Hanna and Yosry A. Attia, Editors

veries obtained by conventional flotation.

EXPERIMENTAL

Material

Galena: A pure sample of galena (Brushy Creek, Missouri) was obtained from the Ward's Natural Sciences Establishment Inc., USA. The sample was crushed in a jaw crusher and dry ground in a steel rod mill and the size fractions— -5 μm (median size 2.4 μm - determined by Cilas Granulomètre 715) and -38+20 μm were prepared by microsieving and wet screening. The samples were dried and packed in plastic bags and no special precautions were taken to prevent oxidation.

Synthetic PbS: Synthetic PbS was precipitated by titrating a lead nitrate solution with stoichiometric amounts of freshly calibrated sodium sulphide solution. The suspension was rinsed several times in 0.1 M $NaNO_3$ medium until the sulphide ion selective electrode showed a constant value. This indicates that the solution has the same concentration level of sulphide ions corresponding to the solubility of PbS in solution [11]. Pure nitrogen gas was used to keep an inert atmosphere during titrations. Lead Sulphide thus prepared was stored in 100 ml volumetric flasks and a desired volume was taken before the experiment. The precipitate was black in colour and in the form of flocs with a high settling velocity. But when subjected to ultrasonic dispersion the solution turned turbid, thus indicating the breakage of flocs and dispersion of fines.

Reagents: All solutions were prepared in deionized water before the experiment. Sodium ethyl xanthate (Hoechst Co., West Germany) was used as collector and methyl isobutylcarbinol (Merck, >97% pure) as frother for flotation tests. The pH was adjusted with dil NaOH or dil HNO_3.

Methods

Aggregations tests

Before an experiment, desired quantities of -5 μm galena or synthetic PbS and coarse galena particles were added to 100 ml of xanthate solution and subjected to ultrasonic dispersion for 15 min. To this suspension 150 ml of xanthate solution was added and the pH adjusted. The suspension was stirred in a plexiglas container (80-120 mm high; 63 mm in diameter) fitted with six baffle plates (each 50 mm x 7 mm x 1.5 mm) and a single bladed paddle stirrer (14 mm x 25 mm x 3 mm). Stirrer speeds upto 1500 rpm were obtained without much air entrainment or vortex formation. At a high speed of 2000 rpm air entrainment was considerably high.

The aggregates of fine-coarse or fine-fine particles formed at high agitations decrease the turbidity of the solution i.e. the concentration of -5 μm particles in suspension. The solution turbidity was measured by a Hach Ratio Turbidity meter.

Electrokinetic measurements

The zeta potentials were determined with Laser Zee Meter Model 501. A known quantity of -5 μm galena or synthetic PbS was added to deionized water or xanthate solution of desired concentration and dispersed in ultrasonic for 15 min. after adjusting the pH. Sufficient care was taken to avoid the side effects of temperature due to sample heating during the measurements.

Flotation tests

Flotation tests were conducted in a Labor Flotation Machine (West Germany). The maximum number of rotations reached in this cell was 300 rpm. After the shear flocculation test the pulp was transferred to the flotation cell. A conditioning time of 2 min was allowed after the addition of frother and air was let in to continue flotation for 2 min. The flotation recoveries thus obtained were compared with the conventional flotation recoveries—i.e. the pulp not subjected to shear flocculation was transferred to the flotation cell and a required amount of xanthate was added. A conditioning time of 15 min and a flotation time of 2 min after the addition of frother, were allowed. This procedure was adopted to evaluate the difference in recoveries between the two methods—i.e. conventional flotation and shear flocculation followed by flotation. Any improvement in flotation recovery with the pulp subjected to shear flocculation could only be due to particle aggregation.

RESULTS AND DISCUSSION

Electrokinetics of galena and synthetic PbS

The electrokinetic behaviour of galena and synthetic PbS is presented in figs. 1-3. In the presence of xanthate both galena and synthetic PbS are negatively charged in the pH range investigated (fig. 1). With increasing pH the negative potential of galena increases while no appreciable change is noticed in the case of synthetic PbS above pH 7. For the xanthate concentrations studied (fig. 2) the negative potential increases with increasing concentration 3.47×10^{-5} - 1.39×10^{-4} moles/l (5-20 mg/l), above which the potential remains constant up to 4.86×10^{-4} moles/l (70 mg/l). A further increase in collector concentration to 6.94×10^{-4} moles/l (100 mg/l) leads to a more negative potential. The trend of the curves for galena and synthetic PbS is similar, the latter being less negatively charged. In the absence of xanthate (fig. 3) both galena and synthetic PbS are negatively charged and the potentials are less negative in comparison to the values observed in the presence of xanthate (fig. 1).

The electrokinetic behaviour of galena was examined by several workers [11-15] and the values reported vary widely. The pzc values of galena in the absence and presence of oxygen were reported to be at pH 2.6 and pH 8.0, respectively. Whereas, for ethyl xanthate concentrations of 10^{-4}-10^{-3} moles/l galena was found to be negatively charged in the pH range 4-12 both

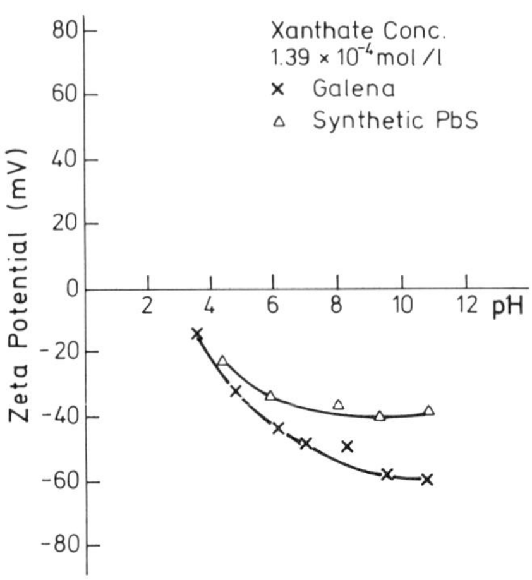

FIG. 1 Zeta potential as a function of pH of galena and synthetic PbS.

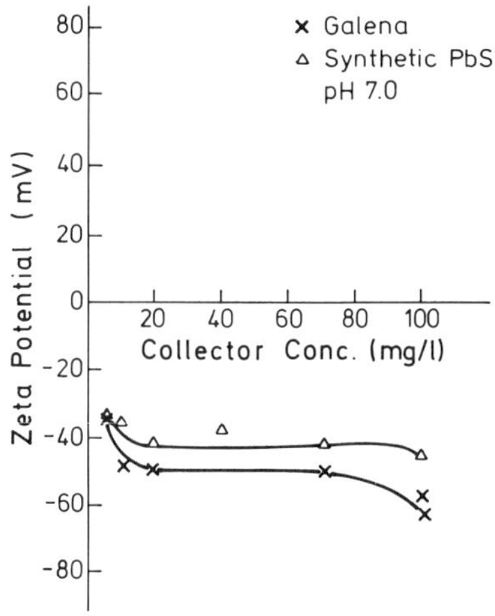

FIG. 2 Zeta potential as a function of collector concentration.

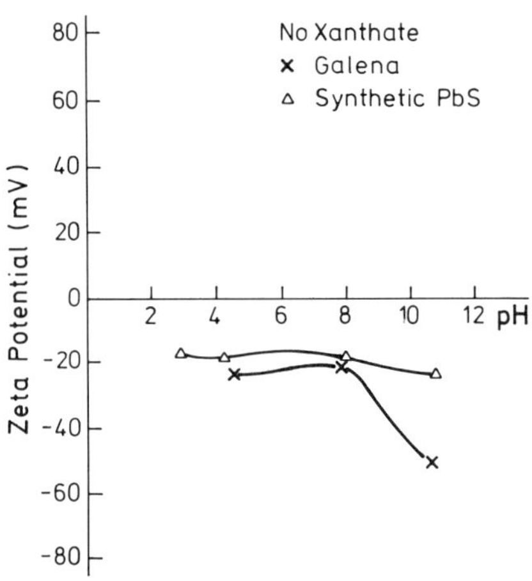

FIG. 3 Zeta potential versus pH of galena and synthetic PbS.

in the presence and absence of oxygen. While zeta potentials were independent of pH in the virtual absence of oxygen ($\sim$ -40 mV) that in presence of oxygen the values varied between $\sim$ -15 mV - -40 mV [12]. A pH_{pzc} = 3 was observed [13] for galena in the absence of xanthate and for similar conditions the zeta potentials were found to be negative throughout the pH range studied [14]. While natural galena was found [15] to be negatively charged with potentials varying from -10 mV to -50 mV in the pH range 2-10, a pH_{iep}5 was observed for synthetic galena (Merck). From the ESCA spectra of synthetic PbS and galena it was found that the former was in a much more oxidized state with sulphates, thiosulphates and molecular sulphur in addition to sulphide. A more detailed discussion of these aspects is given elsewhere [16].

Galena is one of the sulphide minerals which can acquire hydrophobicity under mildly oxidizing conditions. But the species that is responsible for imparting hydrophobicity to the mineral surface is still controversial, although it is generally believed that sulphur is the entity involved. The flotation of sulphide minerals viz., galena, chalcopyrite, pyrrhotite, under mildly oxidizing conditions and in the absence of a collector is well known [17]. However, based on electrochemical and X-ray photoelectron spectroscopic (XPS) studies it was proposed that both the concentration of surface oxides and the degree of sulphur enrichment as the likely species for the hydrophobic nature of the mineral surface [18]. It must be pointed out that the wide variation of results reported by several workers on the electrokinetic behaviour of galena is due mainly to the surface heterogeneity, for example, the effect of oxidation of the sulphide surface and the preparation and pretreatment of the sample, etc.

<u>Shear flocculation of galena</u>

The effect of stirring time on the apparent concentration of -5 μm galena particles in suspension for different collector concentrations is shown in fig. 4. The turbidity values decrease with stirring time for all xanthate concentrations ranging from 0-2.78 x 10^{-4} moles/l. For xanthate concentrations of 6.94 x 10^{-5} and 1.39 x 10^{-4} moles/l where the turbidity values are low i.e. maximum flocculation, the corresponding zeta potentials are -36 mV and -48 mV. A decrease in the apparent concentration of the -5 μm fraction in the absence of xanthate and almost a similar trend for a xanthate concentration of 3.47 x 10^{-5} moles/l, suggest that both the surface coverage (degree of hydrophobicity) and the surface charge are of utmost importance for shear flocculation. The zeta potential of galena in the absence of xanthate at pH 7 corresponds to -20 mV. Since hydrophobicity is a critical factor for shear flocculation the formation of flocs in the absence of xanthate can be expected to be due to the hydrophobic species formed as a result of the surface oxidation. Fig. 5 shows the scanning electron micrograph of the aggregates formed in the absence of xanthate. However, at a high xanthate concentration (6.94 x 10^{-4} moles/l) where the corresponding zeta potential value is ∿-60 mV (see fig. 2), the solution turbidity level remains constant i.e. absence of aggregates. The absence of flocculation

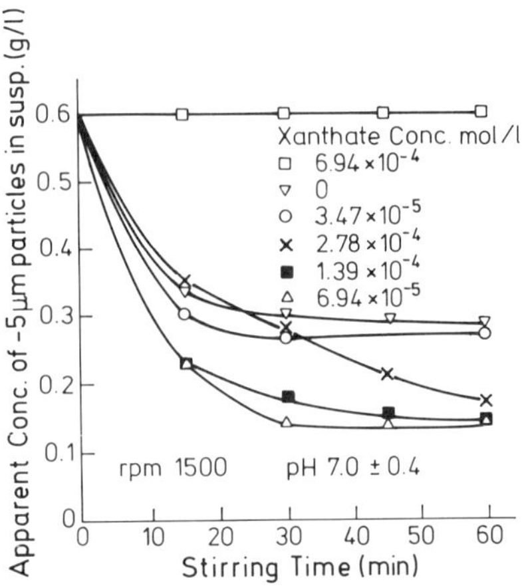

FIG. 4 Effect of collector concentration and stirring time on the concentration of -5 μm particles in suspension; -38+20 μm 4 g/l and -5 μm: 0.6 g/l.

FIG. 5 Aggregates formed in the absence of xanthate after stirring -38+20 μm and -5 μm mixture at 1500 rpm; pH 7; time:60 min.

as seen from fig. 6 may be due to a high negative potential resulting in repulsion between the particles.

In the carrier flotation of wolframite [6-8] better recoveries obtained at pH 6-7 were explained to be due to a lower electrostatic repulsion and favourable reagent adsorption. At high collector concentrations and pH's the mineral surfaces hold high potentials and to overcome the energy barrier it may be necessary to apply higher agitations. But higher stirring speeds (>1000-1500 rpm) may also lead to breakage of aggregates. In the shear

FIG. 6 Effect of high collector concentration on aggregate formation. -38+20 μm and -5 μm mixture at 1500 rpm; pH 7; time of stirring 60 min.

flocculation of hematite the maximum and minimum shear rates were found to be 1200 and 400 rpm, respectively. The larger flocs were found to break at speeds >1200 rpm while no flocs were formed below 400 rpm [3].

Shear flocculation of coarse galena and synthetic PbS

The synthetic PbS sample prepared by potentiometric titrations was used in place of -5 μm galena particles along with coarse galena (-38+20 μm) in order to evaluate the particle size effect on shear flocculation. The synthetic PbS particles can be considered to be different from the natural galena in terms of surface heterogeneity like particle size (≪5 μm), surface oxidation, cleavage planes, etc. It is known that when the difference between the coarse and fine particle sizes is large the fines may flow past the coarse instead of colliding, when subjected to stirring. Further, the hydrodynamic effects increase with particle size and become dominant for particles larger than a few micrometers [19].

The effect of stirring time on the apparent concentration of fines in the shear flocculation of (a). synthetic PbS and coarse galena (-38+20 μm) and (b). -5 μm galena and coarse galena (-38+20 μm) is shown in fig. 7. In the case of former, the turbidity values were found to fluctuate i.e. increasing and decreasing with time even for prolonged periods of stirring. This fluctuation may be due both to the formation and breakage of aggregates. Fig. 8 shows the scanning electron micrograph of the aggregates. Whereas with coarse and fine particles of galena (case b) the turbidity decrease was gradual with time (fig. 7). Fig. 9 shows the aggregates formed

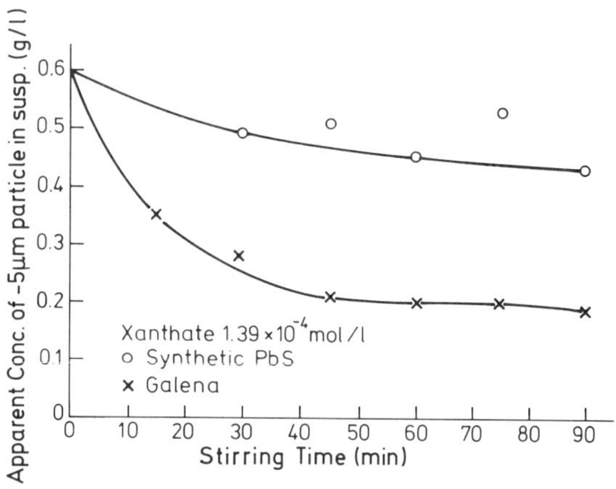

FIG. 7 Effect of stirring time on the concentration of -5 μm particles in suspension in the presence of -38+20 μm galena.

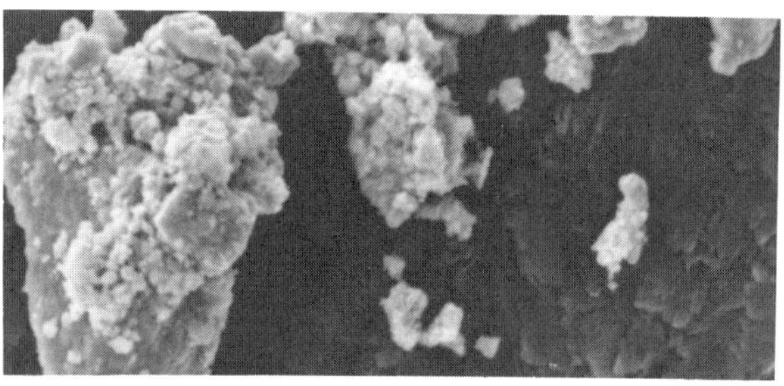

xanthate conc. 1.39×10^{-4} mol/l.

FIG. 8 Aggregate of synthetic PbS (-5 μm:0.6 g/l) and galena (-38+20 μm: 4 g/l); speed 1500 rpm: pH 7; time of stirring: 90 min.

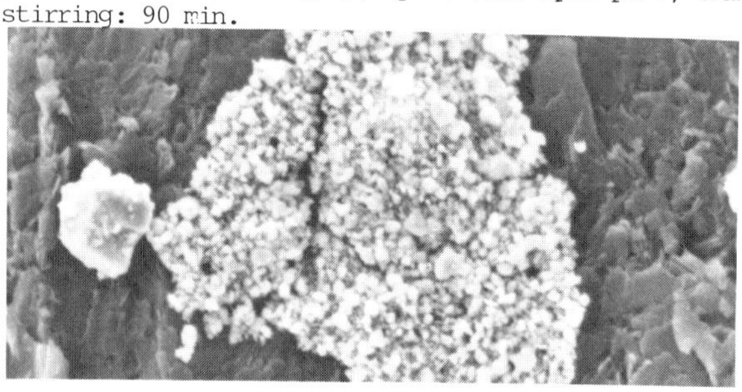

xanthate conc. 1.39×10^{-4} mol/l.

FIG. 9 Aggregate of synthetic PbS (-5 μm:0.12 g/l) and galena (-38+20 μm: 4 g/l) treated with Na_2S; speed 1500 rpm; pH 7; time of stirring: 90 min.

when a mixture of synthetic PbS and coarse galena were pretreated with Na_2S and agitated at 1500 rpm at a xanthate concentration of 1.39 x 10^{-4} moles/l. Two types of aggregates i.e. one between fines and another between fine-coarse (like slime coatings) were observed. The difference between figs. 8 and 9 is in the density of particle coating. This effect is due to the different quantities of fines taken in the two tests. The particles treated with Na_2S attain more negative zeta potentials. The addition of xanthate to sulphidized particles may result in a uniform collector adsorption and under such conditions the aggregates formed due to shear flocculation may be expected to be more stable. The electrokinetic behaviour of synthetic PbS treated with Na_2S and the shear flocculation results are discussed in a forthcoming publication [20].

Flotation

The flotation recoveries of the aggregates of galena formed by shear flocculation are higher in comparison to the recoveries obtained by conventional flotation (fig. 10). By shear flocculation the particle size is increased so that the aggregates formed may have better collision efficiencies with gas bubbles and thereby improvement in fines recovery.

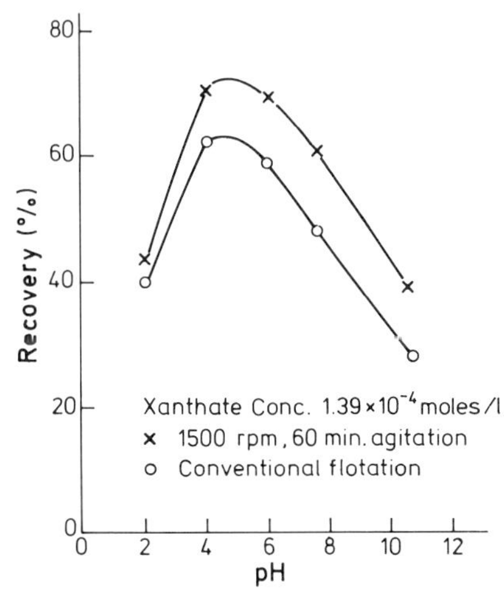

FIG. 10 Flotation recovery as a function of pH.

CONCLUSIONS

From the experimental evidence gathered on the electrokinetics of galena and synthetic PbS and shear flocculation tests, the following conclusions are drawn:

* Among the factors which affect the shear flocculation the sur-

face potential and the degree of hydrophobicity are of utmost importance.

* The aggregate formation in the absence of xanthate is attributed to the surface oxidation products of galena.
* Two types of aggregates—i.e. fine-fine and fine-coarse were predominant depending upon the experimental conditions maintained. For the stirring speeds used the aggregates formed between galena particles were stable in comparison to the aggregates of coarse galena and synthetic PbS.
* The flotation recoveries of aggregates of galena formed by shear flocculation are higher in comparison to the conventional flotation recoveries.
* Like the froth flotation process the shear flocculation mechanism is sensitive to several variables which can be classified into physical, chemical and geometrical.

REFERENCES

1. L. J. Warren, Trans. Inst. Min. Metall., Sect. C: Min. Proc. & Ext. Metall. 84, 99-104 (1975).
2. L. J. Warren, J. Colloid and Int. Science 50, 2, 307-318 (1975).
3. D. W. Fuerstenau, C. Li, and J. S. Hanson in: Proc. Int. Symp. on the Production and Processing of Fine Particles A. J. Plumpton, ed. 7 (Pergamon Press, 1988)pp. 329-335.
4. E. W. Greene and J. B. Duke, Min. Engg. 14, 51-55 (1962).
5. Y. H. Chia and P. Somasundaran in: Ultrafine Grinding and Separation of Industrial Minerals Malaghan G. Subhas, ed. (SME. AIME, 1983) pp. 117-131.
6. W. Hu, D. Wang, and J. Huaai, in: Proc. XIV Int. Min. Proc. Congress. IV-10.1-IV.10.14 (1982).
7. W. Hu, D. Z. Wang, and G. Qu, J. Cent. South. Inst. Min. Metal. 4, 408-414 (1987).
8. W. Hu, D. Z. Wang, and C. Z. Qu, in: Proc. XVI Int. Min. Proc. Congress K. S. Eric Forssberg, ed. (Elsevier, Amsterdam 1988) Part A, pp.445-452.
9. M. Clement, H. Harms, and H. M. Trondle in: Proc. IX Int. Min. Proc. Congress. 1, pp. 179-187 (1970).
10. T. V. Subrahmanyam and K. S. Eric Forssberg, (submitted to Int. J. Min. Proc. Elsevier, Amsterdam).
11. W. Forsling and S. Sjöberg, in: Proc. Konferens i Mineralteknik, Högskolan i Luleå, 14-16 February,1988, pp. 74-87.
12. M. C. Fuerstenau, in: Principles of Flotation R.P. King, ed. 3 (South African Inst. Min. Metall. Monograph Series, 1982) pp. 91-108.
13. A. Yucesoy and B. Yarar, Trans. Inst. Min. Metall., Sect. C: Min. Proc. & Ext. Metall. 97-100 (1974).
14. P. G. Parsonage, in: Flotation of Sulphide Minerals K. S. Eric Forssberg ed. (Elsevier, Amsterdam 1985) pp. 111-139.
15. R. J. Pugh, in: Proc. XVI Int. Min. Proc. Congress K. S. Eric Forssberg, ed. (Elsevier, Amsterdam 1988) Part A, pp. 751-762.
16. T. V. Subrahmanyam, Z. Sun and W. Forsling (Electrokinetic Properties of Galena and Synthetic PbS-in preparation).
17. W. J. Trahar, in: Principles of Mineral Flotation M. H. Jones and J. T. Woodcock, eds. (The Wark Symp. Australasian Inst. Min. Metall. Melbourne, 1984) pp. 117-135.
18. A. N. Buckley and G. W. Walker, in: Proc. XVI Int. Min. Proc. Congress K. S. Eric Forssberg, ed. (Elsevier, Amsterdam 1988) Part A, pp. 589-599.
19. D. H. Everett, Basic Principles of Colloid Science (Royal Society of Chemistry, London, 1988).
20. T. V. Subrahmanyam, Z. Sun, K. S. Eric Forssberg and W. Forsling (Physical, Chemical and Geometrical Variables in Shear Flocculation (in preparation).

THE HYDROPHOBIC AGGREGATION FLOTATION OF RUTILE PARTICLES

SHAOXIAN SONG AND SHOUCI LU
Wuhan Iron and Steel University, Wuhan, P.R.China

ABSTRACT

The hydrophobic aggregation flotation of fine rutile particles has been investigated. The mixture of rutile 10% and amphibole 90% has been separated by hydrophobic aggregation flotation and by conventional flotation for comparison. The size of rutile and amphibole samples is below 20μm. as a result of long time and intense agitation and addition of non-polar oil emulsion together with collector BBA, the flotation recovery of rutile could be remarkably increased. By means of the size analysis, the aggregates formation has been observed. A concentrate with grade 71.4% TiO_2 and recovery 89.00% has been obtained from the rutile-amphibole mixture by hydrphobic aggregation flotation, whereas by conventional flotation the grade and recovery are 26.30% TiO_2 and 45.09% respectively

INTRODUCTION

In our previous papers(1, 2, 3), it has been pointed out that hydrophobic aggregation methods are an efficient and promising way to separate fine and ultrafine mineral particles. The common characteristics of these methods can be concluded as follows; a) selective imparting the surface of mineral particles to be hydrophobic and formation of hydrophobic aggregates; b) intensification of aggregation process by nonpolar oil addition; c) the high energy input during agitaiton.

In this paper a hydrophobic aggregation flotation method is proposed to separate the fine particle rutile-amphibole mixture. Such factors as agitation intensity, agitation time, collector nonpolar oil ratio are examined.

MATERIALS AND EXPERIMENT

Both rutile and amphibole samples for this research were obtained from Dafushan ore deposit and were purified by gravity and magnetic separation. The rutile sample contains TiO_2 92.10%, whereas the purity of amphibole sample is 85%.

The samples were ground until their size is below 20 μm. The size distribution of rutile is shown in Fig.6. It can be seen from the Fig.6 that the d_{50} is about 5.73 μm.

The experiment was carried out in a stirred tank with a four flat baffle. Pulp added reagents was conditioned first in the stirred tank during a given time, then was poured into the flotation cell and floated. The froth product and tailing product obtained by this way were dried, weighed and chemically analyzed.

In our research, the benzyl arsonic acid (BBA) from Zhuzhou reagent factory was used as collector (flocculent), and sodium silicate(C.P.) as modifier.

The size distribution of mineral samples and products was measured by size analyzer of centrifugal type(ZPL-10) made by Dandong optical equipment factory.

Published 1990 by Elsevier Science Publishing Co., Inc.
Advances in Fine Particles Processing
John Hanna and Yosry A. Attia, Editors

RESULTS AND DISCUSSION

As first step of our experiment, the aggregation behaviours of rutile and amphibole and the main factors of aggregation were investigated.

The influence of agitation intensity, represented by r.p. m. and agitation time on recovery of fine rutile particles is shown in Fig.1 and Fig.2. It can be seen that the rutile recovery is strongly dependent on the agitation intensity and agitation time before flotation. The rutile recovery raised gradually with the increase of agitation intensity and time until a high r.p.m. and long agitation time are reached. Furthermore, a close correlation between agitation intensity and time is observed. If a high agitation speed like 3000 r.p.m. is used, a short agitation time (5 min.)is needed to obtain a good recovery. Conversely, if a long agitation time is used, a mild agitation intensity, as 1000 r.p.m., is enough to reach the same recovery. A similar correlation has been obtained by Warren in the case of shear flocculation of scheelite(4).

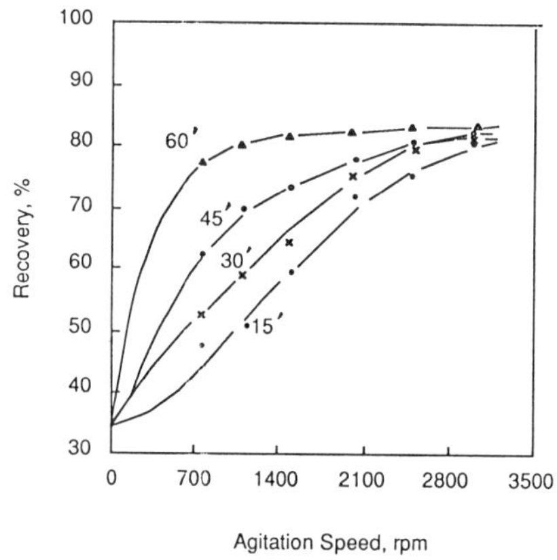

Fig.1. Influence of agitation speed on hydrophobic aggregation flotation of fine particle rutile during giving time. pH=5.4, Na_2SiO_3, 60mg/l, BBA, 10^{-4}mol/l.

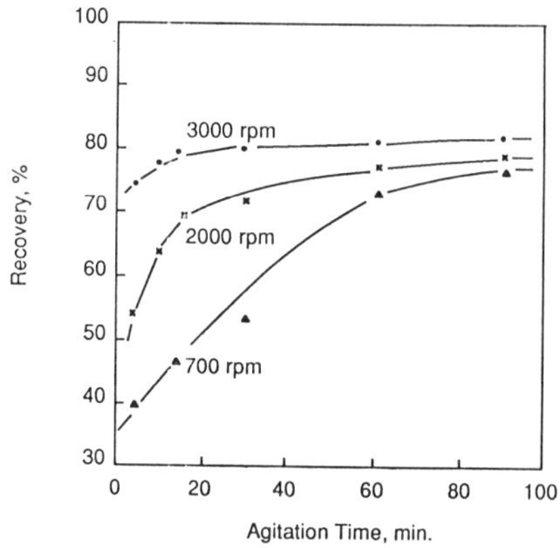

Fig.2. Influence of agitation time on hydrophobic aggregation flotation of fine rutile particles. pH=5.4; Na_2SiO_3, 60mg/l, BBA, 10^{-4}mol/l.

Fig. 3. and Fig.4. show the effect of adding kerosen on hydrophobic aggregation flotation of fine particle rutile when pulp is stirred violently. BBA and kerosen emulsion mixture is produced by the ultrasonic treatment. The emulsion mixture is added into pulp, and the pulp is stirred with giving intensity and duration in the stirred tank. The optimum turbulence with proper emulsion mixture addition ensures a rutile recovery as high as 93% instead of 81% when only BBA is used without non-polar oil. From Fig.3, it can be seen that flotation recovery is gradually increased with the raising of kerosen amount in the emulsion mixture. The result is a good evidence of non-polar oil strengthening acticn on hydrophobic aggregation flotation. Fig.4 shows the influence of agitation time under different BBA kerosen ratio. It can be seen that the addition of kerosen guarantees a stable recovery inereasing about 10% when the agitation time is longer than 10 min.

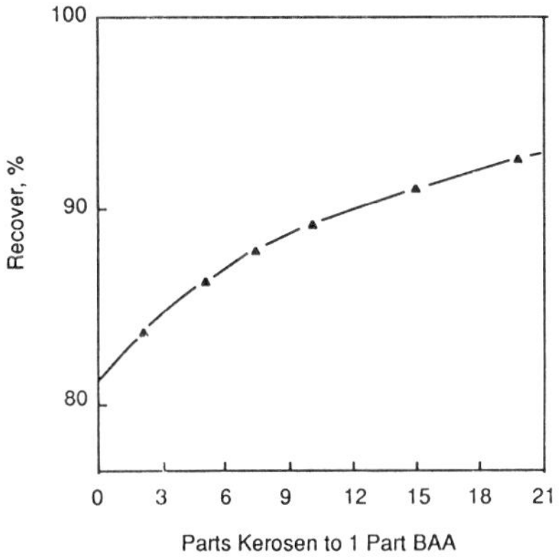

Fig. 3. Influence of BBA and kerosen ratio on the hydrophobic aggregation flotation of fine particle rutile. pH=5.3, Na_2SiO_3, 60mg/l, BBA, 10^{-4}mol/l. agitation speed, 3000 r.p.m., agitation time, 15min.

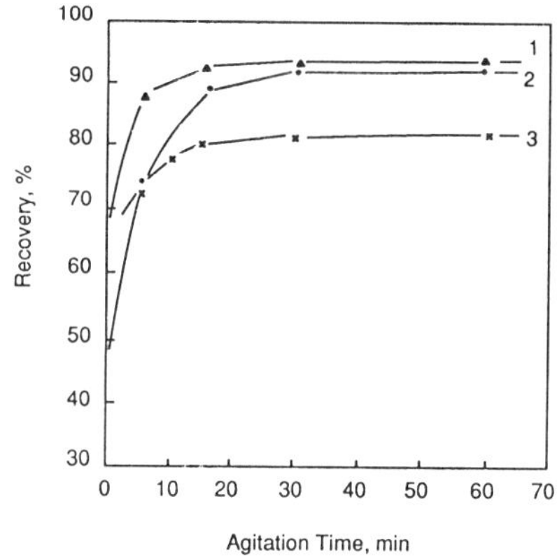

Fig. 4. Affectation of shear field on the hydrophobic aggregation flotation of fine particle rutile in the case of using emulsion mixture of BBA and kerosen. pH=5.3, Na_2SiO_3, 60mg/l, BBA, 10^{-4}mol/l. agitation speed, 3000r.p.m. Ratio of BBA and kerosen, 1, 1 : 20; 2, 1 : 10; 3, 1 : 0.

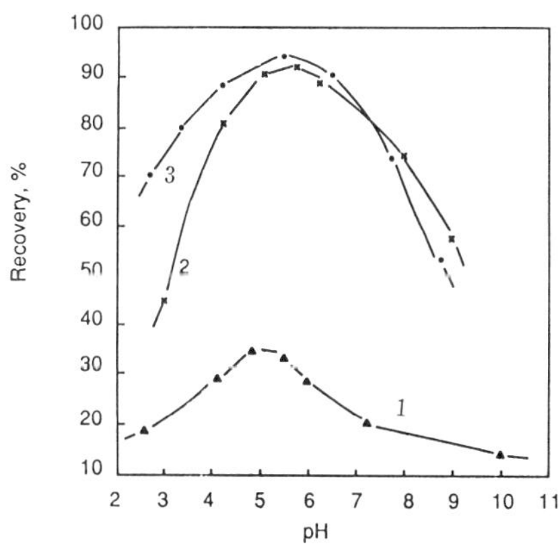

Fig. 5. Comparison of three cases of rutile flotation.
1. Common flotation, Size, −20 μm. Na_2SiO_3, 60mg/l, BBA, 10^{-4}mol/l;
2. Common flotation, Size, −74—+40 μm, BBA, 10^{-4}mol/l;
3. Hydrophobic aggregation flotation, size, 20 μm, Na2SiO3, 60mg/l, BBA, 10^{-4}mol/l, ratio of BBA and kerosen, 1 : 20, agitation time, 15 min., agitation speed, 3000 r.p.m. .

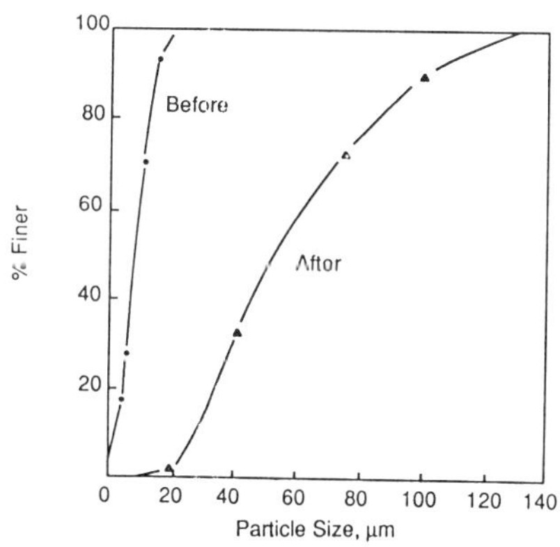

Fig. 6. Size distribution before and after hydrophobic aggregation of fine particle rutile.
Before, pH=5.4.
After, pH=5.3, BBA, 10^{-4}mol/l, Na_2SiO_3, 60mg/l, ratio of BBA and kerosen, 1 : 20, agitation speed, 3000r.p.m. agitation time, 15min. .

A comparison of three cases of rutile flotation is illustrated in Fig. 5. obviously, the conventional rutile flotation of fine particles(-20 μm) does not work well, the recovery of rutile is as low as 36%, whereas for the 74—40 μm size fraction a high recovery is obtained by the conventional flotation. A high recovery 94% can be achieved just by a hydrophobic aggregation flotation technique under pH value between 5—6.

Fig. 6 showe the size distribution before and after hydrophobic aggregation of fine rutile particles. It can be seen from this figure that the particle size is much more enlarged, and -20 μm particles is only 2.67%. It means the aggregates of rutile are produced. Through the hydrophobic aggregation, all fine rutile particles almost gether into aggregates with a size range about +20—-100 μm that is just what required for a successful conventional flotation.

As to the tailing mineral amphibole it was found that amphibole could not be floated by benzyl arsenic acid at all(Fig. 7). To avoid the entrainment and collision of fine amphibole particles with rutile particles or air bubbles, sodium silicate was added as a modifier and dispersant.

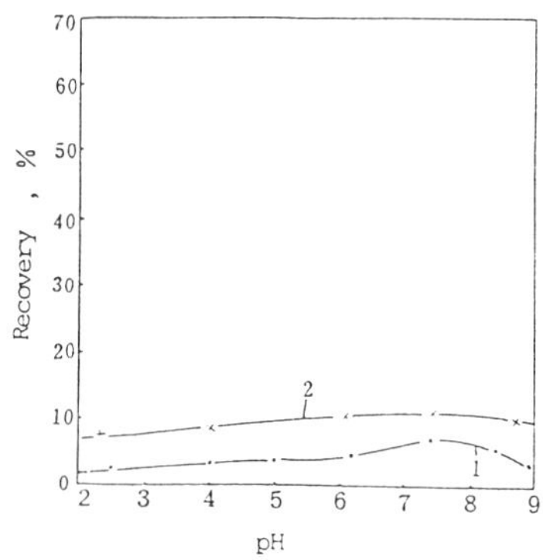

Fig. 7. Flotation behavious of amphibole by BBA.
1. size, -74—$+40$ μm, BBA, 10^{-4}mol/l;
2. size, -20 μm, emulsion mixture of BBA and kerosen(1 : 20), BBA, 10^{-4}mol/l, Na_2SiO_3, 60mg/l, agitation speed, 3000 r.p.m., agitation time, 15 min.

Based on the foregoing described results, a hydrophobic aggregation flotation technique has been worked out for separation of fine particles mixture of rutile and amphibole.

The content of rutile and amphibole in the artificial mixture is 10% and 90% respectively. The dosage of Na_2SiO_3 and BBA is 60mg/l

TABLE I. Separation results by hydrophobic aggregation flotation for artificial mixture of fine particle rutile and amphibole.

	product	mass yield (%)	grade of TiO_2 (%)	recovery of TiO_2 (%)
hydrophobic	feed	100.00	10.00	100.00
aggregation	concentrate	12.47	71.40	89.00
flotation	tailing	87.53	1.25	11.00
conventional	feed	100.00	10.00	100.00
	concentrate	17.14	26.30	45.09
flotation	tailing	82.86	6.63	54.91

and 10^{-4} mol/l respectively. For conventional flotation, the pulp with Na_2SiO_3 and BBA are stirred and floated in the flotation cell. For hydrophobic aggregation flotation, pulp with Na_2SiO_3 and emulsion mixture of BBA and kerosen(ratio is 1 : 20) are stirred strongly (agitation time, 15 min, agitation speed, 3000 r.p.m.)in the stirred tank, and then are poured into flotation cell. The results of both techniques are given in Table I.

It is clear by comparising the results in Table I, that the hydrophobic aggregation flotation technique has been proved to be a very efficient method of fine particles separation for a rutile amphibole mixture, whereas the conventional flotation technique failed to handle such material.

CONCLUSION

1. Hydrophobic aggregation flotation is efficient for fine particles separation. As an example, a mixture of rutile and amphibole(-20 μm) has been separated successfully by this method.

2. A proper combination of high agitation field and addition of non-polar oil emulsion together with collector is essential for hydrophobic aggregation flotation.

3. Fine particles of rutile gather together during such treatment, and form tight and large hydrophobic aggregates, thereby become easy to be floated.

REFERENCES

1. Shouci Lu, Shaoxian Song and Zongfu Dai, Proceedings of 16th IMPC, A, (Elsevier, Amsterdam, 1988), pp. 999-1009.
2. Shouci Lu and Zongfu Dai, Proceeding of the international symposium the production and processing of fine particles, Montreal, (1988).
3. Shouci Lu and Guoqing Li, J. Wuhan Iron and Steel University, NO. 2, 4-12(1984), (in Chinese).
4. L.J. Warren, J. Colloid and Interface Sci. , 50, NO. 2, 307-318(1975).

SELECTIVE FLOCCULATION OF CHRYSOCOLLA FINES WITH ANIONIC POLYACRYLAMIDE/ACRYLATE POLYMER

Y. Ye* and M. C. Fuerstenau**
*Department of Metallurgy and Metallurgical Engineering University of Utah, Salt Lake City, UT 84112; **Department of Chemical and Metallurgical Engineering, University of Nevada, Reno, Reno, NV 89557

ABSTRACT

Selective flocculation of chrysocolla fines has been examined with a variety of flocculants. Of those investigated, anionic polyacrylamide/acrylate polymers function the most effectively. The dissolution of cupric ion from the chrysocolla assumes a key role in flocculation behavior. Both chrysocolla and quartz/gangue are activated by hydrolyzed species of metal ions. Magnesium ion is found to adsorb specifically on chrysocolla and function as an activator under conditions in which significant hydrolysis of Mg^{++} does not occur.

INTRODUCTION

Selective separation of chrysocolla from gangue minerals continues to be one of the greatest challenges facing the mineral industry. Efforts in the past have involved flotation separations utilizing chelating agents (1-7), xanthates (8-10), mercaptan (11), other anionic/cationic collectors (12), sulfidization (10,13-18) and thermal activation (16,19-21). In general, only limited success has been achieved with these techniques. Other approaches need to be examined and selective flocculation appears promising.

Selective flocculation of finely-divided minerals with high-molecular weight polymers is a developing field in mineral processing. Probably the most noteworthy example of the successful application of this technology is the Tilden iron ore. In contrast to the research effort devoted to flotation of chrysocolla, study of selective flocculation of chrysocolla with these polymers is very limited at the present time. Previous investigators have reported that chrysocolla might be readily separated from a binary mixture with nonionic flocculants (22,23). However, it is generally known that nonionic polymers exhibit relatively-low selectivity in selective flocculation separations.

The objective of this study was two-fold: first, to examine the response of chrysocolla to flocculation with various high molecular weight polyelectrolytes and, second, to establish whether chrysocolla can be concentrated from natural ore with this technique.

EXPERIMENTAL MATERIALS

Chrysocolla, Quartz, Ore

Small lumps of pure quartz were crushed to about -48 mesh in a rolls crusher and then cleaned repeatedly with a strong magnet. Following that treatment, the quartz particles were contacted with acidified water

Advances in Fine Particles Processing
John Hanna and Yosry A. Attia, Editors

(pH~3), rinsed with distilled water and then ground in a ceramic mill. After grinding, finely-divided particles, five microns and less in size, were obtained by sedimentation sizing. Size and distribution moduli measured for these particles were about 4 μm and 1.78, respectively.

The chrysocolla sample was from Arizona and was obtained from Wards Natural Science Establishment, Inc. The sample was first crushed into small pieces 1-5 mm in size and, then, carefully hand-picked. Washing of these chrysocolla particles was carried out at neutral pH to avoid significant loss of surface copper ions. After washing, the sample was ground in a ceramic ball mill with distilled water at neutral pH. Similarly to quartz, finely-divided particles less than about five microns were obtained by sedimentation sizing. Measured size modulus and distribution modulus were 3 μm and 1.86, respectively. Analysis of this sample revealed the following:

Component	Wt.%
Cu	27.73
Adsorbed H_2O	6.52
Combined H_2O	15.81

For flocculation of natural chrysocolla ore, the ore was crushed to about -48 mesh and deslimed with tap water. The ore was then wet ground to -500 mesh in a ceramic ball mill with tap water. Major metal elements contained in the sample are as follows:

Elements	Ca	Cu	Fe	Ni	Mg	Mn	Zn	Others
Wt.%	1.22	0.66	0.53	0.28	0.28	0.25	0.02	<0.02

Flocculants

All of the flocculants used in this work are commercial products of American Cyanamid Company. They are Superfloc (SF) 204,212 and 214 (anionic polyacrylamide/acrylate-type polymers). The molecular weight of SF 204 is 4-6 million gm/gm-mole. Both SF 212 and 214 have molecular weights of 12-15 million gm/gm-mole, but the degree of anionic charge is 15~20% for SF 212 and 20~40% for SF 214. Some experiments were conducted with SF 16 and 17 (nonionic), Aerofloc (AF) 550 (anionic) and SF 320 (cationic), but only data which are pertinent to this paper are included. All of the flocculants were prepared bi-weekly as a 0.05% solution and stored at room temperature.

EXPERIMENTAL PROCEDURES

Slurry volume with known solids concentration was one liter. The slurry was placed into a beaker and conditioned by stirring for 10 minutes at desired values of pH. After that time, flocculant was added, and the slurry was stirred gently for another 10 minutes. The slurry was then transferred to a one-liter graduated cylinder for sedimentation. Sedimentation time was controlled at five minutes, after which the pulp liquor containing residual solids particles was pumped into another container. Pumping took about 40 seconds. To prevent suction of flocs into the pulp and bias of experimental results, about 20 mls of pulp liquor were always left over the sediment layer. Both the sediment and the non-settled solids were dried in an oven and analyzed.

EXPERIMENTAL RESULTS

Choice of Flocculant and Critical Flocculant Concentration

The initial experiments in this work were conducted to determine the critical flocculation concentration (cfc) of the flocculants, since it is well known that for a given range of solids concentration, flocculation reaches a maximum at a particular flocculant concentration. This optimal flocculant concentration is often referred to as the critical flocculation concentration or cfc. A reduction in flocculation is expected when the concentration of a flocculant used is either higher or lower than the cfc. This work was carried out with various types of flocculants with chrysocolla and quartz suspensions in separate systems. The results are given in Figure 1. Poor flocculation of both chrysocolla and quartz with cationic SF 320 is revealed. Only about 40% of the chrysocolla particles and 25% of the quartz particles were flocculated under these conditions at a cfc of about 0.3~0.5 ppm even though the zeta potentials of chrysocolla and quartz particles at this pH are negative. Secondly, the data suggest that the separation of chrysocolla from quartz may be possible with both nonionic and anionic SF flocculants. In the case of nonionic SF 127, the cfc for chrysocolla is in a broad range from about 1 ppm to 5 ppm with 80-90% solids flocculated (Figure 1), while the cfc for quartz is about 1 ppm. The quantity of quartz flocculated is reduced from 90% to 30% when the flocculant concentration is increased from 1 ppm to 4 ppm. These results indicate that selective flocculation of chrysocolla from quartz might be achieved with SF 127 utilizing the difference of cfc's for chrysocolla and quartz. However, good selectivity of chrysocolla from quartz could not be achieved from a mixture of the two minerals. Further tests with nonionic SF polymers were, therefore, discontinued.

The cfc of anionic SF 212 is the same for both chrysocolla and quartz under the same conditions, namely, about 0.3~0.5 ppm. At such a cfc, 90% of the chrysocolla and 20% of the quartz are flocculated, respectively (Figure 1). All of the work reported later in this paper was carried out with this type of anionic flocculant.

Effect of pH and Dissolution of Surface Copper Ions

This series of experiments was carried out to determine the pH value suitable for selective flocculation of chrysocolla with anionic SF flocculants. The work was conducted with both single mineral systems and mixed binary chrysocolla/quartz systems. For single mineral systems, flocculation of chrysocolla or quartz was carried out separately as a function of flocculation pH. Typical results are presented in Figure 2 which gives residual solids concentration of chrysocolla and quartz, respectively, as a function of flocculation pH. As shown, chrysocolla particles can be well flocculated with SF 212 in a pH range less than pH 7. When the flocculation pH is greater than 7, flocculation is reduced significantly, which is represented by a sharp increase in residual solids concentration in suspension.

In an alkaline pH range, chrysocolla particles are not flocculated by SF 212. With quartz particles, good flocculation occurs below pH 4 but not above this value.

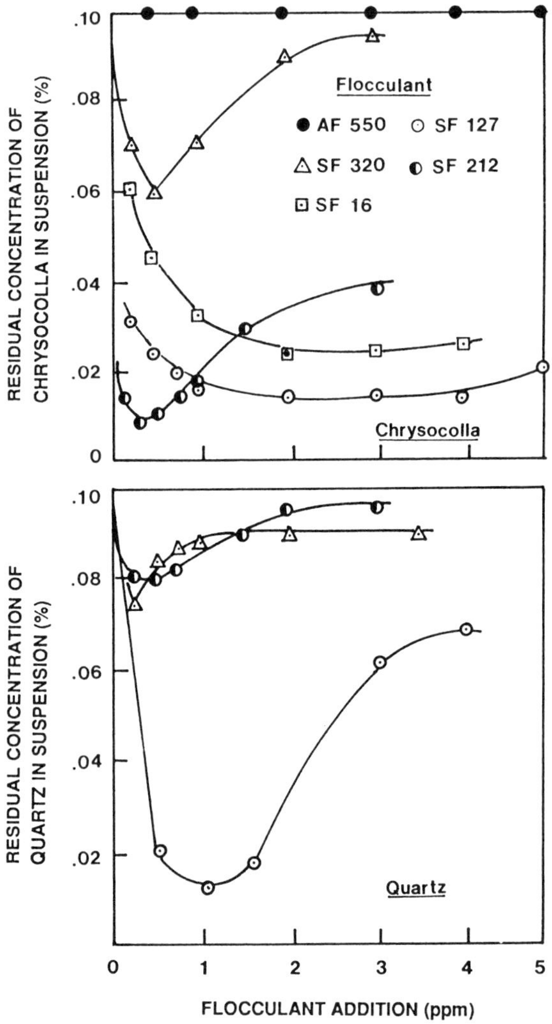

Figure 1. Residual concentration of chrysocolla and quartz particles in separate systems as a function of polymer addition at pH 6.0. (Solids concentration 0.1 wt%.)

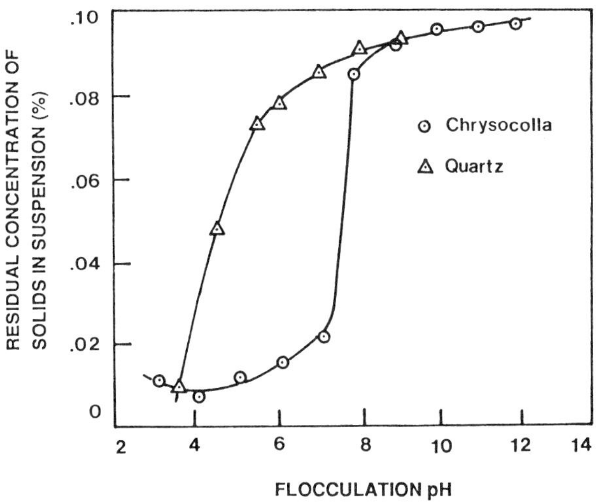

Figure 2. Residual concentration of chrysocolla and quartz in separate systems as a function of flocculation pH. (Solids 0.1%; 1 ppm SF 212.)

With a binary mixture of chrysocolla and quartz, the optimum flocculation pH for maximal recovery of chrysocolla is dependent upon the solids concentration. At higher solids concentration of chrysocolla, flocculation recovery is greatest at about pH 6.5. However, as the solids concentration of chrysocolla is reduced, the optimum pH for maximal flocculation recovery shifts to acidic pH values. As can be noted in Figure 3, optimum flocculation pH shifts from about 6.5 for 0.25% chrysocolla to about 4.7 for 0.025% chrysocolla. The concentration of quartz in these experiments was maintained at 0.25%. The recovery of quartz as a function of pH followed exactly the same pattern as chrysocolla, except that recovery was always lower than that of chrysocolla.

The effect of dissolution of copper ions from chrysocolla on the flocculation of chrysocolla and quartz is significant and is well demonstrated by the data presented in Figure 4. In this group of experiments, the flocculation pH was held constant at pH 5.6, which is the optimum value for highest recovery of chrysocolla with given chrysocolla solids concentration (0.1%). However, in each experiment prior to the addition of flocculant, the pulp pH was first adjusted to an acidic value lower than pH 5.6 and conditioned for about 4 minutes to enhance dissolution of Cu^{++} from the surface. This pH value is referred to as conditioning pH in Figure 4. After conditioning, the pulp pH was then adjusted to 5.6, and the flocculant was added. As can be seen, flocculation recovery of chrysocolla and quartz is improved significantly under these conditions.

Metal Ion Activation

In view of the results obtained with Cu^{++} activation (Figure 4),

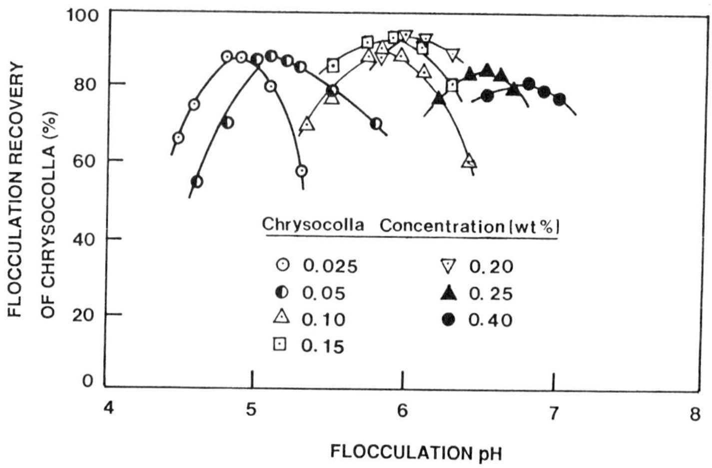

Figure 3. Flocculation recovery of chrysocolla from a binary mixture as a function of flocculation pH. (Quartz concentration 0.25%; 0.5 ppm SF 214.)

experiments were conducted in which Mg^{+2} and Ca^{+2} were added as activators. The flocculation response of chrysocolla and quartz is essentially the same in the presence of Ca^{+2}. See Figure 5. In the case of Mg^{++} additions, however, although the responses of chrysocolla and quartz are nearly the same at pH 10, significant difference in flocculation response is noted at pH 8.5.

Adsorption experiments were conducted next, and the results of these experiments corroborate the flocculation recovery data in Figure 5. As shown in Figure 6, adsorption of anionic SF 204 is much less extensive on quartz than on chrysocolla, and maximal adsorption occurs at pH 10.5. In the case of chrysocolla, maximal adsorption occurs from pH 8.5 to 10.5.

Electrokinetic experiments were conducted to establish the surface characteristics of quartz and chrysocolla in the presence of 1×10^{-3} M Mg^{+2} additions. Zeta reversal occurs at around pH 10.5 which phenomenon has been observed by others (28). (See Figure 7.) With chrysocolla, the value of the zeta potential is reduced in magnitude and is virtually independent of pH from about pH 7 to 10.

Experiments were conducted to establish attainable separations of chrysocolla from quartz from mixtures of these two minerals utilizing Mg^{+2} activation. These results are shown in Figure 8 in which concentrate grade is presented as a function of feed grade of chrysocolla. Flocculation recovery of chrysocolla was maintained at greater than 98 percent in this series. As shown, chrysocolla can be concentrated to 23.2% (6.44% Cu) from a feed grade of 1.8% (0.50% Cu) with single-stage flocculation.

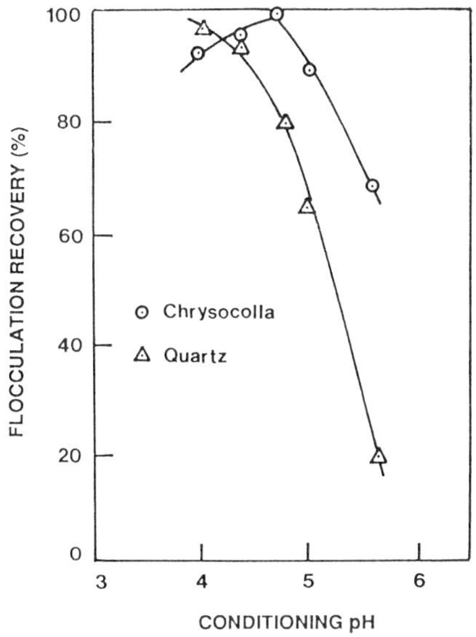

Figure 4. Flocculation recovery of chrysocolla and quartz as a function of conditioning pH. (Solids: 0.1% chrysocolla and 1.0% quartz; 0.5 ppm SF 214; flocculation pH 5.6.)

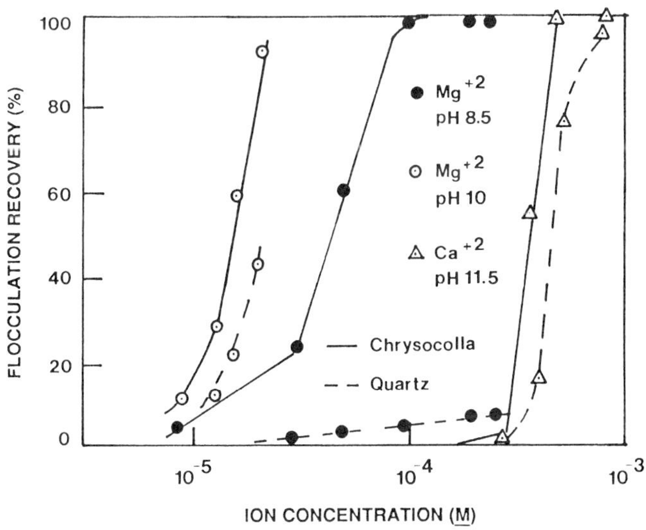

Figure 5. Recovery of chrysocolla and quartz from mixture as a function of ion addition. (0.1% chrysocolla; 1.2% quartz; 0.5 ppm SF 214.)

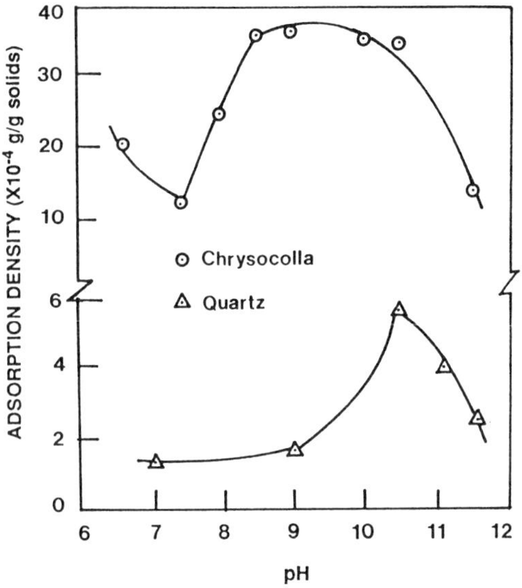

Figure 6. Adsorption density of SF 204 on chrysocolla and quartz as a function of pH. (1×10^{-4} M Mg^{+2}; 5 ppm SF 204; solids 0.1%; particle size 74 x 105 μm.)

Flocculation of Natural Ore

In view of the separation of chrysocolla from quartz that is possible from mixtures of these minerals, experiments were extended to natural ore. In this work cupric ion activation of gangue was observed to have a major impact on selectivity. This fact can be noted in Figure 9 in which flocculation recovery is plotted as a function of pH for various grinding conditions. When the ore was ground at neutral and quite alkaline pH, extensive recovery of gangue was obtained. However, when the ore was ground first at pH 11, flocculation recovery of gangue was minimal in alkaline medium.

In this view, it was found necessary to filter the pulp after grinding, condition the solids in acidified water (pH 5), filter, repulp the solids with fresh water and subject the pulp to flocculation conditions. When these procedures are used, good separation of chrysocolla from gangue is achieved (Figure 10).

Continuous Flocculation

Continuous flocculation of natural ore was conducted after grinding and washing as described above. Pulp containing 2% solids by weight was conditioned with Mg^{++} and SF 204 and pumped into a column, 2 in. (ID) and 20 in. high. The feed inlet on the column was located 3 in. from the bottom. Floccules settled below the feed inlet, while the rougher overflow was pumped to a similar column for cleaner flocculation. Overflow from the cleaner column which represents the tailing in this operation was collected and analyzed. Experimental results obtained with this procedure are given in Table 1. Approximately 80 percent recovery was effected at a concentrate grade of 2.57 percent Cu under these conditions.

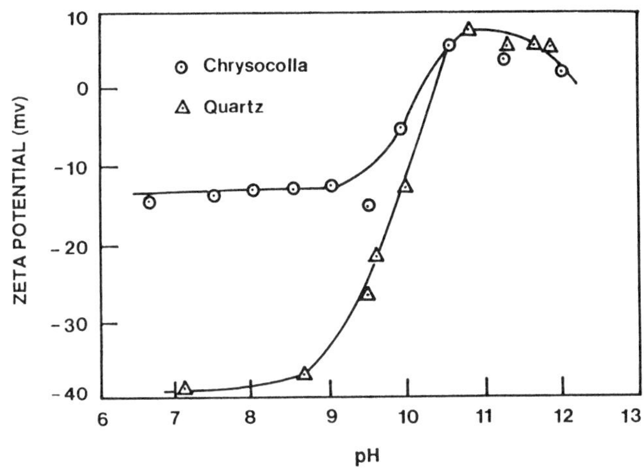

Figure 7. Zeta potential of chrysocolla and quartz as a function of pH in the presence of 1 x 10^{-3} M Mg^{+2} addition.

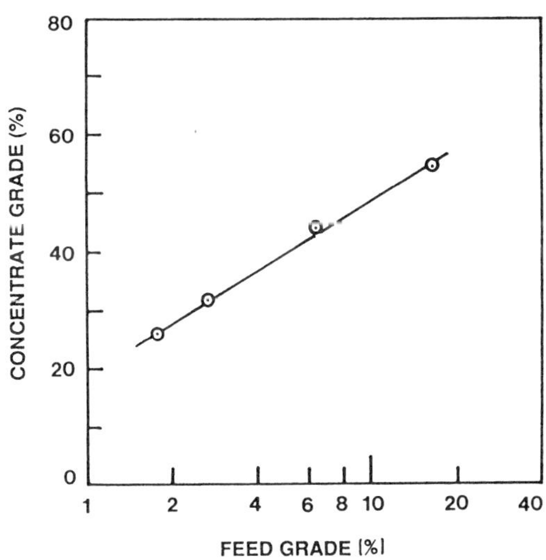

Figure 8. Concentrate grade of chrysocolla from single-stage flocculation versus feed grade. (Flocculation pH 8.5; 2 x 10^{-4} M Mg^{+2}; 0.5 ppm SF 214; flocculation recovery of chrysocolla > 98%.)

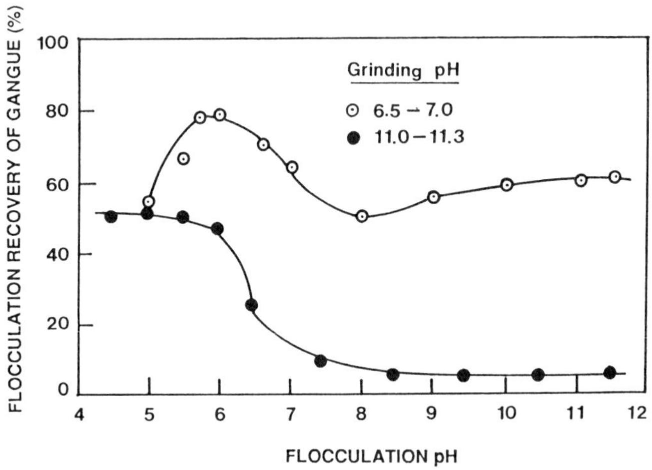

Figure 9. Recovery of gangue as a function of flocculation pH and grinding pH. (1% solids; 0.5 ppm SF 214.)

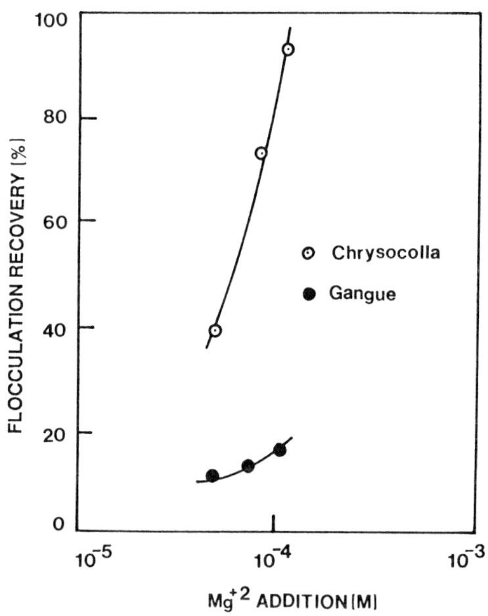

Figure 10. Flocculation recovery of chrysocolla and gangue as a function of Mg^{+2} addition. (Flocculation pH 8.5; 1% solids; 2.5 ppm SF 214.)

Table I. Experimental results obtained with continuous flocculation. Conditions: 1 x 10^{-4} M Mg^{+2}, 1 ppm SF 204, flocculation pH 8.5, feed grade 0.42% Cu, slurry feed rate 4 ml/s.

Table I. Experimental results obtained with continuous flocculation. Conditions: 1×10^{-4} M Mg^{+2}, 1 ppm SF 204, flocculation pH 8.5, feed grade 0.42% Cu, slurry feed rate 4 ml/s.

	Flocculation Recovery (%)	Concentrate Grade (% Cu)
Rougher	61.2	2.56
Cleaner	17.7	2.61

Total recovery 78.9%; Conc. grade 2.57% Cu

DISCUSSION OF RESULTS

Flocculation phenomena observed with the various flocculants used in this investigation were similar to those observed in other studies. That is, the molecular weights of SF 16, 127, 212 and AF 550 are 4×10^6, 15×10^6, $12\text{-}15 \times 10^6$ and 0.25×10^6 gm/gm-mole, respectively, and although the properties of nonionic SF 16 and 127 are the same, the flocculation ability of the former is less than that of the latter due to the difference in their molecular weights. Further, low-molecular weight AF 550 (anionic) does not flocculate chrysocolla in the same concentration range as observed with other flocculants. The higher the molecular weight of the polymer is, the greater is the flocculating ability.

With regard to anionic versus nonionic polymers, the flocculation of chrysocolla with anionic SF 212 at its cfc is superior to that obtained with nonionic 127 even though the molecular weight of SF 212 is less than SF 127. Flocculation results for quartz with these two polymers are just the opposite to those for chrysocolla. This fact suggests that ionic interactions in flocculation phenomena are much stronger than hydrogen bond interactions between the polymers and mineral surface.

Activation of oxide and silicate minerals with metal ions has received extensive study (26-29), and the role of hydrolyzed metal species has been delineated in these studies. Activation studies of chrysocolla, however, have been limited. In previous work the recovery of chrysocolla with amyl xanthate after contact with titanium tetrachloride has been reported (8). The important role of $CuOH^+$ and $Cu(OH)_2$ in chrysocolla flotation with octyl hydroxamate has been demonstrated (3). The results of this investigation also show that metal hydroxy complexes and metal hydroxides are involved in the flocculation of chrysocolla and quartz with polyacrylamide/acrylate flocculants.

The pH range in which hydrolyzed species of Ca^{+2} activate quartz in the presence of anionic flotation collectors is the same as that observed for effective polymer adsorption and flocculation. That is, the pH range in which calcium hydroxide forms on the surface and exhibits a positive charge is the same range that SF 214, a high molecular weight anionic polymer, adsorbs resulting in flocculation. Similar phenomena are also noted when Mg^{+2} is added at pH 10 in which region Mg^{+2} hydrolyzes to $MgOH^+$ and $Mg(OH)_{2(s)}$.

The specific adsorption of Mg^{+2} on chrysocolla is readily apparent from the data in Figures 5 and 7. The zeta potential of chrysocolla in the pH range 7 to 9 in the absence of Mg^{+2} is about -26 mv. In the presence of 1×10^{-3} M Mg^{++}, the magnitude of this potential is reduced to about -13 mv in this same pH range.

Given that the reactions for hydrolysis of Mg^{+2} (30) are,

$$Mg^{+2} + OH^- \quad MgOH^+ \qquad K = 380$$

$$Mg(OH)_{2(s)} \quad Mg^{+2} + 2OH^- \qquad K = 1.82 \times 10^{-11}$$

the activating role that Mg^{+2} exhibits is probably not due to adsorption of hydrolyzed species at pH 8.5. Equivalent flocculation response (say 30% recovery) is effected at about 1.5×10^{-5}M and 4.5×10^{-5}M Mg^{++} at pH 10.0 and 8.5, respectively. If the adsorption of hydroxy complexes and the formation of magnesium hydroxide on the chrysocolla surface were controlling, the Mg^{+2} addition would be expected to be 3.15×10^{-4}M at pH 8.5. The reasons for the specificity of Mg^{++} are unclear at the present time.

Work with mixtures of chrysocolla and quartz demonstrated that selective flocculation of chrysocolla from quartz is possible with high molecular weight anionic polymers. Unfortunately, with a natural ore ground under normal conditions, general activation of chrysocolla and gangue is effected resulting in nonselective flocculation. Elimination of the solution after grinding and conditioning at pH 5 with fresh water did result in flocculation selectivity between chrysocolla and gangue minerals. Such treatment, unfortunately, is not practical in large scale operations.

ACKNOWLEDGMENT

The authors gratefully acknowledge the support provided by the Mining Mineral Resources Research Program of the Department of the Interior administered by the Bureau of Mines.

REFERENCES

1. R.W. Ludt and C.C. DeWitt, Trans. AIME, vol. 184, p. 49, 1949.
2. D.W. Fuerstenau, M.S. Thesis, Montana School of Mines, 1950.
3. H.D. Peterson, M.C. Fuerstenau, R.S. Rickard, and J.D. Miller, Trans. AIME, vol. 232, pp. 388, 1965.
4. J.W. Scott and G.W. Poling, Can. Metal Q., vol. 12, p. 1, 1973.
5. B.R. Palmer, G. Gutierrez and M.C. Fuerstenau, Trans. AIME, vol. 258, p. 257, 1975.
6. J.C. Cecile, French Patent 2355065, 1978.
7. D.R. Nagaraj and P. Somasundaran, Min.Eng., vol. 33, p. 1351, 1981.
8. A.P. Prosser, A.J. Wright and J.D. Stephens, Trans. IMM, vol. 76, p. C233, 1964-65.
9. A.J. Wright and A.P. Prosser, Trans. IMM, vol. 74, p. C259, 1964-65.
10. S. Castro, H. Gaytan and J. Goldfarb, Int. J. Min. Proc., vol. 3, p. 71, 1976.
11. F.F. Aplan and D.W. Fuerstenau, Int. J. Min. Proc., vol. 13, p. 105, 1984.

12. A.C. Gonzalez, G. Gonzalez, J. Laskowski, Trans. IMM, vol. 84, p. C154, 1975.
13. F.W. Bowdish and T.P. Chen, Trans. AIME, vol. 226, p. 21, 1963.
14. F.W. Bowdish and W.S. Stahmamn, Trans. AIME, vol. 238, p. 118, 1967.
15. S. Castro, J. Goldfarb and J. Laskowski, Int. J. Min., vol. 1, p. 141, 1974.
16. S. Castro, H. Sota, J. Goldfarb and J. Laskowski, Int. J. Min. Proc., vol. 1, p. 151, 1974.
17. H. Soto, W. Aliaga and P. Riveros, Trans.IMM, vol. 84, p. C250, 1975.
18. S. Raghavan, E. Adamec and L. Lee, Int. J. Min. Proc., vol. 12, p. 173, 1984.
19. G.A. Parks and C. Kovacs, Trans. AIME, vol. 235, p. 349, 1966.
20. C. Queirolo and S. Castro, Trans. IMM, vol. 85, p. C166, 1976.
21. G. Gonzalez and H. Sota, Int. J. Min. Proc., vol. 5, p. 153, 1978.
22. J. Rubio and J. Goldfarb, Inst. Min. Met., vol. 84, p. C123, 1975.
23. J. Rubio and J.A. Kitchener, Trans. IMM, vol. 84, p. C97, 1977.
24. F.W. Bowdish and T.M. Plouf, Trans. AIME, vol. 254, p. 66, 1973.
25. W.B. Crummett and R.A. Hummel, J. Am. Water Works Assoc., vol. 57, p. 209, 1966.
26. M.C. Fuerstenau, C.C. Martin and R.B. Bhappu, Trans. AIME, vol. 226, p. 449, 1963.
27. M.C. Fuerstenau and D.A. Rice, Trans. AIME, vol. 241, p. 453, 1968.
28. M.C. Fuerstenau, D.A. Eligillani and J.D. Miller, Trans. AIME, vol. 247, p. 11, 1970.
29. M.C. Fuerstenau and B.R. Palmer, Flotation, M.C. Fuerstenau (ed), AIME, New York, vol. 1, p. 186, 1976.
30. J.N. Butler, Ionic Equilibrium, A Mathematical Approach, Addison-Wesley Publishing Company, Inc., Reading, MA, p. 287, 1964.

THERMODYNAMICS OF ADSORPTION OF A HYDROPHOBIC POLYMERIC FLOCCULANT ON COAL, PYRITE AND SHALE MINERALS

S. Yu and Y.A. Attia

Dept. of Materials Science and Engineering, 116 W. 19th Ave., The Ohio State University, Columbus, Ohio 43210

ABSTRACT

A study was performed on the adsorption of a hydrophobic polymer, FR-7A, on individual suspensions of coal, pyrite and shale minerals. A nephelometric method was employed for the determining FR-7A concentration in solutions. From the adsorption isotherms, thermodynamic derivations of ΔG^o_{ads}, ΔG^o_{el}, ΔG_H, ΔH^o_{ads} and ΔS^o_{ads} were made. The study led to a number of findings: 1) adsorption of FR-7A on the minerals followed the Langmuir adsorption model; 2) FR-7A had a higher adsorption affinity to coal than to pyrite or to shale minerals; the higher affinity to coal could be attributed to "hydrophobic bonding"; 3) adsorption affinity of FR-7A on all minerals decreased as pH increased.

INTRODUCTION

Preferential adsorption of a polymeric flocculant on a particular type of mineral or colloidal surface is the key step in achieving selective flocculation separation process for these minerals or colloids. Therefore, adsorption kinetics and isotherms can be very useful in predicting the performance of the selective flocculation separation process.

Adsorption of low molecular weight polymers from solution on solid surfaces has been widely studied, but literature on the adsorption of long-chain polymeric flocculants is relatively scarce [1,2]. The shortage of such studies was due to in part to the difficulty in finding analytical methods which could accurately determine very small concentrations of the high molecular weight polymers. Attia and Rubio [1] developed a nephelometric technique for the determination of very low concentrations (0-10 mg/l) of non-ionic and cationic polymeric flocculants. The method precipitated the polymer with tannic acid in the presence of sodium chloride. Pradip, Attia and Fuerstenau [2] further developed this technique and applied it to the adsorption of polymeric flocculants on some minerals. The results indicated that the nephelometric method was simple, accurate, and allowed for studying the behavior of polymer adsorption on minerals. This method was also employed for studying the adsorption behavior of several other polymers on minerals, including: fluorapatite and quartz in a phosphate system, and coal and shale in a coal slurry system [3].

It is known that polymer adsorption, like other molecular adsorption, is a thermodynamic process. Therefore, the adsorption characteristics can be expressed by some thermodynamic parameters, such as ΔG_{ads} (Gibbs free energy change for adsorption), ΔH_{ads} (enthalpy change of adsorption), and ΔS_{ads} (entropy change of adsorption). The physical meaning of ΔG_{ads} is the indication of adsorption likehood and affinity. The more negative the value of ΔG_{ads}, the higher the adsorption affinity of the adsorbate to the adsorbent. The value of ΔH_{ads} is an indication of the bond type of strength between the polymer's functional groups and the solid's surface cites, e.g.,

Advances in Fine Particles Processing
John Hanna and Yosry A. Attia, Editors

a strong attraction corresponding to a high negative value of ΔH_{ads}. From the sign of ΔS_{ads}, the direction of adsorption/desorption can be identified. For example, a negative value of ΔS_{ads} signifies that adsorption is possible, while a positive value of ΔS_{ads} means the occurrence of a desorption process.

This study aimed at investigating the adsorption behavior of a hydrophobic polymer, FR-7A, which has been demonstrated by Attia and his co-workers [4-7] to be an effective selective flocculant for coal. The adsorption data should provide information to predict the selectivity of the polymer to the mineral components in coal slurry. Furthermore, thermodynamics of polymer adsorption should enable the determination of the mechanisms involved.

MATERIALS AND METHODS

Sample Preparation

The coal samples used in this study were from Upper Freeport seam, supplied by Babcock and Wilcox Research and Development Division, Alliance, Ohio, and Pittsburgh No. 8 seam, obtained from the R & F Coal Company, Lamira Preparation Plant, Warnok, Ohio. Each raw coal sample was crushed and screened to the size range of -10+100 mesh (-1700+150 microns), and then cleaned using a heavy liquid, 1,1,1-trichloroethane with a specific gravity of 1.33, to remove the majority of ash-forming minerals and pyrite (sink product). The clean coal (float product) was washed with acetone to remove the residual solvent on the coal surface, and was then dried in an oven at about 104°C for five hours to evaporate acetone. The clean coal samples were dry-ground to very fine size (the majority of the particles were below 25 microns) using a stainless steel attritor, and were stored in a covered container filled with nitrogen gas at reduced temperature (-10°C) in a refrigerator.

An almost pure pyrite crystal (FeS_2) from Hunzala, Peru, and the argillaceous shale minerals were supplied by Ward's Science Establishment, Inc., Rochester, New York. Each mineral sample was ground separately to 75% less than 25 microns. The samples were stored at -10°C under nitrogen.

Sample Characterization

The characterization data of the mineral samples and the employed instruments are listed in Table 1. The measurements of ash and total sulfur contents followed the ATSM procedures [8]. Fig. 1 shows the measured zeta-potential of the minerals versus pH. The measured pzc's were close to those reported in literature [9-11].

Chemicals

A hydrophobic polymeric flocculant, FR-7A, provided by Calgon Corp., Pittsburgh, PA, was used in this study. The polymer was encapsulated in an oil in water emulsion, and it dispersed well in water. From an observation of the migration of the polymeric colloid clouds toward the anode in the electrophoresis cell, it was concluded that the FR-7A was slightly negatively charged. The molecular weight of FR-7A was believed to be slightly less than one million.

Nephelometric Determination of Polymer Concentration

Since FR-7A solution was naturally turbid, its concentration could be determined from the measurement of its turbidity. A Nephelometer, Model 21,

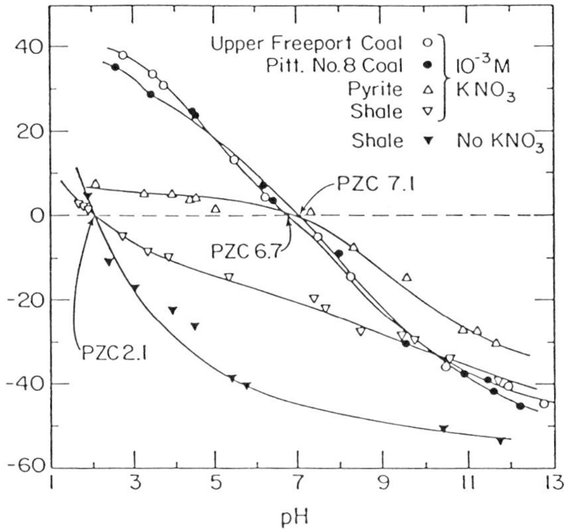

FIG. 1. Zeta-potential of coal, pyrite and shale

TABLE 1. Characterization of the mineral samples

	Coal* UFP	Coal* Pitt	Pyrite	Shale	Instrument for Characterization
Ash content (%)	3.5	4.2	---	---	Isotemp Programmable Ashing Furnace, Model 497, Fisher Sci. Co.
Total sulfur (%)	1.8	1.7	---	---	Sulfur Determinator, Model SC132, LECO Co.
Surface area (m^2/g)	2.62	3.50	3.58	7.42	Accusorb, Model 2100E Micromeritics Instrument Corp.
Area mean diameter (micron)	14.6	20.9	4.12	4.36	Microtrac II Particle Size Analyzer, Leeds & Northrup Co.
Point of zero charge at pH	6.7	7.1	6.7	2.1	Zeta-meter, Laser Zee(tm) Model 501, Penkem, Inc.
Literature Reported pzc	5-8** [9]		6.2-6.9 [10]	1.0 [11]	

* UFP = Upper Freeport; Pitt = Pittsburgh No. 8;
** Bituminous coals.

manufactured by Monitek Inc., Hayward, CA, was employed to for measurement of turbidity of FR-7A solutions. This instrument measured turbidity from 0.01 to 200 Ntu. Calibration curves were established by measuring the turbidity of polymer solutions with known concentrations. The relationship of FR-7A concentration versus turbidity was nearly linear, and it was not affected by solution pH [12]. The detection limit of FR-7A by this method was 0.05 mg/l. From the calibration curves, the polymer concentration of an unknown solution could be determined after measuring its turbidity.

Adsorption Procedure

Adsorption of FR-7A on individual suspensions of coal, pyrite and shale minerals may be defined in a four-step procedure. First, a pre-determined amount of polymer solution was mixed with the slurry by agitation for 5 minutes, and then the mixture was allowed to stand unperturbed a certain time. Subsequently, a solid-liquid separation was conducted. For coal flocs, the separation occurred easily by vacuum filtration using a water-pump. For pyrite and shale minerals, the filtration was performed with a pressure membrane filter. The pore size of the membranes were less than 0.2 microns for those used with shale minerals and 0.45 microns for those used with pyrite particles. The turbidity of the solids-free solution was determined by nephelometry, as described before. The difference in concentration between the initial and residual solutions gave the quantity of polymer adsorbed, and finally the adsorption density was derived using the pre-determined specific surface area of the mineral samples.

EXPERIMENTAL RESULTS

Adsorption Kinetics

The kinetic study (Fig. 2) shows that equilibrium was attained in about 4 to 7 hours for these samples, with a steady adsorption state until the adsorption plateau was reached. The figure also indicates that polymer adsorption in the early adsorption period was faster on coal than on pyrite, and was much faster than on shale. For example, in the pH range of 8.4 to 9.0 and the period of the first hour, the polymer adsorption rate was 0.79 mg/m^2-hr on coal, 0.21 mg/m^2-hr on pyrite, and 0.04 mg/m^2-hr on shale minerals. Thus, the higher adsorption rate on coal favors the selective flocculation process. For adsorption isotherms experiments, 12 hours was taken as the time to achieve adsorption equilibrium.

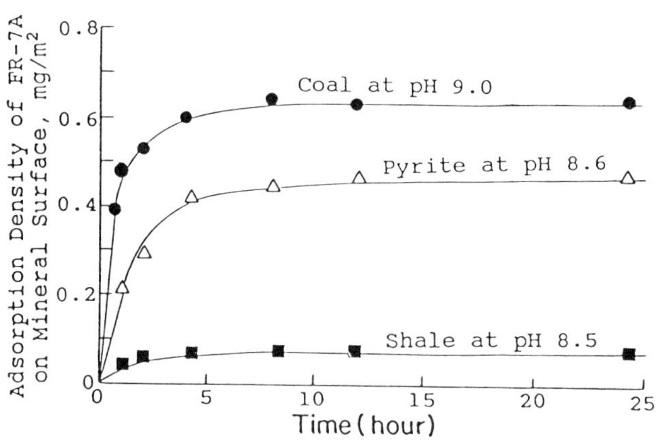

FIG. 2. Adsorption kinetics of FR-7A on coal, pyrite and shale at initial FR-7A concentrations of 100 mg/l and at pH 8.4-9.0

Adsorption Isotherms

Fig. 3 illustrates the adsorption isotherms at various levels of pH. The adsorption density increased with increasing equilibrium concentration

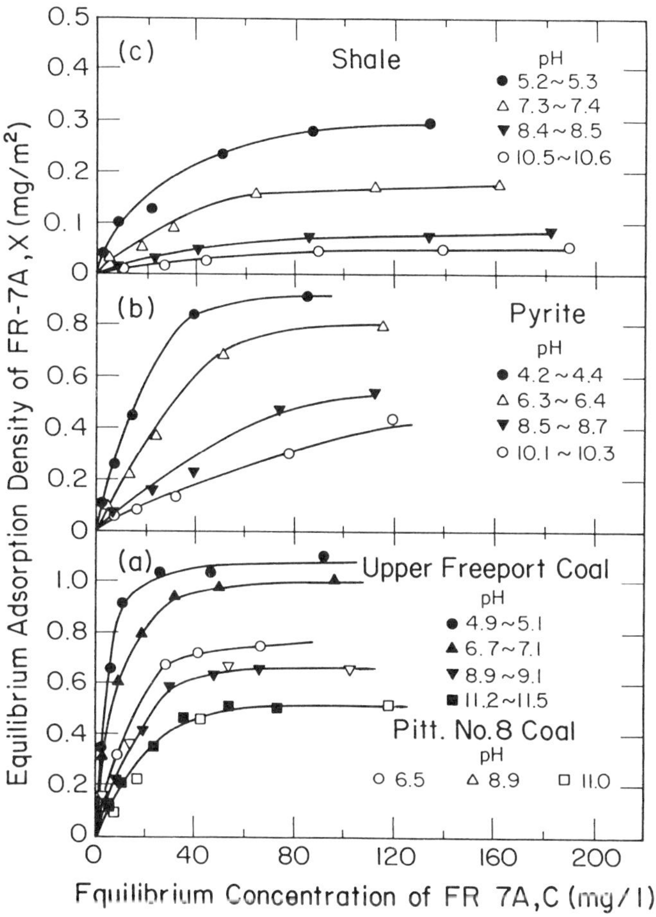

FIG. 3. Adsorption isotherms of FR-7A on coal, pyrite and shale at various levels of pH

of the polymer in the early stages, and then remained constant. This isotherm is very similar to the Langmuir type adsorption model. Namely, at relatively low FR-7A concentrations, only mono-layer coverage occurs on the minerals [13].

The figure also shows that equilibrium adsorption density of FR-7A on coal was about seven times higher than that on shale; namely, the FR-7A had a higher adsorption affinity to coal than to shale minerals. The high adsorption affinity of FR-7A to coal could be attributed to "hydrophobic bond" formation between the hydrophobic polymer molecules and the naturally hydrophobic surface of coal. For shale, adsorption was depressed due both to the natural hydrophilicity and to the relatively high negative zeta-potential of the surface. Referring to the results of adsorption kinetics which show the adsorption rate was much higher for coal than for shale in the early adsorption period, it is believed that the ratio of the polymer adsorption density on coal to that on shale minerals would be much greater if the polymer adsorption and floc formation was conducted for only 5 to 10 minutes, which is usually the time for a selective flocculation process.

However, the equilibrium adsorption density of FR-7A on pyrite was also high. Pyrite, like coal, has also hydrophobic surface in nature, and a pzc very close to coal's. These properties imply a difficulty for separating pyrite from coal in a selective flocculation process using FR-7A as the selective flocculant. To remove pyrite, a selective dispersant should be used to modify the pyrite surface and render it hydrophilic to depress its adsorption of FR-7A. Studies with such a polyxanthate reagent has been reported by Attia and his co-workers [14].

Effect of pH

Fig. 4 shows the effect of pH on the adsorption of FR-7A upon selected minerals at various initial FR-7A concentrations. The figure illustrates that at increasing pH, the equilibrium adsorption density of FR-7A molecules on all the minerals generally decreased. At alkaline pH, the minerals' surfaces are more negatively charged, which resulted in a depression of adsorption of the weekly negatively charged FR-7A molecules because electrostatic repulsion became operative. While at acidic pH, coal and pyrite became positively charged and shale minerals became less negatively charged. As a consequence, polymer adsorption was enhanced due to electrical attraction on all the minerals. However, at neutral pH condition, adsorption of FR-7A on coal was high mainly due to hydrophobic association, while that on shale was low because the electrical repulsion was operative. That meant that selective flocculation using FR-7A would likely occur in a neutral pH range.

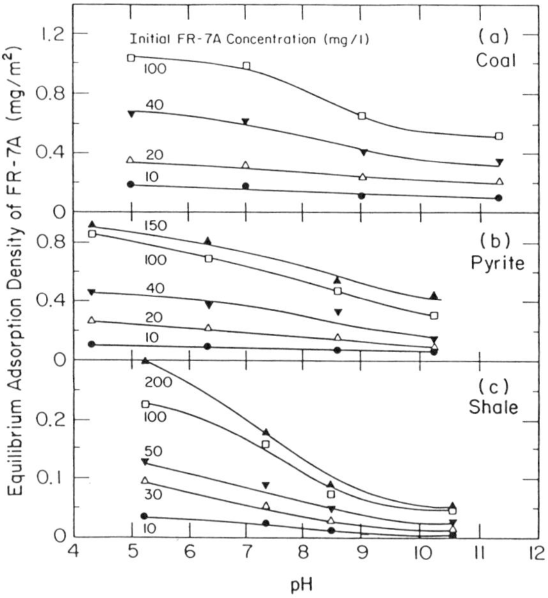

FIG. 4. Effect of pH on the adsorption density of FR-7A on coal, pyrite, and shale

THERMODYNAMICS OF POLYMER ADSORPTION

Determination of Adsorption Equilibrium Constant

The process for FR-7A adsorption can be described as follows:

$$\text{Solvent (adsorbed, } a_1^s\text{) + FR-7A (in solution } a_2\text{)}$$
$$= \text{Solvent (in solution, } a_1\text{) + FR-7A (adsorbed, } a_2^s\text{)} \qquad (1)$$

where a_1 and a_2 are the activities of the solvent water and polymer in solution respectively, and a^s represents the activity at the solid/liquid interface. Letting K be the equilibrium constant for the reaction, we can further derive Eq. (2) following Stumm and Morgan [15], and Pradip, Attia and Fuerstenau [2],

$$K = \exp(-\Delta G^o_{ads}/RT) = \frac{a_2^s a_1}{a_1^s a_2} \qquad (2)$$

where ΔG^o_{ads} is the standard Gibbs free energy change of adsorption of unit mole of monomer molecules of the polymer, R is gas constant, T is temperature. For dilute polymer solutions, the activities in solution can be replaced by equilibrium concentrations, C, and the activities at the surface can be assumed to be related to adsorption densities. Equation (2) can then be re-written as follow:

$$K = \exp(-\Delta G^o_{ads}/RT) = \frac{X}{C(X_o - X)} \qquad (3)$$

where X is the adsorption density and X_o the saturation adsorption density of the polymer on the mineral surface. From Eq. (3), a plot of C/X versus C should yield a straight line:

$$\frac{C}{X} = \frac{1}{X_o K} + \frac{C}{X_o} \qquad (4)$$

The slope and intercept of this relationship would determine both the saturation adsorption density X_o and the equilibrium constant K for the adsorption reaction.

Fig. 5 shows Langmuir plots of FR-7A concentration, C, versus C/X for the three mineral samples at pH range of 8.5 to 9.0. The good linear plots indicate that the adsorption followed the Langmuir adsorption model.

Estimation of Standard Gibbs Free Energy Change

From the obtained K, the Gibbs free energy change for adsorption at standard conditions (usually defined as 25°C and 1 atmosphere), ΔG^o_{ads}, can be calculated [16]. The standard Gibbs free energy change of adsorption is made up of contributions due to specific bonding, ΔG^o_{sp}, and electrostatic bonding, ΔG^o_{el}, ΔG^o_{ads} can be expressed as:

$$\Delta G^o_{ads} = \Delta G^o_{el} + \Delta G^o_{sp} = zF\zeta + \Delta G^o_{sp} \qquad (5)$$

where z is the valency of the polymer's functional group, F is the Faraday constant and ζ is the zeta potential of the mineral surface.

The values for Gibbs free energy change for adsorption of FR-7A on coal

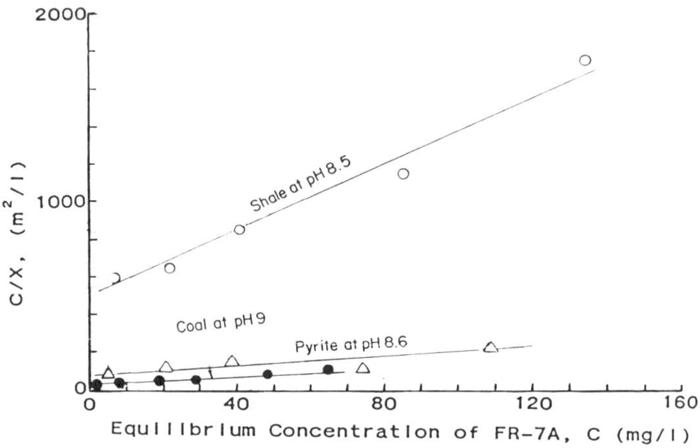

FIG. 5. Langmuir plots for the adsorption of FR-7A on coal, pyrite and shale at pH 8.5 to 9.0

at various levels of pH are shown in Table 2. The magnitudes of K, ΔG^{o}_{ads}, ΔG^{o}_{el} and ΔG^{o}_{sp} per mole of monomer molecules of FR-7A were derived using Eqs. (1) through (5). In the calculation of ΔG^{o}_{el}, the sign of z of FR-7A was taken as negative and the valency was one. At pH higher than pH_{pzc}, the zeta potential of a mineral was negative, and the ΔG^{o}_{el} became positive, which meant that adsorption was depressed. While at pH below pH_{pzc}, the zeta potential was positive, ΔG^{o}_{el} was negative and the adsorption was favored.

Calculation of Entropy Change for Polymer Adsorption

Based on the calculated cross sectional areas of the FR-7A functional groups and the water molecule from the bond lengths and stereo angles [11,17], calculations showed that one functional group of FR-7A molecule, which was random in solution, could replace 5.16 of water molecules, which were arranged in order on a coal surface. The entropy change due to adsorption and replacement of surface water molecules based on ideal mixing can be calculated as follows:

$$\Delta S^{o}_{ads} = -R[n_w \ln(n_w/n) - n_p \ln(n_p/n)] \qquad (6)$$

where n_p and n_w are the numbers of polymer functional groups and water molecules respectively, and the total number of molecules $n = n_w + n_p$. Assuming n_p = 1 mole, we calculated n_w to equal 5.16 moles, and S_{ads} for adsorption of 1 mole monomer FR-7A was calculated at 60.9 J/mol-K. Furthermore, the contribution of the configurational rearrangement to the Gibbs free energy change for adsorption at room temperature, 25°C (298K) was $-T\Delta S^{o}_{ads}$ = -18.2 kJ/mol. This data indicate that configurational rearrangement made the major contribution to the adsorption driving force, ΔG^{o}_{ads} [18].

Estimation of Enthalpy Change of Adsorption

According to Eq. (7),

$$\Delta H^{o}_{ads} = \Delta G^{o}_{ads} + T\Delta S^{o}_{ads} \qquad (7)$$

TABLE 2. Calculation of thermodynamic quantities for the adsorption of FR-7A on coal, pyrite and shale at various levels of pH

pH	X_o Calc.	X_o Meas.	K	ΔG^o_{ads}	ζ	ΔG^o_{el}	ΔG^o_{sp}	ΔH^o_{ads}
	mg/m^2		x10^4	kJ/mol	mv	kJ/mol	kJ/mol	kJ/mol
Upper Freeport coal								
5.0	1.14	1.14	4.49	-26.5	14.4	-1.4	-25.1	-8.3
7.0	1.09	1.00	2.42	-25.0	-3	-0.3	-24.7	-6.8
9.0	0.69	0.68	1.79	-24.3	-26	2.5	-26.8	-6.1
11.3	0.57	0.51	1.19	-23.3	-37	3.6	-26.9	-5.1
Pitt. No. 8 coal								
6.5	0.93	.89	2.79	-25.4	5	-0.5	-24.9	-7.2
8.9	0.76	.60	1.88	-24.4	-22	+2.1	-26.5	-6.2
11.0	0.64	.35	1.10	-23.1	-38.2	+3.8	-26.9	-4.9
Pyrite								
4.3	0.69	.67	2.10	-24.7	6.0	-0.6	-24.1	-6.5
6.3	0.54	.51	1.60	-24.0	1.5	-0.1	-23.9	-5.8
8.6	0.50	.33	1.14	-23.1	-8	0.8	-24.1	-4.9
10.2	0.28	.27	0.85	-22.4	-20	1.9	-24.3	-4.2
Shale								
5.3	0.36	.30	0.94	-22.7	-14	1.4	-24.1	-4.5
7.3	0.23	.20	0.63	-21.7	-22	2.1	-23.8	-3.5
8.4	0.11	.16	0.38	-20.4	-26	2.5	-22.9	-2.2
10.5	0.09	.12	0.28	-19.7	-35	3.4	-23.1	-1.5

the heat of adsorption, or enthalpy change due to the polymer adsorption, ΔH^o_{ads}, can be derived using the values of ΔG^o_{ads} and $T\Delta S^o_{ads}$ determined earlier. Calculated values for ΔH^o_{ads} are listed in Table 2. The data indicate that stronger bonding interactions between the polymer molecules and coal occurs at acidic pH condition. These are due to in part to electrostatic attraction. Also, the interaction between the polymer and naturally hydrophobic coal surface as indicated by ΔH^o_{ads} is greater than that between the polymer and pyrite, and than that between the polymer and shale.

Estimation of Free Energy Change for the Hydrophobic Association "Force"

As mentioned earlier, the high affinity of the polymer to naturally hydrophobic coal as well as pyrite surfaces may be mainly due to the hydrophobic association forces, in addition to van der Waals forces. It is believed that the hydrophobic association effect arises mainly from the configurational rearrangement of water molecules around the hydrophobic species [18-20]. The free energy change due to hydrophobic attraction between a hydrophobic polymer molecule and a "flat" hydrophobic coal surface relative to the size of the polymer's functional groups can be calculated by the hydrophobic energy law developed by Israelachvili and Pashley [19],

$$\Delta G_H = -80.6R\exp(-D/D_o) \quad \text{kJ/mole} \tag{8}$$

where ΔG_H is the free energy change due to the hydrophobic effect, R is the radius of the hydrophobic polymer molecules in nm, D_o is the force decay length and D is the distance between the polymer molecule and the coal surface.

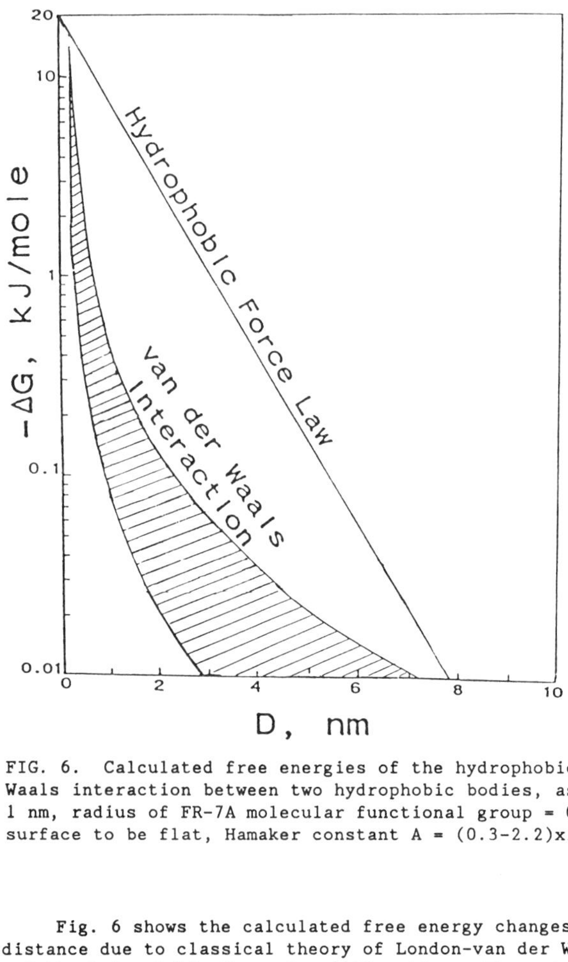

FIG. 6. Calculated free energies of the hydrophobic effect and van der Waals interaction between two hydrophobic bodies, assuming decay length D_o = 1 nm, radius of FR-7A molecular functional group = 0.262 nm, and coal surface to be flat, Hamaker constant A = $(0.3\text{-}2.2)\times10^{-23}$ kJ

Fig. 6 shows the calculated free energy changes for adsorption versus distance due to classical theory of London-van der Waals forces [20], ΔG_{vdw}, and the hydrophobic force law [19], ΔG_H. The value of ΔG_{vdw} was calculated for Hamaker constant A = (0.3-2.2)x10-23 kJ [18], which cover most materials and surfactants. The calculations of ΔG_H were based on the assumptions of D_o = 1 nm, and the calculated radius of FR-7A functional group = 0.262 nm. The calculated maximum free energy change due to hydrophobic effect is -21.9 kJ/mol at D = 0. But, the theory of van der Waals forces is not available for D = 0 between two hydrophobic bodies. Also, we can see from Fig. 6 that when the distance between the two hydrophobic bodies is greater than about 1.5 nm, the predicted free energy due to van der Waals forces is very small. It is difficult to explain how the high adsorption tendency of FR-7A molecules on coal surface is the result of the very small forces (or free energy change). Therefore, the main force for the adsorption is probably due to the hydrophobic association effect. The difference between ΔG^o_{sp} and ΔG_H is due to be the contributions from van der Waals forces, solvation and any possible chemical bonding involved in the adsorption process. It is not possible, however, to determine each item of the free energies at this time.

SUMMARY AND CONCLUSIONS

The adsorption of a hydrophobic polymeric flocculant, FR-7A, on the surfaces of coal, pyrite and shale was studied in detail, using a nephelometric technique to determine the residual concentrations in solution. Based on the obtained adsorption isotherms, thermodynamic adsorption data were derived. The experimental results and thermodynamic calculations have led to a number of conclusions:

1 The adsorption of the hydrophobic polymer on selected minerals follows the Langmuir adsorption model at relatively dilute concentration of the polymer (below about 200 mg/l).

2 The adsorption isotherms indicate that this polymer has a higher affinity to coal than to shale minerals. However, its affinity to naturally hydrophobic pyrite is also quite high, which means that the selectivity of this polymer is poor for pyrite rejection.

3 The thermodynamic derivations show that the strong affinity of the polymer to naturally hydrophobic coal as well as pyrite surfaces is attributable mainly to the hydrophobic association forces, in addition to van der Waals forces.

4 The adsorption of polymer on the minerals is significantly affected by suspension pH. As the pH increases, the equilibrium adsorption density decreases, and the estimated heat of adsorption decreases.

5 The derived enthalpy data show that the interaction between the polymer molecules and the minerals is the strongest with coal, and then with pyrite, and weak with shale minerals.

ACKNOWLEDGEMENTS

The fellowship to S. Yu provided by the Ohio Mining and Mineral Resources Research Institute is gratefully acknowledged.

REFERENCES

1 Y.A. Attia and J. Rubio, Determination of Very Low Concentrations of Polyacrylamide and Polyetheyleneoxide Flocculants by Nephelometry, Brit. Polym. J., 7, pp135-138 (1975).
2 Pradip, Y.A. Attia and D.W. Fuerstenau, The Adsorption of Polyacrylamide Flocculants on Apatities, Colloid & Polymer Sci., 258, pp1343-1353 (1980).
3 Y.A. Attia, Determination of Polymeric Flocculants Concentration by Nephelometry, Flocculation and Dewatering, Eds. B.M. Moudgil and B.J. Scheiner, Enginerring Foundation, AIChE (1990).
4 Y. Attia, S. Yu and S. Vecci, Selective Flocculation Cleaning Upper Freeport Coal with Totally Hydrophobic Polymeric Flocculant, Flocculation in Biotechnology and Separation Systems, Ed. by Y.A. Attia, Elsevier (1987).
5 Y.A. Attia and S. Yu, Production of Super-clean Coals from a High Sulfur Coal by Selective Flocculation, Processing and Utilization of High Sulfur Coals-II, Eds. Y.P. Chugh and R. Caudle, Elsevier (1987).
6 Y.A. Attia and K. Driscoll, Effects of Process Parameters on the Selective Flocculation Cleaning of Upper Freeport Coal, Interfacial Phenomena in Biotechnology and Materials Processing, Ed. by Y.A. Attia, p321-334, Elsevier (1987).
7 Y.A. Attia and K. Driscoll, Effects of Process Parameters on the

Selective Flocculation Cleaning of Upper Freeport Coal, Interfacial Phenomena in Biotechnology and Materials Processing, Ed. by Y.A. Attia, p321-334, Elsevier (1987).
8 ASTM, D3174 "Test for Ash in the Analysis Sample of Coal and Coke", D3177 "Test for Total Sulfur in the Analysis Sample of Coal and Coke", and D2492 "Test for Sulfur Forms Coal", reapproved (1979).
9 F.F. Aplan, Coal Flotation, Flotation - A.M. Gaudin Memorial Volume, 2, Ed. by M.C. Fuerstenau (1979).
10 E.G. Kelly and D.J. Spottiswood, Introduction to Mineral Processing, Wiley Interscience Publication, p101 (1982).
11 D.Z. Wang, Mechanisms and Applications of Froth Flotation Reagents, China Metallurgy Publisher, pp29 (1982).
12 S. Yu, Adsorption of a Hydrophobic Polymer and Its Role in Flocculation and Selective Flocculation of Ultrafine Coal Slurries, M.S. Thesis, The Ohio State University (1990).
13 D.J. Show, Introduction to Colloid and Surface Chemistry, Butterworths, 2nd edition (1970).
14 Y.A. Attia, F. Bavarian and K.H. Driscoll, Use of a Polyxanthate Dispersant for Ultrafine Pyrite Removal from High Sulfur Coal by Selective Flocculation, Coal Preparation, V6, p35-51 (1988).
15 W. Stumm and J. Morgan, Aquatic Chemistry, Wiley (1970).
16 D.W. Fuerstenau, Thermodynamics of Surfaces in Adsorption and Wetting, Principles of Flotation, Ed. by R.P. King, Monograph Series No. 3, SAIMM, Johannesburgh (1982).
17 H.V.V. Lawrence, Elements of Materials Science and Engineering, 4th ed., Addison-Wiley, p520 (1980).
18 C. Tanford, The Hydrophobic Effect: Formation of Micelles and Biological Membranes, Wiley & Sons (1980).
19 J.N. Israelachvili and R.M. Pashley, Measurement of the Hydrophobic Interaction between Two Hydrophobic Surfaces in Aqueous Electrolyte Solutions, J. Coll. & Interface Sci., V98, p500-514 (1985).
20 J.N. Israelachvili and D. Tabor, Prog. Surface Membrane Sci., 7 (1973).
21 V.G. Dashevsky and G.N. Sarkisov, Mol. Phys. 27, p1271 (1974).

SYNTHETIC COPOLYMERS TAILOR-MADE FOR THE pH CONTROLLED SELECTIVE FLOCCULATION OF EXTRAFINE DISPERSIONS OF ILMENITE WITH RESPECT TO RUTILE

V. BERTINI*, A. MARABINI**, M. POCCI***, M. BARBARO**, N. PICCI***, A. DE MUNNO****
*Istituto di Chimica Organica, Università di Genova, Corso Europa 26, 16132 Genova; **C.N.R., Istituto per il Trattamento dei Minerali, Via Bolognola 7, 00138 Roma; *** Dipartimento di Chimica, Università della Calabria, 87030 Arcavacata di Rende (Cosenza); **** Dipartimento di Chimica e Chimica Industriale, Università di Pisa, Via Risorgimento 35, 56100 Pisa.

ABSTRACT

Following a new methodical approach to fully synthetic tailor-made polymeric flocculants for a given mineral dispersion, we prepared several multifunctional polymers for the pH controlled flocculation of titanium minerals active towards ilmenite and selective respect to rutile. They had the structure of radical vinyl copolymers containing 1,2-catecholic functions either free or protected and acrylic acid units. They also had the distinctive property of changing their effect from flocculating to dispersing or inert and vice versa by changing pH, with great advantages for the best fulfilment of the flocculation process. The tendency of the ilmenite particles to gather, as a result of the combined effect of the catecholic and carboxylic functions at acidic pH, is considered an essential condition and a preliminary process for the formation of flocs.

INTRODUCTION

An intense worldwide research concerns the selective flocculation as a fruitful technique for the beneficiation of extrafine minerals, which are untreatable by other conventional techniques and represent either a considerable economic loss or a serious environmental problem. The main bar to the diffusion of such a technique is represented by the low availability of proper selective polymeric flocculants, usually sought among commercial polymers of natural or synthetic origin, modified or not.
We faced this topic seeking a methodical approach to fully synthetic tailor-made flocculants for given minerals and, as several experimental results confirm [1,2], we were successful by conceiving a simplified theoretical model of the morphology of a polymeric flocculant based on only three structural parameters corresponding to three different types of chemical functions that must be present in the polymer.
In this paper we report about the selective flocculation of titanium minerals with tailor-made flocculants containing free or protected catecholic functions able to distinguish between ilmenite and rutile.

EXPERIMENTAL

IR spectra were recorded on a Perkin Elmer 1330 instrument. ^{1}H-NMR spectra were run on Bruker 300 and Varian XL 100 spectrometers with TMS as internal standard. Viscosity measurements ([η]) were made in dioxane at 25°C with an Ubbelohde viscometer. Number average molecular weights ($\overline{Mn}$) were determined by a Knauer membrane Osmometer in dioxane at 36°C.
Reagents, solvents and acrylic acid monomer were from Fluka and purified according to standard procedures. Monomers 1-(4'-styryl)-1-(3',4'-methylenedioxyphenyl) methanol (I) and 1-[3,4-bis(acetyloxy)phenyl]-2-propen-1-one (II) were prepared according to Bertini et al. (3).

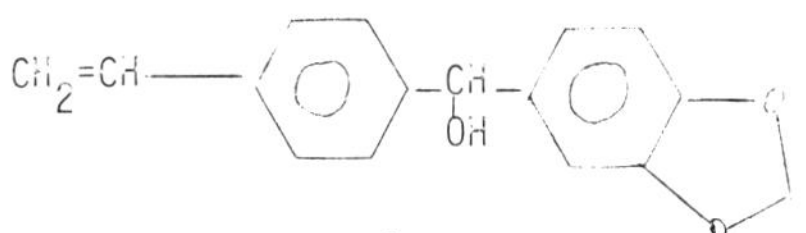

I

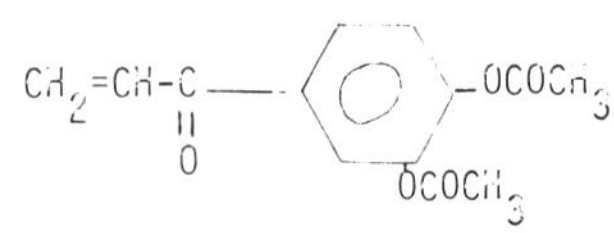

II

Minerals

Rutile (97% TiO_2) from Florida, ilmenite from Risor (Norway) and pure natural quartz were used. Each mineral was ground wet in a porcelain mill using distilled water and fractionated by settling. The fraction with a granulometry lower than 15 micron was collected and used in a slurry form.

Polymers and copolymers

General procedure: Degassed monomers, anhydrous peroxide free dioxane and azobisisobutyronitrile (AIBN) as initiator (0.1% by weight) were introduced in the desired ratios under nitrogen in the polymerization flask and magnetically stirred at 60°C. After a suitable period the mixture was poured into the non-solvent, the precipitated polymer was filtered and vacuum dried overnight at room temperature.
Polymers and copolymers were fractionated through their partial precipitation from stirred 1% dioxane solution in a thermostatic bath at 25°C by slow addition of benzene.
More representative polymerization and copolymerization experiments together with intrinsic viscosities and osmotic number-average molecular weights of the obtained polymers are reported in Table 1.
IR and ^{1}H-NMR spectra of the prepared copolymers fully confirm the disappearance of the vinyl group and the presence of the expected monomeric units.

Chemical transformation of the copolymers II/AA

After preliminary experiments in search of the best reaction conditions, copolymers of acrylic acid with monomer II were transformed into copolymers containing the the 1,2-diphenolic monomeric units (III)

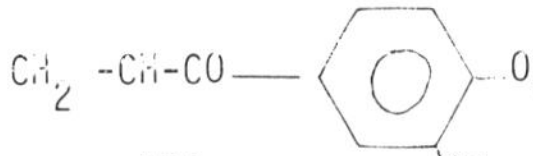

III

Table 1. Representative polymers and copolymers prepared

Monomers M_1	M_2	Mole fraction of M_1 in the feed	Dioxane (ml/mol of monomers)	Time (min)	Polymer (g)	Conversion (%)	Solvent	Non Solvent	[η] ml/g	$\overline{Mn}$
I	AA	0.196	371	1800	3.81	45.2	g, i, j	a, b, c, d, e, k	99.2	184000
II	AA	0.051	273	315	5.24	62.9	d, g, j	b, c, e, f, i, k, l	59.1	80000
III*	AA	0.101**					d, g, i, j,	a, b, c, e, f, k, l	142.6†	420000↓
--	AA	1	250	[illegible]	5.94	76.2	d, g, h, i, j, k, l	a, b, c, e, f	45.6§	360000§

AA = acrylic acid

* monomeric unit of a copolymer III/AA obtained by transformation of a copolymer II/AA

** mole fraction of III in the copolymer

† in methanol at 25° C

↓ in methanol at 36° C

§ at 30° C

a = petroleum ether, b= Benzene, c= diethyl ether, d= THF, e= CH_2Cl_2, f= $CHCl_3$, g= dioxane, h= DMF,
i= methanol, j= 0.1N NaOH, K= water, l= 0.1N HCl

according to the following procedure: the copolymer (1.00 g) dissolved in stirred dioxane (40 ml) was treated under nitrogen with 12N HCl (1.5 ml) at 50°C for 6 hour. After removal of the solvent at reduced pressure, the reaction mixture was dissolved in THF, poured into dichloromethane and the precipitated copolymer was filtered and dried at 30°C under vacuum. The success of the chemical transformation was confirmed by disappearance of the IR absorptions of acetoxy groups at 1770 cm^{-1}. The comparison between the integrals of the ^{1}H-NMR signals of carboxylic group and 1,2-diphenolic residues allowed us to determine the content of III in the deprotected copolymers named III/AA.

Sedimentation tests

Experiments were carried out in a 100 ml graduated cylinder with useful depth of 185 mm.
A volume of slurry corresponding to about 2 g mineral was introduced in the graduated cylinder with 70 ml of distilled water and treated with the desired volume of a 0.1% solution of the polymer in aqueous NaOH 0.01N. The mixture was shaken for 5 minutes, adjusted to the disired pH with standard HCl or NaOH solutions, diluted to volume, shaken for 15 minutes and left to settle for such a time (established on the basis of preliminary tests) that in the absence of polymer and at the same pH and ionic strength, the amount of the mineral still dispersed in the upper 70 ml of the liquid was about 1g, or 50% of the amount used in the test. The upper 70 ml of the dispersion were siphoned off and dried to constant weight, then also the remaining 30 ml were similarly dried and weighed to determine the total amount of the mineral used in the experiment. The obtained weight values were purged of the weights of added non-volatile electrolyte, calculated from the used volumes of HCl, NaOH and alkaline polymer solution, to give the "real weights" of the mineral. As for parallel experiments with and without polymer the weight of the mineral could not be identical, the following normalization relation was used

$$p_o = \frac{p^\circ_{upper} \cdot p_{total}}{p^\circ_{total}}$$

Flocculation power was calculated by the formula

$$F = \frac{p_o - p}{p_o} \cdot 100$$

and dispersing power was calculated by the formula

$$D = \frac{p - p_o}{p_W - p_o} \cdot 100$$

Symbols are:

p°_{upper} = real weight of the mineral contained in the upper 70 ml of the liquid in the experiment without polymer

p°_{total} = real weight of all the mineral used in the experiment without polymer

p_{total} = real weight of all the mineral used in the experiment with the polymer

p_o = normalized weight of the mineral contained in the upper 70 ml of the liquid in the absence of polymer

p = real weight of the mineral contained in the upper 70 ml of the liquid in the experiment with the polymer

p_w = 70% of the real weight of all the mineral used in the experiment with the polymer.

RESULT AND DISCUSSION

The basis of this work is our already mentioned theoretical model which in summary designs for a selective flocculant a multifunctional polymer containing: i) a type of chemical function whose main effect is a specific interaction with the particles of the tested mineral, possibly through a coordination process excluding as many water molecules as possible from the interaction with the particle surface and not tearing off metal ions from it; ii) an adequate polymer chain with strong hydrophobic properties possibly obtainable through a flexible procedure of distribution of the different chemical functions like for instance the radical copolymerization; iii) a type of hydrophilic function dependent on pH as weak acids or bases, matched with the chemical function specifically interacting with the mineral.

According to the model, several free or protected catecholic functions, potentially useful for interactions with titanium minerals especially under acidic conditions, were examined and several vinyl monomers containing them and copolymers with acidic comonomers were prepared and tested. Copolymers of monomers I and II with acrylic acid and analogous copolymers containing the monomeric unit III resulted active towards ilmenite and selective respect to rutile.

Monomers I and II were synthesized and copolymerized in various molar compositions ranging from 5 to 20% with acrylic acid in presence of radical initiators. Copolymers containing the monomeric unit III were obtained from copolymers of II with acrylic acid through hydrolysis of the ester groups. The number average molecular weight ($\overline{Mn}$) of the prepared copolymers, which is known [1,2] not to have a significant effect on the flocculating properties of this type of flocculants, was in the range of 80000 and 420000. Copolymers of I with acrylic acid were easily prepared, but in time their solubility decreased probably owing to a crosslinking reaction involving the alcoholic groups in benzylic position, so they cannot be stored for long time.

As the examined minerals were natural products milled and fractionated for their granulometry, all the flocculation experiments were carried out with the same slurries using each flocculant at the concentrations of 10, 50 and 100 ppm. For comparison purposes, a sample of poly-acrylic acid (PAA) was synthesized and tested in flocculation experiments with both ilmenite and rutile at different pH values (Fig. 1), showing a substantial lack of flocculation at pH higher than 3.

The more significant results obtained in the flocculation experiments of ilmenite and rutile versus pH with the copolymers I/AA, II/AA and III/AA of suitable composition at a concentration of 50 ppm are shown in figures 2, 3 and 4 respectively. In each case the effect of the copolymer is selective between ilmenite and rutile, and the flocculation of ilmenite is strongly dependent on pH. Such property, distinctive of this type of flocculants, allows the copolymer to change its effect from flocculating to dispersing or inert and vice versa by changing pH, with great advantages for the best fulfilment of the flocculation process exploiting the non-flocculating pH conditions.

The copolymers II/AA having molar composition of the monomer II in the 4÷15% range represent the best flocculants for ilmenite selective respect to rutile among all the examined ones, producing well separable stable flocs in good yield by sedimentation and decantation.

Some remarks are worthy of attention: the ilmenite flocculation takes place at fairly acidic conditions where an intensive formation of hydrogen bonds between the particle surface and the carboxylic groups of the copolymer is likely operating; the presence in the copolymer of catecholic functions specifically interacting with the ilmenite particles is essential for the flocculation as revealed by the comparison with the effect of the homopolymer PAA on the same mineral and by the fact that rutile is neither flocculated by PAA nor by the examined copolymers at acidic pH, where it is certainly involved in hydrogen bonding[4]; the same catecholic functions are necessary in the ilmenite flocculants only at rather low concentration, for instance in the copolymer II/AA less than 5% respect to the acrylic matrix.

In the light of such observations it seems unlikely that the main effect of the catecholic functions could be only a help to the hydrogen bonding between the copolymer and the particle. When catecholic functions coordinate to the acidic centers corresponding to crystal lattice coordination vacancies at the surface of the particle, differently from hydrogen bonds, they shield portions of the same surface from water molecules, significantly reducing its direct solvation. By this way the contribution of the copolymer molecules bound to the particle in determining the particle-solvent and particle-particle affinity is enhanced and in consequence of the high content of carboxylic groups in the copolymer, such a contribution is strongly dependent on pH which regulates the carboxylic group dissociation equilibrium. At acidic pH under the effect of a reduction of the particle-solvent affinity the particles may gather and when their copolymer outfits are in contact, the dynamic equilibrium of bond exchanges between flocculant and different particles may produce stable flocs.

CONCLUSIONS

Copolymers I/AA, II/AA and III/AA are good flocculants for ilmenite, selective respect to rutile. Copolymers II/AA are recommended as they produce well separable stable flocs in high yield. As expected the flocculating power is strongly dependent on pH with great advantages for the fulfilment of the flocculation process. Structure and properties of the catecholic functions, which are responsible for specific interactions with the ilmenite particles, suggest their main role in lowering the particle surface solvation and so enhancing the effect of the solvation depending on pH of the polycarboxylic chains bound to the same particle. The tendency of the particles to gather is considered an essential condition and a preliminary process for the formation of flocs, corresponding to a dynamic equilibrium of bond exchanges among flocculant and gathered particles.

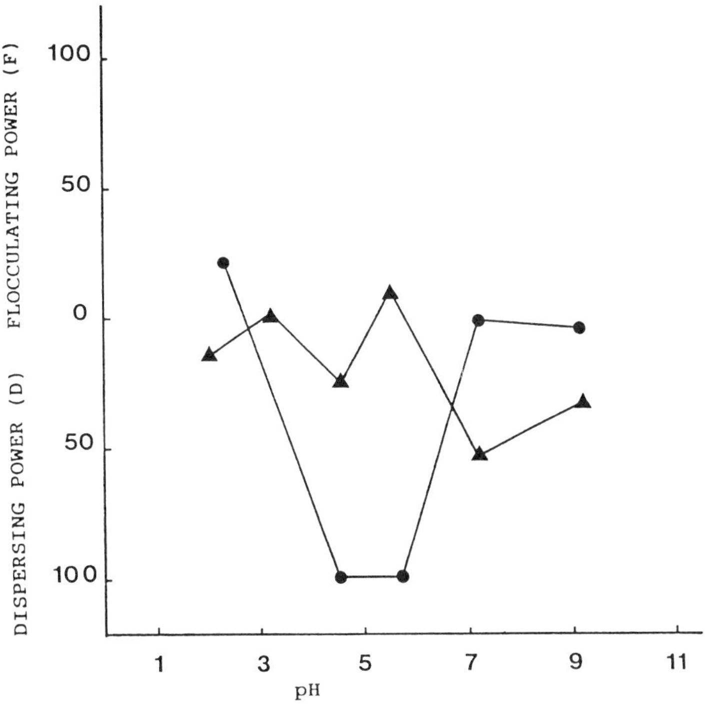

Figure 1. Flocculating and dispersing power of the polyacrylic acid (Table 1) at 50 ppm on ilmenite (●) and rutile (▲) versus pH.

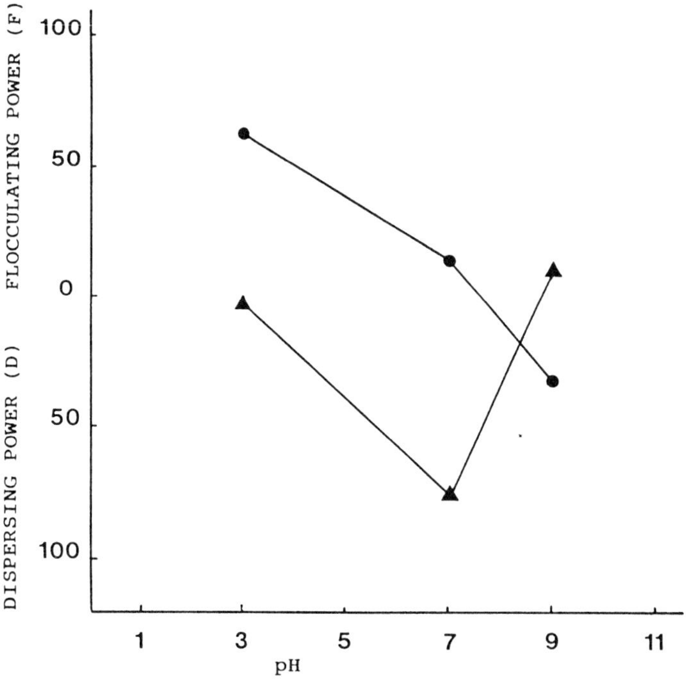

Figure 2. Flocculating and dispersing power of the copolymer I/AA (Table 1) at 50 ppm on ilmenite (●) and rutile (▲) versus pH.

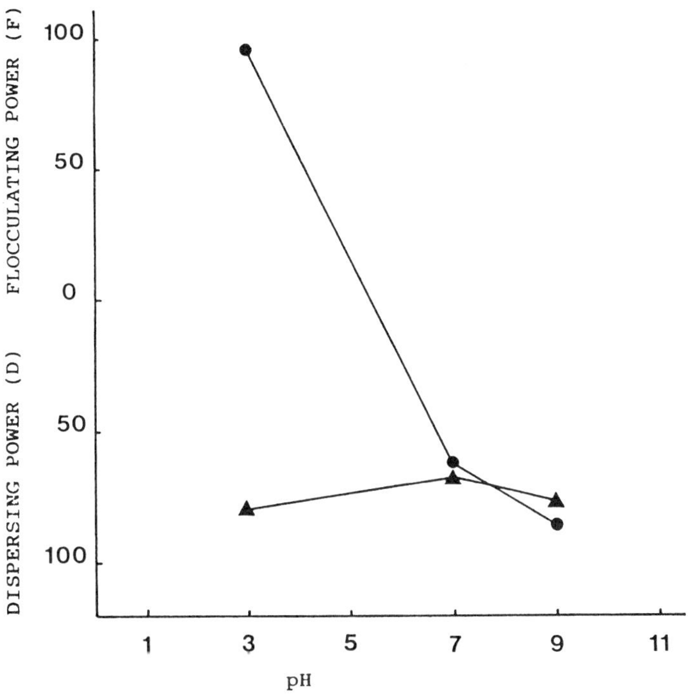

Figure 3. Flocculating and dispersing power of the copolymer II/AA (Table 1) at 50 ppm on ilmenite (●) and rutile (▲) versus pH.

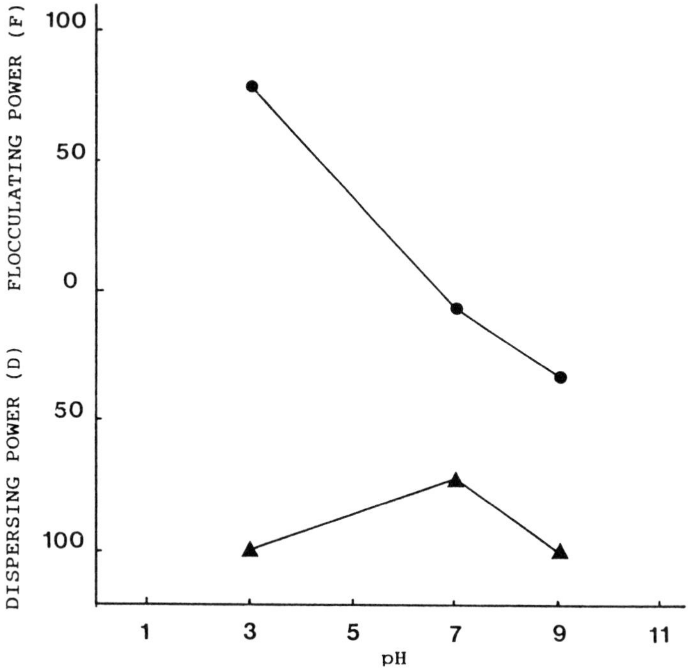

Figure 4. Flocculating and dispersing power of the copolymer III/AA (Table 1) at 50 ppm on ilmenite (●) and rutile (▲) versus pH.

ACKNOWLEDGMENT

This work was supported by CNR-II Progetto Finalizzaio Chimica Fine e Secondaria Sottoprogetto Materiali Polimerici. Special thanks must be extended to Mrs. M. A. Esposito for care and attention paid to flocculation measurements.

REFERENCES

1. V. Bertini, M. Pocci, A. Marabini, M. Barbaro, and A. De Munno, 3-Vinyl-1,2,5-thiadiazole/Methacrylic acid Copolymers as pH Controlled Flocculants for Finely Divided Copper Minerals, **Particulate Science and Technology**, vol. 4, pp. 203-212, (1986).
2. V. Bertini, A. Marabini, A. De Munno, M. Pocci, N. Picci, and M. Barbaro, Design, Preparation and Activity of Tailor Made Copolymers for the pH Controlled Selective Flocculation of Mineral Dispersions, **Flocculation in Biotechnology and Separation Systems**, pp. 247-261, Elsevier Science Publ. B.V., Amsterdam, (1987).
3. V. Bertini, A. Marabini, A. De Munno, M. Barbaro, and M. Pocci, Synthetic Organic Polymers for the Selective Flocculation of Titanium and Iron Ores, Italian Patent Appl. n. 48007-A/84 and corresponding foreign applications.
4. J.E. Gebhardt and D.W. Fuerstenau, Adsorption of Polyacrylic acid at Oxide/Water Interfaces, **Colloids and Surfaces**, vol. 7, pp. 221-231, (1983).

SELECTIVE DESLIMINGS OF FINE IRON ORES BASED ON AGGREGATION BETWEEN MAGNETITE AND HEMATITE

QUN XU, * M.J. ZHANG, J.K LOU, ** AND P. SOMASUNDARAN*

*Langmuir Center for Colloid and Interfaces, Henry Krumb School of Mines, Columbia University, New York, N.Y. 10027, USA; **Beijing General Research Institute of Mining & Metallurgy, 1 Wenxing St., Xizhimenwai, Beijing, P.R. China*

ABSTRACT

Hematite fines were found to form aggregates with magnetite particles in the same slurry, without the help of any flocculants. Because of this aggregation, fine iron ores containing hematite and magnetite can be successfully concentrated by multi-stage deslimings, provided that the ore is finely ground and silicate gangues properly dispersed. The effectiveness of this technique has been tested on a commercial scale(100 ton/h). The increased magnetic attraction between magnetite and hematite due to the terrestrial magnetization of magnetite and its size effect are revealed as responsible for the aggregation, considering the fact that the hematite-magnetite system is already in relatively week dispersion because of their low zeta potentials.

INTRODUCTION

A large part of the iron resources in the world occurs as low-grade finely disseminated hematite ores with magnetite often in association. This is particularly true in the case of ores in China and the United States. To make use of such resources, some simple and efficient concentration methods are indispensable because of the large quantities and low price of iron ores. Desliming and selective flocculation by use of natural and synthetic polymers has been studied extensively for separating hematite fines from gangues[1,2], but selective flocculation of real ores is achieved with difficulty because the bridging mechanism responsible for polymer flocculation makes the process very sensitive to changes in the chemical conditions in the solution as well as on the surface of the mineral particles. This is especially true if the ore is a complex one or contains soluble species. As a result, few cases of selective flocculation have been reported on a commercial scale.

An innovative method for recovering fine iron ores is reported in this paper. It was found that selective desliming of fine hematite ores is easily achieved if certain amount of magnetite(either associated in the same ore or extraneously added) is present in the feed, and the silicate gangue is properly dispersed(by pH adjustment and use of such common dispersants as sodium silicate and sodium hexametaphosphate). In this process, the sedimentation of iron ore particles is enhanced by the formation of aggregates among hematite and magnetite particles even without any flocculants. For

Advances in Fine Particles Processing
John Hanna and Yosry A. Attia, Editors

commercial tests (100 ton/h), an iron concentrate assaying >65%Fe was produced at 85% recovery from a feed containing 40%Fe by fine grinding and multi-stage deslimings with caustic soda and sodium silicate added to the ball mills and the conditioning tanks to disperse the silicate gangues. Results obtained using this technique are presented here and the mechanism of aggregation between hematite and magnetite is discussed.

MATERIALS AND METHODS

Pure hematite, magnetite and augite(silicate gangue) were hand-picked at Bai-yun E-bo Mine in Inner Mongolia, China. These minerals were further enriched by gravity means to yield 95% pure samples. Ore sample(A) of 42% Fe with magnetite and hematite in an approximate ratio of 1:1 was taken from Baotou Concentrator in China. Because of the complex mineralogical composition(114 minerals were commonly found in the deposit) and fine dissemination, various conventional concentration methods and their combinations attempted during the last two decades could produce only a concentrate assaying 55-58%Fe at a recovery of 60-70%. Most of the hematite was lost in the tailings as fine particles.

Another ore sample(B) of 31.5% Fe was prepared from ore sample A by removing magnetite using low intensity magnetic separation at H=800 Oe. X-ray diffraction analysis showed hematite, augite, quartz, fluorite and carbonates as the main minerals. This sample was ground in a steel ball mill to 98% passing 400 mesh in tap water(hardness 140 ppm) and then dewatered and dried at 40 ^{o}C for future use.

Selective desliming of pure minerals and ore sample B were carried out in a 40-ml settling tube using 3-gram batches. The pulp was agitated in a 100 ml conditioning tank under 20% pulp density for 15 minutes(impeller diameter 15 mm, speed 1500 rpm) and then diluted to 11% pulp density for desliming. The experiments were conducted in distilled water. For selective desliming tests of ore sample A, 90 gram samples were ground with sodium silicate and sodium hydroxide in a steel ball mill at 50% pulp density. The slurry was then transferred to a 500-ml glass beaker and diluted to 11% for desliming by settling-and-syphoning method. All the experiments with this sample were conducted in tap water(hardness 140 ppm).

RESULTS AND DISCUSSION

Selective Desliming

The results of desliming tests with pure mineral mixtures are shown in Figure 1. The addition of fine magnetite markedly increases the recovery of hematite in the settled products. For each test, the silicate gangue mineral, i.e. augite, was fixed at 1 gram. The total amount of magnetite and hematite, with their ratio at different levels, was maintained at 2 grams. Magnetite was almost exclusively found in the settled products. The aggregates of magnetite and hematite under microscope were found to be in the 100 - 200 micrometer size range.

Desliming results with ore sample B in the presence of added fine magnetite and ferrite powder as aggregation media are shown in Figure 2. It is seen that both magnetite and the ferrite powder can increase the recovery of hematite. The -1.5 μm ferrite powder, manufactured for making permanent magnets, was of the magnetic domain size and thus was a collection of magnetic particles. The effectiveness of ferrite powder for desliming purposes suggested that magnetic force was important in the aggregation process.

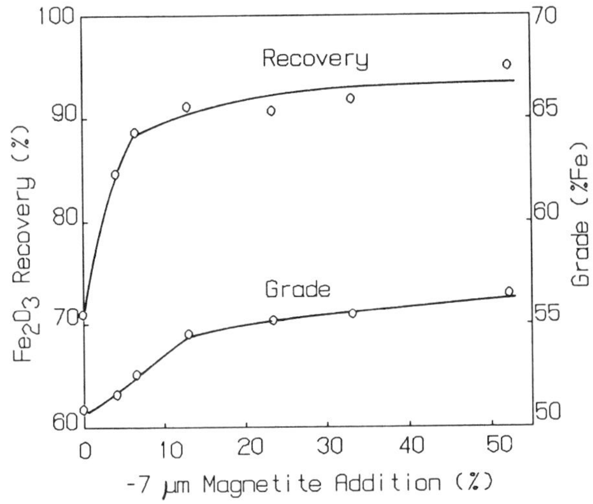

FIG. 1. Effect of fine magnetite addition on desliming of pure mineral mixtures. Augite -20+10 μm; hematite -20+10 μm; NaOH 200 mg/l, Na_2SiO_3 500 mg/l, pH11.5.

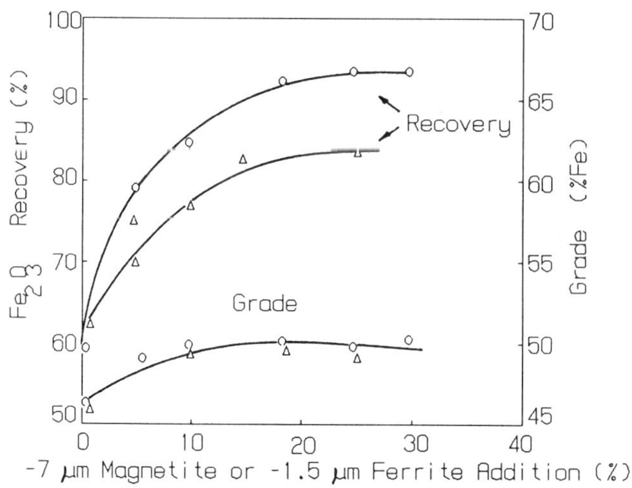

FIG. 2. Effect of fine magnetite and ferrite powder addition on desliming of ore sample B. —o— , -1.5 μm ferrite powder as aggregation medium; —△— , -7 μm magnetite as aggregation medium; NaOH 200 mg/l, Na_2SiO_3 500 mg/l, pH11.5.

The fineness of magnetite was found to be an important parameter for improving recovery in desliming experiments. It is seen from Figure 3 that finer the magnetite, higher is the iron recovery, while the grades remain the same.

Figure 4 shows that agitation intensity was not a critical factor in this case. This rules out shear flocculation or hydrophobic flocculation[3,4], which require strong agitation to provide particles with energy to overcome the potential barriers, as a possible mechanism for aggregation between magnetite and hematite.

The effect of sodium hydroxide and sodium silicate on desliming of ore sample A is shown in Figure 5. It is seen that sodium silicate alone(not NaOH) is effective for achieving selective desliming at high dosage by simultaneously increasing the recovery and grade. In practice, however, sodium hydroxide is used together with sodium silicate so that the pH can be easily adjusted to above 9 and excessive addition of sodium silicate is avoided. Experiments with pure augite showed that the role of sodium silicate and pH adjustment is to eliminate the nonselective coagulation effects of such cations as Fe^{3+}, Ca^{2+}, and Mg^{2+}. Other dispersants such as sodium hexametaphosphate or humates were also found to be effective for selective desliming.

Results obtained for the five stage desliming of ore sample A in tap water(hardness 140 ppm) are given in Table 1. Locked-circuit tests(slime II, III, IV and V returned to their preceding desliming stages), yielded a concentrate of 65.52%Fe at 86.87% iron recovery. Reagent consumption was observed to be at 3.3 kg/t NaOH and 5.6 kg/t sodium silicate for locked-circuit tests.

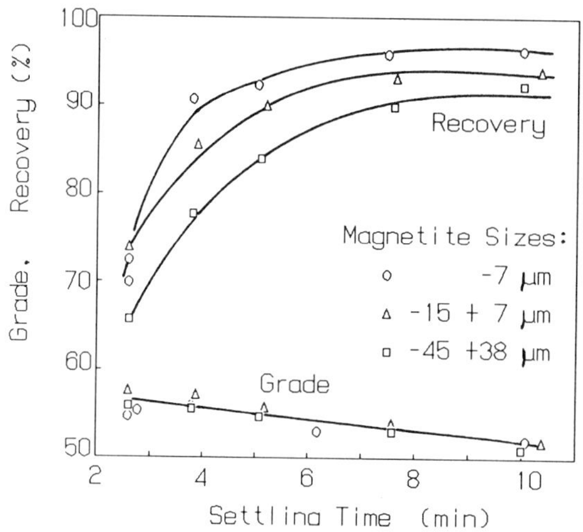

FIG. 3. Effect of magnetite fineness on desliming of ore sample B. Magnetite addition = 20% of the total feed; NaOH 200 mg/l, Na_2SiO_3 500 mg/l, pH11.5.

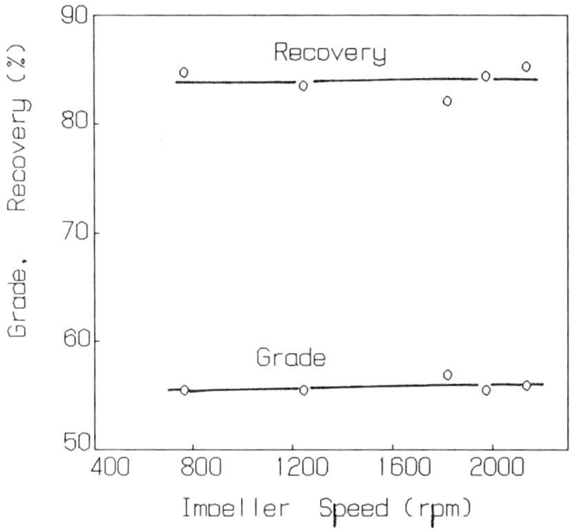

FIG. 4. Effect of pulp agitation on desliming of ore sample B. Agitation time = 15 min; magnetite(-7 μm) = 20% of the total feed.

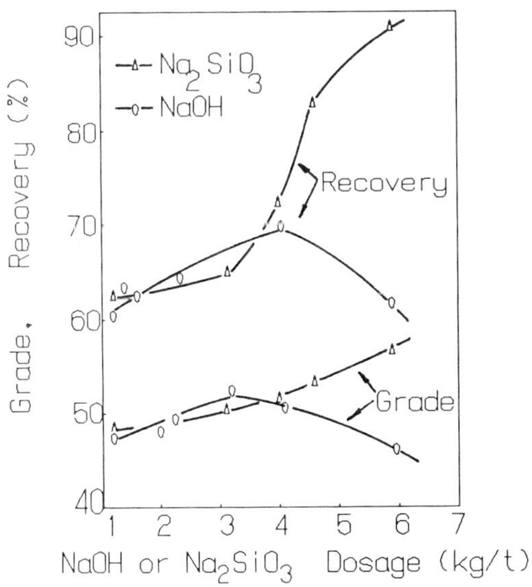

FIG. 5. Effect of NaOH or Na_2SO_3 addition on desliming of ore sample A. Tap water, hardness 140 ppm.

TABLE I. Results for five-stage desliming of ore sample A.

Product	Yield % Wt	Grade %Fe	Recovery %
Slime I	25.05	10.20	6.07
Slime II	10.61	14.10	3.56
Slime III	5.50	19.60	2.56
Slime IV	3.34	26.50	2.10
Slime V	2.45	46.95	2.73
Concentrate	53.05	65.80	82.97
Feed	100.00	42.07	100.00

Mechanism of Aggregation Between Magnetite and Hematite

It is known that flotation and surface characteristics of magnetite are essentially the same as those of hematite[5-7]. The aggregation between magnetite and hematite therefore can not be explained solely on the basis of their surface properties. However, considering the fact that aggregation is normally obtained if the zeta potential is less than 14 mV [8], the zeta potential values of magnetite and hematite($\approx$ -30 mV as shown in Figure 6) make them in weak dispersion[9] and aggregation may therefore be possible by providing some extra-force. Magnetic attraction between magnetite and hematite is considered to be the reason here as suggested by the effectiveness of ferrite powder as aggregation medium (Figure 2).

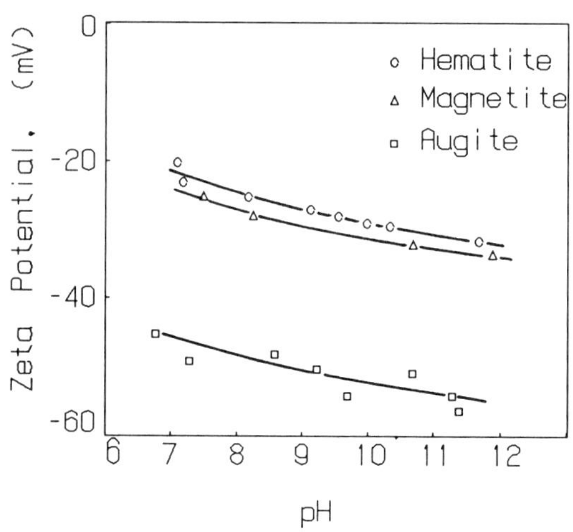

FIG. 6. Zeta potentials of pure minerals as a function of pH, 10^{-3} M KNO_3.

To understand the magnetic attraction between magnetite and hematite particles, a model spherical magnetite particle with uniform terrestrial magnetization, M, was selected and the coordinates were set as shown in Figure 7, with M in z axis direction. The particle is hypothetically sliced into laminae perpendicular to the z axis. Each of the laminae can be equalized by equivalent surface current theorem[10] into a circle carrying electric current dI:

$$dI = M \sin\theta \, dS \tag{1}$$

An electric current circle of this type gives magnetic induction, dB, on z axis at point P as[11]:

$$dB = \mu r^2 dI/[2(r^2 + (l-z)^2)^{3/2}] \tag{2}$$

Integrating the contributions of each laminae yields H at point P:

$$H = B/\mu = M[5(R/l)^3 + 3(R/l)]/6 \tag{3}$$

and

$$\text{grad}H = dH/dl = -M[5(R/l)^4 + (R/l)^2]/2R \tag{4}$$

$$H\text{grad}H = -M^2[(25(R/l)^7 + 20(R/l)^5 + 3(R/l)^3]/12R \tag{5}$$

Where M - Terrestrial magnetization of magnetite
R - Radius of spherical magnetite particle
l - Distance from the center of the magnetite sphere to point P on z axis ($l > R$)

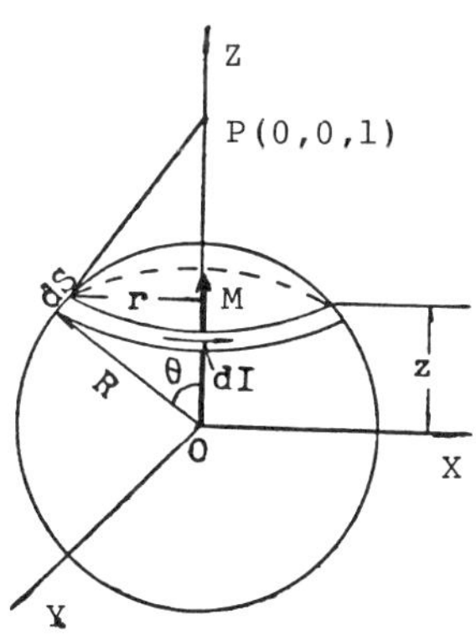

FIG. 7. Model magnetite particle for magnetic calculation.

The calculated results of eq.(4) and (5) are shown in Figures 8 and 9, respectively. It is seen that both | gradH| and | HgradH| near the surface of a magnetite particle increases significantly when the particle size is less then a certain value. This phenomenon is similar to the "High Gradient Effect" of steel wool or other magnetic matrices in high gradient magnetic separators and is in agreement with the experimental results that finer magnetite can attain better desliming results(Fig. 3). Terrestrial magnetization for calculation is taken as low as 1040 A/m(0.2 emu/g), but the | gradH| at the surface of a 10 μm magnetite particle can be as high as 6.24×10^9 A/m^2(7.8 Tesla/cm). This is of the same order as the value of up to 10 Tesla/cm that can be reached in high gradient magnetic separators[12]. Also taking into account the relatively weak electrical repulsion between magnetite and hematite because of their low zeta potentials, we may conclude that the increased magnetic attraction in the fine particles size region is sufficient to bring the particles into aggregation. Further support to this consideration is that aggregation of hematite fines(1.2 μm) can be obtained by weak geomagnetic fields[13].

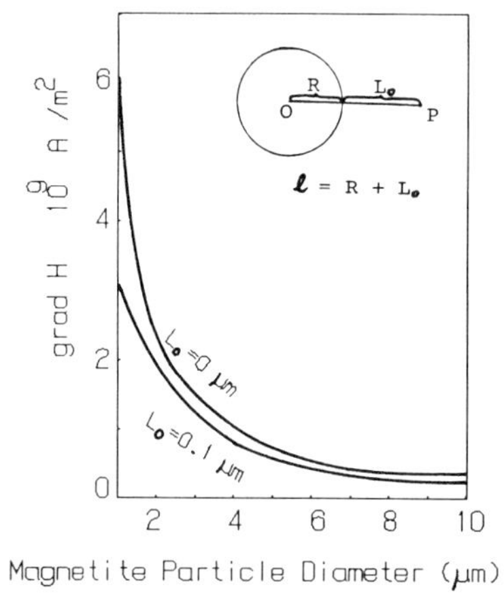

FIG. 8. Calculated | gradH| by eq.(4). M=1040 A/m.

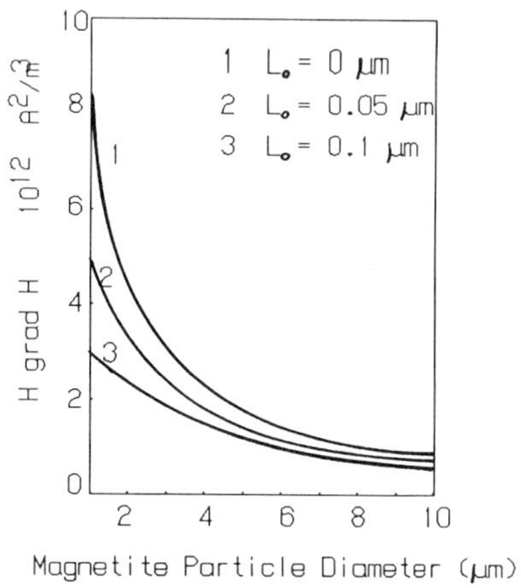

FIG. 9. Calculated | HgradH| by eq.(5). M=1040 A/m.

Because of the magnetic nature of the attraction between the magnetite and the hematite, the desliming process is quite insensitive to the chemical properties of the slurry. Selective separation is therefore possible even with complex ore systems and in the presence of hard water(140 ppm). We have tested iron ores containing magnetite and hematite from various sources and selective desliming results were always achieved. This is attributed to the fact that the magnetic properties are intrinsic to these minerals and therefore the aggregation is to a large extent independent of the origin of the minerals. Also because of the magnetic nature of aggregation, applied magnetic fields were found in our experiments as beneficial to the desliming process by enhancing the settling rate of the aggregates. But no applied field was used for all the tests reported in this paper.

CONCLUSIONS

A selective desliming process for recovering fine iron ores without using any flocculants has been developed. The main feature of the process includes use of fine magnetite to aggregate hematite particles. Aggregation is revealed to be related to the terrestrial magnetization of magnetite. Magnetic attraction on hematite is higher in the fine particle size region due to the "High Gradient Effect". This aggregation is analogous to the capturing process of hematite fines to wire wool or ball bearing matrices in high gradient magnetic separators. This technique is especially suitable for fine iron ores containing magnetite and hematite, but it can also be used for hematite-only ores by adding extraneous magnetite that can be recycled.

REFERENCES

1. A.F. Colombo, in: Fine Particles Processing. P. Somasundaran ed. (AIME, New York, NY, 1980) 2, pp.1034-1056.
2. D.F. Bagster and J.D. McIlvenny, Int. J. Miner. Process., 14, 1(1985).
3. L.J. Warren, Trans. Inst. Min. Metall., Sect. C, 84, 99(1975).
4. W.Z. Xing and J.B. Xu, Act Metallurgical Sinica, 19: B1(1983), (in Chinese, with English abstract).
5 I. Iwasaki, S.R.B. Cooke, and Y.S. Kim, Proc. Int. Symp. on Fine Particle Processing. 1, Ch. 34, 652(1962).
6. I. Iwasaki, S.R.B. Cooke, and H.S. Choi, Trans. AIME, 217, 237(1960).
7. K. Nakatsuka, J. Matsuoka, and J. Shimoiizaka, Proc. 9th Int. Miner. Process. Congr., Prague, 1, 251(1970).
8. P. Somasundaran, in: Fine Particles Processing. P. Somasundaran, ed. (AIME, New York, NY, 1980) 2, p. 947.
9. A.F. Colombo, in: Proceedings of XIth IMPC, Ente Minerario Sardo, ed. 470(1975).
10. P. Lorrain and D.R. Corson, Electromagnetic Fields and Waves. Third Edition, W.H. Freeman and Company, New York, 1988, pp. 362-380.
11. D.T. Paris and F.K. Hurd, Basic Electromagnetic Theory. McGraw-Hill, New York, 1969, p. 220.
12. B.A. Wills, Mineral Processing Technology, 4th edition, Pergamon Press, 1988, p.617.
13. M. Ozaki, T. Egami, N. Sugiyama, N., and E. Matijevic, E., *J. Colloid and Interface Sci* 126, 212(1988).

PART 6.

SEPARATION OF FINE PARTICLES BY FLOTATION

High Speed Photographic Investigation of Coal Flotation

R.F. Batchelder* and C.C. Li**

*United States Department of Energy, Pittsburgh Energy Technology Center, PO Box 10940, Pittsburgh, Pennsylvania 15236

**University of Pittsburgh, Department of Electrical Engineering, Pittsburgh, Pennsylvania 15260

ABSTRACT

Individual bubble-particle collisions were studied using a high speed video recorder. An algorithm for determining a particle's trajectory in three-dimensional space during a collision with an air bubble through a frame by frame analysis of the recorded video sequence is described. Evidence that the moment of bubble-particle attachment can be detected from the particle trajectory is examined.

BACKGROUND

Froth flotation of coal is a process that selectively separates particles according to their surface properties. The particles are dispersed in an aqueous solution and contacted with rising air bubbles, resulting in multitudes of bubble-particle collisions. Particles that remain attached to an air bubble are buoyed to the surface and skimmed off. Particles that do not attach to air bubbles remain in suspension and are removed as tailings. The separation selectivity arises from the fact that for hydrophobic particles (e.g., coal) an air bubble attachment is energetically favorable, while the reverse is true for hydrophilic particles (e.g., ash-forming mineral particles) [1].

A simplified illustration of a bubble-particle collision is shown in Figure 1. The relative motion between the bubble and particle causes the fluid to move along streamlines around the bubble. Both viscous and inertial forces affect the trajectories of particles. Viscous forces tend to force the particle to travel along these fluid streamlines. If the density of the particle is greater than the density of the fluid, the inertial force will cause the particle to travel in a straight line. If the inertial force predominates over the viscous force (ie. such as would occur with a large, dense particle), the particle will tend to cross the streamlines as they curve around the bubble and a bubble-particle collision will result. In most cases, after a collision, the particle will slide along the surface of the bubble under the influence of fluid flow. The sliding time is the time between the moments of collision and separation. Figure 2 defines the concept of sliding time. At the moment of collision, a thin fluid film exists between the bubble and particle. Physical attachment between the bubble and particle occurs if this liquid film thins and ruptures, leaving a particle-gas interface. As illustrated in Figure 3, the induction time is defined as the time required for this liquid film to thin and rupture. From our definition of physical attachment, it follows that attachment will take place only when the induction time is less than the sliding time (see Figure 2). Thus, it is the induction time distribution of the feed particles that determines the selectivity. If the induction time of all the particles is much less than the average sliding time, all the particles will attach. For long induction times (compared to average sliding time), the particles will fail to attach. Selectivity in flotation exists when the induction time for one group of particles is less than the average sliding time, and the

Advances in Fine Particles Processing
John Hanna and Yosry A. Attia, Editors

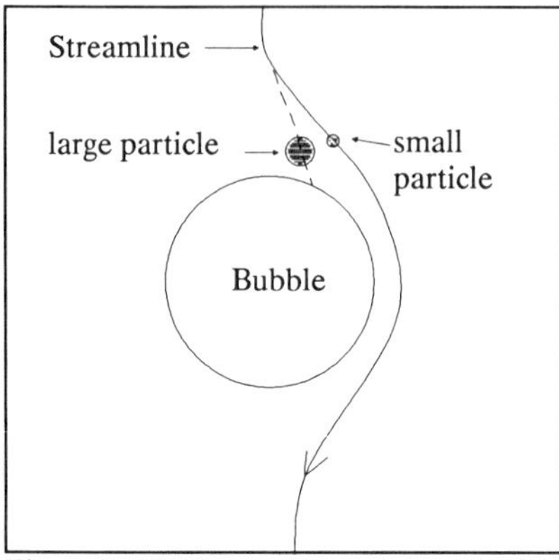

Figure 1. Fluid streamlines around bubble

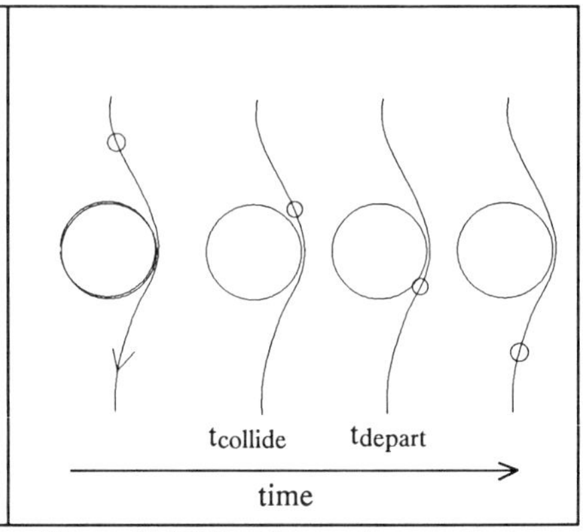

Figure 2. Sliding time = $t_{depart} - t_{collide}$

induction time for a second group of particles is more than the average sliding time, as illustrated in Figure 4. The key to increasing the selectivity of a flotation separation is to widen the difference in induction times for the two particle groups by use of surface conditioning.

OBJECTIVE

Traditional research on flotation is done with a small flotation cell. The experimenter observes the effects of changes in process variables on the results of a batch flotation test. The experimenter searches for the optimum conditions (relative to selectivity and yield) by covering wide ranges of the process variables. Each measurement of flotation yield is the averaged result-set for a multitude of bubbles, particles and approach velocities, each of which is only statistically described. For this reason, it is difficult to answer questions such as: "What is the most efficient bubble size?", "What is the particle size effect?", and "How much turbulence is optimum?". These and other questions must be answered by studying the bubble-particle collision which is the most elementary facet of flotation,

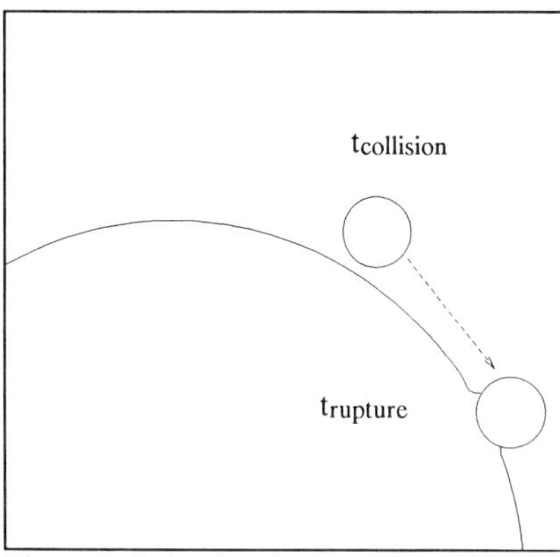

Figure 3. Induction time = $t_{rupture} - t_{collision}$

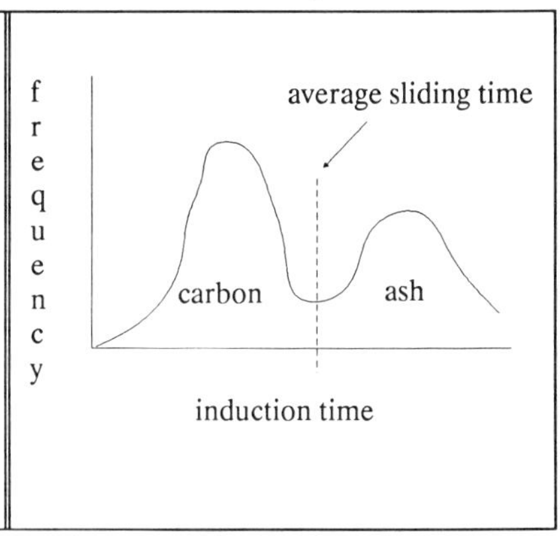

Figure 4. Induction time distribution

because this is the event which determines selectivity. In this study we control the variables for each collision, including the bubble size, particle size, and approach velocity. Our data will be used to develop a mathematical model for use in predicting performance, and for directing changes that will improve separation efficiency and rejection of pyrite and mineral matter.

APPROACH

The intent of this study is to measure the induction times for individual coal particles. This information will provide an induction time distribution, which can be used to predict a sample's response to flotation. The goal was to develop techniques to measure the induction time for an individual particle by filming the bubble-particle collision using high speed photography, followed by a quantitative analysis of the event.

It was hypothesized that as a particle slides along the bubble surface, its velocity would undergo a perturbation when the separating fluid film ruptured. If this could be detected on playback of the film, the induction time could be directly measured.

Bubble Cell Development

Two Crossing Streams Our first attempt was to reproduce the work of Brown[2], where a stream of bubbles rises through the field of view and a stream of particles falls through the field of view. The operation of the cell resulted in very few visible collisions because neither the trajectory of the bubble or the trajectory of the particle could be controlled.

Tapered chamber flow cell Figure 5 illustrates the next stage of cell development. In an attempt to fix the position of the bubble, we designed a tapered (flaring) flow cell in which a vertical velocity gradient exists, with the velocity greater at the top of the cell. When the bubble was introduced, it sought a position where its buoyant force would equal the fluid drag force. Particles were then introduced into the fluid stream from an overhead mixing chamber. The particle-bubble collisions are recorded on high speed video tape. Figure 6 is an image recorded with this cell. It can be seen that the bubble is unrestrained and about to collide with a particle. An attached particle can be observed on the bottom of the bubble. The following problems were encountered with the tapered flow cell:

1) absence of any control over the particle-bubble collision angle
2) oscillation of bubbles with diameters > 1 mm
3) trackability of particles only when their trajectories occurred on the periphery of the bubble.

Stereo flow cell Figure 7 illustrates the stereo flow cell configuration. An adjustable, constant, downward flow of fluid was maintained through the cell. An air bubble was held stationary on the end of a tube. Particles were directed from the mixing chamber through a tube towards the bubble. Adjustment of the tube position allowed some control over the bubble-particle collision angle. Restraining the bubble on the end of a tube greatly simplifies the control of the bubble-particle collision.

The recording of bubble-particle collisions with the previous cells was conducted with a single camera which captured one planar projection of the event. Therefore, it was possible to measure only two components of the particle's velocity. It was impossible to

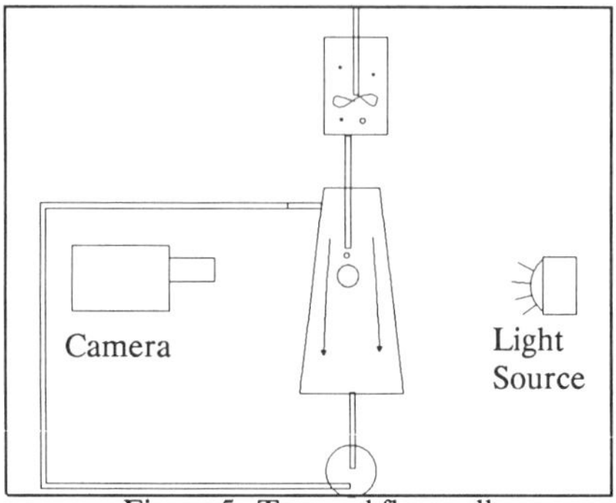

Figure 5. Tapered flow cell

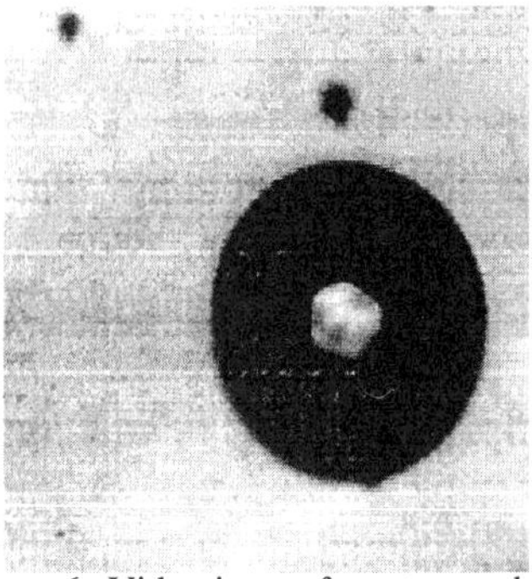

Figure 6. Video image from tapered cell

track a particle's true trajectory around a bubble without knowing its position in three-dimensional space. To correct this, two cameras were utilized with their optical axes oriented 90 degrees apart (see Figure 8). A SP2000 high speed video recorder combines the input from both cameras in a split frame as shown in Figure 9.

Optical Equipment

The bubble-particle collisions are recorded on a Kodak SP2000 high speed video recorder shown in Figure 10. The SP2000 is capable of recording up to 2000 frames/second onto magnetic tape. The event can then be analyzed during replay at slow speed or at a single frame at a time. The Kodak SP2000 has provisions for two simultaneous camera inputs, which are displayed in a split frame on replay. Two Vivitar 100 mm (f = 2.8) macro lens were used on the video cameras. Two Mole cool lights (600 watt) were utilized as back lights. Pieces of velum were placed on the cell between the bubble and the light to act as diffusers.

Figure 11 is a diagram of the hardware configuration. An ITI PCVision Plus frame grabber installed in a PC/AT personal computer was used to analyze the images. The frame grabber digitizes the image into an array of 512 pixels x 480 pixels x 256 grey levels.

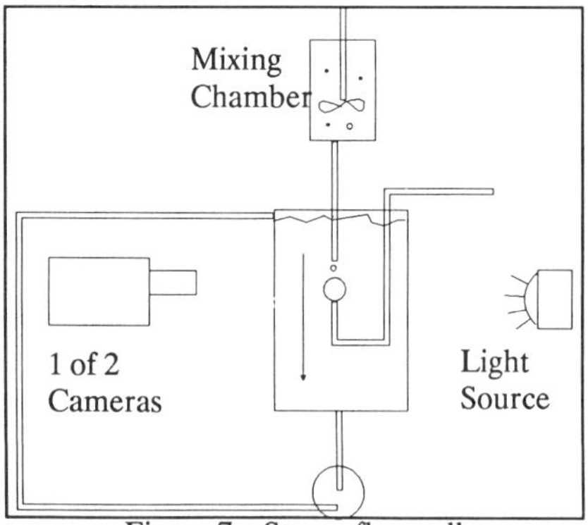

Figure 7. Stereo flow cell

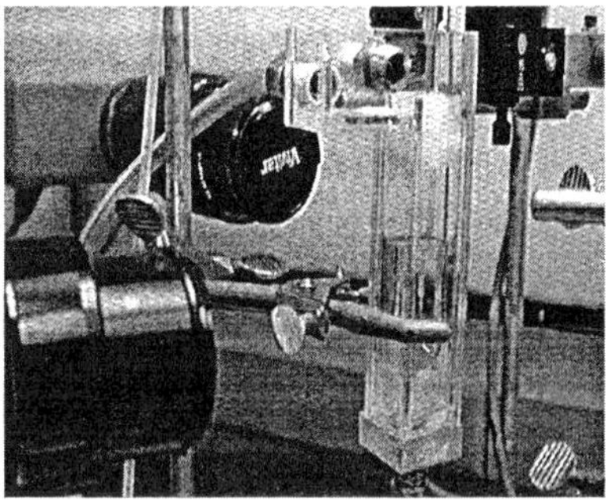

Figure 8. Stereo flow cell

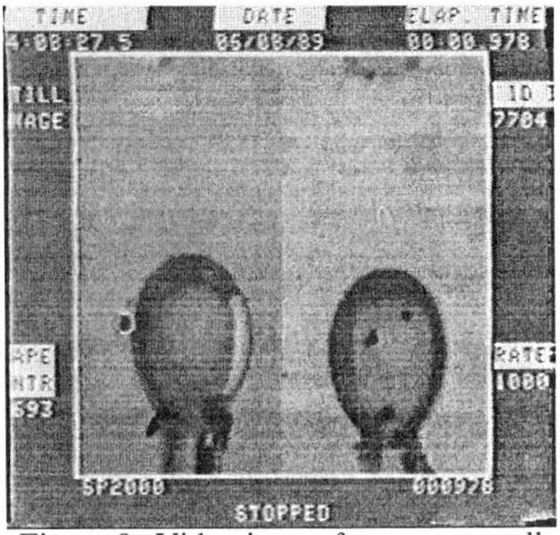

Figure 9. Video image from stereo cell

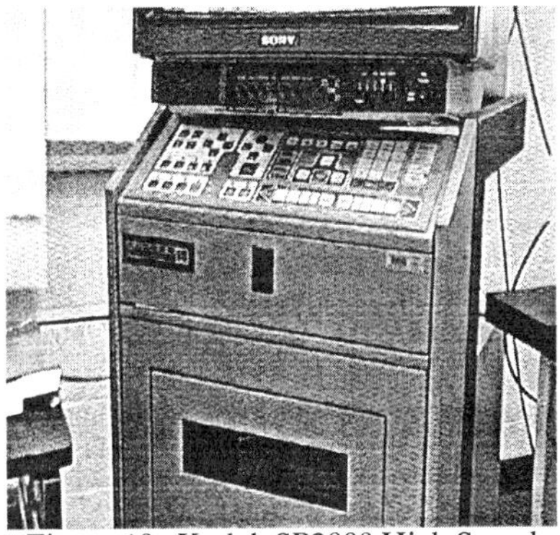

Figure 10. Kodak SP2000 High Speed Video Recorder

Interactive particle tracking program

An image analysis program referred to as interact, which allows the user to track particles in three-dimensional space, was written in the C programming language. Figure 12 is a representation of the interact control screen with which the user controls the processing sequence by selecting the appropriate button with the mouse. A typical session is described below:

1) The desired starting frame is selected from the SP2000 (i.e., frame 0).
2) The interact program is started.
3) The frame 0 is grabbed and digitized.
4) The two projections of the bubble contour are traced automatically after the user selects the grey level threshold. The contour tracing algorithm as described on page 143 of Pavlidis [3] was used. The video image with the bubble contour traced in white is shown in Figure 13. The bubble contour coordinates are stored in a disk file for later analysis.
5) The user locates the interior of the particle in both projections using the mouse and the program automatically traces the particle contour, identifies the interior points, calculates the centroid locations, and calculates the cross

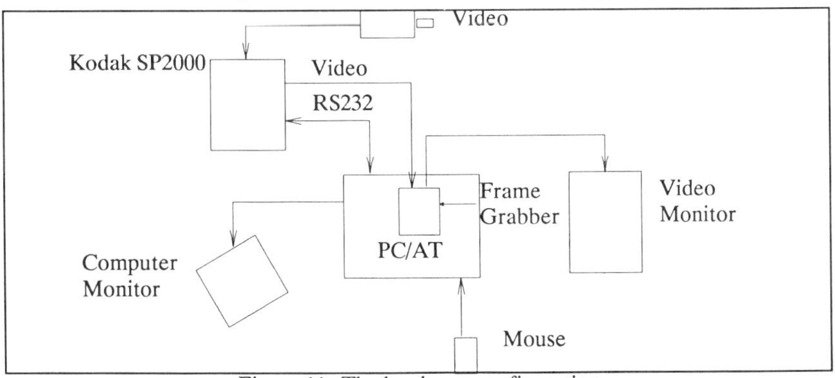

Figure 11. The hardware configuration

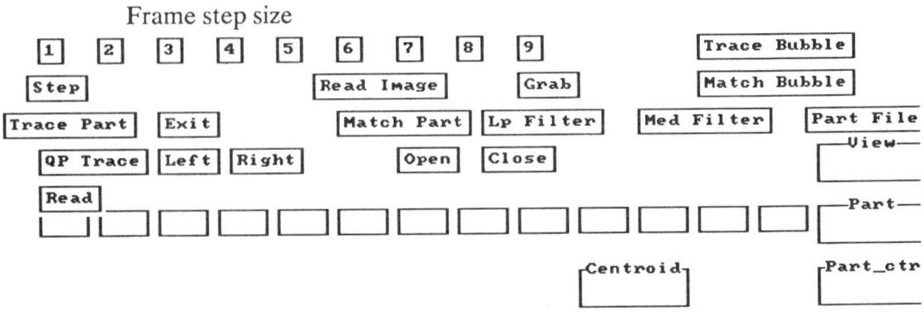

Figure 12. Software control panel

sectional areas. The video image with the particle contour filled in white is shown in Figure 14. The particle centroid locations and particle cross sectional areas are appended to a disk file for later analysis. We used the contour filling algorithm as described on page 174 of Pavlidis [3].

6) The user specifies the number of frames to advance. The computer sends a command to the SP2000 to advance to the specified frame.
7) The new frame is grabbed and digitized.
8) The bubble contours from the starting frame are superimposed on the current frame to measure any image registration offset. Correction for this offset is made by translation so that the current image is aligned properly with frame 0.
9) Steps 2-8 are repeated until a sufficient number of collisions have been recorded.

This sequence is followed except where the particle image intersects with the bubble image. In this situation a uniform background does not exist for the particle image, making automatic contour tracing impossible. In this event the interact program allows the user to interactively translate and rotate a superimposed contour of the same particle from a previous frame using the mouse. When the two images are manually superimposed, the particle's centroid is appended to the data file.

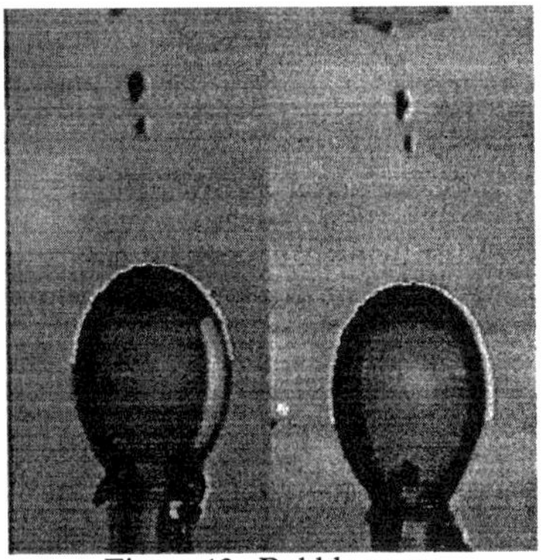

Figure 13. Bubble trace

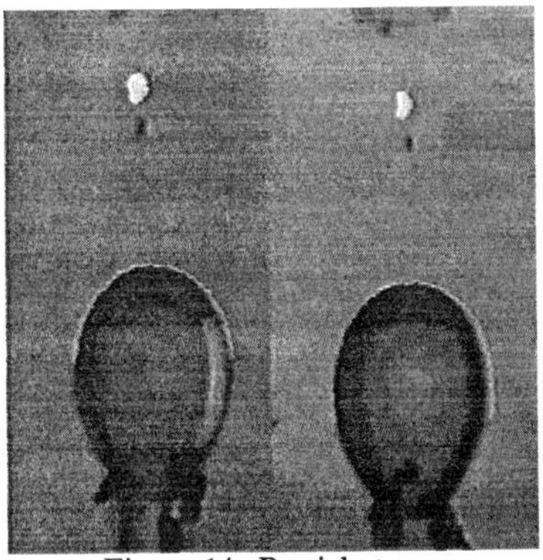

Figure 14. Particle trace

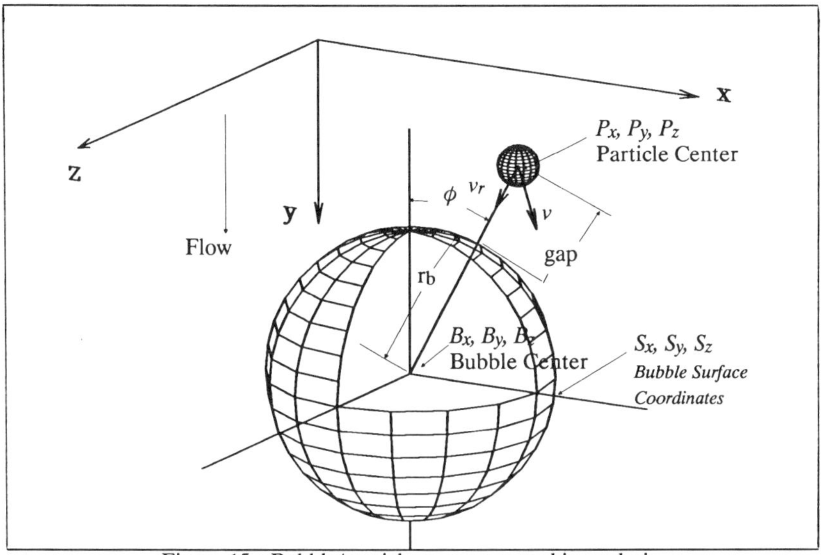

Figure 15. Bubble/particle geometry used in analysis

Calculations

The dimension data is converted from pixels to mm with the calibration factor determined by referencing the known diameter of the tube supporting the bubble.

The interact program outputs two data files as shown in Table I and Table II. Table I contains a sample of several lines from the particle centroid output file.

Table I. Particle centroid locations from the interact program.

Elapsed Time ms	Particle Number	Particle centroid location			Particle Cross Sectional Area	
		P_x mm	P_y mm	P_z mm	Left View 10^{-4} mm^2	Right View 10^{-4} mm^2
0	1	1.63	.89	3.50	56	91
4	1	1.63	1.02	3.50	51	99
.	.	.	.	.	.	.
56	1	1.70	2.14	3.77	61	85
60	1	1.70	2.17	3.81	61	85

Table II lists a sample of the bubble surface (contour) points (343 total points), S_x and S_y, from the left camera projection; the xy plane projection. The center was taken as the midpoint of the horizontal chord of maximum length drawn through the bubble. In this example the center of the bubble is located at $B_x = 3.00$ mm, $B_y = 5.23$ mm, and $B_z =$ 6.25 mm. Since the bubble is not spherical, the bubble radius r_b, varies with the angle ϕ. The angle ϕ and the bubble radius r_b are calculated by conversion to polar coordinates (see Figure 15).

Table II. Bubble contour coordinates at $S_z = 0$ (see Figure 15).

S_x mm	S_y mm	ϕ	r_b mm
3.00	3.92	0.0	1.31
3.26	3.96	11.8	1.31
.	.	.	.
3.80	4.44	45.0	1.29
4.03	5.23	90.0	1.03
.	.	.	.
3.65	6.55	153.4	1.47

Figure 15 defines the nomenclature used in the calculations. The center of the bubble is denoted by (B_x, B_y, B_z) and the center of the particle at the time t_i by (P_{xi}, P_{yi}, P_{zi}). The bubble center was computed from the bubble contours in two projections.

Using a 3-point difference, equation 1 gives the magnitude of the particle velocity,

$$v_i = \frac{\sqrt{(P_x^{i+1}-P_x^{i-1})^2+(P_y^{i+1}-P_y^{i-1})^2+(P_z^{i+1}-P_z^{i-1})^2}}{t_{i+1} - t_{i-1}} \tag{1}$$

at time t_i.

Equation 2 defines r_p (i.e. the distance between the center of the particle and the center of the bubble).

$$r_p^i = \sqrt{(P_x^i-B_x^2)+(P_y^i-B_y^2)+(P_z^i-B_z^2)} \tag{2}$$

Equation 3 gives the angle between the upward vertical line from the bubble center and to the radial line from the bubble center to the particle center at the time t_i,

$$\varphi = 90 + \frac{180}{\pi} \tan^{-1} \frac{P_y^i - B_y}{\sqrt{(P_x^i-B_x^2)+(P_z^i-B_z^2)}} \tag{3}$$

Equation 4 defines the gap (i.e. the distance between the particle center and the bubble surface).

$$gap^i = r_p^i - r_b \tag{4}$$

Where r_b is the distance from the bubble center to the bubble surface along the radial line to the particle center at time t_i. Because the bubble is symmetric about the y axis, r_b can be computed from 2-dimensional projection of the bubble contour as listed in Table II.

RESULTS

The stereo flow cell was filled with a 10^{-3} M solution of sodium lauryl sulfate. The particle mixing chamber was charged with 100 x 140 mesh Pittsburgh seam coal. The trajectories for hundreds of particles were determined. For sake of this discussion the trajectories of two particles are addressed. Figure 16 shows the particle velocity profiles for a particle/bubble collision where particle attachment resulted. Figure 17 shows the particle velocity profiles for a bubble-particle collision where a particle attachment did not result.

DISCUSSION

We were able to successfully track particles during the collision process in 3 -dimensional space. The collision angle can be accurately determined by monitoring the gap. Analysis of Figure 16 reveals that the collision occurred at an angle of around 10 degrees (see point a) where the gap can be seen to reach a constant value approximately equal to the particle radius. The original hypothesis, that the point of film rupture can be detected from the particles's velocity profile, can not be confirmed at this time, however Figure 16 appears to indicate a possible velocity perturbation (see point b) at 80 degrees. This perturbation could be the result of the fluid film rupture. Examination of Figure 17 reveals that for the case of no attachment, a velocity perturbation is not observed. If this hypothesis can be verified, a direct measurement of the induction time could be achieved.

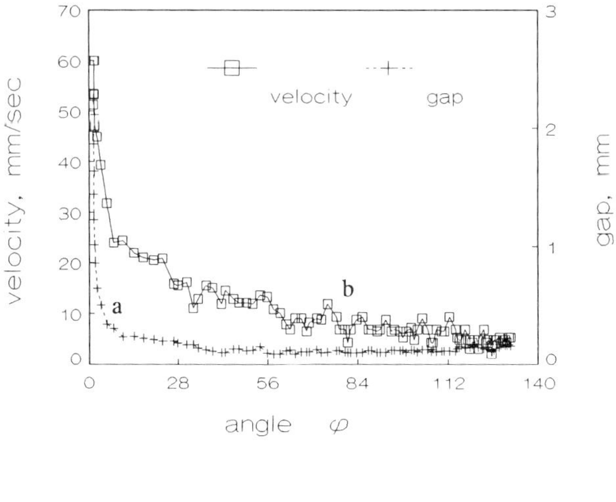

Figure 16 Particle attachment

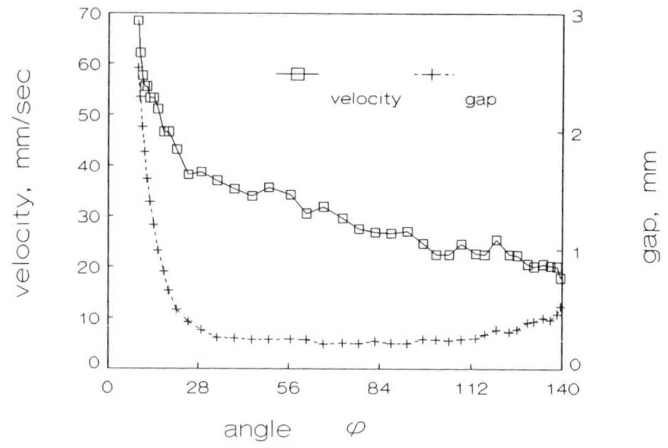

Figure 17 No particle attachment

CONCLUSIONS

1) We were able to construct a stereo cell where three-dimensional trajectories of particles during particle/bubble collisions could be measured.
2) We developed software which allowed the user to automatically trace the contour of the bubble and particles.

DISCLAIMER

Reference in this paper to any specific commercial product, process, or service by trade name, trademark, manufacturer, or otherwise does not necessarily constitute or imply its endorsement, recommendation, or favoring by the United States Department of Energy.

REFERENCES

1. A. Scheludko, Adv. Colloid & Interface Sci. 1, 4 (1967).
2. D.J. Brown, Fuel Soc. J. 16, 22 (1965).
3. T. Pavlidis, Algorithms for Graphics and Image Processing (Computer Science Press 1982).

A STUDY OF SURFACTANT/OIL EMULSIONS FOR FINE COAL FLOTATION

Q. YU, Y. YE, AND J. D. MILLER
Department of Metallurgical Engineering
University of Utah
Salt Lake City, Utah 84112

ABSTRACT

Emulsification studies show that the combination of an oil-soluble surfactant and water-soluble surfactants with oil can produce a stable, water-dispersive emulsion. In particular, with the addition of a nonionic wetting surfactant, emulsified oil is highly dispersed in water with an oil/water droplet size of 5 to 35 µm. Photography demonstrates that such droplets can quickly adsorb on and spread at the coal/water interface, significantly improving the flotation response of coal.

INTRODUCTION

Emulsification is usually referred to as the dispersion of one liquid such as oil in another immiscible liquid such as water. Such an emulsion is thermodynamically unstable due to increased interfacial area, and the separation of two liquids is expected to be spontaneous. To stabilize the emulsion, mechanical energy or a surfactant is needed.

Mechanical emulsification can be done with jets or ultrasonic devices [1]. Emulsions thus produced usually are only stable over a short period of time. Surfactants adsorb at the oil/water interface, lowering the surface free energy and thus stabilizing the emulsion. In addition, adsorption of surfactants at the oil/water interface often changes the characteristics of the electric double layer, which significantly reduces the coalescence of oil droplets by collision. Emulsions thus produced are stable over a long period of time and convenient for transport and utilization.

With such stabilized emulsions and their ability to facilitate the oil wetting of coal particles, emulsified oils have been found to significantly improve coal flotation recovery even at a reduced dosage when compared to direct oil addition during fine coal flotation [2-14]. In this way, further research in this area is warranted due to an increased demand for fine coal flotation technology.

In this study, we selected three differet wetting surfactants, anionic, cationic and nonionic. It was found that the nonionic surfactant provided the best stability to the surfactant/oil emulsion and the best dispersion characteristics in water. Significant improvement in coal flotation was also achieved with such an emulsion. The preliminary results are concisely presented in this paper.

EXPERIMENTAL

Oil Emulsification

In this phase of the work, an alkane oil, an oil-soluble emulsifier (Span 80, HLB = 4.3) and a wetting surfactant (either anionic, cationic or nonionic) were mixed together at different weight ratios. After mixing,

Advances in Fine Particles Processing
John Hanna and Yosry A. Attia, Editors

the emulsions were placed into test tubes. One-hundred twenty-four hours later, visual examination was carried out to check the prepared emulsions for signs of separation. Wetting surfactants used were OT-100 (anionic, HLB = 14.3), CTAB (cationic, HLB = 12.3) and FCA (nonionic, HLB = 12.6). Table I gives a list of molecular structure of the three surfactants.

Table I. List of Molecular Structures of the Three Reagents Used in this Study.

Reagent	Molecular Structure	Functional Group
Surfactant		
Anionic OT	$CH_3-(CH_2)_7-O-C(=O)-CH_2$ $CH_3-(CH_2)_7-C-C(=O)-CH-SO_3Na$	$-CH-SO_3^- \mid Na^+$
Cationic CTAB	$CH_3-(CH_2)_{14}-CH_2-N(CH_3)_3Cl$	$-N^+(CH_3)_3 \mid Cl^-$
Nonionic FCA	$CH_3-(CH_2)_{12}-(OC_2H_5)_{67}-OH$	$-(OC_2H_5)-OH$
Oil-Soluble Emulsifier		
Span 80	$C_{17}H_{33}COC_6H_8O(OH)_3$	$R \mid -COOC_6H_8O(OH_3)$

The photographs shown in Figures 1-3 represent the observed experimental results as well as the weight percentage of the three liquids used for emulsion perparation. It is evident from these photos that, with the anionic surfactant, a stable emulsion was obtained only when the blending weight percentage was 70% oil, 15% surfactant and 15% emulsifier Span 80, or when the weight ratio of water-soluble surfactant to oil-soluble emulsifier was 1:1. With an increase or decrease in the weight ratio of surfactant to emulsifier from 1:1, the emulsions prepared were unstable. For the cationic surfactant, emulsions were stable only when the weight ratio of surfactant to emulsifier was greater than 1:1. Below such a ratio, emulsions were unstable. For the nonionic surfactant, however, stable emulsions were obtained at any weight ratio of surfactant to emulsifier as clearly shown by the photo in Figure 3, and, in fact, a stable emulsion was obtained even without such an oil-soluble emulsifier (Tube No 7, Figure 3). From these results, it is clear that the nonionic surfactant provides the greatest flexibility for oil emulsification over both the anionic and cationic surfactants investigated.

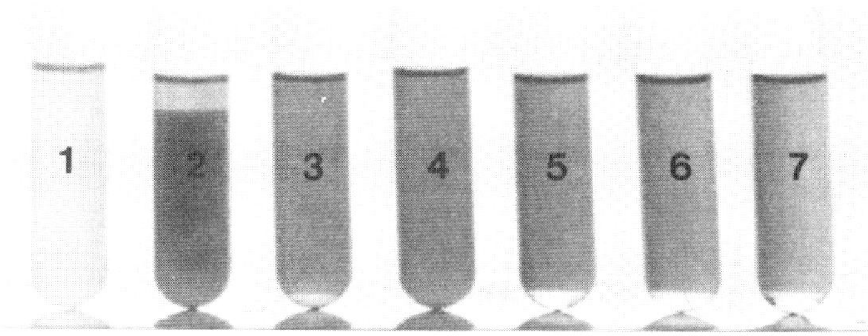

Tube No.	1	2	3	4	5	6	7
	Weight Percentage						
Oil	100	70	70	70	70	70	70
Surfactant	0	5	10	15	20	25	30
Emulsifier	0	25	20	15	10	5	0

FIG. 1. Anionic surfactant/oil emulsions prepared at different compositions and as observed after 124 hours of storage.

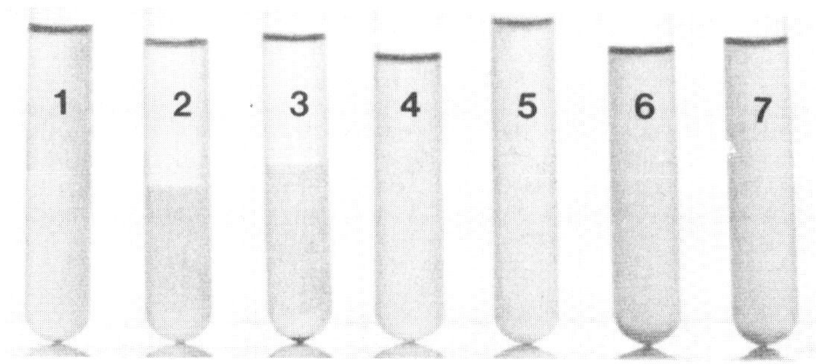

Tube No.	1	2	3	4	5	6	7
	Weight Percentage						
Oil	100	70	70	70	70	70	70
Surfactant	0	5	10	15	20	25	30
Emulsifier	0	25	20	15	10	5	0

FIG. 2. Cationic surfactant/oil emulsions prepared at different compositions and as observed after 124 hours of storage.

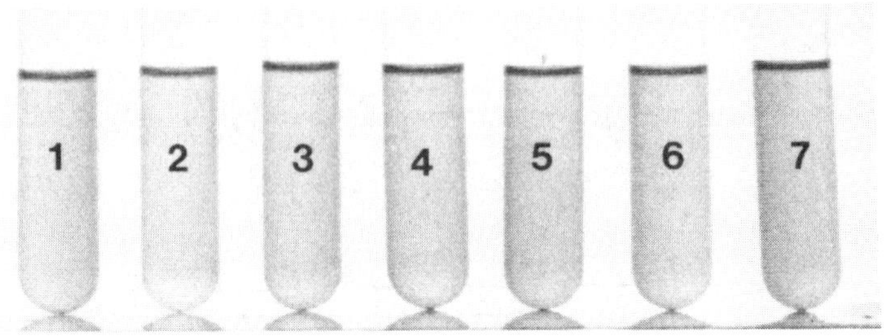

Weight Percentage							
Tube No.	1	2	3	4	5	6	7
Oil	100	70	70	70	70	70	70
Surfactant	0	5	10	15	20	25	30
Emulsifier	0	25	20	15	10	5	0

FIG. 3. Nonionic surfactant/oil emulsions prepared at different compositions and as observed after 124 hours of storage.

Dispersion of Oil Emulsions in Water

Aside from creating a stable surfactant/oil emulsion, another critical criterion for application of the emulsion is its dispersion characteristics in water. If the emulsified oil can be readily dispersed in water, a better wetting of the oil collector at the coal surface can be achieved to improve flotation. Generally, unmodified oil cannot be dispersed in water, as shown in Figure 4. Wetting of coal particle surfaces by the oil is mostly accomplished by mechanical mixing during conditioning of the coal suspension.

In the case of dispersion tests for oil emulsions in water, 1 mℓ each of stabilized anionic, cationic and nonionic surfactant/oil emulsion was injected into a beaker containing 150 mℓ distilled water. All three surfactant/oil emulsions were prepared with 70% oil, 15% surfactant, and 15% emulsifier. This condition, according to previous tests, provides a stabilized state for all three emulsions.

Most important, however, is that these emulsified oils can be dispersed in water. Further, it was observed that the nonionic surfactant/oil emulsion provided for the best dispersion with immediate dispersion upon injection; anionic surfactant/oil emulsion was next, and the cationic surfactant/oil emulsion had the poorest dispersion characteristics. These results are shown in Figures 5-7.

After injection, the nonionic surfactant/oil emulsion was dispersed almost instantaneously in water as a uniform dispersion. Figure 8 presents a photo taken 5 minutes after the injection of the emulsion into the water. As shown in Figure 8, a complete dispersion has been obtained.

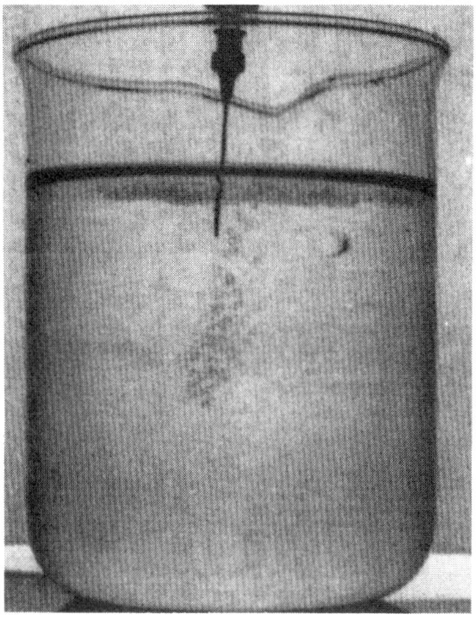

FIG. 4. Lack of dispersion during injection of oil into water.

FIG. 5. Injection of 1 mℓ anionic surfactant/oil emulsion into 150 mℓ water.

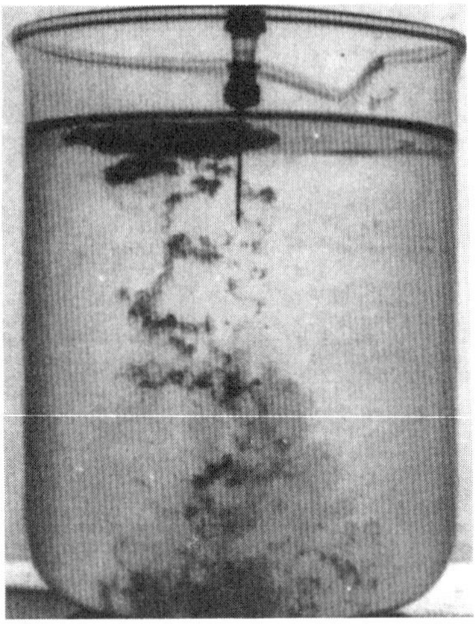

FIG. 6. Injection of 1 mℓ cationic surfactant/oil emulsion into 150 mℓ water.

FIG. 7. Injection of 1 mℓ nonionic surfactant/oil emulsion into 150 mℓ water.

FIG. 8. Spontaneous dispersion of 1 mℓ nonionic surfactant/oil emulsion in 150 mℓ water.

Microphotographic Examination

The water-dispersion of the nonionic surfactant/oil emulsion was examined under an optical microscope. One photograph typical of the dispersion is given in Figure 9. From this work, it was found that the majority of emulsion droplets formed during dispersion are in a size range of 5 to 35 µm. In this way, a better oil coating at the coal particle surfaces with a reduced reagent addition is anticipated.

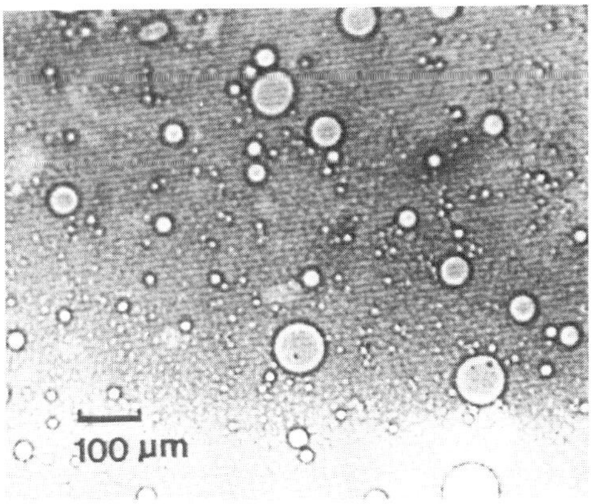

FIG. 9. Microphotograph of the dispersed droplets of the nonionic surfactant/oil emulsion in water.

Adsorption of Emulsion Droplets at the Coal/Water Interface

It was shown from this study that the prepared nonionic surfactant/oil emulsion is rapidly dispersed in water and that the emulsion droplets quickly adsorb on and spread at the coal/water interface. In this test, one drop of the nonionic surfactant/oil emulsion was carefully introduced at the coal/water interface. As soon as the emulsion drop contacted the coal surface, it immediately spread at the coal/water interface to form a thin film oil coating. This result is shown in Figure 10. As a comparison, contact of the hydrocarbon oil droplet at the coal/water interface is also given in Figure 10. It is clearly shown that the nonionic surfactant/oil emulsion provides for significantly enhanced oil spreading at the coal/water interface, while the spreading of oil droplets at the coal surface is quite limited.

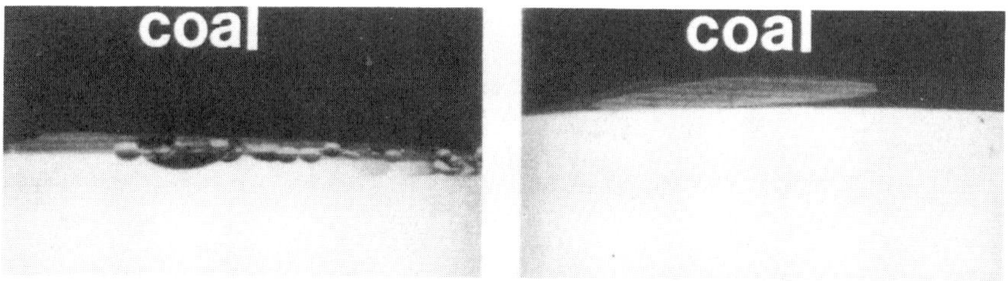

Oil Only　　　　Surfactant/Oil Emulsion

FIG. 10. Spreading of hydrocarbon oil and nonionic surfactant/oil emulsion at coal/water interface.

Flotation Experiments

Bench-scale flotation experiments were carried out for three different coal types; one high-volatile bituminous coal and two difficult-to-float sub-bituminous coals. The tests were done with a 4-ℓ flotation cell, at 10% solids and 4 ℓ/min air flowrate. The flotation time was controlled at five minutes. Again, the emulsions were prepared with 70% oil, 15% surfactant, and 15% emulsifier. Figure 11 gives the combustible recovery of the high-volatile bituminous coal versus reagent addition for oil, anionic surfactant/oil emulsion, cationic surfactant/oil emulsion, and nonionic surfactant/oil emulsion. From the figure it is evident that a 20 to 30% increase in flotation recovery was obtained at the same level of reagent addition when the nonionic surfactant/oil emulsion was used, as compared with oil-only addition. Improved flotation recovery with the anionic surfactant/oil emulsion was also obtained; however, the cationic surfactant/oil emulsion did not lead to any improvement in the flotation recovery.

Table II presents the flotation results for the two difficult-to-float sub-bituminous coals. As can be seen from the table, at the same ash content in the clean coal products, a 10 to 15 percent increase in flotation yield was achieved when the nonionic surfactant/oil emulsion was used instead of oil only, even though the level of addition of the emulsified oil was 30% less. The improved flotation performance with emulsified oil is obviously significant.

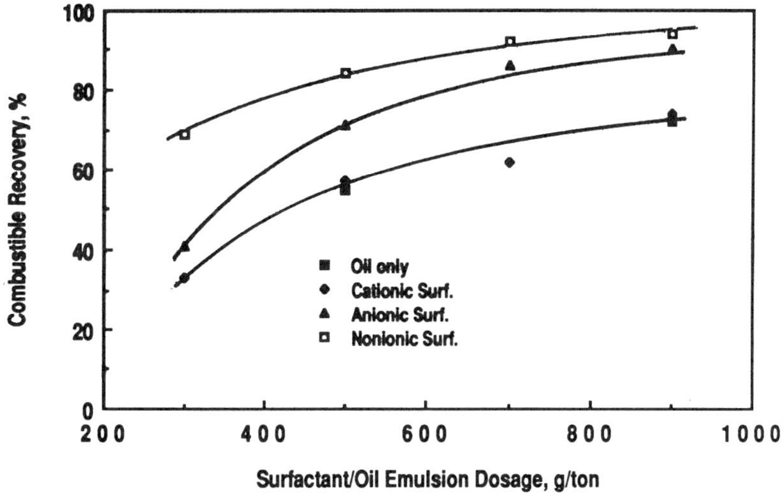

FIG. 11. Combustible recovery versus reagent addition for single-stage bench-scale flotation of a high-volatile bituminous coal.

Table II. Bench-Scale Flotation Results of Two Difficult-to-Float Sub-Bituminous Coals with Oil Only and with the Nonionic Surfactant/Oil Emulsion

				Clean Coal	
Sample	Reagent	Dosage (kg/ton)	Feed (Ash %)	Yield (%)	Ash (%)
SUB No. 1	Oil Only	4.42	13.43	71.4	8.45
	Emulsified oil	3.12	13.58	87.6	8.58
SUB No. 2	Oil Only	3.55	17.5	77.4	10.3
	Emulsified oil	2.65	16.8	86.4	10.9

DISCUSSION

In general, hydrocarbon oils can be emulsified with the addition of an oil-soluble emulsifier such as Span 80, as used in this current work. Nevertheless, such a prepared emulsion has very poor dispersion characteristics in water. Consequently, this way, a water-soluble surfactant should be introduced into the system so that the stabilized oil/water emulsion can better be dispersed in water.

Most of the hydrocarbon oils can be prepared to form a water-dispersive emulsion by using surfactants with a proper HLB value. This value is calculated as

$$HLB = (W_a \cdot HLB_a + W_b \cdot HLB_b)/(W_a + W_b) \quad (1)$$

where HLB_a and HLB_b are the HLB values of the water-soluble wetting surfactant and oil-soluble emulsifier, respectively, while W_a and W_b are their corresponding weight fractions in the emulsion. In this work, the hydrocarbon oil used was emulsified with a HLB value of about 9.3 using an anionic surfactant. For the cationic surfactant, the oil was emulsified with a HLB value of 8.3-12.3. For the nonionic surfactant, however, the oil was emulsified with a HLB value of 5.7-12.6, a much broader range than was required for either the anionic or cationic surfactants. The nonionic surfactant, therefore, provides for better flexibility than the other two ionic surfactants. In addition to the emulsion stability, the emulsion prepared with the nonionic surfactant had the best dispersion characteristics in water.

Since wetting surfactants are mostly adsorbed at the oil/water interface with their polar groups pointing toward the water, it is quite usual that the surfaces of emulsified oil droplets prepared with anionic, cationic, and nonionic surfactants carry a negative, positive, and neutral charge, respectively. In this way, when emulsified oil droplets collide with coal surfaces, the nonionic surfactant/oil emulsion droplets are expected to have a strong tendency to spread at the coal surfaces due both to hydrogen bonding from the surfactant's polar groups and to hydrophobic bonding phenomena. Such spreading will push interfacial water from the coal surface and lead to the stabilization of an oil film at the coal surface. Figure 12 schematically shows the wetting of the coal/water interface by a nonionic surfactant/oil emulsion droplet. The wetting phenomena associated with the interfacial water displacement by the oil emulsion leads to the stabilization of an oil film rather than attached discrete microdroplets. Improvement in bubble attachment at the surface of coal particles, and hence flotation performance, is thus expected.

FIG. 12. Schematic showing the wetting of the coal/water interface by a nonionic surfactant/oil emulsion droplet.

From these same considerations, wetting of the coal surface by anionic surfactant/oil emulsion and cationic surfactant/oil emulsion is much weaker than nonionic surfactant/oil emulsion, because the polar nature of the ionic surfactants may provide for hydrogen bonding or coulombic attraction at the coal surface. This type of adhesion will result in a tendency for a residual water film to be retained between the coal and oil. Consequently, improvement in flotation performance with ionic surfactant/oil emulsion is less than that with the nonionic surfactant/oil emulsion, as has been observed in flotation experiments.

CONCLUSION

Experimental results show that the surfactant/oil emulsion prepared with a nonionic surfactant gives the best stability to the surfactant/oil emulsion and provides for the best dispersion characteristics of the surfactant/oil emulsion in water.

Significant improvement in coal flotation can be achieved with the surfactant/oil emulsion due to enhanced adhesion/spreading of the oil at the coal/water interface.

REFERENCES

1. S. C. Sun, L. Y. Tu, and E. Ackerman, Min. Engin., 7, 656 (1955).
2. A. R. Burkin and J. V. Bramley, J. Appl. Chem., 11, 300 (1961).
3. A. R. Burkin and J. V. Bramley, J. Appl. Chem., 13, 417 (1963).
4. J. M. W. Mackenzie, Trans. Am. Inst. Min. Engrs., 244, 393 (1969).
5. J. M. W. Mackenzie, Trans. Am. Inst. Min. Engrs., 247, 202 (1970).
6. W. W. Wen and S. C. Sun, Trans. Am. Inst. Min. Engrs., 262, 174 (1977).
7. J. A. Beardsley, Coal Min. Process., July, 64 (1978).
8. S. S. Wang, C. J. Scanlon, and M. J. Scanlon, U.S. Patent 4,340,467 (1980).
9. W. W. Wen and S. C. Sun, Separation Sci., 16, 1491 (1981).
10. J. Sabik, Inst. J. Min. Proc., 9, 245 (1982).
11. R. O. Keys, U.S. Patent 4,504,385 (1982).
12. M. J. Scanlon et al., World Coal, Feb., 54 (1983).
13. R. O. Keys, U.S. Patent 4,606,818 (1984).
14. R. O. Keys, U.S. Patent 4,678,561 (1984).

SEQUENTIAL SEPARATION OF CARBONATE AND SILICEOUS GANGUE MINERALS DURING PHOSPHATE ORE PROCESSING

I. ANAZIA AND JOHN HANNA
Mineral Resources Institute, The University of Alabama
Box 870204, Tuscaloosa, AL 35487-0204

ABSTRACT

A unique fatty acid flotation process is described for sequential separation of carbonate and siliceous gangue minerals from low grade phosphate ores. The process involves a first stage of selective carbonate/phosphate separation followed by phosphate/silica separation during the second stage. In the first stage, a carbonate rich froth is removed without specific depression of the phosphate minerals and without conditioning of the pulp with the fatty acid collector prior to flotation. In the second stage, the same collector was selectively used to produce a market grade phosphate concentrate in the froth, in the presence or absence of a silica depressant. Selectivity was achieved using commercial grade fatty acid collectors, frothers and pH modifiers. The factors affecting carbonate/phosphate and phosphate/silica separation are discussed. Tests on a high-MgO siliceous phosphate matrix from south Florida, yielded phosphate concentrates analyzing 31% P_2O_5, 0.7% MgO and 4% acid insoluble matter with a P_2O_5 recovery of 80%.

INTRODUCTION

The MRI "no conditioning" process for carbonate gangue removal from phosphate ores has been demonstrated to be effective on sedimentary carbonate-phosphate ores from a variety of sources [1,2]. For non-siliceous dolomitic and/or calcareous phosphates (i.e. low in acid insoluble matter), carbonate/phosphate separation by the MRI process involved a single step selective fatty acid flotation of the carbonate gangue to produce market grade phosphate concentrates. Ores containing significant amounts of siliceous material required an additional phosphate/silica separation step after carbonate gangue flotation.

Details of the selective fatty acid carbonate/phosphate separation technique and its applications to various ores have been reported elsewhere by the authors [3,4]. The present work being discussed in this paper deals mainly with the development of the two step carbonate/phosphate/silica separation process. This includes studies on the effect of pH, type of pH modifiers, silica depressants and collector dose on the grade, recovery and flotation efficiency of high grade phosphate concentrates with low MgO and silica content.

EXPERIMENTAL

Ore Tested

A high-MgO phosphate matrix from the W.R. Grace Company, Four Corners Mine in Bartow, Florida was tested. The run-of-mine material

Advances in Fine Particles Processing
John Hanna and Yosry A. Attia, Editors

analyzed 12.1% P_2O_5, 24.1% CaO, 8.9% MgO, and 29.0% acid insoluble matter (insol).

Examination of the sample using polarized light and scanning electron microscopy revealed that it consisted of apatite, dolomite, feldspar and quartz. The apatite was brown to black and occurred as smooth pebbles, as pebbles with inclusions of quartz and dolomite, as pebbles with partial rims or depressions filled with carbonate, as phosphate cementing dolomite and quartz, and as fossil shark teeth. The dolomite occurred as soft agglomerates of rhombohedral grains (typically <30 microns) with quartz and apatite pebble inclusions; and as carbonate rims (described above). The quartz occurred as rounded grains and as inclusions in apatite or dolomite agglomerates.

X-ray diffraction analysis and cell parameters "a" and "c" indicated that the apatite mineral in the matrix was francolite having the following empirical formula:

$$Ca_{9.62}\ Na_{0.273}\ Mg_{0.106}\ (PO_4)_{4.976}\ (CO_3)_{1.024}\ F_{2.41}$$

Based on McClellan and Lehr's [5] analytical method, the theoretical composition of the francolite mineral was 36.28% P_2O_5, 55.34% CaO, 4.7% F, 4.63% CO_2, 0.87% Na_2O and 0.44% MgO. The X-ray diffraction pattern of the carbonate mineral in the samples matched the pattern for dolomite. However, the refractive indices for the carbonate mineral were between the literature values for dolomite and calcite.

Flotation Feed: The matrix was slurried in tap water and screened at 35-mesh (417 microns), and the minus fraction was deslimed at 150-mesh (104 microns). The plus 35-mesh material was roll crushed to minus 35-mesh and deslimed at 150-mesh. The resulting 35 x 150 mesh products of both steps were mixed and stored wet until used. Chemical analysis of this feed was approximately 10% P_2O_5, 1% MgO, 11% CaO, and 65% acid insoluble matter (insol). Thus, most of the MgO in the matrix was removed in the minus 150-mesh slime fraction.

Point counting and sink/float separation of the 35 x 150 mesh feed indicated that the apatite particles were not completely liberated from the dolomite and quartz. About 56% of the phosphate occurred as free particles analyzing 31.4% P_2O_5, 0.6% MgO and 3% insol. About 40% of the phosphate occurred as locked carbonate/phosphate particles analyzing 21.4% P_2O_5, 5.4% MgO, and 4.7% insol. About 92% of the siliceous impurities occurred as free particles.

Flotation Procedure

Flotation tests were carried out in a model D-1 Denver laboratory flotation machine with the impeller speed set at 1000 rpm. The deslimed material (35 x 150 mesh) was pulped in the flotation cell at 18% to 24% solids using Tuscaloosa tap water. Unless otherwise stated, oleic acid was used as collector and pine oil as frother. Test products were analyzed and recoveries calculated.

Carbonate Flotation: Batches of 300-600g of wet deslimed feed were mixed to the required pulp density. The pH of the pulp was adjusted to about pH 5.5 and maintained throughout the carbonate flotation step. Unless otherwise stated, sulfuric acid was used for pH adjustment. The collector used was a pine oil (frother) and fatty acid mixture, emulsified with several drops of dilute NaOH prior to addition to the pulp in three equal increments. Immediately after each collector addition, the pulp was aerated and the carbonate froth was collected for about 1.5 minutes. Thus, no collector conditioning time was allowed. The flotation products were analyzed for MgO, P_2O_5, and insol.

Phosphate Flotation: The cell product obtained from the carbonate flotation step was conditioned at the required pH for three minutes in the presence or absence of sodium silicate to depress siliceous gangue. This was followed by another three-minute conditioning with oleic acid collector and flotation of the phosphate minerals, leaving the silica-rich tails in the cell product. No frother was added in this step.

RESULTS

Carbonate Flotation

A number of fatty acid collectors including pure oleic acid and commercial tall oil products were tried and found to be effective reagents for carbonate flotation. Pure oleic acid was, however, selected for testwork involving a three factor, two level (2^3) full factorial design to optimize the carbonate flotation step, with respect to collector dose, pH and % solids. The factors included in the design and the factor levels tested are given in table 1. In addition to replicate center points, star points were included in order to estimate non-linear factor responses. Also, included in table 1 are the product grades and recoveries as well as the MgO Separation Efficiencies (MgO S.E.) for the nineteen experiments conducted. The MgO S.E. is a response function, similar to the coefficient of separation equation by Gaudin [6]. The MgO S.E. was used to quantify the degree of carbonate/phosphate separation as: MgO S.E. = $R_e - R_f$ where: R_e is the MgO recovery in the carbonate froth and R_f is the P_2O_5 recovery in the same carbonate froth.

Statistical analysis of the data indicated that pH and collector dose factors and cross factors had predominant effects on MgO S.E., compared to the effect of pulp solids. Based on the MgO S.E. model developed, the response surface contours at 8, 16, and 24% solids were plotted separately and stacked as shown in figure 1. The contours show that the region of best response shifts toward lower pH and higher collector dose, as percent solids is increased. In addition to a higher collector dose, a lower pH implies higher pH regulator consumption. However, because flotation processing of dilute slurries is usually uneconomical, a pulp solids higher than 8% was preferred. Hence, 16% was considered to be the best of the three pulp densities tested. From the response surface figure for 16% solids, a pH of 5.5 and a collector dose of 1.5 kg/T oleic acid were considered the best conditions to use in further tests.

Effect of Collector Conditioning: The effect of conditioning the pulp with the oleic acid collector prior to aeration was studied under conditions stated above--pH 5.5, 1.5 kg/T oleic acid and 16% pulp solids. The results are shown in figure 2. The conditioning times shown are the totals for the 3-stage collector additions, i.e., 3-minute conditioning time entailed a 1-minute conditioning after each collector addition.

As shown in figure 2, collector conditioning had a deleterious effect on the selectivity of the carbonate/phosphate separation. Figure 3 shows that even at "zero" conditioning time the dolomite content of each froth decreased, while the apatite increased after each collector addition.

Phosphate Flotation

Effect of pH: Results of tests done to study the effect of pulp pH on the phosphate flotation step are shown in table 2. In these tests,

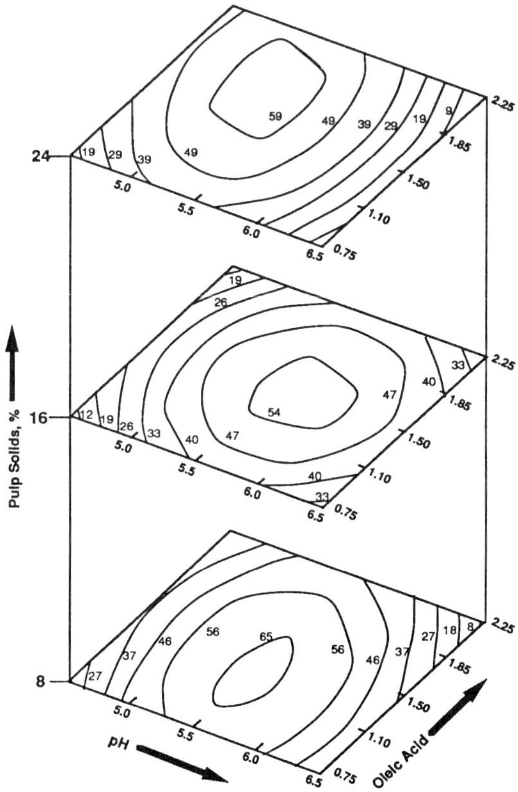

Figure 1. MgO Separation Efficiency Contours of Oleic Acid Dose at Various Pulp Solids Percentages

Table 1. Experimental Design for Optimization of Carbonate Flotation Parameters

* Exp't	[A] Collector Dose, kg/T	[B] pH	[C] % Solids	MgO S.E., %	MgO Grade, %	MgO Recovery, %
(1)	0.75	4.5	8	21.4	16.90	21.7
A	2.25	4.5	8	31.9	15.80	32.3
B	0.75	6.5	8	38.6	2.72	72.1
AB	2.25	6.5	8	5.7	1.60	88.5
C	0.75	4.5	24	17.7	15.85	17.9
AC	2.25	4.5	24	48.5	16.89	49.8
BC	0.75	6.5	24	1.9	2.12	78.1
ABC	2.25	6.5	24	-2.0	1.20	95.4
M	1.50	5.5	16	58.2	11.88	63.0
M	1.50	5.5	16	65.3	12.34	70.9
M	1.50	5.5	16	60.6	11.55	66.9
M	1.50	5.5	16	58.5	12.15	62.6
M	1.50	5.5	16	58.3	13.46	61.1
MA+	2.25	5.5	16	27.0	4.01	83.0
MA-	0.75	5.5	16	40.9	7.69	48.8
MB+	1.50	6.5	16	14.1	2.39	66.9
MB-	1.50	4.5	16	15.6	14.50	15.9
MC+	1.50	5.5	24	55.5	9.49	66.7
MC-	1.50	5.5	8	58.6	13.60	61.1

* Listed in Standard Yates Order

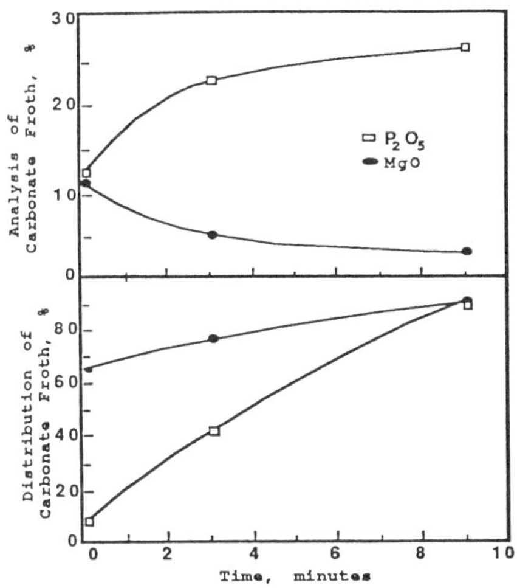

Figure 2. Effect of Collector Conditioning Time on Carbonate Flotation

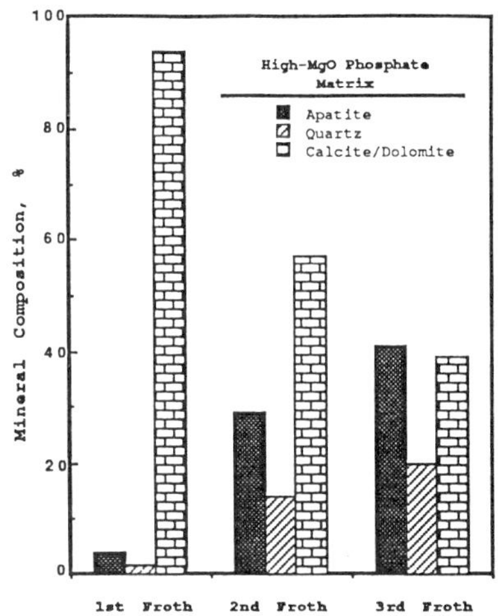

Figure 3. Effect of Collector Stage Addition on the Mineralogic Composition of the Carbonate Froth

Table 2. Effect of pH on Phosphate Flotation

Pulp pH	Product	Weight, %	Analysis, %			Distribution, %		
			P_2O_5	Insol	MgO	P_2O_5	Insol	MgO
6	Carb. Froth	7.4	13.02	8.04	10.21	10.2	0.9	72.9
	Phos. Conc.1	24.3	29.69	6.54	0.86	76.1	2.4	20.2
	Phos. Conc.2	2.5	27.04	11.04	0.24	7.2	0.4	0.6
	Tailing	65.8	0.93	96.28	0.10	6.5	96.2	6.3
	Composite Feed	100.0	9.48	65.79	1.04	100.0	100.0	100.0
7								
	Carb. Froth	4.0	10.52	7.81	13.05	3.9	0.5	55.4
	Phos. Conc.1	20.3	24.68	22.33	0.76	46.0	7.2	16.3
	Phos. Conc.2	15.3	21.03	33.08	0.77	29.6	8.1	12.4
	Tailing	60.4	3.70	87.73	0.25	20.5	84.2	15.9
9	Composite Feed	100.0	10.88	62.89	0.95	100.0	100.0	100.0
	Carb. Froth	4.5	5.09	7.38	16.81	2.3	0.5	63.4
	Phos. Conc.1	21.7	12.51	60.06	0.47	27.1	20.8	8.6
	Phos. Conc.2	53.3	10.04	65.02	0.41	53.2	55.3	18.5
	Tailing	0.5	8.58	71.37	0.55	17.5	23.4	9.5
	Composite Feed	100.0	10.06	62.67	1.18	100.0	100.0	100.0

Conditions

Carbonate Flotation: ph 5.5, 3 steps, with 0.5 kg/T oleic acid for each step.

Phosphate Flotation: 3-minute conditioning (after carbonate flotation) with 0.5 kg/T Na_2SiO_3 (SiO_2/Na_2O = 3.25) followed by flotation of Phosphate Conc 1. Then 0.25 kg/T oleic acid with 2 minute conditioning for Phosphate Concentrate 2.

each cell product obtained after carbonate flotation at pH 5.5 was first conditioned for two minutes with 0.5 kg/t of sodium silicate depressant at the desired pH level. This was followed by another 3-minute conditioning at the same pH with 0.25 kg/t oleic acid as phosphate collector. Subsequent flotation yielded phosphate-rich froths and clean silica cell products.

The data in table 2 show that the grade and recovery of P_2O_5 in the phosphate concentrate decreased with increase in pulp pH from 6 to 9. The best flotation results were obtained at pH 6, at which point about 83% of the phosphate was recovered as concentrate analyzing about 29.4% P_2O_5 and 7.0% insol. The low insoluble content of the phosphate concentrate indicates a high degree of selectivity in the phosphate/silica separation step. However, at higher pulp pH's of 7 and 9, this selectivity was vitiated, as evidenced by the high acid insol and low P_2O_5 analyses of phosphate concentrates obtained under those conditions. For example, the phosphate concentrates recovered at pH 7 analyzed about 23% P_2O_5 and 27% insol. At pH 9, most of the silica floated indiscriminately along with the phosphate values, indicating that selectivity in phosphate/silica separation was poor at alkaline pH's.

Collector/Depressant Effects and the Role of pH: Tests co-varying the three factors of pH, collector and depressant doses were randomly conducted on the carbonate flotation cell product using a 2^3 full factorial design, containing replicate center points and star points to estimate non-linear factor responses. The factor levels are shown in table 3, along with the response for each test. As was done for the carbonate flotation step, in addition to the use of P_2O_5 grade and recovery as response variables, P_2O_5 separation efficiency (P_2O_5 S.E.) was used as a response function that quantified the degree of phosphate/insol separation as follows: P_2O_5 S.E. = $R_{pf} - R_{if}$. Where: R_{pf} is the P_2O_5 recovery in phosphate froth and R_{if} is the insol recovery in the same phosphate froth. Figures 3, 4 and 5 have been plotted based on the models developed for the three response variables.

Statistical analysis of the data revealed that pH was the most significant factor in determining product grade and recovery and therefore P_2O_5 S.E. It is interesting to note that in the first experiment in table 3 a high grade phosphate concentrate with reasonable P_2O_5 recovery was obtained at pH 5.8 without the use of sodium silicate as a depressant, and without additional oleic acid to collect the phosphate minerals. This indicates that the residual collector from carbonate flotation was sufficient to produce relatively clean products at slightly acidic pH's without the use of a specific silica depressant. In general, however, the use of sodium silicate appeared to improve phosphate flotation, indicating its possible role as activator for phosphate minerals during fatty acid flotation under the test conditions used.

Figure 4 shows that the numerical values of P_2O_5 separation efficiency contours, obtained in test runs with 1.0 kg/t sodium silicate, were virtually the same as those obtained for tests run with 0.5 kg/t sodium silicate. Thus, increase of sodium silicate dose above 0.5 kg/t was neither beneficial nor deleterious to selective phosphate/silica separation. However, the values of P_2O_5 S.E. obtained for 0.5 kg/t sodium silicate were higher than those obtained without the use of sodium silicate. Since P_2O_5 S.E. is an expression incorporating grade and recovery, grade/recovery contours may more clearly show the collector/depressant effects and the role of pH.

The grade/recovery contours for tests run with no sodium silicate and 0.5 kg/t sodium silicate are plotted in figures 5 and 6 respectively.

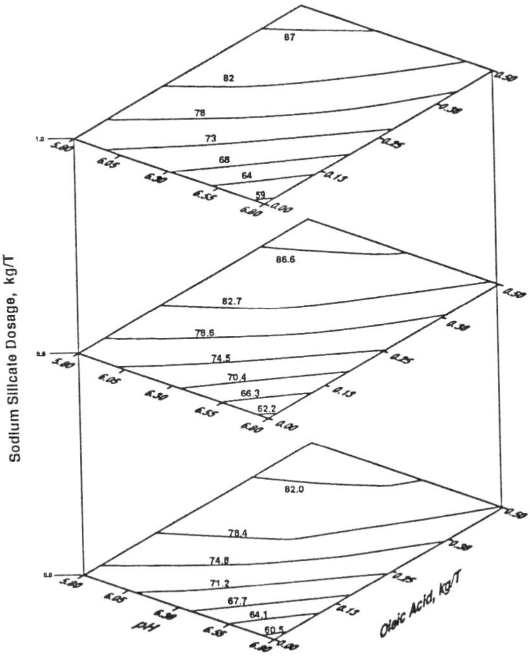

Figure 4. P_2O_5 Separation Efficiency Contours of Oleic Acid Dose vs pH at Various Sodium Silicate Dosages

Table 3. Experimental Design for Optimization of the Phosphate Recovery Step

Exp't	A:Collector Dose, kg/T	B:Depressant Dose, kg/T	C:pH	P_2O_5 Sep'n Effic'y	P_2O_5 Grade, %	P_2O_5 Rec'y, %
(1)	0.00	0.0	5.8	66.9	30.60	68.7
A	0.50	0.0	5.8	89.3	30.74	91.2
B	0.00	1.0	5.8	81.8	30.21	83.9
AB	0.50	1.0	5.8	87.8	30.70	89.2
C	0.00	0.0	6.8	60.7	23.02	73.5
AC	0.50	0.0	6.8	71.8	25.81	77.4
BC	0.00	1.0	6.8	51.3	25.11	56.6
ABC	0.50	1.0	6.8	86.3	29.39	89.2
M	0.25	0.5	6.3	68.3	29.01	71.0
M	0.25	0.5	6.3	84.8	29.90	87.1
M	0.25	0.5	6.3	85.8	*30.20	87.6
M	0.25	0.5	6.3	87.3	29.13	89.4
M	0.25	0.5	6.3	84.8	29.70	87.5
MA+	0.50	0.5	6.3	83.6	*30.49	86.1
MA-	0.00	0.5	6.3	77.1	30.40	78.6
MB+	0.25	1.0	6.3	76.2	29.91	78.5
MB-	0.25	0.0	6.3	85.1	*29.73	87.8
MC+	0.25	0.5	6.8	80.5	27.32	86.7
MC-	0.25	0.5	5.8	79.4	26.99	81.3

* Tests for which reflotation was done.

In a general sense, within the sample space, the values for recovery appeared to be insensitive to pH while the values for grade seemed insensitive to collector dose. In examining both figures, it is clear that the use of sodium silicate contributed marginal increases in grade and significant increases in recovery.

Effects of Type of pH Modifier: As mentioned above, good grade phosphate concentrates were obtained with P_2O_5 recoveries, at about pH 6, without the use of a silica depressant and without collector addition other than that used in carbonate flotation. Tests were then conducted to better understand the role of pH and type of pH modifier, during the phosphate flotation step, under these conditions (i.e. at pH 6, without addition of collector and silica depressant). Results of these tests are shown in table 4.

In the first test recorded in the table, no pulp pH modifier was used. The pulp was merely conditioned for three minutes, after carbonate flotation, thereby allowing the pH to increase from 5.5 to about 5.8 prior to phosphate flotation. In the other tests shown in the table, the pH was maintained at pH 6 with the indicated modifier, the quantity of which amounted to only a few drops of a 5% solution.

In the test conducted without pulp pH adjustment, over 60% of the phosphate values were floated by the residual collector remaining in the pulp from the carbonate flotation step. The high P_2O_5 content of 30.5% and the low insol analysis of the phosphate concentrates obtained indicate that the siliceous materials were effectively depressed at about pH 6.

Conditioning the carbonate-free pulp with various pH modifiers, to maintain a constant pH of about 6, improved phosphate flotation recovery but the grade of products depended on the type of pH modifier used. Control of pH with strongly alkaline sodium silicate produced a good phosphate concentrate containing about 30% P_2O_5, 6.5% insol and 0.86% MgO with a P_2O_5 recovery of 76%.

The outstanding results obtained with sodium silicate may be attributed to its dual function as pH modifier and as a selective silica depressant. Further comparison of the results obtained with and without sodium silicate indicate that this reagent had enhanced phosphate flotation recovery from 60-76%.

DISCUSSION

In conventional phosphate beneficiation practice, the widely used "double float" froth flotation technique developed for upgrading siliceous phosphatic ores, has generally been ineffective for beneficiating high dolomite/limestone phosphate ores. The fatty acid collectors used to float phosphate minerals (in alkaline circuits pH 8-10) also collected carbonate minerals. Therefore, efforts were made to develop selective carbonate or phosphate collectors such as phosphoric and polycarboxylic acid esters, n-alkylamine propionic acids, and n-substituted sarcosine. In addition, specific depressants such as diphosphonic or fluosilicic acids were used to achieve adequate selectivity in the fatty acid flotation circuits [7].

The carbonate flotation results obtained in this investigation have demonstrated the premise of "instant" fatty acid flotation of carbonate gangue minerals, under slightly acidic conditions, without the specific use of a phosphate depressant. Selective fatty acid flotation of carbonate gangue from francolite was achieved at pulp pH of 5.5 (figure 4) using any strong inorganic acid for pH regulation. This suggests that

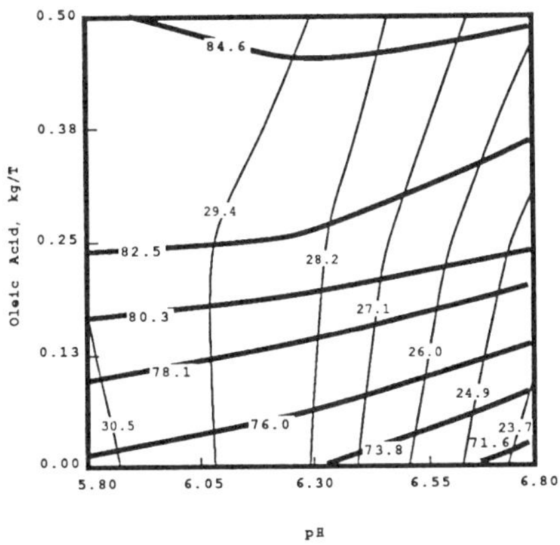

Figure 5. P_2O_5 Grade/Recovery Contours for Rougher Phosphate Concentrates Obtained without the Use of a Silica Depressant; (——— Grade; ▬▬▬ Recovery)

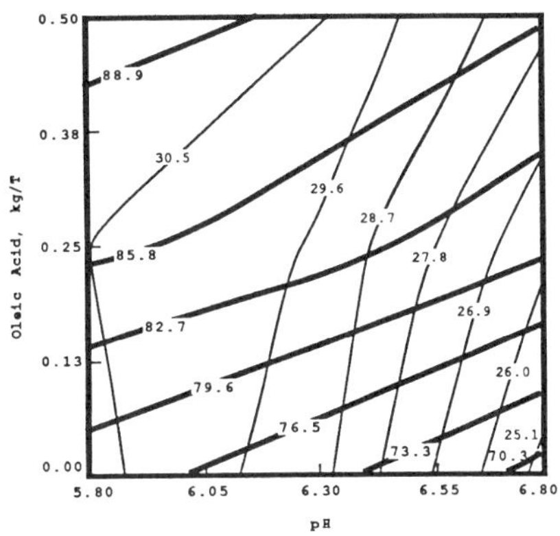

Figure 6. P_2O_5 Grade/Recovery Contours for Rougher Phosphate Concentrates Obtained with 0.5 kg/T Na Silicate as Silica Depressant; (——— Grade; ▬▬▬ Recovery)

Table 4. Screening of pH Modifiers for the Phosphate Recovery Step of the MRI Process

pH Modifier	Product	Weight, %	Analysis %			Distribution %		
			P_2O_5	Insol	MgO	P_2O_5	Insol	MgO
None	Carbonate Froth	6.8	12.73	8.17	10.62	9.1	0.9	68.7
	Phosphate Conc.	19.1	30.45	2.91	1.11	60.3	0.8	20.0
	Tailings	74.1	3.97	87.24	0.16	30.6	98.3	11.3
	Composite Feed	100.0	9.62	65.76	1.06	100.0	100.0	100.0
Na_2SiO_3	Carbonate Froth	7.4	13.02	8.04	10.21	10.2	0.9	72.9
	Phosphate Conc.	24.3	29.69	6.54	0.86	76.1	2.4	20.2
	Tailings	68.3	1.89	93.17	0.11	13.7	96.7	6.9
	Composite Feed	100.0	9.48	65.79	1.04	100.0	100.0	100.0
NaOH	Carbonate Froth	4.7	8.87	7.00	14.03	4.2	0.5	60.3
	Phosphate Conc.	22.3	28.16	9.94	0.97	64.4	3.4	23.7
	Tailings	73.0	4.20	85.65	0.24	31.4	96.1	16.0
	Composite Feed	100.0	9.78	65.07	1.08	100.0	100.0	100.0
NH_4OH	Carbonate Froth	5.4	13.48	6.76	10.96	6.7	0.6	60.0
	Phosphate Conc.	23.6	30.32	6.28	0.98	67.1	2.3	23.7
	Tailings	71.0	3.92	86.40	0.22	26.2	97.1	16.3
	Composite Feed	100.0	10.66	63.22	0.98	100.0	100.0	100.0

Test Conditions: See text on screening of pH modifiers.

the hydrogen ion concentration is more important in determining collector adsorption than the anionic species of the pH regulator.

Unlike phosphate minerals (apatite and francolite) the carbonate minerals (dolomite and calcite) react readily with the inorganic acids. This results in preferential dissolution or removal of mixed surface contaminants on the carbonate particles and exposure of fresh clean surface sites suitable for fatty acid adsorption. As a result also, CO_2 microbubbles are generated on the carbonate particle surfaces, leading to enhanced oleic acid adsorption at the solid/liquid/gas interface. These are favorable conditions for particle/bubble attachment and "instant" flotation of carbonate minerals. On the other hand, the reaction of the acids with the apatite (francolite) particles may produce high concentrations of orthophosphate ionic species and the formation of highly structured phosphate-rich water layers surrounding the mineral particles. The phosphate-rich water layer is reported to be strongly hydrogen bonded to apatite and therefore depresses its flotation [8]. Prolonged conditioning of the pulp appears to remove these depressant layers from the apatite surface and results in increased flotation, as shown in figure 5. Under slightly acidic conditions, Johnston and Leja [9] reported that oleic acid can adsorb much faster on dolomite than on apatite because of the evolution of CO_2 microbubbles on the carbonate surface and/or the presence of hydrogen-bonded phosphate ionic species on the apatite surface. This mechanism finds support from our recent adsorption and zeta potential studies on dolomite and francolite minerals at pH's of 4 to 6. Fourier Transform IR Spectra, FTIR, have shown that oleic acid is physically adsorbed on both minerals. The adsorption kinetics was faster for dolomite than francolite. Thus, the selectivity in carbonate/phosphate separation, in slightly acidic media, may be attributed to preferential dissolution of the mineral surface layers, which results in changes in adsorption kinetics and preferential flotation of the carbonate minerals.

As experienced in commercial practice, the results in table 2 show that oleic acid flotation in the alkaline pulp produced a rougher phosphate concentrate with a high silica content. In practice, such a product is further processed by strong acid scrubbing (de-oiling) to remove fatty acid coatings, followed by cationic flotation of the siliceous gangue (double flotation) in order to produce a marketable grade product of 30-31% P_2O_5, 0.7-1.0% MgO and 8-10% insol.

The lack of selectivity in fatty acid phosphate/silica separation under alkaline conditions may be due to slime coating or surface contamination of the silica particles by the precipitated Ca-/Mg- fatty acid salt complexes formed in the pulp. Such complexes may not exist in slightly acidic pulps as indicated by the high selectivity of phosphate/silica separation achieved at pH 6 in the absence of sodium silicate (table 4). The high grade phosphate concentrate with the low silica content recovered indicates a strong depressing effect of acidic pH's on the siliceous material during fatty acid flotation of the phosphates. Addition of sodium silicate improved the P_2O_5 recovery in the phosphate concentrate and increased the rejection of siliceous gangue in the tailing product. The exact mechanism by which sodium silicate actually depressed silica or activated phosphatic particles is not well understood. Some investigators suggested that activation of apatite (francolite) may be due to enhanced adsorption of oleic acid in the presence of colloidal hydrosilicic acid species prevailing under acidic pH's [10]. Other investigators suggested that the polyvalent cations which can interfere with flotation may interact with the soluble silicates species forming insoluble compounds, thereby reducing interference by those cations [11,12].

CONCLUSIONS

The research has demonstrated that the MRI "no-conditioning" process is effective for selectively floating the carbonate minerals from a high-MgO phosphate matrix obtained from the W.R. Grace Company Four Corners Mine. In the phosphate flotation step that followed, high quality phosphate concentrates containing about 31% P_2O_5, 0.7% MgO, and 3% insol were obtained by a unique phosphate flotation procedure. The technique obviates the need for de-oiling and silica flotation steps used in current phosphate practice.

The selectivity in carbonate/phosphate separation, in slightly acidic media, may be attributed to preferential dissolution of the mineral surface layers, which results in changes in adsorption kinetics and preferential flotation of the carbonate minerals.

ACKNOWLEDGEMENTS

Financial support for this work was provided by the Florida Institute of Phosphate Research (Grant #86-02-066). Any opinions, findings and conclusions or recommendations expressed in this work are those of the authors and do not necessarily reflect the views of the Florida Institute of Phosphate Research. The authors also wish to acknowledge the cooperation of the Florida phosphate producers in supplying the samples used in this study.

REFERENCES

1. I.J. Anazia and J. Hanna, Min. and Metall. Proc., pp. 196-202, (November, 1987).
2. I.J. Anazia and J. Hanna Int'l J. Mineral Processing 23, pp. 311-314, (1988).
3. J. Hanna and I.J. Anazia, "Carbonate/Phosphate Flotation Separation by the MRI No-Conditioning Process", 118th AIME/SME National Meeting, Las Vegas, Nevada, Feb. 27-March 2, 1989, Preprint #89-144.
4. J. Hanna, et al., "Process for Separating Carbonate and Non-Carbonate Salt-Type Minerals", U.S. Patent pending Serial No. 07/395,996, August 21, 1989.
5. G.H. McClellan and J.R. Lehr, American Mineralogists 54, pp. 1374-1391, (1969).
6. N.S. Schulz, Trans. SME/AIME 247, pp. 81-87, (1970).
7. B.M. Moudgil and P. Sumasundaran: Advances in Mineral Processing, P. Sumasundaran, ed. (AIME/SME), pp. 426-441, (1986).
8. M. Bertolucci, F. Jantzef, and D.L. Chamberlain, Interaction of Liquids at Solid Substrates, R.F. Gould, ed. (ACS), Advances in Chemistry Series, Washington, D.C., 1968, pp. 124-132.
9. D.J. Johnston and J. Leja, Trans., Canad. Inst. of Min. and Metall. 87, pp. 237-242, (1978).
10. V.I. Klassen and V.A. Mokrousov, An Introduction to the Theory of Flotation, J. Leja and G.W. Poling eds., (London, Butterworths, 1963) pp. 321-335.
11. M.A. Eigeles, "Theoretical Basis of the Flotation of Non-Sulfide Minerals", Metallurgizdat, (1950).
12. N.A. Ianis, "The Effect of Alkali Regulators on the Adsorption of Sodium Silicate by Calcium Minerals", Proc. 2nd Sci Tech. Sess, Metallurgizdat, 1952.

SPLIT FLOTATION OF CALCITE FROM WOLLASTONITE AND MICROCLINE - THE CALCITE RICH WOLLASTONITE ORE OF NORTHERN SWEDEN

R. Sivamohan and Huang Fugen
Division of Mineral Processing, Technical University of Luleå,
S951 87 Luleå, Sweden.

ABSTRACT

A detailed laboratory research program to find possible ways to selectively float the above cited complexly and finely interlayered ore, was carried out. Entrainment experiments of different size fractions of this ore showed that it could be an important contributor to the flotation of slime. Split flotation experiments have shown that substantial improvements in selectivity as well as a >25% reduction in collector consumption are possible compared to normal flotation. Starvation flotation gives flotation concentrates of narrow size distributions. Reverse flotation of calcite enables a simple flotation circuit because of the availability of the calcite specific collectors, namely, LF AK100 and Berol 860. The experiments carried out on mixtures of coarser size fractions with the -10μm fraction in different proportions have shown that the coarser particles, in the presence of ultrafine particles, do not float as fast as they float in the absence of such particles.

INTRODUCTION

The broad objective of this study has been to thoroughly probe the possibilities of obtaining separate calcite and wollastonite concentrates from the finely and complexly interlayered calcite rich wollastonite-microcline-Fe garnet ore of Northern Sweden at high recoveries with acceptable grades. The ore samples used in the present work assay, on average, ~65% $CaCO_3$ and ~15% $CaSiO_3$.

Achieving successful concentration between very finely sized particles of different minerals is difficult. In the flotation of such particles, there are two major problems, the reduced rate of flotation [1,2,3,4,5,6] as well as the increased degree of entrainment of particles of such sizes [7]. Several methods have been suggested to improve selectivity during very-fine particle flotation. Among them are column flotation [8] and shear-flocculation [9,10,11,12,13].

In the experimental study reported in this summary article, we carried out, amongst others, entrainment experiments of the different size fractions; split conditioning and split flotation of the different size fractions and flotation of mixtures of -10μm particles with the coarser fractions in different proportions.

EXPERIMENTS

Material

The calcite rich wollastonite-microcline ore material obtained for testing was supplied by Jokkmokk Mineral Company. Visual examination (i.e. the observation of colour and shape of different minerals) showed that the bulk material varied greatly in mineral contents. Large blocks of the material were fractured by hydraulic pressure and 300 kgs of the fractured species were crushed to -2.4 mm with jaw crushers. The crushed ore was ground in a laboratory stainless steel rod mill. The average calcite content of the commminuted material was ~65% $CaCO_3$. The wollastonite content was ~15% $CaSiO_3$.

Advances in Fine Particles Processing
John Hanna and Yosry A. Attia, Editors

Equipment

The AGITAIR Model LA500 flotation machine with a 1.35 liter cell was used. In all experiments, a near-constant pulp level was kept during the flotation by letting in additional water as necessary. The air flow rate was kept constant at 3.0 liters/min.. The impeller speeds used were 900 and 600 rpm. In this summary, only the results with 900 rpm are discussed.

The machine used in wet high intensity magnetic separation was a Jones P40. It is the smallest (up to 500 kg/h in continuous operation; 50 kg/h in batch operation) in a series of wet HIMS equipment manufactured under licence by KHD Humboldt Wedag AG. The comminuted material was divided into two parts for magnetic concentration. One part was deslimed before being subject to magnetic concentration while the other part was not deslimed.

Methods:

Entrainment Experiments

Entrainment is the mechanism by which the hydrophilic particles are recovered in the float product. The not-deslimed non-magnetic product, obtained after high intensity magnetic separation of the ore, was divided into six size fractions, 100-75μm, 75-53μm, 53-38μm, 38-20μm, 20-10μm and -10μm, and used as feed. In the entrainment experiments, only the frother, not the collector, was added. The conditions for the entrainment experiments reported in figure 1 were as follows: MIBC, 1 drop; pH 6; feed weight 500 g.

Split Flotation

Split flotation is defined as the process of flotation where a fractionated feed is used. Three broad groups (groups 1, 2 and 3) of split flotation experiments were carried out on the six size fractions mentioned above. The three groups were: test feeds with the same number of particles (group 1); test feeds of the same material weight (group 2); and test feeds with variable amounts of -10μm material (group 3). The collectors used in the first two groups (1 and 2), were LILAFLOT AK100 and AEROPROMOTOR AP845E. The former is an alkali carboxylic salt produced by Kenograde, Stockholm. The latter is an amine produced by Cyanamid, Rotterdam. In the group 1 experiments, each size fraction tested contained the same number of particles. Thus it was possible to observe the effects of flotation strictly as a function of particle size. The same number of bubbles were also used during flotation. The procedure used to determine the material weight in different fractions that would give a constant number of particles in all the fractions has been described elsewhere [14].

Table I gives the calculated conditions. From table I, it can be seen that the weights of the fractions, 100-75μm and -10μm, that would give the same number of particles as that in 500 grams of the non-magnetic product, are 1625 and 1.3 grams respectively. It is impractical to bench float a feed as high as 1625 grams of the 100-75μm fraction, or, a feed as little as 1.3 grams of the -10μm fraction, in a 1.35 liter cell. Therefore, in group 1 experiments, only the middle four size fractions were floated. In group 2 experiments, the weight of the feed to flotation was kept constant at 500 grams (table II). In group 3 experiments, Berol 860, an alkali carboxylic salt, from Berol Chemical Company, Sweden, was used as collector and the feed weight was kept constant at 250 grams (table III).

The amounts of the collectors added in all of the three groups of experiments, were decided in such a way that the amounts of the collectors added per unit surface area were always the same. The calculated amounts are given in tables I-III. One drop of undiluted MIBC was added in the experiments of groups 1 and 2. No frother was added in group 3 experiments. H_2SO_4 was used to adjust the pH to 6.0±0.2 in the groups 1 and 2 experiments. In group 3 experiments, pH was adjusted to 9.5±0.2 with sodium carbonate. Also, in group 3 sodium silicate was used as a dispersant at

$7.25X10^{-4}$ mol/l. The results of these three groups of split flotation experiments are found in figures 2-4, respectively. For the two coarsest size fractions, 100-75μm and 75-53μm, the split flotation was also carried out with an increased amount of Berol 860. The results of these experiments are also found in figure 4.

Table I. Conditions for Group 1 Tests (feeds with constant number of particles)

	non-mag.	75-100 (μm)	53-75 (μm)	38-53 (μm)	20-38 (μm)	10-20 (μm)	-10 (μm)
Number of particles in the non-mag. divided by that in a fraction, on a unit volume basis	1	3.25	1.65	0.69	0.17	0.057	0.0027
Amount needed in each fraction to give the same number of particles as that in 500 g of non-mag. (g)	500	1625	824	347	86	24	1.3
Specific surface area (m^2/g)	0.58	0.11	0.15	0.18	0.26	0.47	2.77
Total surface area (m^2)	290	178.3	123.6	62.46	22.36	13.16	3.60
Collectors (g/t)							
AK100	40	7.6	10.6	12.4	18.0	32.4	192
AP845E	20	3.8	5.3	6.2	9.0	16.2	96

Table II. Conditions for Group 2 Tests (feeds with constant material weight)

	Non-mag.	100-75 (μm)	75-53 (μm)	53-38 (μm)	38-20 (μm)	20-10 (μm)	-10 (μm)
Specific surface area (m^2/g)	0.58	0.11	0.15	0.18	0.26	0.47	2.77
Test Feed (g)	500	500	500	500	500	500	500
Total surface area (m^2)	290	55	75	90	130	235	1385
Collectors(g/t)							
AK100	40	7.6	10.6	12.4	18.0	32.4	192
AP845E	20	3.8	5.3	6.2	9.0	16.2	96

Normal Flotation

Normal flotation is defined as the process of flotation where the feed has a very wide size distribution. Above mentioned tables I and II contain one experiment each which involve the non-fractionated non-magnetic charge. Normal flotation, in this article, refers to these particular flotation experiments. Figure 5 includes the results of these experiments.

Table III. Conditions for Group 3 Tests(feeds with variable amounts of -10μm material)

The ratio of each size fraction to -10μm frac. (wt%)		$CaCO_3$ in mixture (%)	Total surface area in 250 grams (m^2)	added Berol 860 (g/t)
100-75μm	100 : 0	67.95	28	59
	90 : 10	69.69	94	200
	80 : 20	71.47	161	341
	70 : 30	73.15	227	482
	60 : 40	74.44	294	625
75-53μm	100 : 0	72.17	38	81
	90 : 10	73.18	105	220
	80 : 20	75.27	168	358
	70 : 30	76.34	235	501
	60 : 40	77.61	300	641
53-38μm	100 : 0	74.11	45	95
	90 : 10	75.69	110	236
	80 : 20	76.65	173	374
	70 : 30	77.06	240	512
	60 : 40	78.18	305	650
38-20μm	100 : 0	76.64	65	138
	90 : 10	76.89	128	274
	80 : 20	77.69	190	406
	70 : 30	78.94	253	542
	60 : 40	80.62	318	676
20-10μm	100 : 0	79.27	118	251
	90 : 10	79.94	175	374
	80 : 20	81.10	233	496
	70 : 30	81.65	290	620
	60 : 40	82.03	348	742
-10μm		83.54	693	1472

Starvation Flotation

Starvation flotation refers to a flotation environment where low amounts of collectors are only added so that the total flotation of the desired mineral is not achieved in one single flotation step. Instead the flotation of the desired mineral is achieved in two or more flotation steps with the repeated addition of the collector. It is generally believed, for example in the Swedish mineral industry, that such a flotation process has its advantages such as reduced collector consumption and improved selectivity. We probed the possibility of starvation-floating the non-magnetic product obtained from the comminuted deslimed material. Calcite (the major mineral component) was the mineral targeted to float (reverse flotation). The frother was undiluted MIBC. H_2SO_4 was used to adjust the pH. The collectors used were AK100 and AP845E. The results of the magnetic concentration as well as the subsequent flotation of the non-magnetic product are given in table IV. Also this table shows that the coarseness of the particle size distribution increases from Flot1 product to Flot3 product. Flot1 is much finer than Flot2 and Flot3. Also, Flot3, Flot4 and Sink products contain almost the same size distribution.

Table IV Results of the WHIMS and flotation experiments

Product	Weight (%)	$\frac{CaCO_3}{CaSiO_3}$ (%)	$\frac{\text{Calcite}}{\text{Wollastonite}}$ Dist.(%)	D_{50} (μm)
Non-mag.	67.6	70.2 / 17.0	72.0 / 77.9	61.4
Flot1	21.9	87.7 / 7.8	29.1 / 11.6	29.0
Flot2	23.4	88.1 / 6.8	31.3 / 10.7	65.0
Flot3	10.5	69.3 / 14.9	11.0 / 10.6	98.3
Flot4	1.8	13.8 / 31.3	0.4 / 3.8	79.7
Sink	10.1	1.6 / 60.4	0.2 / 41.2	85.0
Midd.	7.5	54.6 / 20.5	6.2 / 10.4	50.3
Mag.	8.1	11.3 / 0.0	1.4 / 0.0	77.0
slime	16.8	80.4 / 10.2	20.4 / 11.7	10.8
Feed	100.0	66.0 / 14.7	100.0 / 100.0	

RESULTS and DISCUSSION

Entrainment studies

The recovery-time curves at different size fractions of figure 1 show that the entrainment generally increases as the particle size decreases. For example, for the fraction 100-75μm, almost no calcite could be recovered by entrainment. In contrast to this, most of the calcite could be recovered by entrainment in the flotation of the -10μm fraction (d_{50}=3.2μm). After six minutes of flotation, ~85% calcite could be recovered by entrainment in the flotation of fraction -10μm (fig. 1).

Split flotation studies

The results of the group 1 experiments where each size fraction tested contained the same number of particles, are given in figure 2. The results of the group 2 experiments where the weight of the feed to flotation was kept constant, at 500 grams, in all size fractions, are found in figure 3. Comparison of the results in fig. 2 with those of in fig. 3 show, for example, that better grades are obtained when a lower material weight is floated under otherwise similar conditions. A detailed comparison is found elsewhere [14].

The collector Berol 860 has proven to be a more suitable collector in floating calcite. The use of Berol 860 leads to considerably better grades at comparable recoveries and times of flotation than that could be obtained with AK100 and AP845E (fig. 3&4). Because of this better performance shown by Berol 860, it was later used in the flotation studies of the mixtures of various coarse fractions with -10μm fraction in different proportions.

When the results of the flotation experiments involving the mixtures of coarser and -10μm fractions are considered (fig. 6), it is seen that at a given calcite recovery the calcite concentrate grade decreases as the amount of the -10μm fraction is increased. For example, at 96% of calcite recovery the calcite concentrate grade decreases from 93% $CaCO_3$ to 89% $CaCO_3$ when the amount of -10μm increases from 10 wt% to 40 wt% in the mix-

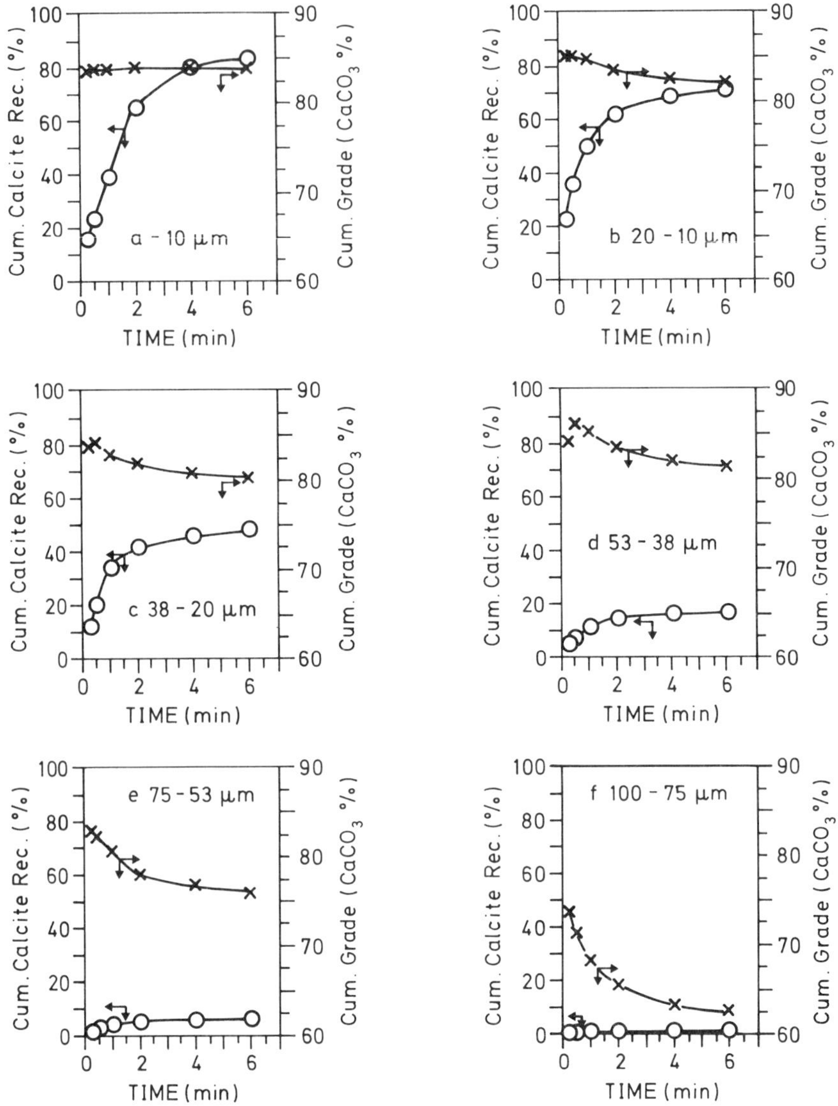

Fig.1. Grade-time and Recovery-time curves of the entrainment experiments of the six size fractions.

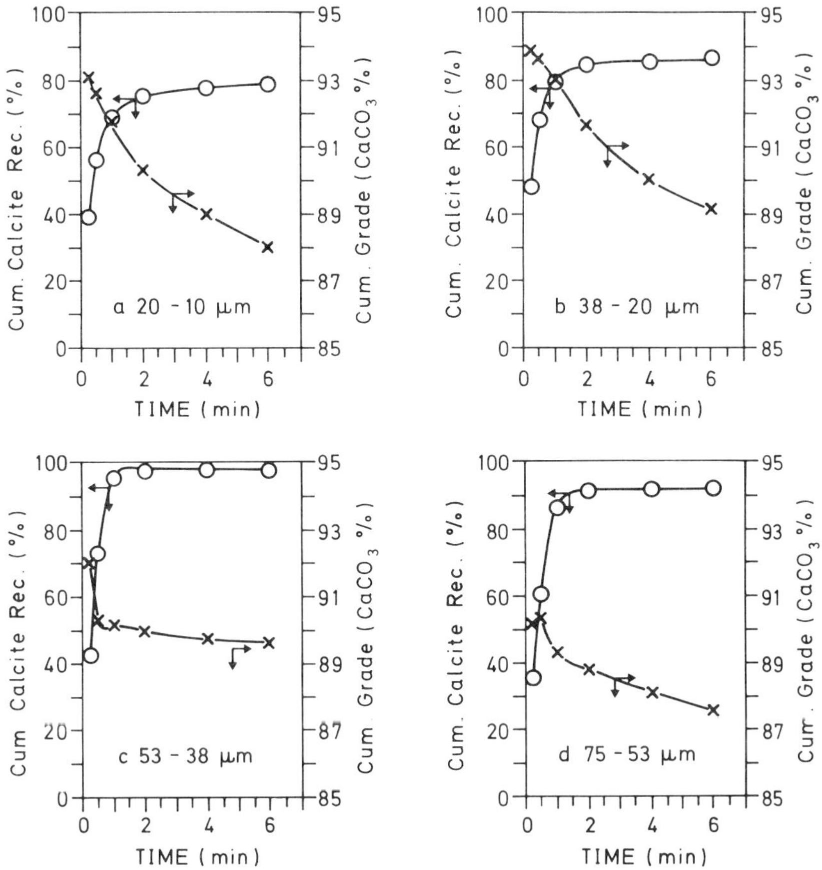

Fig.2. Grade-time and Recovery-time curves of the split flotation with AK100 and AP845E, (The number of the particles is same.).

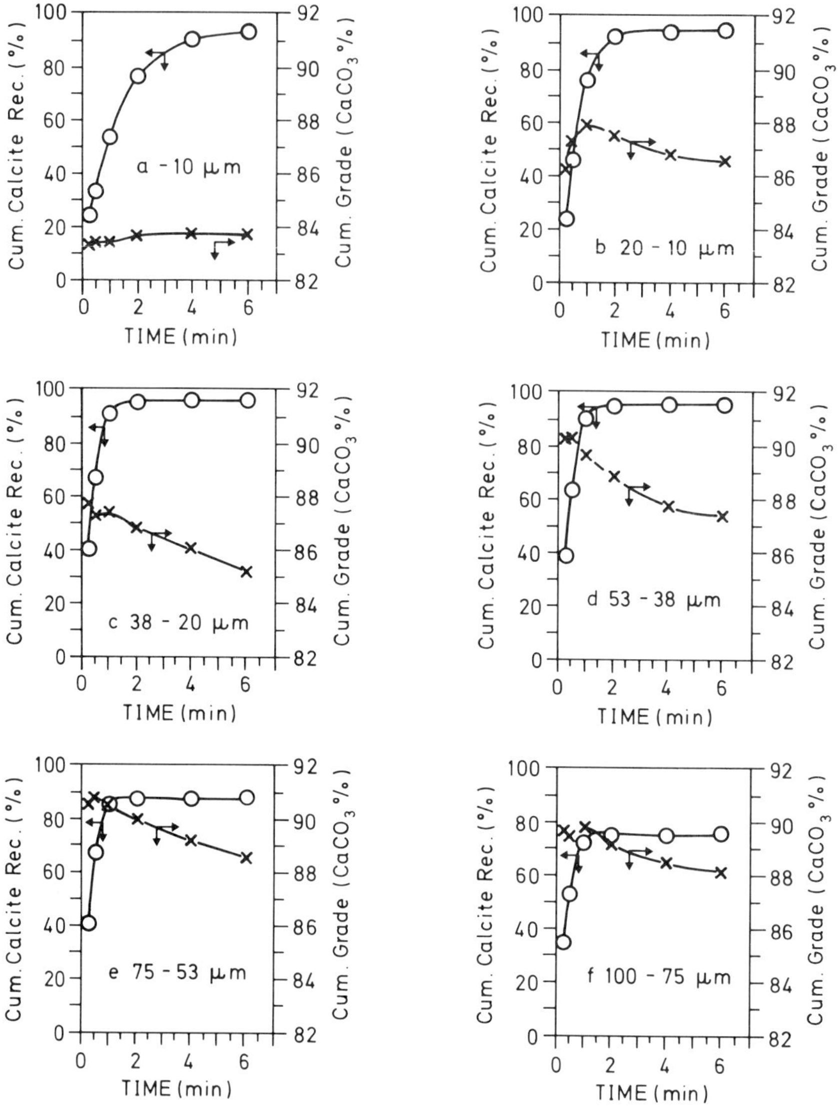

Fig.3. Grade-time and Recovery-time curves of the split flotation with AK100 and AP845E, (The weight is same.).

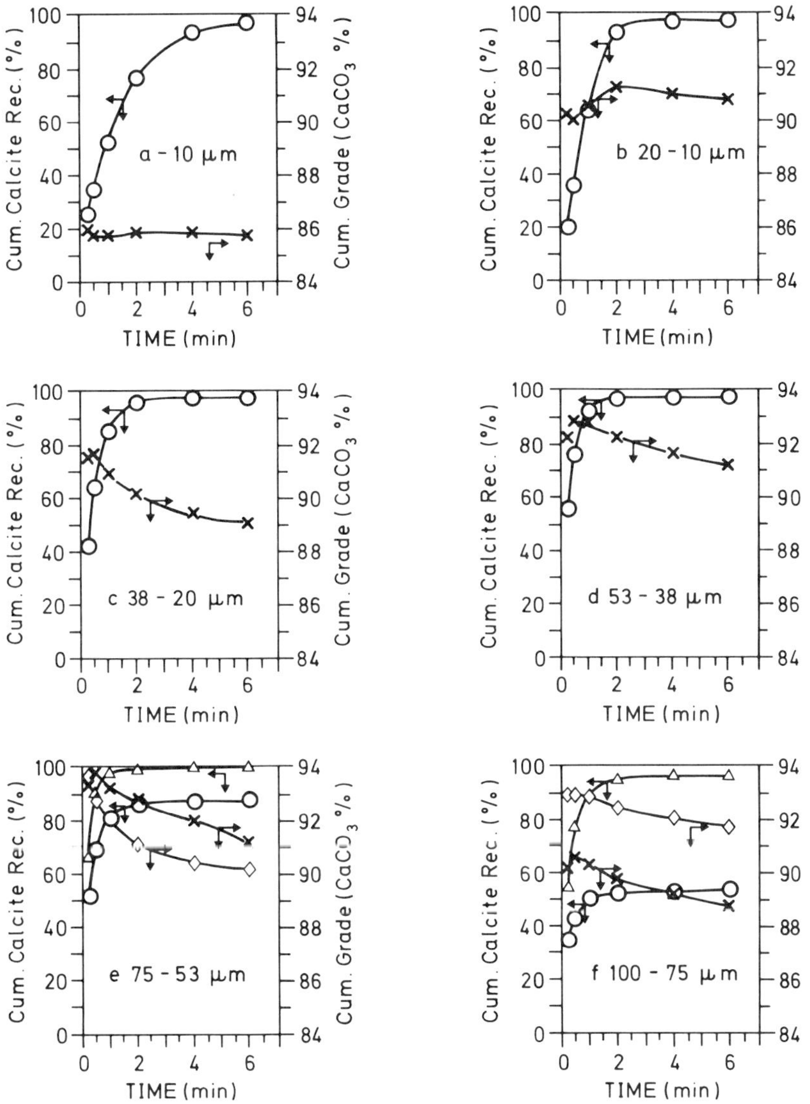

Fig.4. Grade-time and Recovery-time curves of the split flotation with Berol 860. ○ , × :2.1g/t of Berol 860; △ , ◇ : 4.2g/t of Berol 860.

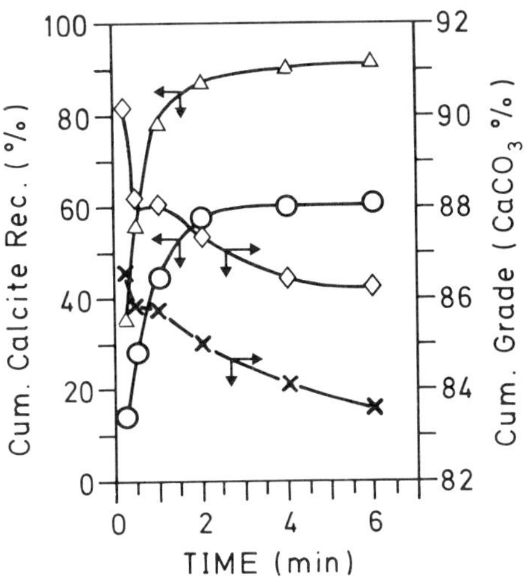

Fig.5. Grade-time and Recovery-time curves, ○ , × :the result of the non-magnetics flotation, △ , ◇ : the total result of the split flotation (The weight is same.).

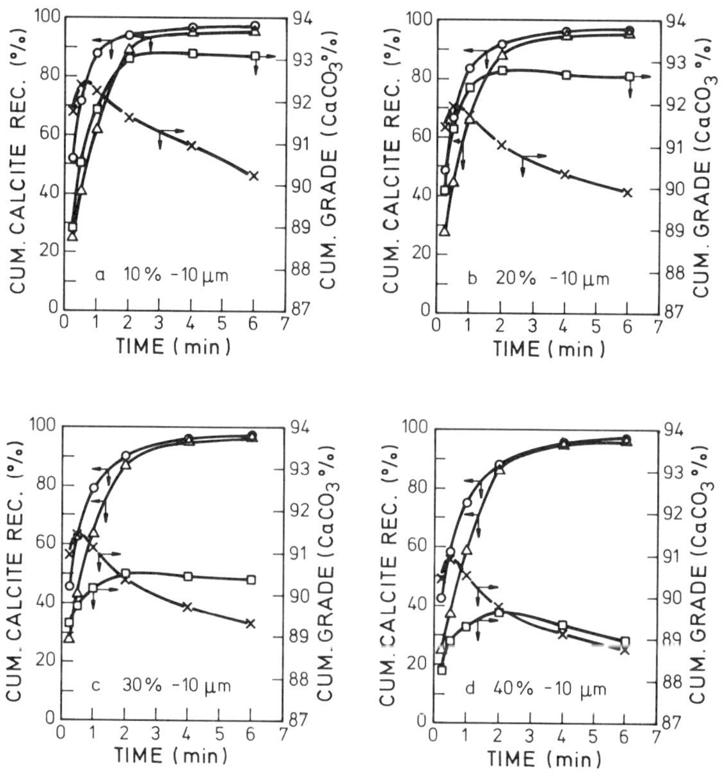

Fig. 6. Grade-time and recovery-time curves with new Berol 860. coarse size fraction: 53-38μm; o : calculated recovery; x : calculated calcite grade; Δ : experimental recovery; □ :experimental calcite grade.

tures with 53-38μm fraction.

From the results of the split flotation of coarser size fractions as well as the -10μm fraction with Berol 860 as collector, the expected results for the flotation of different mixtures are calculated. The calculating procedure is described elsewhere [15]. Figure 6 gives the calculated results for the 53-38μm fraction, in addition to the experimental ones. The results for other size fractions can be found elsewhere [15]. It is seen in figure 6 that the calculated calcite recoveries are much higher than the experimental ones at the initiation of flotation of 53-38μm in all four proportions. As the flotation time increases the difference between the calculated and experimental recoveries becomes less and less. The calculated calcite concentrate grades are also higher than experimental ones. However, as the flotation time increases, the situation reverses (fig. 6). The higher calculated calcite recovery obtained at the beginning of flotation, compared to the experimental ones, is due to that in the presence of ultrafine particles, the coarser particles fail to float as fast as they float in the absence of such particles. One probable reason could be that, in the mixture, the number of ultrafine particles is much more than the number of coarser ones and this would reduce the chances of coarser particles colliding with bubbles. This leads to lower calcite grades and recoveries at the beginning of flotation.

Normal flotation studies

Comparison of the results of the split flotation of group 2 (fig. 3) with those of the normal flotation involving the non-magnetic product (fig. 5) reveals that for the six minutes of flotation, the calcite concentrate grades and recoveries of the flotation of the six size fractions are all higher than those of the normal flotation. Put the concentrates of the flotation of the six size fractions of group 2 together, then the total grades and recoveries of the flotation of the different size fractions at different times are obtained (fig. 5). It is seen that the grade and recovery are much higher in split flotation than in normal flotation with AK100 and AP845E as collectors. After only one minute of flotation, the calcite grade and recovery are 2.3% and 33.6% higher in split flotation than in normal flotation, respectively. Obtaining an optimal physicochemical condition for floating a narrow size fraction, as in split flotation, should be much easier than in normal flotation.

However, in the split flotation of the fraction -10μm with AK100 and AP845E the calcite concentrate grade is only ~1% higher than the grade of the feed (fig. 3). It is possibly due to that high non-selective recovery by entrainment of the ultrafinely sized particles leads to calcite concentrates with low grades. Thus, if the amounts of the collectors added for split floating the -10μm fraction are excluded, since the addition of collectors hardly leads to any selective flotation, then the total amounts of the collectors needed for split floating the other size fractions would be less by over 25% compared to normal flotation.

Starvation flotation studies

The starvation flotation of calcite from the non-magnetic product yields two concentrates very rich in calcite(table IV). Flot1 and Flot2 concentrates both assay ~88% $CaCO_3$ with recoveries ~29% and ~31% respectively. Moreover, the products of the flotation have narrow size distributions, which is advantageous in the upgrading flotation because it may not be possible to have a physicochemical environment that would suit the particles of many different sizes as that found in normal flotation [14]. Trahar [17] wrote that if the adsorption density depends on particle size then we should not be able to measure the surface areas of particles of different size distributions by any method which depends on adsorption. He further wrote that such methods were widely accepted and used and therefore there was no relationship between particle size and adsorption density. Surface area measurements by adsorption method is not a valid argument against a relationship between particle size and adsorption density. The theory used in the method of surface area measurement of particulate masses by gas adsorption at a very low temperature is the BET for multilayer adsorption. It is clear from this theory that the adsorption measurements to determine the surface area corresponds to the conditions where the particles are already covered with the first monolayer. Adsorption of the second, the third and other monolayers take place only on the energetically homogenised, gas molecules covered surface of the particles. Therefore such an adsorption event is not at all influenced by the surface energy of the particles themselves.

Developing a strategy for processing the ore material under study

In order to develop a reasonable mineral processing strategy for this material, we attempted to upgrade the various products found in table IV by flotation. In the upgrading flotation of Flot1, Flot2 and Slime products, silicate was used as dispersant and Berol 860 was used as collector. The same chemicals used for the starvation flotation of the non-magnetic product were used for the upgrading of Flot3 and middling products as well. These two products were reground prior to upgrading flotation. The Flot4 and Sink products have rather a high grade in wollastonite. A wollastonite selective reagent was tried in order to upgrade these two products. The other major mineral present in Flot4 and Sink products is microcline, a K,Al silicate. Satisfactory results were obtained. A detailed description of this part of the study is found elsewhere [16].

CONCLUSIONS

1. The entrainment experiments show that in the flotation of finely sized calcite, entrainment can be an important contributor to the calcite recovery. The degree of entrainment generally increases with decreasing particle size.
2. It may not be possible to have a physicochemical environment that would suit the particles of many sizes as that would be found in normal flotation. Split flotation overcomes this limitation. Possibilities to reduce the collector consumption by >25% while obtaining substantial improvements in both total grade and recovery exist in split flotation with AK100 and AP845E.
3. Starvation flotation helps to optimize the physicochemical conditions in cleaner flotation by giving rougher products of narrow particle size distributions.
4. Experiments with Berol 860 show that, in the flotation of mixtures, the coarser particles fail to float as good as they float in the absence of ultrafine particles.
5. The mineral processing strategy developed in this work for the finely and complexly interlayered calcite rich wollastonite, microcline and Fe-garnet ore of northern Sweden gives an overall calcite recovery of 89.4% at a grade of 90.7% $CaCO_3$ as well as a wollastonite product assaying 53.0% $CaSiO_3$ at a recovery of 92.2%.

REFERENCES

1. D. Reay and G.A. Ratcliff, Canadian J. Chemical Eng., 51, 178-185 (1973).

2. G.L. Collins and G.J. Jameson, Chemical Eng. Sci. 31, 985-991 (1976).

3. G.J. Jameson, S. Nam and M. Moo Young, Minerals Sci. Engng. 9, 103-118 (1977).

4. R.J. Gochin, IMM(U.K) Trans. 92, C52-58 (1983).

5. Z.W. Jiang and P.N. Holtham, IMM(U.K) Trans., 95, C187-194 (1986).

6. R.H. Yoon and G.H. Luttrell, Coal Preparation 2, 179-192 (1986).

7. J.P. Bisshop and M.E. White, IMM(U.K) Trans., C191-194 (1976).

8. K.V.S. Sastry (Editor), Column Flotation'88 - Proc. Inter. Sym. on Column Flotation, (SME, USA 1988) 315 pp.

9. L.J. Warren, J. Colloid Interface Sci. 50, 307-318 (1975).

10. L.J. Warren, IMM(U.K) Trans. 84, C99-104 (1975).

11. P.T.L. Koh and L.J. Warren, IMM(U.K) Trans., 86, C97-100 (1977).

12. P.T.L. Koh and L.J. Warren, in: J. Laskowski (Ed.), Proc. 13th. Int. Miner. Process. Congr., (Elsevier, Amsterdam 1979) 1, 294-315.

13. R. Sivamohan and J.M. Cases, Int. J. Miner. Process., (1989), in press.

14. F. Huang and R. Sivamohan, Minerals Engineering, U.K., (1989), in press.

15. F. Huang and R. Sivamohan, in: G.S.Dobby (Ed.) Proc. Int. Symp. on the Processing of Complex Ores, CIM, Canada, (Pergamon Press 1989).

16. R. Sivamohan and F. Huang, Miner. & Metal. Process., May, 69-72 (1989).

17. W.J. Trahar, Int. J. Miner. Process. 8, 289-327 (1981).

OPERATING PARAMETERS IN THE COLUMN FLOTATION OF ALABAMA OIL SHALE

C.W. SCHULTZ AND JOHN B. BATES
Mineral Resources Institute, The University of Alabama
Box 870204, Tuscaloosa, AL 35487-0204

ABSTRACT

A factionally designed experiment performed in a one meter column flotation cell identified the important factors affecting the flotation of eastern oil shales. These initial tests were performed in a batch mode, and as a result, the effect of certain operating variables was observed.

Several series of tests were performed in the same cell operating in a continuous, i.e. equilibrium, mode. In the course of optimizing the separation, operating conditions were established which are outside the normal realm of column cell operations. Preferred conditions included positioning the pulp-froth interface below the midpoint of the column and introducing feed into the froth phase rather than into the pulp.

A third generation of experiments were performed in a variable height column. This configuration has allowed independent evaluations of froth height, feed positioning, spray water addition and froth drainage.

INTRODUCTION

The Mineral Resources Institute of The University of Alabama is developing a beneficiation system for the eastern (Devonian) oil shales. One element of the program is evaluation of the potential of column flotation for this purpose.

The investigation to date has proceeded in three stages. In the first stage a factorial array of tests were performed in a batch mode. The data from those tests (1) established the effects of a basic set of variables which included air flow rate, frother concentration, air sparger pore size, and percent solids. The data were used to set baseline conditions for subsequent tests.

Batch tests unfortunately do not permit evaluation of parameters which are peculiar to column cells (i.e. pulp height and froth depth). Therefore, in the second stage of the program tests were performed in a continuous (i.e. equilibrium) mode. In the course of the second stage testing, it became apparent that our preferred operating conditions were well outside the range which is considered normal for column operation [2,3,4]. Typically, the pulp-froth interface was maintained at about 1/3 of the total column height. Feed was introduced into the froth at a point slightly above the midpoint of the column.

To gain a greater understanding of the effect of operating parameters, a third stage of testing was undertaken. A column cell was fabricated in sections to permit independent control of pulp height, froth depth, and feed position.

Advances in Fine Particles Processing
John Hanna and Yosry A. Attia, Editors

This paper presents the observations made in the second and third stage of the research program and offers the authors' interpretation of those observations.

EQUIPMENT AND PROCEDURE

The arrangement of the column cell and auxiliary equipment for continuous flow testing is shown schematically in figure 1. The feed sump (1) is filled with a sufficient volume of prepared sample to permit a large number of tests to be performed (typically 12). Past experience has shown this is necessary due to high sample variability and variability in the size distribution resulting from ultra fine grinding.

The feed slurry is maintained at about 20% solids and is constantly recirculated and stirred. Sample is metered from the circulating pipe by a peristaltic pump (2). The feed slurry was diluted with reagentized water (3) by a second peristaltic pump (4). Wash water (5), also reagentized is supplied through a third peristaltic pump (6). While this feed system may seem unduly complex it does permit us to independently vary either the wash water rate or the net solids content of the cell. In the tests reported here the feed rate and net percent solids were constant at 12.5 gms/min and 3.3 percent respectively.

Diluted feed enters the column through 6.35 mm diameter copper tubing and is discharged upwardly at the center of the column. Tailings are discharged through flexible tubing which can be adjusted so as to control the position of the pulp-froth interface.

The column is 7.62 cm internal diameter and 109.2 cm high. It is made from lucite tubing and is fitted with a 5 cm diameter fritted glass air sparger.

In performing a series of tests both the concentrate and tailing are allowed to discharge continuously. The system is allowed to equilibriate for a period of 30 minutes after the pulp and froth reach operating levels. Concentrate and tailing samples are taken simultaneously for timed intervals (five to fifteen minutes, depending on the volume of sample desired). After sampling a change in operating conditions is made and the system is again allowed to equilibriate. Operating in this fashion a series of six tests can be performed in a day.

Samples are analyzed by measuring their specific gravity with a helium pycnometer. Specific gravity is converted to Fischer Assay oil yield using a previously established relationship. Because the Fischer Assay procedure has a high variability or experimental error the use of the specific gravity relationship has the effect of reducing the experimental error in our flotation tests. All the data reported here were calculated from specific gravity measurements.

STAGE TWO RESULTS

Figure 2 shows the effect of feed size on the grade-recovery relationship obtained for an Alabama oil shale. The grade-recovery curves were developed by varying the air flow rates from 4.3 to 6.0 standard liters per minute (SLPM). The curves also illustrate that the Alabama shale liberates in such a manner that high recoveries are possible at relatively coarse sizes but substantially finer grinding is required to achieve satisfactory concentrate grade.

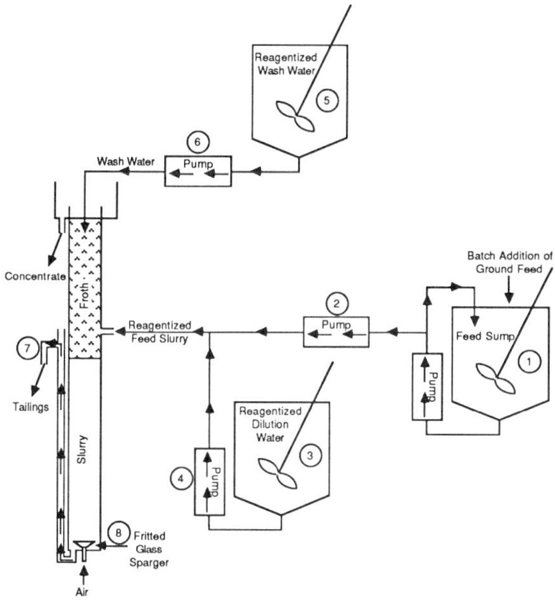

Figure 1. Schematic of Continuous Column Cell Flotation Circuit

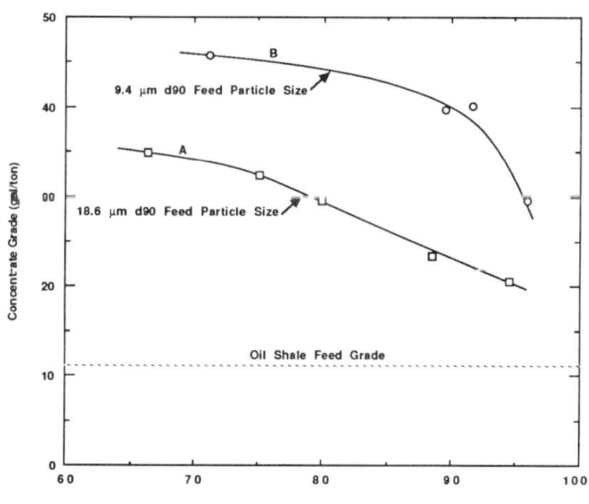

Figure 2. Fischer Assay Oil Grade Versus Recovery Using Column Flotation for Alabama Oil Shale, Effect of Feed Particle Size

The effect of increased frother concentration on the grade-recovery relationship is illustrated in figure 3. As might be predicted the increase from 30 ppm to 42 ppm Dowfroth resulted in increases in recovery and decreases in grade at all air rates. The greater effect on recovery however results in an overall more favorable grade-recovery relationship.

A surprising effect was noted when an attempt was made to increase kerogen (oil) recovery by raising the position of the pulp froth interface. The interface was raised from 35.6 cm to 71.71 cm with the feed inlet fixed at 63.5 cm above the sparger. The results are shown in figure 4. Note that at the lowest air flow rates there was indeed an increase in recovery. At the highest air rate the recoveries are approximately equal but the lower interface (greater froth depth) yielded a higher grade and a significantly more favorable grade-recovery curve.

The implications of this data is that considerable recovery can occur in the froth phase, and, in conditions of high air flows, the rate of collection in the froth phase is equal to that in the pulp phase.

At this point in the investigation it became clear that we needed more flexibility in our ability to evaluate column operating parameters. For this purpose a new column cell was designed. The column is built in sections so that the froth depth and pulp height could be varied independently. The column was also equipped with several access ports so as to allow changes in the feed inlet position.

STAGE THREE RESULTS

The first series of tests performed with the variable height column were designed to gain insight to the effect of feed position and to assess the effect of increased column height. The first pair of grade-recovery curves are shown in figure 5. The lower grade-recovery curve was obtained at our base line conditions, i.e. column height 109.2 cm, pulp-froth interface 35.6 cm and feed inlet at 63.5 cm. The curve is considerably below that shown in figure 4 due to a coarser grind (d_{90} = 12.2 as compared with d_{90} = 9.4 microns in figure 4). The upper curve in figure 6 resulted from raising the feed inlet to 88.9 cm while all other conditions were held constant. The effect is similar to that noted in figure 4. The increased height of the feed position resulted in increased recovery in the froth phase with an accompanying decrease in concentrate grade. There is a distinct improvement in the overall grade-recovery relationship.

Figure 6 compares the base case with observations made with the column height at 134.6 cm. Note that both grade and recovery are markedly improved by the increased column height. The effect of raising the feed position is obscured, and except for the points at the middle air flow level there appears to be no effect or perhaps even a reversal of the effect noted in the 43 inch column.

The effect of air sparger pore size at two column heights is shown in figure 7. At the 109.2 cm column height the finer sparger pore size yields a significantly improved grade-recovery curve. At a column height of 134.6 cm however the effect of sparger pore size is negligible. While the data are shown as two curves it is difficult to say that the individual points are not all a part of the same grade-recovery curve. All test shown in figure 8 were conducted with the feed inlet at 63.5 cm and the pulp-froth interface at 14 inches.

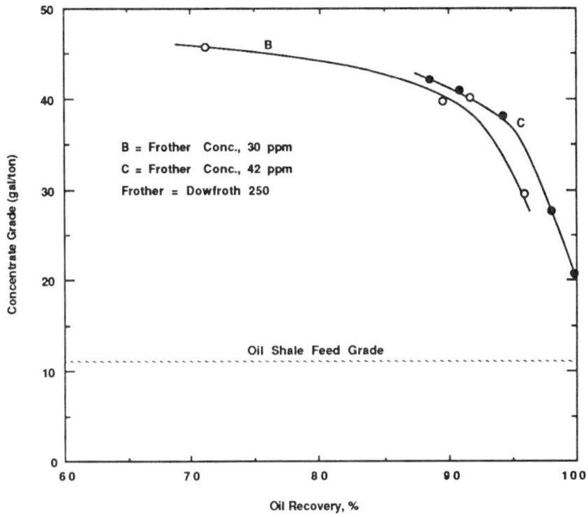

Figure 3. Fischer Assay Oil Grade Versus Recovery Using Column Flotation for Alabama Oil Shale, Effect of Frother Concentration Added

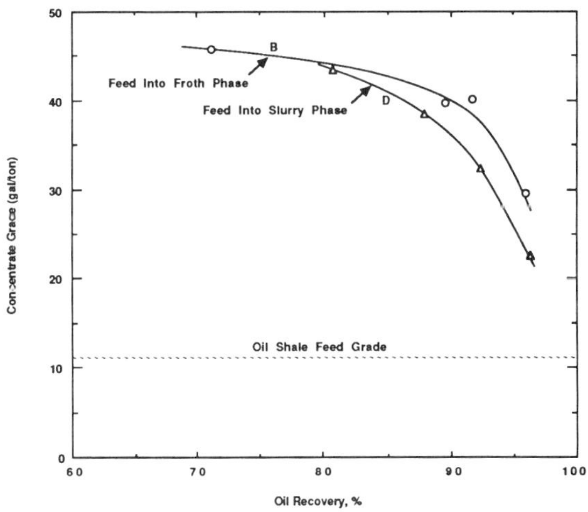

Figure 4. Fischer Assay Oil Grade Versus Recovery Using Column Flotation for Alabama Oil Shale, Effect of Entrance Location of Feed Slurry

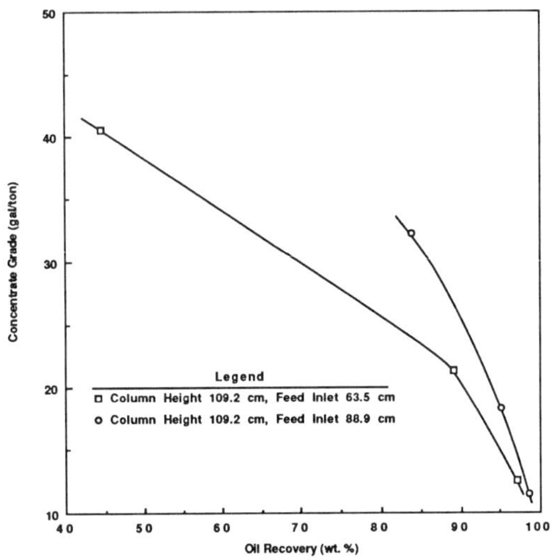

Figure 5. The Effect of Column Height and Feed Inlet Position on Grade - Recovery Relationships

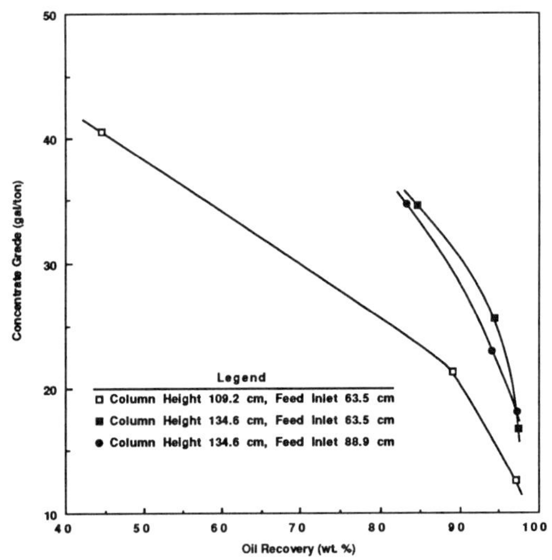

Figure 6. The Effect of Column Height and Feed Inlet Position on Grade - Recovery Relationships

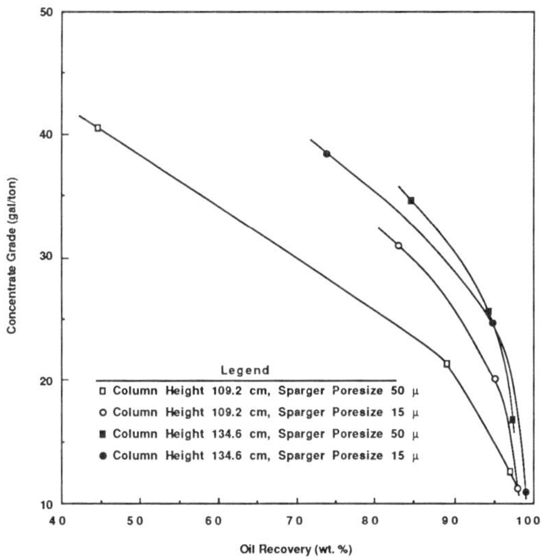

Figure 7. Effect of Column Height and Air Sparger Pore Size on Grade and Recovery of Alabama Oil Shale

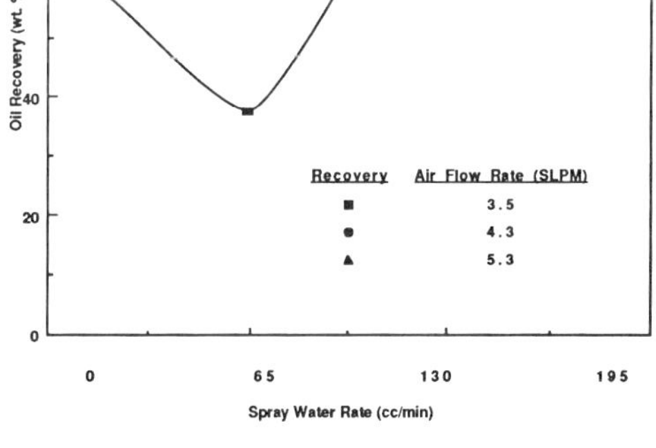

Figure 8. Effect of Air Flow and Spray Water Rates on Kerogen Recovery

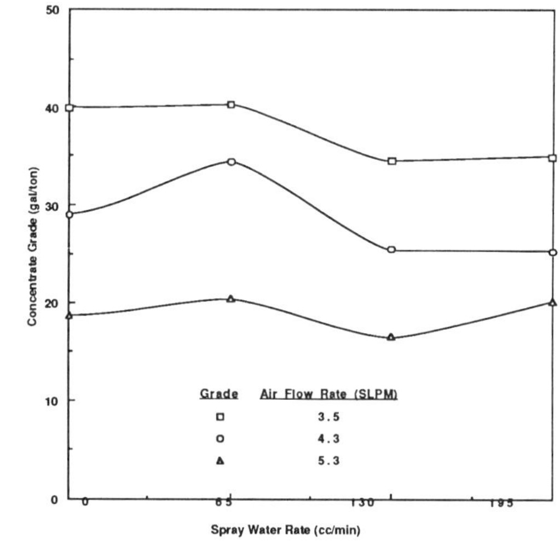

Figure 9. Effect of Air Flow and Spray Water Rates on Concentrate Grade

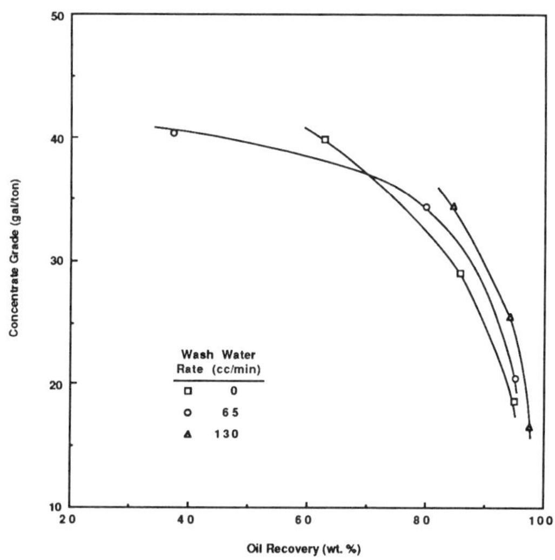

Figure 10. Effect of Wash Water Rate on Grade and Recovery of Alabama Oil Shale

Figures 8 and 9 show the effect of wash water rate on recovery and grade respectively. In this series of tests the dilution water (as shown in figure 1) was adjusted to maintain the net solids content in the cell constant at 3.3%. This was done because earlier work had shown that the solids concentration had a marked effect on flotation performance and our objective here was to isolate the effect of spray water.

Figure 8 shows a strong interaction between air rate and spray water rate in their effect on kerogen recovery. Less interaction is seen in the effect on concentrate grade. As with all other variables the effect of wash water rate is dampened at high air flow rate.

The data from figures 8 and 9 are summarized in figure 10 as grade-recovery curves. The optimum is achieved at a wash water rate of 130 cc per minute which is the base line rate used in all previous tests.

SUMMARY AND CONCLUSIONS

The data presented in this report indicate that some benefit is derived from lower than normal levels of the pulp-froth interface. It is also demonstrated that high recoveries can be achieved in the froth phase. These observations were made outside the range of normal column operation but are consistent with the data of Yang [5,6].

Figures 6 and 7 show that the effects of feed inlet positions and air sparger pore size are obscured in the taller (53 inch) column. This is perhaps due to having reached a limiting grade-recovery relationship, that is, having approached the theoretical washability curve for shale ground to that size.

If that assumption is correct, then the limiting condition has been reached at a relatively short column height (i.e. low L/D ratio).

The interpretation that can be made of the data presented here is limited by the range of conditions which were tested. All tests were performed at low feed rates and low solids levels. It is expected that increases in these factors may well require greater column height.

Our future plans include extending the range of operating test conditions and instrumenting the column in order to aid in our interpretation of the resulting data.

REFERENCES

1. Schultz, C.W., John B. Bates, and M. Misra, "Column Flotation of Eastern Oil Shales".
2. Sastry, Kal V.S., and Kevin D. Loftus, "Mathematical Modeling and Computer Simulation of Column Flotation," Column Flotation 88. Proceedings of an International Symposium on Column Flotation. K.V.S. Sastry, Editor, Society of Mining Engineers, Littleton, CO, 1988, pp. 57-68.
3. Attia, Y.A. and Shaning, Yu, "Feasibility of Separation of Coal Flocs by Column Flotation," Column Flotation 88, Proc. etc, pp. 249-254.
4. Ynchousti, R.A., J.D. McKay and D.G. Foot, "Column Flotation Parameters-Their Effects," Column Flotation 88. Proceedings, etc., pp. 157-172.

5. Yang, D.C., "Static Tube Flotation for Fine Coal Cleaning," Proceedings of the Sixth International Symposium on Coal Slurry, Combustion and Technology, 1984, pp. 582-597.
6. Yang, D.C., Private Communication

PART 7.

FINE PARTICLE PROCESSING WITH MULTIPLE PHYSICAL, CHEMICAL AND BIOLOGICAL PHENOMENA

GRINDING AND FLOTATION CHARACTERIZED WITH THE PARAMETER ACTION

R. VARBANOV
Central Laboratory of Mineral Processing, Bulgarian Academy of Sciences, 1, Ave. A. Ivanov, P.O. Box 32, 1126 Sofia Bulgaria

ABSTRACT

The product S = E.t, called Action (E - energy, t - time) is well known in physics. It takes several forms such as: Hamilton's integral $S = \int_{t_1}^{t_2} L.dt$, Max Planck's formula h = E.t and Kinetic momentum $\bar{K} = m.\bar{v} \times \bar{r}$. These forms have dimension Youle. Sec . All natural processes correspond to the principle of the smallest Action (S = E.t = min). All artificial processes including the technological ones are in the most cases rather far from this minimum, but with the technical progress they approach this minimum gradually. Pontryagin's maximum principle widely used for the optimization of processes is a modification of the principle of the smallest Action.

In the area of grinding or flotation the specialists use the parameters power and energy consumption and the time of grinding or flotation. If one juxtaposes power $P = P.t^0$, energy $E = P.t^1$ and Action $S = E.t = P.t^2$, the essential difference is evident. When they treat grinding and flotation, the best performance criterion is neither P nor E but S. For the same initial and final conditions, the best performance of grinding or flotation process corresponds to the smallest value of the parameter Action S = E.t (E - energy consumption, t - time of flotation or grinding). If the time is the shortest at the equal other conditions, the performance is the best one - increased capacity, decreased production cost etc.

INTRODUCTION

The parameter Action is well known and widely used in physics. Most fundamental phenomena in Nature as well as in Science are closely connected with this parameter. We would like citing its three most used forms in Science.

The first form is Hamilton's integral.

$$S = \int_{t_1}^{t_2} L.dt \tag{1}$$

where: S - parameter Action, J.s.
L = T - K, Lagrange's function (T - potential energy, K - kinetic energy of the system considered), J.
t - time, s.

This type of integral exists for both the conservative systems (wuthout friction) as well as for the systems with dissipation of energy (with friction, with irreversible effects etc.).

The second form is Max Planck's formula:

$$E = h.\nu \text{ or } h = E.t \text{ since } \nu = 1/t \tag{2}$$

Advances in Fine Particles Processing
John Hanna and Yosry A. Attia, Editors

where: h - elementary quantum Action, J.s.
E - energy, J.
ν - frequency, s^{-1}.

The Quantum theory and the Quantum statistics largely use the parameter h.

The third form represents the Kinetic momentum:

$$\vec{K} = m.\vec{v} \times \vec{r} \tag{3}$$

where: m - mass of the system, kg.
$\vec{v}$ - velocity, m/s.
$\vec{r}$ - radius vector, m.

According to the principle of the smallest Action the processes including the technological ones should correspond to S = E.t = minimum.

With the realization of the technical progress the technological processes developed and projected by the specialists approach gradually this minimum.

Pontryagin's maximum principle widely used for the optimization of the processes is a modification of the principle of the smallest Action.

In the areas of grinding or flotation, the specialists use for estimation and characterization the power and energy consumption and the time of grinding or the time of flotation, but do not use the parameter action. Let us juxtapose the three parameters: power, energy and Action in order to elucidate the essential differences between them: $P = P.t^0$, $E = P.t^1$ and $S = E.t = P.t^2$. In grinding and flotation, the best parameter for estimation, characterizing and comparison is the Action (S) as cumulative parameter. For the same initial and final conditions, the best performance of grinding or flotation process would be that gives the smallest value of the parameter S = E.t (E - energy consumption for unit mass of ore, t - time of flotation or grinding.).

Theory

Let us consider a relationship between recovery, parameter Action and some general parameters (recovery (t), flotation rate constant K, revolution n etc.).

Let us begin with linear kinetic equation of flotation:

$$\varepsilon(t) = 1 - e^{-Kt} \tag{4}$$

where: $\varepsilon(t)$ - recovery
K - flotation rate constant
t - flotation time

It is possible to obtain the time of flotation applying the operation in to both sides of equation (4):

$$t = -\frac{\ln[1-\varepsilon(t)]}{K} \tag{5}$$

Using Newton's law for the tangential tension:

$$\tau = \mu \cdot \frac{\Delta V}{\Delta y} \tag{6}$$

where: τ - tangential tension
μ - dynamic viscosity of the aerated pulp
$\frac{\Delta V}{\Delta y}$ - velocity gradient

For the friction force F_{fr} applied onto the rotor of a

flotation machine can be determined from formula (6) by multiplying both sides by R^2_{rot} and assuming the friction surface area $A = R^2_{rot}$ and $\Delta y = R_{rot}$.

$$F_{fr} = \mu . v . R_{rot} \tag{7}$$

where: v - velocity of rotor.
R_{rot} - rotor radius.

The assumptions $A = R^2_{rot}$ and $\Delta y = R_{rot}$ are reasonable from the hydrodynamical point of view and these are only both estimations with good approximations.

Further for the power consumption of the rotor $P = v . F_{fr}$ follows:

$$P = \mu . v^2 . R_{rot} \tag{8}$$

For the energy consumption of the rotor E could be obtained:

$$E = P.t = -\mu . v^2 . R_{rot} . \frac{\ln[1 - \varepsilon(t)]}{K} \tag{9}$$

respectively for the parameter Action it follows:

$$S = P.t^2 = \mu . v^2 . R_{rot} . \left\{ \frac{\ln[1 - \varepsilon(t)]}{K} \right\}^2 \tag{10}$$

It is known that the power consumption of the rotor of a flotation machine could allways be measured. This way we have two possibilities - to measure and calculate the parameter Action.

In the case that we replace the velocity of rotor using the equation:

$$v = \omega . R_{rot} = \frac{\Pi . n}{30} . R_{rot} \tag{11}$$

for F_{fr} could be obtained:

$$F_{fr} = \mu . \frac{\Pi . n}{30} . R^2_{rot} \tag{12}$$

and further for the power follows:

$$P = \mu . \left(\frac{\Pi . n}{30} \right)^2 . R^3_{rot} \tag{13}$$

The energy consumption will obtain the form:

$$E = - \frac{\mu . \Pi^2 . n^2 . R^3_{rot} . \ln[1 - \varepsilon(t)]}{900 . K} \tag{14}$$

and finally for the parameter Action we shall obtain the equation:

$$S = \frac{\mu . \Pi^2 . n^2 . R^3_{rot}}{900} . \left\{ \frac{\ln[1 - \varepsilon(t)]}{K} \right\}^2 \tag{15}$$

The parameter action could be defined for the elementary act of flotation as well, this is for the attachment of a particle to the bubble. For this purpose we could use the results from the theory of flotation [1,2,3,4].

The changing of the energy during the attachment is equal to:

$$E = \Pi . r^2_p . \gamma . (1 - \cos \theta)^2 \tag{16}$$

where: p - particle radius
θ - particle equilibrium contact angle,
γ - surface tension on the interface liquid/air.

The time of contact given by Evans's equation:

$$t_c = \sqrt{\frac{\pi \cdot m}{2 \cdot \gamma}} \qquad (17)$$

where: m - mass of particle

permits the following equation to be obtained for the parameter action for elementary act of flotation:

$$S = E.t = \pi \cdot r_p^2 \cdot \gamma \cdot (1 - \cos \theta)^2 \cdot \sqrt{\frac{\pi \cdot m}{2 \cdot \gamma}} \qquad (18)$$

Discussion

We would like discussing the following question. Is the product S = E.t really giving additional information for flotation or grinding process. Compared to the information obtained with the values for power, energy and time cousidered and used separately.

Two flotation processes are compared at the same ore, reagent regimes, initial and final conditions and results. Only the flotation machines are different in the two cases. If the energy consumption and the time of flotation for one of the processes have smaller values compared to the other one, the conclusion will be clear. If the energy consumption and the time of flotation change in different directions (the energy value smaller, but the time of flotation longer for the first one compared to the second one) what chould be the final decision? The parameter Action gives the answer - the process of the shorter time of flotation should be be better (the parameter Action is much more sensitive from the time S = P.t - time is raised to the second power, than from the power, which is raised to the first power).

Principally, the following objections could arise. The final conclusion depends on economical factors as well on parameters discussed above - the price of energy, the investment cost, the production coste etc. The time usually has not a definite price directly measured and evaluated, but let us once again concentrate our attention on the following well known and established facts:

1. Why do people make more fast cars, trains, planes, processes etc?
2. Why do the equations (1), (2) and (3) are so largely used in Science and Technique?
3. Why time is money?

Conclusions

1. The parameter Action could be measured and calculated both for the technological flotation process and for the elementary act of flotation.
2. The most essential and important parameter for the flotation and grinding is Action. It combines the information included in the energy and time and gives additional, more rich information.

REFERENCES

1. C. Nutt. Adhesion of solid particles to flat interfaces

and bubbles. Chem. Engng. Sci. 12, pp. 133-144 (1960)
2. A. Sheludko, B. Radoev and A. Fabrikant, On the theory of flotation 11. Adhesion of particles to bubbles. Annuaire de L'Universite de Sofia, 63. pp. 44-54 (1968).
3. A. Sheludko, B. Toshev and D. Bojadjiev, Attachment of particles to a liquid surface Capillary theory of flotation. Chem. Soc. Farad. Trans. 1, 12, pp. 2815-2828 (1975).
4. L. Evans, Ind. Engng. Chem. 46, pp. 2420-2450 (1954).
5. R. Varbanov, Colloid and Polymer Sci. 263, pp. 75-80 (1985).

UPGRADING FINE-GRAINED IRON ORES:
(i) General Review (ii) Agbaja Iron Ore.

G.G.O.O. Uwadiale,
National Metallurgical Development Centre,
P. M. B. 2116,
Jos.
NIGERIA.

ABSTRACT

Methods of upgrading fine-grained iron ores are reviewed - a fine-grained ore deposit defined as one with mineral matter so finely disseminated within the gangue matrix that crushing and grinding, to effect liberation, only produce 'fines' that respond poorly to conventional beneficiation processes.

The review, which covers the broad spectrum of industrial, pilot plant and laboratory techniques, as well as novel propositions, embraces such unit processes as: (i) Flotation - the Yang process of direct flotation of minus 500 mesh iron ore without a desliming step; (ii) Reverse flotation/flocculation-flotation; (iii) Selective flocculation; (iv) Electrolytic coagulation; (v) Oil agglomeration; (vi) Magnetizing reduction; and (vii) Oil extraction, Caustic digestion, Acid leaching.

The processing techniques of froth flotation, flocculation with starch, flocculation with polyacrylamides, electrolytic coagulation, oil agglomeration and magnetizing reduction were utilized on Nigeria's largest iron ore deposit, Agbaja iron ore. The deposit, with 95% of the ore grains <5μm, yielded very good results only with the magnetizing reduction and oil agglomeration techniques. Concentrates with >60% Fe assays and >83% Fe distributions were obtained. The positive response to these two processing techniques, and poor response to others, are discussed.

INTRODUCTION

Although the terms 'coarse-grained', 'intermediate-size', and 'fine-grained' are not assigned definite or specific dermacative values (micrometers, or other measuring units) in minerals processing, a fine-grained iron ore is often construed as one in which mineral matter is so finely disseminated within the gangue matrix (or vice versa) that crushing and grinding, to effect liberation, produce minute particles that respond poorly to conventional beneficiation equipment and/or processes (froth flotation, magnetic separation, gravity separation, etc).

On modification, or alteration, of the surface potentials of these poorly-responsive 'fine particles', however, concentration of the desired mineral(s), from gangue, may take place. Changes in surface potentials may be effected by: (i) the transformation of a non-magnetic mineral to the magnetic form, rendering it assessible to magnetic separation, (e.g. rhombohedral α-hematite to cubic crystalline γ-hematite; or goethite/hematite to magnetite), (ii) collector adhesion to the fine minerals' surfaces, rendering them amenable to oil agglomeration (iii) addition of electrolyte to ore pulp, to decrease particle electrostatic repulsion and promote selective coagulation, (iv) other techniques.

In this paper, a general review of processes that have been directed at upgrading fine-grained iron ores is presented. These processes include (i) froth flotation, (ii) reverse flotation/flocculation-flotation, (iii) selective flocculation, (iv) electrolytic coagulation, (v) oil agglomeration, (vi) magnetizing reduction, and (vii) oil extraction, caustic digestion, acid leaching. They have been conducted at different scales (laboratory, pilot-plant, industrial-,) with varying degrees of success, over the past years.

I. GENERAL REVIEW

Froth Flotation: The size limit to which hematite/goethite samples are often pulverized, prior to froth flotation, is 100 x 125 mesh (150 x 125μm), with desliming at 20μm[1,2,3]. However, some iron ore deposits are so finely disseminated that the excavated samples have to be ground to <5μm to effect adequate liberation of ore minerals from gangue.

Although froth flotation of finely ground sulphide minerals (<5μm size) is well reported in the literature[4-8], a similar situation dose not hold for the iron oxides and silicates. The difference seems to lie in the specificity of the collectors used in floating sulphide minerals as well as the type of bonds formed between metal sulphide and these collector compounds as against similar relations between the oxides and silicates, and the collectors employed in their selective flotation.

The works of Fuerstenau et al.[9], Reghavan and Fuerstenau[10], Houot[11] and Yang[12] stand clear as the finest iron ore (or composited iron oxide and silicate) flotation in the literature. The generally preferred close alternative is the selective flocculation-desliming-flotation technique[13-17].

Fuerstenau et al.[9] studying samples of iron ores, including a -44μm group, found hydroxamate a better collector than oleate, while Reghavan and Fuerstenau[10] in their microflotation studies, with hydroxamate, effected a separation of 0.2μm hematite from 1.8μm quartz. An industrial flotation of -44μm iron ore, by Hanna Mining Company (U.S.A), is mentioned in a recent paper by Houot[11].

Yang[12] reports a unique process involving direct flotation of minus 500 mesh iron ore without a desliming step. A critical ingredient in the process is the conditioning reagent made from Na_2SiO_3, $Fe(NO_3)_3.9H_2O$ and H_2SO_4 at a mixing ratio of 5:1:1. Tests on fine-grained iron ores, including oxidized and partially oxidized samples, gave very high Fe_T assays and recoveries (Average: 65% assay, and 87% recovery).

With phosphoriferous iron ores, direct selective flotation is simple, provided the constituent minerals are not fine-grained - as illustrated by the work of Kihlstedt[18] on a series of Scandinavian iron ores. Phosphorus in the crude (as apatite) varied from 0.017 - 3.1%P, while Fe_T was 16.7 - 38.7%. Concentrates, assaying 65% Fe and as low as 0.005%P were obtained.

However, with very fine-grained phosphoriferous ores, direct selective flotation has not been effective, and selective oil agglomeration has been employed[31,32].

Reverse Flotation/Flocculation-Flotation

Because of the poor efficiency of direct flotation systems in handling fine-grained iron ores (prior to the work of Yang[12]) the technique of reverse flotation, (flocculation-flotation), was developed in the early sixties principally by the U.S. Bureau of Mines. The historical background and technical details of the process development and operation at laboratory, pilot plant and full industrial levels are subjects of various papers by Colombo[15,16,19], Frommer[13,14], Numela[17] and other investigators[20].

The process involves grinding the fine-grained ores to -37µm in aqueous medium, dispersing the pulp at pH of around 11 with sodium silicate (sodium hexametaphosphate, sodium tripolyphosphate or combinations of these), activating the siliceous minerals with calcium ions while depressing the iron minerals with starch, tapioca flour, dextrin, etc., and then desliming, followed by flotation of the activated siliceous gangue with an anionic collector. Some systems do not employ calcium ion activation of the siliceous minerals. In such systems, cationic flotation follows the desliming step and terminates the operation[15,16,19].

Selective Floccultation

The initial stage here is similar to that in the reverse flotation process, but instead of subsequent flotation step after desliming, the flocculated mineral is re-dispersed in order to release entrained gangue, and then re-flocculated. The process of flocculation-desliming-dispersion and re-flocculation is continued until a clean concentrate is obtained.

The detailed principles and applications of the process are discussed in reviews by Somasundaran[21,22], Read et al.[23,24] and Yarar and Kitchener[25]. The technique does not, however, seem to have been extensively utilized in iron minerals concentration as only the laboratory[24,26] studies of Read is frequently encountered. The investigations involved using partially hydrolysed polyacrylamides of varying anionic characters, to separate synthetic mixtures of -20µm hematite and silicate.

The anionic character of the flocculant was found to have a profound effect on selective flocculation of the mineral mixture. Whereas the strongly anionic A70 promoted selective flocculation of hematite, the mildly anionic A130 showed no selectivity at all, while the weakly anionic A100 preferentially flocculated the silicate.

Electrolytic Coagulation

Laboratory experiments have been reported in which suspensions of mineral mixtures were effectively concentrated by selective coagulation in electrolytic media. Pugh and

Kitchener[27], and Pugh[28] effected separations of mixtures of very fine quartz-rutile and quartz-hematite suspensions with or without an electrolyte. For the quartz-hematite investigation, 0.05-0.2μm particles of naturally occurring minerals were suspended in 0.01-0.015M NaCl solution at pH 9. Selective coagulation of the hematite took place under these conditions leaving the quartz stable in suspension; but increasing the electrolyte concentration to 0.1M resulted in coagulation of the quartz. The experiments were conducted on 1:1 quartz-hematite mixtures. With 9:1 mixtures, selective coagulation was not so successful.

Oil Agglomeration

Puddington et al.[29-32] developed certain observations of a flocculating effect, on addition of a third immiscible phase to a suspension of ore minerals, or coal, in water, into a processing technique which has found a virtually universal application in the concentration of fine-grained minerals and coals. The technique has been effectively used to concentrate cassiterite, illmenite, baryte, calcite, coals and iron ore bodies containing apatite.

Sparks and Sirianni[32], and Sirianni and co-workers[31] report the beneficiation of the phosphoriferous Snake River iron ore deposit (Canada) by the selective agglomeration method where all conventional processing techniques failed. The ore was finely disseminated and required grinding down to 4μm for liberation of ore-minerals from gangue. Recovery and grade of the iron extracted were in the order of 91% and 66%, respectively, from a raw ore assaying 44% Fe_T. The phosphorus content was reduced from 0.34% in feed to 0.02% in final concentrate.

In certain phosphoriferous ore deposits, where the extremely fine grain size of the phosphate mineral(s) results in their not being readily identified, even with the use of qualitative SEM with EDAX, selective oil agglomeration may not be effective in dephosphorizing the ore[33-35]. With such iron ore deposits, dephosphorization - by lime injection into the converter (after blast furnace operation) - is often proposed.

Magnetizing Reduction

This process appears to be the most effectively used in the treatment of iron ores (particularly fines) which are not responsive to conventional beneficiation operations (froth flotation, gravity concentration and direct magnetic separation). It involves the conversion of non-magnetic iron minerals to the magnetic form by reducing the ore at elevated temperatures. The resulting artificial magnetite is concentrated from the non-magnetic gangue by magnetic separation.

$$3Fe_2O_3 + H_2 \longrightarrow 2Fe_3O_4 + H_2O - \text{heat} \quad (1)$$

$$3Fe_2O_3 + CO \longrightarrow 2Fe_3O_4 + CO_2 + \text{heat} \quad (2)$$

In most magnetizing reduction systems, the ore is reduced at a particular temperature and then quenched in water, prior to magnetic separation. But in the oxidation-reduction technique, designed for heat economy, both reduction and oxidation stages are involved[39], resulting in the production of -hematite:

$$4Fe_3O_4 \quad + \quad O_2 \longrightarrow 6Fe_2O_3(\gamma) \quad + \quad heat \qquad (3)$$

Cooling the γ-hematite is effected to recover its sensible heat which is used to preheat and dry-feed ore, resulting in a cyclic process as regards heat requirements.

The process requires only the initial fuel to dry and reheat the ore to reaction temperature; thereafter it is self-sustaining.

The development of high intensity magnetic separators[36-38] which efficiently concentrate particles down to -44μm, has rendered magnetizing reduction highly attractive in the beneficiation of fine-grained,non-mangetic,iron ores.

Oolitic Ores: Comparing the magnetizing reduction results of Ramin[40], Asswan *(Egypt) and Jdaidet-Yabus *(Syria) iron ores with results obtained, on the same deposits, by other processing techniques, yields the glaring superior scientific performance of the magnetizing reduction technique in the beneficiation of fine-grained iron ores.

The Ramin iron ore - principally a 1-2μm goethite deposit - was previously concentrated by gravity methods[41] and only resulted in a product assaying 42% Fe, with a recovery of 75%. But magnetizing reduction gave concentrates assaying 84-95.7%Fe, with recoveries 93-95.7%. (However, magnetizing reduction, here, involved total reduction to metallic iron, instead of magnetite).

With the Asswan iron ores, flotation concentrate gave a recovery of 58.5%, with an Fe grade of 49% whereas typical magnetizing reduction experiments resulted in recoveries of 86-89.3% Fe with assays of 53-54.68%. A similar increase in values of assays and recoveries was observed with the Jdaidet-Yabus ores.

Oil Extraction, Caustic Digestion, Acid Leaching

Shergold and Mellgren developed the oil extraction technique and designed a laboratory apparatus with which they concentrated quartz and hematite, in iso-octane, in the presence of dedecylamine and sodium dodedcyl sulphate collectors[42-44] The fundamentals of the technique are similar to those of the oil agglomeration process but the experimental procedure is different. A critical difference is the high oil requirement in the oil extraction process.

Caustic digestion of finely disseminated silica from iron ores has been proposed by Tiemann et al.[45-47] and Herzog and Backer[48] as a viable processing route when conventional techniques fail. In this process, quartz, silicates and other impurities dissolve at the boiling point of caustic soda or several degrees lower, depending on operating variables. The iron mineral is filtered from the solution, the silica precipitated with lime, and caustic soda is recycled.

Reeve[49] developed an acid leaching method for treating low grade iron ores at Appleby-Frodingham. The process involves the use of hydrogen chloride gas or aqueous hydrochloric acid

---

* Personal Communications.

to distill iron from the ores as pure ferric chloride at 300-350°C. The ferric chloride so produced is hydrolysed by steam at the same temperature range to produce ferric oxychloride, and finally by further steam treatment at 550°C to produce pure ferric oxide, with the regeneration of HCl gas for use in chloridizing further supplies of ore.

II. AGBAJA IRON ORE

This is the largest iron ore deposit in Nigeria. The principal constituent mineral is goethite, accompanied by minor hematite, siderite, quartz and kaolinite. Fe assay in the crude deposit is 46.5%. Particle grain sizes range from submicron to 500μm, and to achieve significant liberation, ore samples must be pulverized to <5μm, as 95% of the particles lie in this size region.

Concentration investigations, on the ore deposit, were conducted via the processes of (i) electrolytic coagulation, (ii) flocculation with starch, (iii) flocculation with polyacrylamides, (iv) oil agglomeration, and (v) magnetizing reduction.

Experimental

In the electrolytic coagulation, selective flocculation and oil agglomeration investigations, a charge of 50g Agbaja iron ore +50ml water +0.07g Na_2SiO_3 was ground in a ceramic pot for 6 hours. The pulp was emptied into the glass liquidizer unit of a National Panasonic blender, diluted to the 0.8 litre mark, and agitated for 10 minutes.

The underlisted conditions applied in respective tests:

Electrolytic Coagulation: Agitation was conducted around neutral pH, with or without 10ml of 10% NaCl.

Flocculation with Starch: 10-20ml of 2.5% AnlaR Starch (BDH) was added to the pulp at the eighth minute of agitation.

Flocculation with Polyacrylamides: The pulp was agitated with 5ml of 0.5M NaCl solution, and the pH was varied from one polyacrylamide to another. The polyacylamide was slowly added, two minutes before termination of agitation, as 0.25% solution.

Oil Agglomeration: 5ml oleic acid was added to the ball mill charge. 7ml kerosene and 0.07g Na_2SiO_3 dispersamt are added to the agglomeration cell (blender) at every separation stage.

Magnetizing Reduction: Known wights and sizes of ore and coal, in a poreclain crucible, were placed in an electric furnace and heated to particular temperatures. On roasting for one hour, the samples were quenched in water and concentrated in a Davis tube magnetic separator.

Results and Discussion

The techniques of electrolytic coagulation, and selective flocculation (by starch or polyacrylamides used in this investigation) are not viable processing routes for Agbaja iron ore. The results indicate poor selectivity, as evidenced by virtually identical partitioning of iron (assays, and yields) to the concentrates and tailings. The consequence of this poor selectivity, and an equally poor flotation response[33] (by the Yang[12] and conventional flotation methods), is that Agbaja iron ore will not be amenable to beneficiation by the selective flocculation-desliming-flotation technique which, fundamentally, consists of the processes of selective flocculation and flotation.

Electrolytic Coagulation: Chemical assay of concentrates showed that electrolytic coagulation did not yield good results in many investigations (assay: 47-56% Fe; recovery: 45-68% Fe).[33]

Microscopic examination of some mill product, after 6 hours grinding, showed that 90% of the material was finer than 1μm (one micrometer), with about 96% liberation. With these finer sizes, however, coagulation rates and sedimentation were slow. It took over two hours to complete each investigation. The mudlines were difficult to observe, particularly during the washing stages, and accurate siphoning of suspension (tailing) was not possible.

Starch Flocculation: From the results of previous investigators, if substantial amounts of <2μm iron oxides are present in the ore, incomplete selective flocculation and a higher iron content in the slime fraction, can be expected[15,16,19]. The ground slurry used in these tests contained mineral particles that were 90% less than one micrometer. Selective flocculation did not take place.

Flocculation with Polyacrylamides: The poor selectivity of polyacrylamides in the flocculation of Agbaja iron ore is attributed to the extreme fineness of the ground ore, and also to the broad flocculating properties of polyacrylamides. Although Read[26] and Yarar and Kitchener[25] report that the selective flocculation process "should work as well on 0.1μm as on 20μm particles,"[28] there is no experimental work encountered in the literature that supports this assertion of flocculating predominantly 0.1μm particles. On the contrary, particles less than 2μm are reported to cause poor selectivity in the flocculation of iron ores by starch[16]. The mechanism of adsorption of starch and polyacrylamide on hematite and silica is physical (hydrogen bonding and electrostatic interaction[25,28,50], and the flocculation properties of the two reagents, with respect to size of minerals treated, may also be similar.

Of greater significance than the size factor, however, is the broad aggregating properties of most flocculants, including polyacrylamides. The adverse effects of such broad flocculating properties is particularly conspicuous in natural ore pulps which, invariably, consist of many minerals with, sometimes, relatively close points of zero charge. In such pulps, flocculants which aggregate minerals over a wide range of pH, produce poor beneficiation.

The reaction of minerals to polyacrylamide-based flocculants at pH ~6.0, and in distilled water, is shown in the investigation of Slater et al.[50]. At pH ~6, an anionic polyacrylamide flocculates hematite, alumina, flourite, calcite, galena, kaolinite, mica, and talc, with the exception of quartz. With the non-ionic polyacrylamide, only aged quartz will be left in suspension, and with the cationic polyacrylamide, only alumina will be left unflocculated.

An attempt to selectively flocculate hematite from a pulp containing these minerals with the use of polyacrylamides will obviously be difficult. However, previous workers[25,26,51] have oversimplified the subject by conducting investigations on cautiously selected synthetic pairs of pure minerals with widely divergent points of zero charge and/or specific gravities. Such situations are unlikely in natural ore deposits.

Oil Agglomeration: Agbaja iron ore was upgraded, with high Fe grades and recoveries, by the oil agglomeration technique. Concentrate assays of 65-67% Fe and recoveries of 90-95% were obtained[33,34]. This is readily achieved on wet-grinding the ore to liberation point, <5μm, and subjecting to oleic acid (collector) and kerosene (bridging liquid) at pH 8.0-10.0. At these pH values, soluble oleate ions predominate, in a well dispersed pulp.

Typical results obtained in the oil agglomeration investigation are shown in Table 1. At pH 11.00 no agglomerates are formed; the pulp is permanently dispersed.

Table 1: Selective Oil Agglomeration of Agbaja Iron Ore. pH region of highest Fe assays and recoveries.

pH	Wt., g	Wt., %	Assay, %Fe	Distribution, %Fe
8.0	40.20	76.35	58.21	89.77
8.5	41.82	79.96	60.52	94.08
9.0	43.53	70.06	65.02	89.33
10.0	32.69	60.11	64.05	72.50

Magnetizing Reduction: The highest Fe assays in concentrates, and the highest recoveries, were obtained from samples roasted at 500-700°C, prior to quenching and magnetic separation (Table 2). XRD investigations showed that concentrates in this temperature range, principally, contain magnetite(35).

A particular advantage of the magnetizing reduction processing method is that under the reduction-roasting conditions, finely disseminated iron oxide and quartz grow and coalesce into larger, separate, grains[52] which eliminates the fine-size constraints associated with many beneficiation unit operations.

Table 2: Magnetizing Reduction of Agbaja Iron Ore: roasting temperatures with high magnetite recoveries.

Temp., °C	Product.	Wt., g	Wt., %	Assay, % Fe	Distribution, % Fe
500	C	14.79	67.70	59.37	84.40
	T	7.05	32.30	22.32	15.60
	Head	21.84	100.00	46.21	100.00
600	C	13.19	69.70	60.00	87.34
	T	6.09	30.30	20.00	12.66
	Head	20.07	100.00	47.88	100.00
700	C	12.20	62.10	60.67	83.77
	T	7.44	37.90	19.26	16.23
	Head	19.64	100.00	44.98	100.00

CONCLUSION

A number of processes have been proposed for upgrading fine-grained iron ores. Of these, the reverse flotation (flocculation-flotation) and magnetizing reduction methods have been widely employed at all levels - laboratory, pilot plant, and industrial - with success.

The oil agglomeration technique has also generated very good results, but has been utilized principally at the laboratory level. The long duration of grinding, for processed samples, is economically untenable in the direction of its elevation to industrial scale.

Similar extremely slow processing speed also casts a negative mark on electrolytic coagulation of natural ores. Although leaching and cementation operations go on for lengthy periods of time in the copper processing industry, faster alternative benefication methods, in the direction of specific final goals, exist in the iron ore industry.

The broad flocculating property of polyacrylamides holds and adverse effect in natural ore deposits, where ore minerals and (some) gangue may lie relatively close in points of zero charge. Selection, by previous workers, of minerals (usually two), of widely divergent PZCs and/or specific gravities, does not depict the accurate processing methodology of natural ores.

Ore deposits which require grinding to <2μm, to effect liberation, cannot be processed by flocculation with starch as well stated by previous investigators.

REFERENCES

1. Kulkarni, R.D., and Somasundaran, P., Trans AIME, 262, 1977, pp 120-124.
2. Iwasaki, I., Cooke, S.R.B., Harraway, D.H., and Choi, H.S., Trans AIME, 223, pp 97-108.
3. Cooke, S.R.B., Iwasaki, I., and Choi, H.S., Trans AIME, 217 1960, pp 76-83.
4. Trahar, W.J., and Warren, L.J., Int. J. Miner. Proc. 1976, 3 pp 103-131.

5. Gaudin, A.M., Groh, J.O., and Henderson, H.B., AIME Tech. Publ., 1931, 414, pp 3-23.
6. Trahar, W.J., Int. Jour. Miner. Proc. 1976, 3, pp 151-166.
7. Hornsby, D., and Leja, J., Surf. & Colid. Sc., 12, 1982, (Ed., Egon Matijevic), pp. 217-313.
8. Collins, D.N., and Read, A.D., Minerals Sc. & Engng. 3, April 1971, pp 19-31.
9. Fuerstenau, M.C., Harper, R.W., and Miller, J.D., Trans AIME, 247, 1970, pp 69-73.
10. Raghavan, S., and Fuerstenau, D.W., AIChE Symp. Series 150, 71, 1975, pp 59-67, Somasundaran, P., and Grieves, R.B., Eds.
11. Houot, R., Int. Jour. Miner. Proc., 1983, 10, pp 183-204.
12. Yang, D.C., Beneficiation of Mineral Fines, 1979, AIME, pp 295-308 (Somasundaran, P., & Arbiter, N., Eds).
13. Frommer, D.W., Mining Engineering, April 1964, pp 67-71, 80.
14. Frommer, D.W., Blast Fur. Steel Plant, May 1969, pp 380-389.
15. Colombo, A.F., and Frommer, D.W. Flotation - A.M. Gaudin Memorial, 1976, pp 1285-1304 (Fuerstenau, M.C. Ed.)
16. Colombo, A.F., Fine Particle Processing, 1980, pp 1034-1056. (Somasundaran, P., Ed.)
17. Numela, W., ibid., pp 63-74.
18. Kihlstedt, P.G., Proc. Int. Mineral Dressing Congress, Stockholm, 1957, pp 559-570.
19. Colombo, A.F., Beneficiation of Mineral Fines, (SME of AIME), 1979, pp 237-252. (Somasundaran, P. and Arbiter, N., Eds.)
20. Durand, M., Gauthier, F., and Guyot, R., Proc. 6th Int. Miner. Process. Congr., Cannes, 1963, Pergamon, pp. 385-394.
21. Somasundaran, P., Beneficiation of Mineral Fines, AIME, 1979, pp. 183-196. (Somasundaran, P., and Arbiter, N., Eds)
22. Somasundaran, P., Fine Particles Processing, AMIE, 1980, pp. 947-979. (Somasundaran, P., Ed).
23. Read, A.D., and Hollick, C.T., Mineral Sc. Engrng, 8, 1976, pp. 202-213.
24. Read, A.D., and Whitehead, A., Tenth Int. Process Congr., Lond., 1973, IMM, pp. 949-957.
25. Yarar, B., and Kitchener, J.A., Trans. IMM, Sect. C., March 1970, pp. C23-C33.
26. Read, A.D., Trans. IMM, Section C., March 1971, pp. C24-C31.
27. Pugh, R.J., and Kitchener, J.A., J. Colld. Int. Sc., 38, (3), March 1972, pp. 656-657.
28. Pugh, R.J., Colld & Polymer, Sc., 252, 1974, pp. 400-406.
29. Smith, H.M., and Puddington, I.E., Can. J. Chem., 38, 1960, pp. 1911-1916.
30. Farnand, J.R., Smith, H.M., and Puddington, I.E., Can. J. Chem., Engrng. 39, 1969, pp. 94-97.
31. Sirianni, A.F., Coleman, R.D., Goddhue, E.C., and Puddington, I.E., Can. Min. Metall. Bull., 1968, pp. 731-735.
32. Sparks, B.D., and Sirianni, A.F., Int. Jour. Min. Proc. 1, 1974, pp. 231-241.
33. Uwadiale, G.G.O.O., Univ. of Strathclyde, 1984, 341 pp.
34. Uwadiale, G.G.O.O., 1990, Min. & Met. Processing, AIME, (in press).
35. Uwadiale, G.G.O.O., and Whewell, R.J., Met. Trans B., 19B, 1988, pp. 731-735.

36. De Vaney, F.D., Int. Jour. Min. Proc. 1, 1974, pp 675-688.
37. Stone, W.J.D., ibid., pp. 733-743.
38. Hopstock, D.M., and Colombo, A.F., Fine Part Processing (AIME), 1980, 2, pp. 1242-1260 (Somasundaran, P., Ed.)
39. Stephens, Jr., F.M., Langston, B., and Richardson, A.C., J. Metals, 5, 1953, pp. 780-785.
40. Weissberger, S. and Zimmels, Y., Int. J. Miner. Processing 11, 1983, pp. 115-130.
41. Boskovitz-Rohrlich, Vo., Mitzmager, A., and Mizrahi, J., Mining Magazine, 108, 1963, pp. 325-331.
42. Mellgren, O., and Shergold, H.L., Trans. IMM, 75, 1966. pp. C267-C268.
43. Shergold, H.L., and Mellgren, O., Trans. IMM, 78, 1969, pp. C121-C132.
44. Shergold, H.L., and Mellgren, O., Trans. IMM, 81, 1971, pp. C60-C68.
45. Tiemann, T.D., Trans. AIME, 223, 1962, pp. 173-178.
46. Stone, R.L., and Tiemann, T.D., Trans AIME, 229, 1964, pp. 217-222.
47. Metha, A.J., and Tiemann, T.D., Trans AIME, 260, 1976, pp. 101-103.
48. Herzog, E., and Backer, L., Proc. Int. Miner, Proc. Congr., Cannes, 1963, pp. 171-184.
49. Reeve, L., J. Iron Steel Inst., 1955, pp. 26-40.
50. Slater, R.W., Clark, J.P., and Kitchener, A.J., VIII Int. Miner. Proc. Congr., Leningrad, 1968, Paper C-5, 7 pp.
51. Read, A.D., British Polymer, J., 1972, 4, pp. 253-264.
52. Dhalem, D.H., and Sollenberger, C.L., Proc. 6th Int. Miner, Processing Congr., Cannes, 1963, pp. 407-420.

PROCESSING OF HEMATITIC IRON ORES

JOHN HANNA AND IBEZIM J. ANAZIA
Mineral Resources Institute, The University of Alabama,
Box 870204, Tuscaloosa, AL 35487-0204

ABSTRACT

The beneficiation characteristics of low grade Birmingham red iron ores of the Big Seam and Ferruginous Sandstone seams were investigated using flotation and magnetic separation after reduction roasting. Samples of Big Seam ore responded more readily to concentration than the Ferruginous Sandstone ores. The impurities inherent with the iron oxide minerals limited the grade of concentrates recoverable from these ores to a maximum of about 61% Fe if reasonable iron recoveries are sought.

Flotation concentrates assaying 60-61% Fe and about 0.1% P with attendant iron recoveries of about 63% were achieved from Big Seam ore charges while those obtained from the low phosphorus Ferruginous Sandstone samples analyzed 54-56% Fe. Reductive roasting and magnetic separation of ground charges of the two ore samples yielded concentrates assaying 60-61% Fe. Iron recoveries were better than those achieved by flotation, but the magnetic products of Big Seam ore were high in phosphorus (0.2-0.3% P).

INTRODUCTION

The recovery of upgraded iron ore concentrates from Alabama's Birmingham District ores using various beneficiation schemes has been studied since about 1895 [1]. As late as the 1960's, a number of physical separation techniques were proposed for treating the ores. However, the iron concentrates produced rarely contained more than 56% Fe even when the pulps consisted of ore ground to minus 50 microns [2,3]. Only during the past several decades, with the success of extremely fine grinding processes in the Lake Superior region, has it become apparent that adapting such technology to the characteristics of Birmingham District ores might facilitate production of higher grade concentrates [4,5]. Therefore, this investigation focused on developing new or improved beneficiation techniques for recovering concentrates suitable for modern blast furnace smelting. Among the techniques applied were flotation, reductive roasting and magnetic separation, magnetic flocculation and hydrocyclone concentration.

EXPERIMENTAL

Samples Tested

Based on information from previous investigations, five samples of red iron ore were obtained for this investigation. Two samples were from the Big Seam (BS-1 and BS-2) and three were from the Ferruginous Sandstone formations (FS-1, FS-2 and FS-3). Chemical analyses of the samples are given in table 1.

Advances in Fine Particles Processing
John Hanna and Yosry A. Attia, Editors

Table 1. Chemical Analyses of Birminghan Red Iron Ores

Sample	Assay, % Fe	Insoluble	CaO	P
FS-1	35.5	44.5	0.1	0.10
FS-2	35.0	50.0	2.0	0.09
FS-3	23.6	39.0	12.0	0.12
BS-1	34.7	24.2	11.1	0.36
BS-2	37.7	13.5	16.2	0.34

FS = Ferruginous Sandstone ore
BS = Big Seam ore

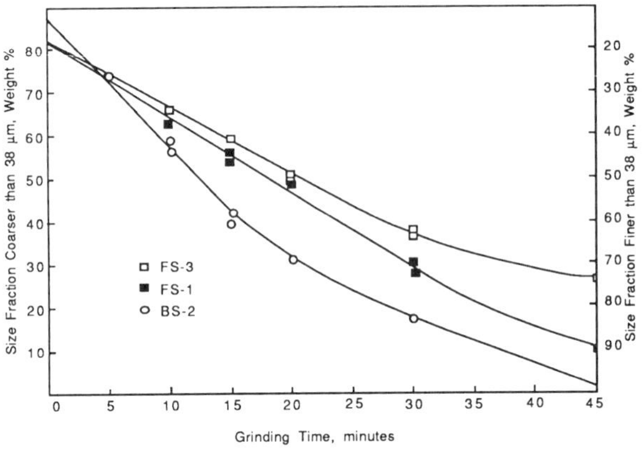

Figure 1. Effect of Grinding Time on Size Reduction of Ferruginous Sandstone Samples at 50 % Solids

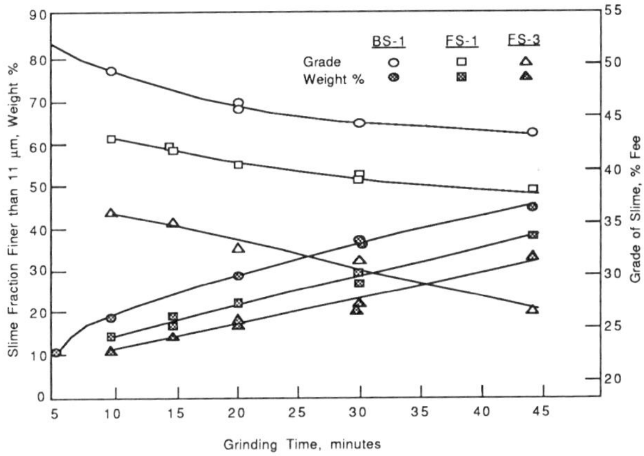

Figure 2. Quantity and Quality of Slime Produced from Ground Big Seam and Ferruginous Sandstone Samples as a Function of Grinding Time

The major iron-bearing minerals in the red iron ores are microcrystalline hematite, commonly associated with variable amounts of hydrated iron oxides (goethite-limonite). With the exception of a few occurrences of specular hematite, the bulk of the iron minerals are in oolitic and fossiliferous structures and are poorly crystallized [6]. These structures also host a large number of mineral impurities such as clays, calcite, apatite, and colloidal silica. Detailed mineralogic studies of Hagni and Cooper have indicated that carbonate-fluorapatite is the major phosphorus-bearing mineral in the Big Seam ores [7]. Most of the apatite particles were in the size range of 5-30 microns and were finely disseminated in the host hematite and calcite particles.

From a beneficiation standpoint, the red iron ores may be classified into three main types, (1) siliceous, e.g. Upper Ferruginous Sandstone, (2) calcareous-siliceous, e.g. Big Seam and Irondale, and (3) siliceous and/or calcareous-clay, e.g. Lower Ferruginous Sandstone. The Ferruginous Sandstones, low in phosphorus, belong to the first category. These resources consist essentially of minus 300 micron sand grains coated with and cemented together by iron oxide with little or no carbonate. The Big Seam ores are high in phosphorus and contain large amounts of the carbonate minerals, sand grains coarser than 300 microns, and some clay minerals. Where part or all of the carbonate minerals have been leached out or replaced by iron oxides, the ores contain higher amounts of clay minerals (type 3). Most of the soft red ores belong to this category.

Apparatus and Techniques

The dry samples were stage crushed to minus 2 mm (10 mesh) using jaw and roll crushers. Wet grinding was carried out in a 30 cm x 20 cm diameter stainless steel rod mill operated at 78% of critical speed using twenty-six 1.7 cm stainless steel rods as the grinding medium.

The flotation feed, unless otherwise stated, was prepared by rod mill grinding of the minus 2 mm 10-mesh ore for 20-30 minutes at 50% solids using Tuscaloosa tap water. The pulps were either floated directly or they were deslimed at 11 microns before floating the sand fractions. Flotation was conducted using the reverse flotation technique [8,9].

A standard flotation procedure was followed in which the charges were treated with 0.5 kg/ton of $CaCl_2$ followed by flotation of the activated quartz gangue with 2 kg/ton of Pamak 4 (distilled tall oil) using 1 kg/ton of cauticized starch as the iron oxide depressant. The rougher froth was cleaned twice. After analysis of the test products, the middling products were composited either with the cleaner siliceous froth product (tailing) or the cell product (iron concentrate) depending upon the iron analysis.

Reduction roasting was carried out in an externally heated steel alloy rotary drum furnace patterned after the furnace designed by Dean and Davis [10] using either natural gas, bituminous coal, waste coal or lignite as the reducing agents. Most of the tests were made on one-kilogram charges although a few tests were made on larger feed charges. After heating the charges at the desired temperature and for a predetermined length of time, each sample was cooled to about 250°C followed by quenching in water. When natural gas was used as the reductant, the retort was swept with the same gas during cooling. When a solid reductant was used, a neutral atmosphere of nitrogen was maintained in the retort while cooling. The quenched charges were wet ground at 50% solids in the rod mill using 0.9 kg/ton each of NaOH and Na_2SiO_3.

Grinding for 45 minutes yielded ground pulps containing about 95% of material finer than 38 microns.

Split portions of the reduced charges were separated in an Eriez Davis Tube (Model EDT) magnetic separator. The other portions were used for magnetic flocculation tests, where the dispersed pulps were mildly stirred using a stainless steel 3-blade impeller with a 51-micron teflon coated magnetic bar attached on each blade to flocculate the magnetic iron oxide. Upon termination of the stirring, the floccules settled rapidly, usually in one minute or less. The suspended nonmagnetic constituents were separated from the settled floccules by decantation at 44 micron. The floccules then were washed three times by repulping and decantation.

RESULTS AND DISCUSSION

Grinding Studies

Tests were made on raw and reduced charges of the Big Seam and Ferruginous Sandstone ores to determine their grinding characteristics and the maximum grade of iron products recoverable. The results of these investigations have been reported by the authors [11,12] and a summary of the major conclusions are as follows:

Significant differences were observed in the grinding characteristics of the Big Seam and Ferruginous Sandstone samples as shown in figure 1. These differences were attributed to the wide variability in particle size and hardness of the constituents of the red ores.

Mild grinding of these ores produced substantial quantities of sand-size grains and generated iron-enriched slimes finer than 11 microns. Fine grinding to essentially minus 38 microns generally improved liberation of the iron-enriched particles, but generated excessive quantities of low grade slime as shown in figure 2.

Liberation studies showed that the iron bearing minerals in particle sizes finer than 38 microns encapsulate substantial quantities of submicron-sized colloidal siliceous and clay-type minerals in addition to the phosphorus bearing mineral, apatite. Thus, indiscriminate fine grinding of the red ores has only limited merit for improving mineral liberation or recovering high grade products. This is particularly true in preparing raw ore pulps for flotation and supports the established axiom of the need to employ stage grinding and/or intermediate grinding in the minus 106 micron ranges if excessive slime production and over-grinding of the coarse iron oxide particles are to be avoided. This is more evident with the raw Big Seam ores. On the other hand, fine grinding of the reduced ores improves the iron recovery by magnetic separation and minimized slime production. However, the maximum grade of the magnetic product was limited to 61% Fe, indicating incomplete liberation of the reduced ores. This may be attributed to the high temperature phase transformation and/or chemical bonding of the host iron minerals and the encapsulated impurities.

Flotation Studies

Tests were made to evaluate the merit of separately floating the "coarse" and "fine" fractions of the Big Seam and Ferruginous Sandstone ores. Flotation results of tests made on sample BS-2, after grinding for different lengths of time, are shown in table 2. Concentrates assaying 59-60% Fe with iron recoveries of 51-70% were produced by floating the "coarse" plus 38 micron fractions. The "fine" minus 38 plus 11 micron fractions yielded slightly lower grade concentrates of 57-58% Fe with iron recoveries in the range of 70-85%. However, the overall iron recovery and grade of the composite iron concentrates declined with the fineness of the grind. The decline in the iron recovery is mainly due to the production of excessive fines, or slimes, during fine grinding and their possible adverse effects on reverse flotation of iron minerals. The lower grade concentrates produced from the finely ground pulps may be related to the increased liberation of the non-floatable clay minerals remaining with the iron minerals in the pulp that limited the quality of products obtained. Therefore, the fine grinding and reverse flotation approach was not pursued for upgrading the Big Seam BS-2 sample. As shown in table 2, the optimum flotation results were obtained by mild grinding of the ore for about 20 minutes. The deslimed ore gave good grade iron concentrates with reasonable recoveries from the coarse- and fine-sized fractions.

Table 3 shows the phosphorus analyses of the flotation products obtained from the 20-minute grind of sample BS-2. The flotation concentrates were generally lower in phosphorus compared to the corresponding feed analysis. The concentrates recovered from the coarse fraction contained the lowest phosphorus at 0.11%, while the middling products in general gave the highest phosphorus analysis. The grade of the composite flotation concentrate of 59.3% Fe and 0.17% P was good, but the iron recovery was low because of the heavy iron losses in the slimes. Regrinding and upgrading the middlings and slime products might recover additional iron enriched low phosphorus concentrates.

Flotation tests also were made on charges of "coarse" and "fine" fractions of Ferruginous Sandstone samples FS-1 and FS-3. The results of typical tests are given in tables 4 and 5 respectively. The composite iron concentrates recovered by flotation of both fractions were somewhat lower in grade than those produced from sample BS-2. This is attributed to their high clay content which reported with the iron minerals in the concentrate cell product after silica flotation. The presence of substantial quantities of clay minerals was verified by SEM and EDS examination of the flotation concentrates obtained from the two sandstone samples.

Influence of Grinding Additives on Flotation: Various types of organic and inorganic compounds were previously tested to modify the grinding characteristics and control the production of slimes from the Big Seam ore [9,10]. The tests showed that certain additives had generally improved grindability of the ore and in some cases selectivity controlled the quality and quantity of the minus 11 micron slimes produced. For example, dispersants such as silicates, phosphates or some lignin sulfonate products were unselective in producing large amounts of finer-sized products during grinding. Other reagents, such as certain anionic or cationic surfactants, inorganic salts and polymer products, favored selective grinding of the ores which resulted in production of iron-enriched or iron-deficient slime fractions [11]. Because many of the reagents tested are also used as flotation reagents, tests were made to investigate their effects on the reverse flotation step.

The flotation tests were conducted on the deslimed ground pulps prepared using the additives under the conditions described in table 6. The test data using over 34 various reagents indicated that some of the grinding additives had enhanced flotation selectivity and/or iron recovery in the concentrates. Selected examples of the more effective reagents are given in table 6. The data shows that addition of 0.25-1.0 kg/t anionic surfactants such as Na-oleate, Aerosol OT and Cyanamid 899 R

Table 2. Effect of Feed Size on Flotation of Sample BS-2

Grind					Coarse (+37μ)			Fine (37 x 11μ)		
						Fe			Fe	
Time, minutes	Size, microns	Weight, %	Assay, %	Flotation Product	Weight, %	Assay, %	Distrib'n, %	Weight, %	Assay, %	Distrib'n, %
5	+ 38	73.6	34.4	Concentrate	29.9	59.1	51.4	60.0	56.9	85.5
	- 38 + 11	16.0	40.0	Middling	22.8	39.6	26.3	10.6	31.2	8.1
	- 11 (slime)	10.4	54.1	Tailing	47.3	16.2	22.3	22.3	8.9	6.4
	Head	100.0	37.3	Flotation Feed	100.0	34.4	100.0	100.0	40.0	100.0
10	+ 38	57.2	33.4	Concentrate	32.5	60.1	58.6	44.6	58.0	69.2
	- 38 + 11	24.2	37.7	Middling	25.2	38.4	28.9	23.2	37.4	23.3
	- 11 (slime)	18.6	47.8	Tailing	42.3	9.9	22.5	32.2	8.8	7.5
	Head	100.0	37.1	Flotation Feed	100.0	33.4	100.0	100.0	37.7	100.0
20	+ 38	34.1	34.9	Concentrate	34.6	59.8	59.1	45.2	58.5	80.4
	- 38 + 11	38.7	33.0	Middling	18.5	42.5	22.4	9.8	36.3	10.7
	- 11 (slime)	27.2	46.1	Tailing	46.9	13.8	18.5	45.0	6.7	8.9
	Head	100.0	37.2	Flotation Feed	100.0	35.0	100.0	100.0	33.0	100.0
30	+ 38	17.4	37.4	Concentrate	44.8	58.4	69.9	42.9	58.0	74.2
	- 38 + 11	52.2	33.5	Middling	18.4	43.9	21.4	13.0	37.3	14.4
	- 11 (slime)	30.4	44.8	Tailing	36.8	8.8	8.7	44.1	8.5	11.4
	Head	100.0	37.6	Flotation Feed	100.0	37.4	100.0	100.0	33.5	100.0

Test Conditions : 1 kg charge of minus 10 mesh ore, rod mill ground at 50% solids with 1 kg/ton each of NaOH and Na_2SiO_3; flotation at pH 11.2 with 2 kg/ton Pamak 4 after gangue activation with 0.25 kg/ton $CaCl_2$ and addition of 1.1 kg/ton pearl starch to depress the iron oxides; in tap water.

Table 3. Metallurgical Balance of Grinding and Reverse Flotation of Coarse and Fine Fractions of Sample BS-2

Size, microns	Flotation Product	Weight, %	Analysis, %		Distribution, %	
			Fe	P	Fe	P
	Concentrate	11.8	59.8	0.11	19.0	3.9
	Middling	6.3	42.5	0.39	7.2	7.4
	Tailing	16.0	13.8	0.41	5.9	19.9
Coarse (+37μ)	Flotation Feed	34.1	34.9	0.30	32.0	31.2
	Concentrate	17.5	58.5	0.21	27.5	11.1
	Middling	3.8	36.3	0.90	3.7	10.4
	Tailing	17.4	6.7	0.63	3.1	33.2
Fine (37x11μ)	Flotation Feed	38.7	33.0	0.47	34.3	54.7
	Combined Conc.	29.3	59.3	0.17	46.5	15.0
	Combined Feed	72.8	33.9	0.39	66.3	85.9
	-11 μm Slime	27.2	46.1	0.17	33.7	14.1
Slimes (-11μ)	Composite Head	100.0	37.2	0.33	100.0	100.0

Test Conditions : 20-min rod mill grind of 1000 grams of minus 10 mesh ore at 50% solids using 1 kg/ton each of NaOH and Na_2SiO_3
Floated at pH 11.2 using 0.25 kg/ton $CaCl_2$, 1.1 kg/ton pearl starch, 2 kg/ton Pamak 4 and 0.01 kg/ton pine oil; in tap water.

Table 4. Results of Reverse Flotation of Coarse and Fine Fractions of Sample FS-1

Size, microns	Flotation Product	Weight, %	Analysis, % Fe	Distribution, % Fe
	Concentrate	29.2	51.5	42.7
	Middling	0.5	33.9	0.5
	Tailing	19.7	2.2	1.3
Coarse (+37µ)	Flotation Feed	49.4	31.7	44.5
	Concentrate	19.8	55.3	31.2
	Middling	-----	----	-----
	Tailing	10.7	2.6	0.8
Fine (37x11µ)	Flotation Feed	30.5	36.9	32.0
	Combined Conc.	49.0	53.0	73.9
	Combined Feed	79.9	33.7	76.5
	-11 µm Slime	20.1	41.0	23.5
Slimes (-11µ)	Composite Head	100.0	35.2	100.0

Test Conditions : 20-min rod mill grind of 1000 grams of minus 10 mesh ore at 50% solids using 1 kg/ton each of NaOH and Na_2SiO_3.
Floated at pH 11.2 using 0.25 kg/ton $CaCl_2$, 1.1 kg/ton pearl starch, 2 kg/ton Pamak 4 and 0.01 kg/ton pine oil; in tap water.

Table 5. Results of Reverse Flotation of Coarse and Fine Fractions of Sample FS-3

Size, microns	Flotation Product	Weight, %	Assay, % Fe	Distribution, % Fe
	Concentrate	5.4	54.0	15.3
	Middling	6.5	29.8	10.1
	Tailing	37.1	4.8	9.3
Coarse (+37µ)	Flotation Feed	49.0	13.6	34.7
	Concentrate	4.9	53.9	13.8
	Middling	10.1	28.8	15.2
	Tailing	19.2	3.8	3.8
Fine (37x11µ)	Flotation Feed	34.2	18.3	32.8
	Combined Conc.	83.2	15.5	67.5
	Combined Feed	10.3	54.0	29.1
	-11 µm Slime	16.8	37.0	32.5
Slimes (-11µ)	Composite Head	100.0	19.1	100.0

Test Conditions : 20-min rod mill grind of 1000 grams of minus 10 mesh ore at 50% solids using 1 kg/ton each of NaOH and Na_2SiO_3.
Floated at pH 11.2 using 0.25 kg/ton $CaCl_2$, 1.1 kg/ton pearl starch, 2 kg/ton Pamak 4 and 0.01 kg/ton pine oil; in tap water.

during grinding improved both flotation and grinding results. High grade flotation concentrates analyzing 60-61% Fe and iron-enriched slimes assaying 57-58% Fe were recovered from the Big Seam ore. Also, addition of $CuSO_4$ enhanced flotation selectivity and improved iron grade and recovery in the concentrate and produced smaller amounts of iron-enriched slimes. Some polymers, such as PAMS and PA-18, showed certain selective dispersion/flocculation and/or activation/depressant effects on the iron oxide or gangue minerals during desliming and flotation.

Moreover, phosphorus analysis of the flotation products indicated that many of the additives used had reduced the phosphorus content of the iron flotation concentrates to less than 0.2% P, compared to about 0.3-0.4% P in the feed material. In several tests, however, the phosphorus content in the concentrate was reduced to about 0.1% as shown in the results presented in table 7. In the test with NH_4F, the iron flotation concentrate analyzed 61% Fe and 0.1% P. Iron recovery was about 63% and the phosphorus rejection from a composite of the siliceous froth and the minus 11 micron slime was 71%. The above results demonstrate the potential benefits of using grinding aids during mild grinding and flotation in recovering high grade flotation concentrate from the Big Seam ore BS-2. Meanwhile, the recovery of low-phosphorus (< 0.1%P) iron concentrates is a continuing challenge in the processing of the red ores. However, the promising results of grinding additives indicate that it may be possible to selectively remove a substantial part of the phosphorus from the iron concentrates. Further research on this aspect of the problem is warranted.

Flotation of Reduced Ore: Based on the fact that reduction roasting of the red ores proved to minimize slime generation and iron losses during very fine grinding, tests were made to study the flotation response of reduced red ores. The roasted charges, after grinding in the rod mill to essentially minus 38 microns, were directly beneficiated without prior desliming by anionic flotation of the siliceous gangue minerals. Generally, the iron concentrates produced from the reduced charges were of intermediate grade of 55-58% Fe and gave relatively high iron recoveries of about 88% for the Big Seam samples, and 68-75% for the Ferruginous Sandstone samples. The improved iron recoveries from the reduced ores over those reported for non-reduced samples may be attributed to changes in the physical and chemical behavior of the slime-forming minerals during the roasting step. However, flotation of reduced ore does not appear to have any particular merit, in as much as the quality of the iron concentrates obtained is not better than that obtained by treating the raw ore.

Magnetic Separation of Reduced Ore

The magnetic separation of reduced low-grade Birmingham red iron ores to recover high grade iron oxide concentrates has been extensively studied [1,2,8,13,14]. However, because most of the earlier studies were conducted on feeds ground to a size range of 150 to 75 microns, many researchers concluded that recovery of magnetic concentrates assaying substantially more than 56% Fe was virtually impossible even if the ores were ground finer than 50 microns [13,14,15].

Therefore, in the current investigation, magnetic separation tests were made on extremely fine ground charges of the reduced Big Seam and Ferruginous Sandstone ores. The tests were made using coal, lignite ore and natural gas as the reducing agents. The reduced charges were finely ground to substantially minus 38 microns and separated using the Davis Tube tester. The magnetic concentrates recovered showed a maximum grade of 61% Fe with attendant iron recoveries of 85-97%. The

Table 6. Effect of Grinding Additives on Flotation Characteristics of Big Seam Sample BS-2

		Iron Concentrate			Minus 11 μm Slime	
			Fe		Fe	
Additive	kg/ton	Weight, %	Assay, %	Distribution, %	Assay, %	Distribution, %
Copper Sulphate ($CuSO_4$)	0.25	43.5	58.8	67.7	54.3	5.4
	1.0	38.7	60.8	60.0	51.5	2.4
Sodium Oleate (SO)	0.25	35.4	59.9	55.2	57.0	8.4
	1.0	34.3	60.6	52.9	58.2	13.0
Sodium Di-octyl sulfosuccinate (Aerosol OT)	0.25	39.0	60.1	58.7	56.0	8.8
	1.0	39.0	59.4	60.4	53.6	6.1
Cyanamid 899R (Petroleum Sulfonate)	0.25	34.9	60.5	49.6	56.6	15.0
	1.0	33.0	60.4	45.0	57.6	28.8
Poly 2-Acrilamido 2-methylpropane Sulfonic Acid (PAMS)	0.25	34.4	59.1	54.7	45.8	21.2
	1.0	35.0	61.2	55.6	42.2	17.6
PA-18	0.25	43.9	58.0	67.8	52.6	3.4
	1.0	37.6	58.5	58.5	53.8	10.5
NaOH Only	1.0	39.0	57.9	62.5	50.6	4.0

Test Conditions: 1 kg charges of minus 2 mm ore, rod mill ground for 10 minutes with the quantity of reagent specified and 1kg/ton NaOH, deslimed at 11μ m, then floated at pH 11.2 using 0.3 kg/ton $CaCl_2$, 1.3 kg/ton pearl starch, 2 kg/ton Pamak 4, and 0.01 kg/ton pine oil.

Table 7. Phosphorus Distribution in Flotation Products of Sample BS-2

				Assay, %		Distribution, %	
Additive	kg/ton	Product	Weight, %	Fe	P	Fe	P
NH_4F	1.0	Concentrate	39.1	60.8	0.10	63.1	12.6
		Middling	8.8	43.5	0.57	10.1	16.2
		Tailing	46.0	15.4	0.46	18.7	68.1
		Flotation Feed	93.9	37.0	0.32	91.9	96.9
		-11 μm Slime	6.1	50.5	0.16	8.1	3.1
		Head	100.0	37.8	0.31	100.0	100.0
XFS-4272	1.0	Concentrate	26.7	60.2	0.13	42.7	10.5
		Middling	12.4	45.5	0.45	15.0	16.9
		Tailing	42.2	16.8	0.43	18.8	55.0
		Flotation Feed	81.3	35.4	0.34	76.5	82.4
		-11 μm Slime	18.7	47.4	0.31	23.5	17.6
		Head	100.0	37.6	0.33	100.0	100.0
NaOH (Reference Test)	1.0	Concentrate	39.9	58.3	0.18	62.5	21.4
		Middling	10.3	44.7	0.50	12.4	15.4
		Tailing	45.8	15.7	0.45	19.3	61.6
		Flotation Feed	96.0	36.5	0.34	94.2	98.4
		-11 μm Slime	4.0	53.2	0.13	5.8	1.6
		Head	100.0	37.2	0.33	100.0	100.0

Test Conditions: 1 kg charges of minus 2 mm ore, rod mill ground for 10 minutes with the quantity of reagent specified and 1kg/ton NaOH, deslimed at 11μ m, then floated at pH 11.2 using 0.3 kg/ton $CaCl_2$, 1.3 kg/ton pearl starch, 2 kg/ton Pamak 4, and 0.01 kg/ton pine oil.

results of typical tests conducted on various ores are given in table 8. The data demonstrated that, by fine grinding, it is possible to increase the limit of the grade of magnetic concentrate from 56% Fe to as much as 61% Fe.

Other test results on sample BS-2, given in table 9, indicate that grinding to about 92% minus 38 microns was optimum. Further grinding to 90% minus 25 microns incrementally increased grade of iron concentrate but the iron recovery declined. Inability to recover high grade concentrates from the minus 25 micron material confirmed our earlier conclusion that the intimate association and fine dessimination of micron-sized gangue minerals in the iron oxide particles limited the maximum grade of concentrates that could be recovered. The limit appears to be about 61% Fe regardless of the fineness of grind. In table 9, the phosphorus analysis of the magnetic concentrates produced of 0.26-0.4% P, which are much higher than the 0.1-0.2% P reported for the flotation concentrates in table 7.

The poor phosphorus rejection of 28-40% in the reductive roasting and magnetic separation tests, compared to 80-90% by flotation, may be attributed to undesirable high temperature phase transformations occurring during reduction that result in chemical bonding of some of the phosphorus bearing minerals with the iron oxide minerals. This conclusion is supported by SEM and EDS examinations of concentrates from magnetic flocculation and magnetic separation tests.

Magnetic Flocculation of Reduced Ores

Generally, the iron oxides produced by reduction roasting are easily magnetized in weak magnetic fields and retain much of the magnetism for long periods of time. This property was used to selectively flocculate the magnetic iron particles while leaving the non-magnetic gangue minerals in suspension [16,17]. Table 10 shows the results of magnetic flocculation tests made on the reduced ores BS-2, FS-1 and FS-3.

Compared to the Davis Tube magnetic separation results reported earlier, the magnetically flocculated concentrates were lower grade rougher products. Mild agitation of the flocculated material did not improve its grade, indicating that the nonmagnetic impurities were strongly entrapped in the flocs. However, vigorous agitation of the flocculated rougher concentrate in a hydrocyclone produced clean concentrates analyzing 60% Fe with iron recovery of 65-75%. As in the case of magnetic separation, the magnetic flocculation concentrates were high in phosphorus (table 10).

SUMMARY AND CONCLUSIONS

The feasibility of applying fine grinding and fine particle processing technology to the low-grade Birmingham red iron ores, to recover high grade iron ore concentrates was investigated. The following conclusions were reached:

(1) In general, indiscriminate fine grinding of raw (unroasted) red ore generated excessive amounts of iron-rich slimes which interfered with subsequent separation steps, particularly flotation. Removal of these slimes improved separation but resulted in high iron losses due to lack of technology for treating such fine sizes. Thus, any improvement

Table 8. Bituminous Coal Reduction and Magnetic Separation of Red Iron Ores

Type of Ore	Product	Weight, %	Fe Analysis, %	Fe Distribution, %
BS-1	Magnetic Concentrate	71.5	57.5	98.0
	Nonmagnetic Tailing	28.5	3.0	2.0
	Feed	100.0	42.7	100.0
BS-2	Magnetic Concentrate	67.4	60.9	97.5
	Nonmagnetic Tailing	32.6	3.3	2.5
	Feed	100.0	42.2	100.0
FS-1	Magnetic Concentrate	52.6	61.0	84.6
	Nonmagnetic Tailing	47.3	12.4	15.4
	Feed	100.0	38.0	100.0
FS-2	Magnetic Concentrate	59.3	59.7	97.2
	Nonmagnetic Tailing	40.7	2.2	2.8
	Feed	100.0	34.7	100.0
FS-3	Magnetic Concentrate	36.8	59.7	89.6
	Nonmagnetic Tailing	63.2	4.4	10.4
	Feed	100.0	26.7	100.0

Test Conditions : 1 kg charge of minus 2 mm ore, 1 hour reduction at 650°C with 100 grams of coal, ground 45 minutes in the rod mill at 50% solids to 95% minus 38 μm, Davis Tube separation.

Table 9. Effect of Grinding on Magnetic Separation of Reduced Sample BS-2

Size of Grind, μm	Product	Weight, %	Assay, % Fe	Assay, % P	Distribution, % Fe	Distribution, % P
80% minus 75	Magnetic Concentrate	67.9	56.6	0.40	90.1	
	Magnetic Middling	5.5	54.8		7.1	
	Nonmagnetic Tailing	26.6	4.5		2.8	
	Feed	100.0	42.6		100.0	
82% minus 38	Magnetic Concentrate	68.3	59.0	0.34	92.8	68.4
	Magnetic Middling	3.0	58.4	0.39	4.1	3.4
	Nonmagnetic Tailing	28.7	3.2	0.37	3.1	28.2
	Feed	100.0	43.0	0.34	100.0	100.0
92% minus 38	Magnetic Concentrate	55.9	61.1*	0.26	79.9	44.0
	Magnetic Middling	11.4	60.2	0.45	16.0	15.6
	Nonmagnetic Tailing	32.7	5.3	0.41	4.1	40.4
	Feed	100.0	42.7	0.33	100.0	100.0
90% minus 25	Magnetic Concentrate	44.4	61.5	0.26	63.6	
	Magnetic Middling	21.2	59.6		29.3	
	Nonmagnetic Tailing	34.4	8.9		7.1	
	Feed	100.0	42.9		100.0	

Test Conditions ; 1 kg charge of minus 2 mm ore, 1 hour reduction at 650°C with 100 grams of coal, ground for different times in the rod mill at 50% solids, Davis Tube separation.

*Composites of magnetic concentrate and magnetic middling = 60.9% Fe.

Table 10. Magnetic Flocculation of Red Iron Ores

Ore and Products		Weight, %	Assay, % Fe	Assay, % P	Distribution, % Fe	Distribution, % P
BS-2	Magnetic Sediment	72.7	55.9	0.04	97.2	85.6
	Nonmagnetic Slime	27.3	4.3	0.18	2.8	14.4
	Head	100.0	41.8	0.34	100.0	100.0
FS-1 *	Magnetic Sediment	67.2	53.4		96.7	
	Nonmagnetic Slime	32.8	3.7		3.3	
	Head	100.0	37.1		100.0	
FS-2 **	Magnetic Sediment	69.9	50.8		97.8	
	Nonmagnetic Slime	30.1	2.6		2.2	
	Head	100.0	36.3		100.0	

* Average of four experiments
** Average of ten experiments

in liberation gained by fine grinding is generally offset to some extent by slime interference and low iron recovery.

(2) Anionic flotation of the gangue minerals from the deslimed sands and/or size-fractionated ore charges gave concentrates assaying 60% Fe for the Big Seam samples and 53-56% Fe for the Ferruginous Sandstone samples.

(3) Flotation of deslimed plus 11 micron sand fractions of the Big Seam ore, after selective mild grinding with additives to minimize slime formation, yielded concentrates assaying 60-61% Fe and 0.1% P with iron recoveries of 60-63%. In this process a phosphorus rejection of about 71% was achieved.

(4) Roasting with natural gas, coal, or lignite as reductants, followed by ultrafine grinding to 92% minus 38 microns and wet magnetic separation, yielded concentrates analyzing about 60-61% Fe with iron recoveries of 75-93%. However, the magnetic products of Big Seam ore were high in phosphorus (0.2 - 0.3% P).

(5) Ultrafine grinding of the Big Seam ore to 90% minus 25 microns failed to yield much better grade of iron concentrates or to improve iron recoveries. The inherent nonferrous mineral impurities in the iron oxide minerals limited the maximum grade of concentrates that could be recovered to about 61% Fe regardless of the extent of fine grinding. Compared to flotation of unroasted ore, roasting and magnetic separation was an inferior processing approach.

ACKNOWLEDGMENTS

This investigation was funded by grants from the Office of Surface Mining and the Bureau of Mines, U.S. Department of the Interior, under contract numbers G5195002, G5105056 and G1105056. The authors express their appreciation to those agencies for their support of the study. The authors express their appreciation and gratitude to Professor Carl Rampacek, former Director of the Mineral Resources Institute (MRI) for his continuous encouragement, unfailing advice and assistance in the preparation of this report. The cooperation and help provided by other MRI staff are greatly acknowledged.

REFERENCES

1. W.B. Phillips, Trans. AIME, 25, (1895), 23 pp.
2. J.R. Thoenen, (1953) pp. 64-71.
3. T.A. Simpson, et al., Assessment of Iron Ore Availability in Alabama and the Southeastern Appalachian Region, Final Report, NSF Grant AER 77-16114, (University of Alabama), (1978) pp. 102.
4. J.W. Willar and G.A. Dawe, Min. Cong. J., (1975) pp. 40-48.
5. H.S. Hanna and C. Rampacek, Mining Eng. (April 1982), pp. 395-403.
6. S.R.B. Cooke, U.S. BuMines Bull. 2391, (1936), pp. 104-109.
7. R.D. Hagni and M. Cooper, U.S. BuMines, open files, 1982.
8. J.B. Clemmer, et al., U.S. BuMines R.I. 3799, (1945), 42 pp.
9. R.E. Perry, et al., U.S. BuMines R.I., 6123, (1962), pp. 13.
10. R.S. Dean and C.W. David, U.S. BuMines Bull. 425, (1941), pp. 144-145.

11. H.S. Hanna and I.J. Anazia, 55th Colloid and Surface Science Symposium, Cleveland, Ohio, (1981).
12. H.S. Hanna and I.J. Anazia, MRI Technical Report Series T.R. #9, (University of Alabama), (1982), pp. 126.
13. O. Lee, B.W. Gandrud, and F.D. Devaney, U.S. BuMines Bull. 278, (1927) pp. 75.
14. W.H. Coghill and G.D. Coe, 1946, U.S. BuMines Bull. 464, (1946).
15. E.G. Davis and I.L. Feld, U.S. BuMines R.I. 7627, (1972), pp. 10
16. V. Hencl and J. Svoboda, 1981, Proceedings 13th Int'l Mineral Processing Congress, Warsaw, June 1979, (et al., Elsevier, N.Y., 1981), pp. 472-488.
17. W.E. Lamont, et al., U.S. BuMines R.I. 7728, (1973), pp. 15.

ORE AND COAL PROCESSING WITH THE TURBOCHARGER ELECTROSTATIC SEPARATOR

R. Ciccu, G. Alfano, P. Carbini, M. Ghiani, N. Passarini, R. Peretti, A. Zucca

Department af Mining and Minerals Engineering, CNR Centre
University of Cagliari, Cagliari, Italy
ENIChem-ANIC SpA, Milan, Italy

ABSTRACT

The features of the Turbocharger pilot plant set up at the Department's laboratories are illustrated. The separation results are presented for different operating conditions, such as nature and temperature of the target surface, rotor spinning velocity, feed rate, electrostatic field intensity, electrode geometry and position of selection splitters. The effect of particle size is also underlined. Expected improvements with apparatus development and technique refinement are also pointed out. The problems connected with the commercial application of the system are dealt with in the conclusions.

INTRODUCTION

Electric separation using contact electrification has long been recognized as a suitable technique for ore and coal beneficiation. However industrial applications are very few and restricted to those cases where dry methods are strongly recommended, e.g. salt separation |1|. Other fields of potential interest are phosphate upgrading |2| in arid regions where deposits are mainly located, and coal cleaning |3| to avoid using large amounts of water, especially for the finest sizes.

The reason for the lack of development lies mainly in the poor knowledge of the principles of static electrification of mineral particles and consequently in the insufficient understating of tribocharging phenomena which often appear as random and inconsistent |4|.

Contact electrification of mineral particles may be the result of

Advances in Fine Particles Processing
John Hanna and Yosry A. Attia, Editors

different concomitant physical or physico-chemical processes involving electron and/or ion exchanges between the solid phases in contact although the prevailing mechanism under carefully controlled environment conditions seems to be related to electron transfer according to the energy band model of semiconducting solids |5|.

It has been found that mutual particle charging of two different minerals is generally always consistent, whereas in the case of contact against a third surface satisfactory charging is only achieved if the latter's electrophysical characteristics are intermediate between those of the two components to be separated.

However the real situation is generally more complex than that idealized by the theory since minerals are not simple semiconductors nor do they possess a defined lattice as is the case of coals and oolithic phosphates, for instance.

Therefore further research is needed in order to elucidate those aspects not yet sufficiently investigated and construct a suitable model of general validity for tribocharging phenomena.

As for equipment, the latest step in the development of this technology is the Turbocharger separator which was specifically conceived with a view to industrial applications |4|.

Its distinct feature is a relatively high particle contact force so as to obtain strong and consistent charging, in line with the findings of basic research which indicate that charge per particle is almost proportional to impact or sliding energy until saturation is reached. The technical solutions adopted in the design of the charging device have been suggested by experience gained in the development of the previous air stream models, in order to overcome the drawbacks arising with economically burdensome pneumatic circuits.

The new separator is characterized by the following features and performance:

- short residence time inside the charging device, thus increasing throughput capacity, by allowing impingement rather than sliding contacts;
- intensive frictional actions in a turbulent medium where inter-particle rubbing is enhanced;
- limited air flow rate through the device and into the separation chamber in order to minimize fine particle draft, facilitating at the same time the recovery of solids fron the main air stream;
- interchangeability of the charging surface for the specific separation problem;
- possibility of target surface heating at the temperature corresponding to optimum selectivity.

EXPERIMENTAL

The pilot plant installed at the Department's laboratories is depicted in the diagram of figure 1.

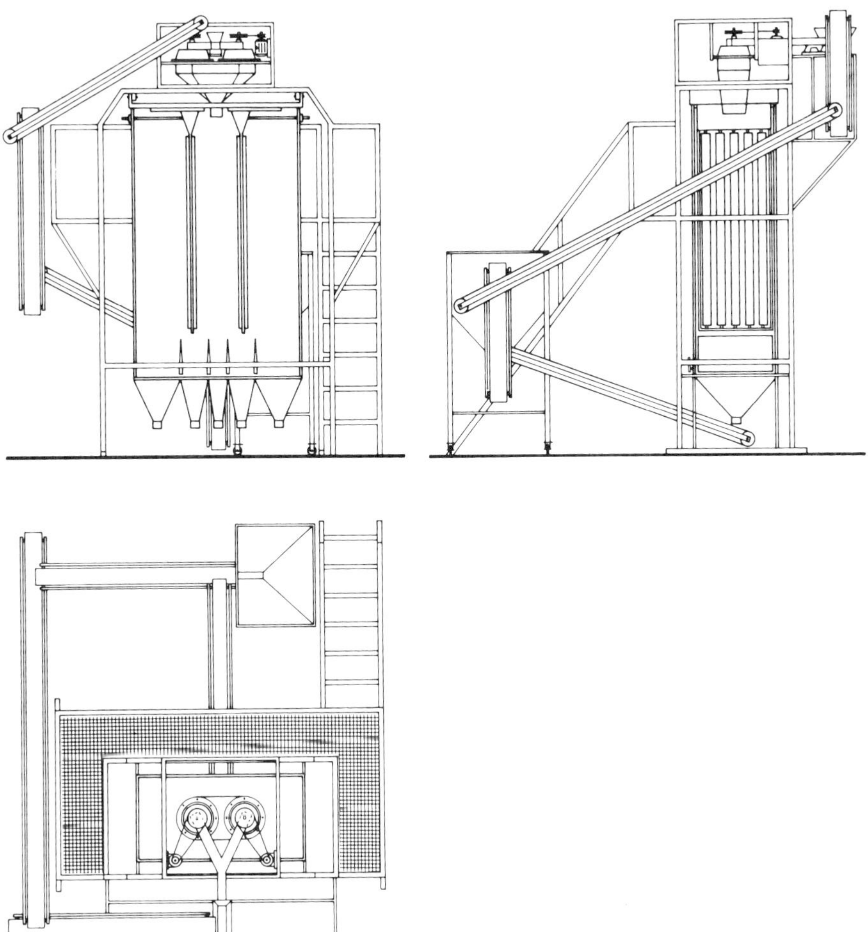

Figure 1. Front, side and top view of the Turbocharger pilot plant with twin-rotor charging device and self-cleaning rotating pipe electrodes.

Briefly, the modular charging unit consists of a rotor spinning around a vertical axle inside a frustum encasement placed at the top of a funnel-shaped receptacle provided with flux straightening vanes.

Mineral particles, ground to suitable sizes, are fed through a circular opening around the rotor axle, accelerated by the centrifugal force along the radial blades and thrown against the surrounding target surface; they are then drawn by the air vortex and collide with the incoming flux undergoing repeated impingements.

Charged material falls by gravity into the separation chamber through the discharge slot and is separated by means of splitters at the common edges of collecting bins.

The pilot plant is completed by the auxiliaries for feed hoisting, middlings recycling and final product recovery, consisting of a series of movable conveyor belts.

In order to assess the performance of the separator as a function of the different geometric settings and operating variables, systematic tests have been carried out resorting to a mixture of two pure components (coal and limestone) hand-sorted from the run-of-mine.

The results allowed to determine the influence of the single variables providing useful information for optimum setting in the case of industrial ore beneficiation.

In order to reproduce the conditions of continous run, a sizeable sample of 50:50 test material was loaded into the head hopper and fed into the plant without interruption. After each variable adjustment according to the experimental plan, time was allowed for stabilization; samples were then taken from each of the current rougher products and from the circulating load. Each sample was weighed and analyzed for ash and mineral matter, thus obtaining the relevant data for separation efficiency calculations.

The following parameters have been explored over a range covering a broad sphere of interests:

- H.T. voltage from 20 to 40 kV;
- electrode spacing from 25 to 35 cm;
- target surface temperature from 20 °C to 160 °C;
- rotor speed from 120 to 140 rad/s;
- splitter position adjustment by 4 cm steps;
- feed rate from 0.8 up to 4.5 kg/min, corresponding to 240 and 1,350 kg/h per metre of chamber width.

Each parameter was varied independently keeping the others at their average value.

Furthermore the products of a prolonged test have been screened and each class analyzed for studying the efficiency as a function of particle size. The results are illustrated and discussed below.

RESULTS AND DISCUSSION

High Tension Voltage

The effect of voltage is represented by the curves of Figure 2. As anticipated, separation selectivity increases with voltage while at the same time the circulating load progressively diminishes. In fact particle paths become more consistent with increasing field intensity, electrostatic and mass-related forces being better balanced for improved selectivity. On the contrary, draft, inertial and gravity forces do prevail when field intensity is too weak thus increasing the probability of erroneous or insufficient separation. Peak point does not yet appear, suggesting that higher voltage might in this case give better results. For a given feed material optimum voltage varies with grinding size since particle charge and mass-related forces are both strongly affected.

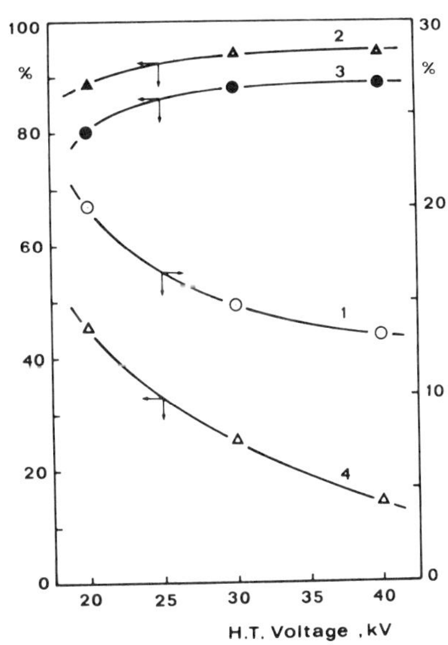

Fig. 2 Separation results versus HT voltage.
1 - Mineral Matter content
2 - Fuel Recovery
3 - Ash Removal
4 - Circulating Load

Electrode Spacing

The influence of this parameter only partially corresponds to that produced by voltage, although both have a direct effect on the static field. In fact two aspects often ignored deserve careful consideration. The first is related to the field intensity which is not constant in the inner space between the two opposite sets of electrodes, reaching a maximum in the vicinity of the active electrode and a minimum near the grounded one and following a close-to-exponential function for the intermediate points. Therefore reducing electrode spacing should result in a marked improvement.

However the favourable effect of increased field strength can be offset by the second aspect. Actually, as electrode spacing is diminished, particle crowding is enhanced thus favouring electrostatic agglomeration phenomena wich are of course detrimental for separation selectivity, unless feed rate is reduced. The finer the grinding size,the more marked the effect.

The combined result of both counteracting effects of electrode spacing variation is clearly reflected by the curves of Figure 3 showing the presence of an optimum point corresponding to the best compromise between the advantage

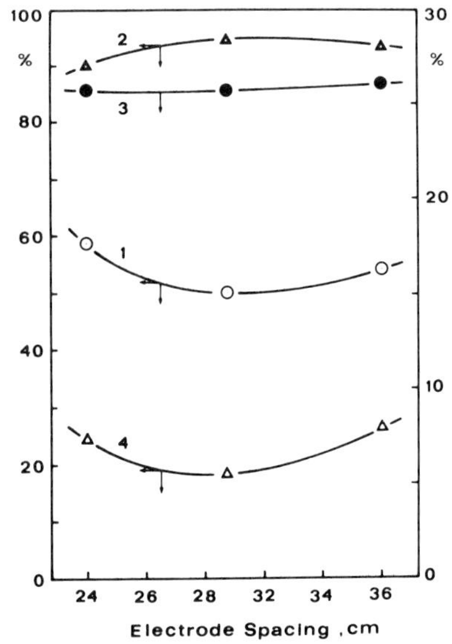

Fig.3 Separation results versus electrode spacing
1 - Mineral Matter content
2 - Fuel Recovery
3 - Ash Removal
4 - Circulating Load

of a stronger field and the disadvantage of particle crowding. This optimum point shifts, depending on grinding size and feed rate: electrodes should be spaced farther apart for finer sizes and increasing throughput, suitably adjusting either the voltage or the chamber height or both, in order to achieve sufficient particle deflection and hence satisfactory separation results.

A three-electrode combination will next be experimented in an attempt to improve separation efficiency.

Rotor speed

Although the explored range was very narrow, the effect of increasing rotor speed was a marked improvement in fuel recovery at constant splitter setting while the variation of ash content was negligible. It appears that saturation charge, i.e. the maximum charge isolated particles can hold in air, is achieved first for limestone whereas higher energy is required for coal. The outcome is shown by the curves of Figure 4.

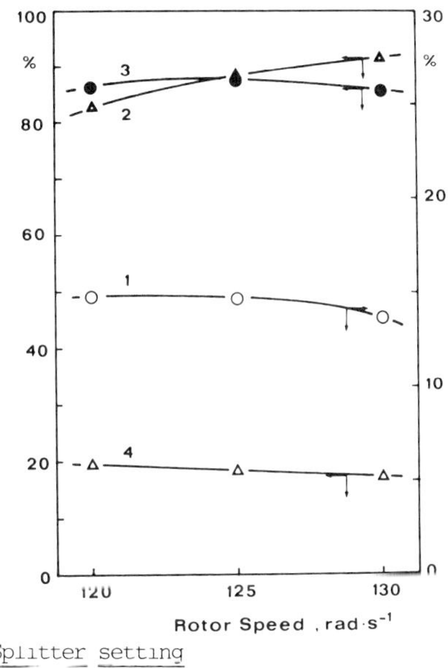

Fig. 4 Effect of rotor speed on separation results.
1 - Mineral Matter content
2 - Fuel Recovery
3 - Ash Removal
4 - Circulating Load

Splitter setting

The study of this parameter yields information on the dispersion of the end points of particle paths thus allowing to assess the extent of charging consistency. Two series of tests have been performed: in the first, the distance of the inner splitters from the central symmetry plane was increased thus increasing the circulating load; in the second, the inner splitters were both moved either to the right or the left, maintaining the same relative distance, thus allowing to explore the sensitivity of separation results to selection system setting. Outer splitters were always kept in a fixed position.

The curves of Figure 5 show that circulating load increases resulting in an improvement of product quality and fuel recovery as central splitters are symmetrically opened farther apart.

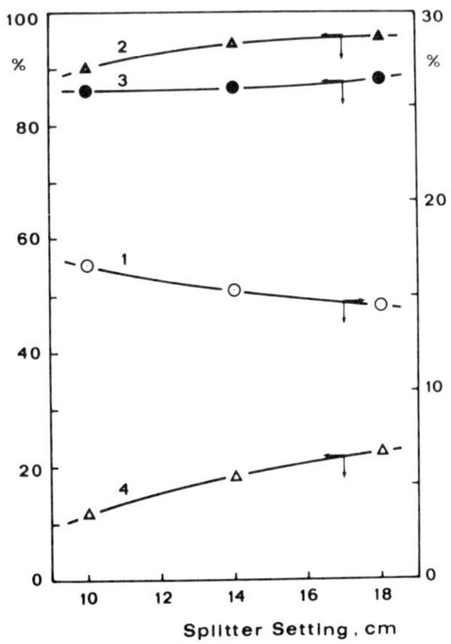

Fig. 5 Effect of inner splitters setting (diverging)
1 - Mineral Matter content
2 - Fuel Recovery
3 - Ash Removal
4 - Circulating Load

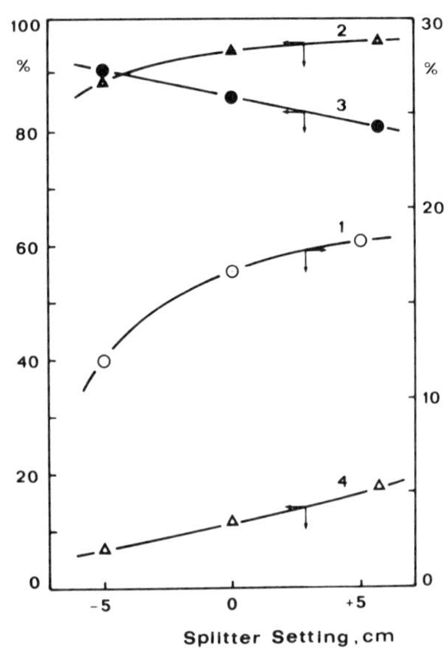

Fig. 6 Effect of inner splitters setting (parallel)
1 - Mineral Matter content
2 - Fuel Recovery
3 - Ash Removal
4 - Circulating Load

Conversely, if both splitters are shifted parallel to each other, better product quality is progressively obtained as the splitters are moved towards the electrode where coal is collected. Fuel recovery deteriorates and more mineral matter is rejected in the waste. Circulating load also slightly decreases (Figure 6).

Target surface temperature

The optimum temperature for coal/limestone separation seems to be around 80 °C. In fact peak fuel recovery and ash removal and corresponding minimum ash content and percent circulating load have been observed when target surface is moderately heated, as shown by Figure 7.

Since the effect of heating varies greatly for the different mineral substances, conclusions only refer to the specific case.

However variations are not very important, separation being quite satisfactory even at room temperatures. Judging from the curves it seems that heating is slightly more effective for limestone than for coal particles.

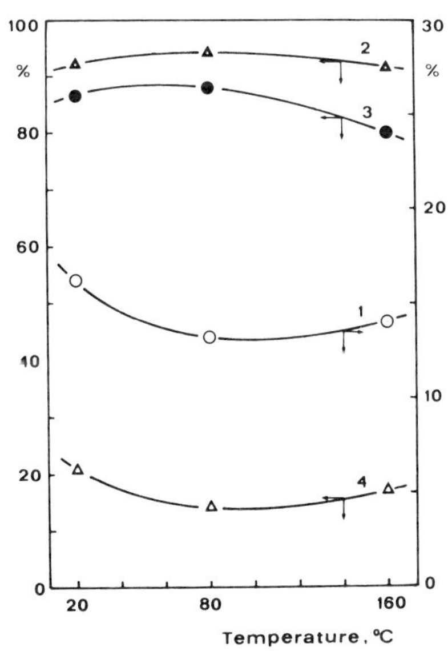

Fig. 7 Coal/Limestone separation results as a function of target surface heating temperature.
1 - Mineral Matter content
2 - Fuel Recovery
3 - Ash Removal
4 - Circulating Load

Feed rate

Considerable attention should be given to feed rate because treatment capacity is one of the most important conditions for industrial applicability and processing cost per tonne depends greatly thereon.

Generally, electric separation results with the classic drum-type separators tend to deteriorate as throughput increases since conductance charging (either pure static or electrodynamic) is poorer and less selective.

The reverse is the situation with the Turbocharger, due to the fact that charging is enhanced at higher collision probability as happens at increasing feed rate, until clogging occurs.

As the curves of Figure 8 clearly show, the ash content of the cleaned product diminishes although with decreasing gradient, while both fuel recovery and mineral matter removal seem little affected. Percent circulating load tends to increase suggesting incipient agglomeration.

Close extrapolation indicates that feed rate could be increased well beyond the extreme point without impairing efficiency. Besides, this seems corroborated by the fact that no drawbacks like irregular material flow through the separation chamber were observed during the high rate tests.

Since circulating load is about 30 %, a throughput capacity of slightly less than 2 t/h per metre of chamber width has been demonstrated as feasible, and might be further increased resorting to taller electrodes.

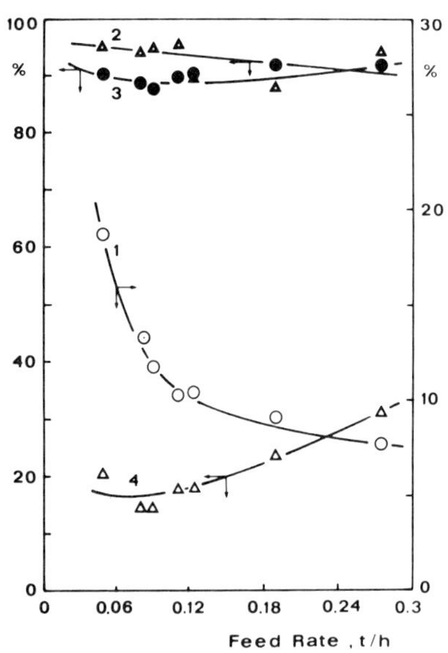

Fig. 8 Separation results at different feed rates

1 - Mineral Matter content

2 - Fuel Recovery

3 - Ash Removal

4 - Circulating Load

Grinding size

Separation efficiency for the various size classes has been studied by screening the end products of a two-stage separation and analyzing the individual size fractions. Results are shown in Figure 9.

As can be seen, efficiency deteriorates somewhat, though still remaining acceptable, for the finest fractions, as commonly found in all dry processes. Further research is being conducted in order to improve both the product quality and fuel recovery, possibly by modifying the rotor design.

FURTHER ILLUSTRATIONS

The efficiency of the Turbocharger system has also been experimented on a number of run-of-mine coals and crude ores |6| obtaining very interesting results. The most significant are summarized in the following table.

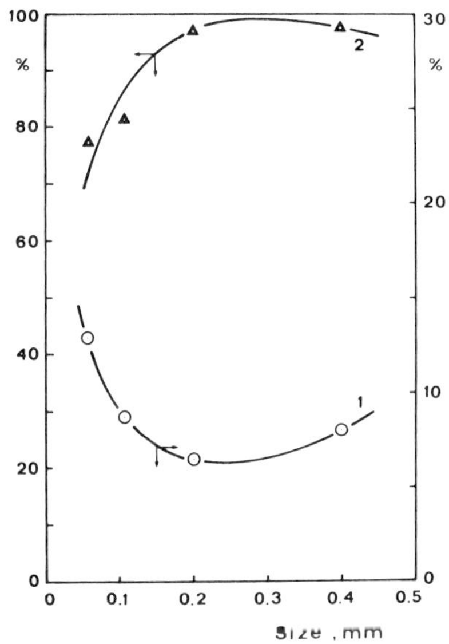

Fig. 9 Separation results of the Coal/Limestone mixture for each class of product screening.
1 - Mineral Matter content
2 - Fuel Recovery

Table 1. Results of electric separation with the Turbocharger separator. Two stage test flowsheet: Rougher followed by Cleaner and Scavenger

COALS	Feed Assay		Product Assay		Percent removed		Heat Rec.
	Ash %	Pyr.S %	Ash %	Pyr.S %	Ash	Pyrite	%
A	11.75	2.28	4.77	0.96	64.7	63.4	93.8
B	16.95	0.60	3.01	0.16	93.1	92.0	57.1
C (1)	6.27	0.15	2.18	0.06	79.5	73.6	66.7

(1) washed commercial coal

PHOSPHATES (*)	Feed Assay P_2O_5 %	Product Assay P_2O_5 %	Gangue removed %	P_2O_5 Rec. %
A	23.7	29.4	71.9	85.3
B	19.9	29.0	80.4	88.4
C	21.4	29.1	78.0	84.1

(*) After middlings recycling simulation

IRON ORE	Feed Assay Fe %	Feed Assay SiO_2 %	Product Assay Fe %	Product Assay SiO_2 %	Quartz removed %	Fe Recov. %
	69.53	0.59	69.82	0.17	76.0	84.0

BARITE ORE	Feed Assay $BaSO_4$ %	Product Assay $BaSO_4$ %	Gangue removed %	$BaSO_4$ Rec. %
	36.3	95.75	97.4	96.9

CONCLUSIONS

Despite the need for further technological improvement, the results of pilot-scale experiments fend further weight to the prospects of industrial application of the Turbocharger system, already envisaged in the earlier stages.

Separation selectivity looks quite satisfactory even relatively high feed rates and for the finest size classes on condition that tribocharging properties of the components to be separated are either naturally well differentiated or can be modified through suitable external actions.

Furthermore power consumption is relatively moderate (even in those cases where heat treatment is required) since it can be limited to target surface heating. The unit cost of processing is acceptably low also on account of limited wear and easy operation monitoring.

REFERENCES

1. R. Dutta, Electrostatic Separation of Potassium-Magnesium Salts from Crude Mineral Salt, Industrial Minerals, N. 206, pp. 71-73, November 1978.

2. G. Alfano, P. Carbini, R. Ciccu, M. Ghiani, R. Peretti and A. Zucca, La separation tribo-electrique des minerais phosphates, Industrie Minerale-Mines et Carrieres, Jan-Feb 1989.

3. M. Carta, C. Del Fa', R. Ciccu. L. Curreli and M. Agus, Technical and Economic Problems connected with the Dry Cleaning of Raw Coal and in particular with Pyrite Removal by means of Electric Separation, Proc. VIIth Int. Coal Prep. Congr., pp. 33, Sydney, Australia, 1976.

4. G. Alfano, P. Carbini, R. Ciccu, M. Ghiani, R. Peretti and A. Zucca, Progress in Triboelectric Separation of Minerals. Proc.XVIth Int. Min. Proc. Congr., part A, pp. 833-844, Stockholm, Sweden, 1988.

5. M. Carta, R. Ciccu, C. Del Fa', G. Ferrara, M. Ghiani and P. Massacci, Improvement in Electric Separation and Flotation by Modification of Energy Levels in Surface Layers, Proc. Xth Int. Min. Proc. Congr., pp. 28, London, U.K., 1973.

6. G. Alfano, M. Carta, R. Ciccu and C. Del Fa', Electric Separation of Finely Divided Particles in Gaseous Stream or in Vacuo, Conf. Rec. IEEE Ann. Meet. pp. 955-965, Chicago, USA, 1984.

BIOMETALLURGY FOR MANGANESE AND COPPER ORES.

Authors: L. Toro, C. Abbruzzese (*), F. Vegliò and B. Paponetti
Dipartimento di Chimica, Ingegneria chimica e Materiali, University of L'Aquila, 67040 Monteluco di Roio, Italy.
(*) Istituto per il Trattamento dei Minerali (CNR), via Bolognola 7, 00138 Rome, Italy.

ABSTRACT

The paper discusses the biotreatment of Italian ores of potential commercial interest, using oxidizing and reducing microorganisms in mixed and pure cultures. The study focused on copper and manganese ores.

Bioleaching tests were carried out in Erlenmeyer flasks, a L.K.B. modified microfermenter and a newly designed reactor. In particular the development during the time of the chalcopyrite bioleaching of catalyzed by Thiobacillus ferrooxidans is proposed. The main biochemical and chemical reactions concerned in the production of soluble and insoluble species were investigated during the culture growth particularly the time when the principal reactions begin and their duration.

Experimental runs were also performed using urea in place of ammonium sulphate.

Manganese bioleaching was carried out utilizing various microrganisms coming from the natural habitat; interesting results as regards rates and yields were obtained by mixed cultures. Tests of some strains isolated from these cultures were carried out.

INTRODUCTION

The study of the bioleaching processes is usually performed using different reactors; stirred tanks for continuous bioleaching CuFeS2 concentrates by Grudev et al. were used [1]. On the other hand Pachuka tanks reactors are commonly employed for extracting Ni, Co, Cu ed Au [2,3] as well as for microbic desulphurization of coal [4,5].

In this work a new reactor, employed in preliminary tests of copper bioleaching, is presented.

Copper bioleaching.

Thiobacillus ferrooxidans is capable of promoting the oxidation of various metal sulphides in ores; in this process metal dissolution depends on the "growth" of bacterial cultures [6].

The solubilization of copper and iron is linked to the oxidation of sulphide first to sulphur and then to sulphate [7,8]. In particular the ferric iron in solution is not very stable in the redox potential/pH operative conditions and trends to reprecipitate. The reprecipitation of ferric iron, can reduce the surface open to the attack of the growing number of microorganisms [9].

Ammonium salts are the routine nitrogen source in metal bioleaching [10]. The use of urea instead of ammonia in such processes offers clear theoretical advantages since it contributes both to the nitrogen and carbon supply for microbial cells [11]; in particular this substrate can be of interest in a context where carbon dioxide, which is easily lost at low pH values, behaves as a limiting nutrient for autotrophic microorganisms [12].

This work proposes a development scheme of calchopyrite bioleaching of a italian concentrate in unlimiting CO2 and O2 concentrations.

Manganese Bioleaching.

Manganese deposits in northern Latium (Italy), worked since the beginning of the century, represent an interesting land-based source of manganese.

From the processing aspect, pyrometallurgy cannnot readily employed owing to the large amount of silica and alkaline earths in the ore. And, of course pyrometallurgy is costly [13].

On the other hand, the hydrometallurgical processes for extracting manganese from ores containing MnO2 make use of SO2 [14];furthermore certain processes using carbohydrates, formaldehyde or carbon monoxide as reductants in aqueous ammoniacal media have been reported [15]. These chemical reagents at high concentrations can become dangerous pollutants.

The recovery of manganese by bioleaching with different microorganisms can offer an interesting alternative process for the future [16,17].

The present research has been devoted to the comparison also of a biometallurgical process in lab-scale using limiting and unlimiting oxygen conditions for the solubilization of manganese from low grade manganese ores.

EXPERIMENTAL

Ores

Chalcopyrite. The 400 mesh (<38 micron) fractions of a concentrated mineral from Fenice Capanne (Tuscany, Italy) have been used [18].

Manganese dioxide. The manganese ore samples were obtained from the Casale Castiglione area (Province of Viterbo) where the deposit has been partially exploited by SAMIM [13].

Microorganisms

Copper bioleaching

A strain of *Thiobacillus ferrooxidans* has been isolated from surface lagoon waters near Cotilia (Rieti, Italy) in 9K medium iron-free and selected in presence of increasing amounts of chalcopyrite [19].

Manganese bioleaching

Several samples were collected from the surface of natural waters near L'Aquila (Italy). During the first phase they were used as mixed cultures; then after isolation in malt-agar they were reinoculated as pure cultures and some of these were identified from the taxonomic point of view [21].

Microrganisms Analysis

The isolation of microrganisms was performed in Petri plates with maltagar medium.

The taxonomic identification of the microrganisms employed was carried out by Api Systems.

Reactors

The reactors utilized in the work reported here were the New Brunswick Shaker G 25, the Microfermenter LKB 1601 and the Biomet Pilot Plant.

Shaker G25

This apparatus permits batch bioleaching with shaking at between 40 and 400 rpm, temperatures between 5 and 60 C above ambient and reaction volumes up to 400 ml. With this apparatus it is not possible to control evaporation of the reactive systems owing to the incubation temperature effect nor to run tests at constant pH and/or controlled redox potential.

L.K.B. 1601

The L.K.B. 1601 microfermenter is a well known stirred reactor capable to carry out batch tests in appropriate experimental conditions.

"Biomet" pilot plant

The "Biomet" pilot plant was specifically designed and built for continuous and batch bioleaching tests of different minerals (see figure 1).

The use of this plant forms the final stage of lab-scale experimentation in our laboratory.

The R1 reactor in which the mineral is dissolved by microorganisms can be considered the very heart of the plant. The mineral can be fed in a continuous or semicontinuos manner through a hole in the reactor hole. The feeder consists of a hopper with vibrator (T1).

Gas normally consists of ambient air, compressed as required. The oxygen and/or carbon dioxide content of the air can be increased as deisired by mixing these with air in mixer M1. Gas flows are measured and indicated by flow-meters. They are controlled individually by self-regulating diaphragm valves. Air (enriched or normal) is fed into the recycling circuit of the reactor.

Mineral salts are dissolved in the growth broth, stored in tank D5 and fed to the reactor (R1) feed pump (G4).

By maintaning a very high external recycle via centrifugal pump (G1) an adequate transfer of material is ensured without the help of mechanical stirrers. G1 is fed from the conical bottom of the reactor (R1) and delivers the liquid to an exchanger with a water-cooled spiral and heating coil (HT2).

Unit E1-HT2 keps the temperature constant about thirty degrees via a TIRC probe fixed inside the reactor.

Air is introduced into the recycled flow at the exchanger outlet. The inlet to the reactor is on the side, low down, just above the conical bottom. It takes the form of a length of pipe which terminates in an upturned elbow in a central position.

Reaction conditions are analysed and controlled by three electrodes which measure conductivity, pH and dissolved oxygen. It is considered necessary to measure conductivity to compensate for the inevitable loss of water by evaporation. When conductivity rises above a certain level, namely the solution becomes too concentrated, the on-off rheostat valve Cyl is switched on and distilled water is fed from tank SR1 to restore optimum conditions.

pH control is essential for the proper progress of the bioleaching reaction. There is a tendency for pH to fall. When it drops below the setpoint, the control operates the on-off rheostat valve pHyl and 2% NaOH is fed in from D2 for a short time to restore the optimum pH . Finer adjustement is possible with valve pHy1 by feeding 1:5 H2SO4 from tank SR3 with manual regulation of valve Rp16.

It is useful to meter dissolved oxygen to guarantee that the bacteria population has the amount of oxygen needed not only to prevent anoxia but also to ensure that the oxygen does not become a limiting substrate.

The apparatus can be operated at different reaction volumes by opening and closing various of the three overflows provided. In this way it is possible to have volumes of 20, 40 and 60 litres within the reactor which, added to the volume of about 30, 50 and 70 l.

The pulp overflows into a sump from which it is extracted by ejector PJ1 whose motive fluid is recirculating of neutralizer R2.

The neutralizer also receives reactor cooling water coming from the water-cooled spiral and NaOH for neutralization coming from an alcaline-tank . The latter is regulated manually by means of valve.

The discharge, by overflow terminates in a collection tank whence it is sent to the sewer via an immersion pump.

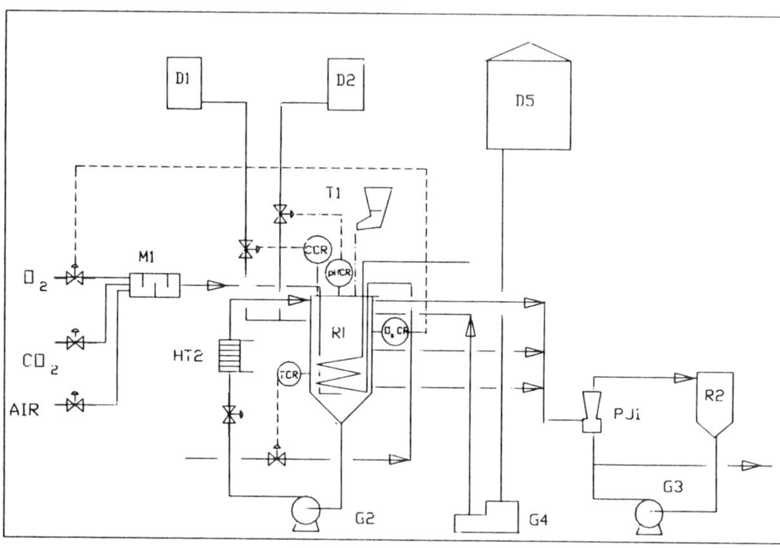

Figure 1: A scheme of the "Biomet" pilot plant with the principal parts of this apparatus (the explanations are defined in the text).

Batch tests

New Brunswich shaker.

300 ml Erlenmeyer flasks containing 100 ml of systems were inoculated 10% (v/v) of suspension. Incubation was performed at 35°C in a rotatory shaker at 250 rpm (revolutions per minute).

L.K.B. 1601 microferementer.

A volume 9 l of reactive system for manganese bioleaching was inoculated 1 l of microbial suspension in active growth. Incubation was performed at 35°C at 200 rpm (revolutions per minute) for seven days.

Biomet pilot plant

A volume 54 l of reactive system for copper bioleaching was inoculated 6 l of microbial suspension in active growth. Incubation was performed at 35°C for a weeck.

Chemical analyses

Liquors

Samples for analyses of the ions were collected after different times of incubation.

Ferrous iron in clear decantation overnatants after biological attack has been determined by titration with 0.1 N $KMnO_4$ [22].

Total iron copper and manganese in solution were determined by Atomic Absorption Spectrophotometry (Perkin Elmer 3030).

Sulphates in the overnatants has been measured by turbidimetry with $BaCl$ and UV-visible Spectrophotometry [23].The sucrose was assayed according to Barham D. & Trinder P., Analyst, 97, 142, (1972) [23].

Solid residues

The recovery of the solid residues at different times during the process was obtained with the Millipore filtration system fitted with membranes mod.HAWG 047 S2.

The microanalysis was carried out at different times with a scanning electron microscope (S.E.M.) for detecting the elemental composition on treated minerals. Jarosite standards were prepared according to Vaschetti [24].

For determination of the precipitates one gram samples of solid residues of chalcopyrite bioleaching at different times were treated with 5 N HCl for 30 minutes at room temperature. The proceedings of the concentrate of $CuFeS$ was calculated in two different ways both making the difference between the leaching residues with the 5 N HCl treated residues and the difference between the starting chalcopyrite with that biologically treated by the measure of copper solubilization.

Urease activity was assayed according to Schlegel and Kaltwasser [25].

RESULTS AND DISCUSSION

Chalcopyrite bioleaching.

The experimental results obtained in pilot plant are compared with tests carried in orbital shaker in limiting CO_2 and O_2 concentrations.

The results of batch tests in terms of copper solubilization with 10%, 20% and 30% of pulp concentrate of the chalcopyrite are presented. Significant extraction of copper can be attributed to the activity of the microorganisms as can be deduced from the figure 2 a. The copper in solution resulted increasing versus time according to the content of pulp in orbital shaker while different was the extractive behaviour in pilot plant as is shown in figure 2 b.

This apparatus resulted proper for processing contents of pulp up to 20%; on the other hand the solubilization rate maintained higher in comparison to that in orbital shaker (see figures 3a and 3b). Probably this different behaviour can be ascribed to the different oxygen availability.

The biochemical and chemical reactions involved in the production of soluble in insoluble species during a bioleaching of chalcopyrite process are indicated in table I.

Tab. I. The biochemical and chemical reactions involved in the production of soluble in insoluble species during a bioleaching of chalcopyrite process.

$$CuFeS_2 + 5/4\ O_2 + 5\ H^+ \xrightarrow{bacteria} Fe^{3+} + 2\ S + Cu^{2+} + 5/2\ H_2O \quad (1)$$

$$S + H_2O + 3/2\ O_2 \xrightarrow{bacteria} SO_4^{-2} + 2\ H^+ \quad (2)$$

$$Fe^{+3} + 3\ H_2O \longrightarrow Fe(OH)_3 + 3\ H^+ \quad (3)$$

$$Fe(OH)_3 + 4/3\ SO_4^{-2} + Fe^{+3} + H_2O + 2/3\ K^+ \Longrightarrow 2/3\ K[Fe_3(SO_4)_2(OH)_6] + H^+ \quad (4)$$

$$2\ Fe^{+3} + 2\ H_2O \longrightarrow Fe_2(OH)_2^{+4} + 2\ H^+ \quad (5)$$

$$Fe^{+3} + SO_4^{-2} \longrightarrow FeSO_4^+ \quad (6)$$

$$FeSO_4^+ + SO_4^{-2} \longrightarrow Fe(SO_4)_2^- \quad (7)$$

$$SO_4^{-2} + H^+ \longrightarrow HSO_4^- \quad (8)$$

$$2\ Fe(OH)_3 + FeSO_4^+ + K^+ + SO_4^{-2} \longrightarrow KFe_3(SO_4)_2(OH)_6 \quad (4')$$

$$CuFeS_2 + 2\ Fe_2(SO_4)_3 \longrightarrow CuSO_4 + 5\ FeSO_4 + 2\ S \quad (9)$$

Tab.II - Experimental results in batch tests with 10% of pulp content.

(days)	0	1	2	3	4	5	7	9	10
copper (g/l)	.76	2.08	2.92	3.56	4.0	4.72	5.75	6.76	7.76
iron III)(g/l)	.39	.07	.42	.5	1.04	1.38	2.06	3.03	3.66
pH	2.29	3.22	2.8	2.26	2.08	1.98	1.85	1.75	1.63
sulphate(g/l)	1.8	1.8	1.99	5.97	8.7	10.18	11.37	12.56	16.27
RW(g)	9.65	9.74	9.57	9.44	9.28	9.2	9.02	8.82	8.61
SRW(g)	.03	.49	.63	.64	.76	.79	.91	.98	.95
FeRW(g)	.0	.13	.18	.20	.23	.28	.29	.29	.31
KRW(g)	.0	.0	.0	.02	.03	.04	.05	.06	.07

where: RW is weight (g) of the solid residues, SRW is weight (g) of the 5 N HCl soluble in the solid residues, FeSW is weight (g) of the 5 N HCl soluble iron in the solid residues and KSW is weight (g) of the 5 N HCl soluble potassium in the solid residues, respectively.

The analytical results of some "marker species" see table II suggest that the whole process in orbital shaker and in the presence of 10% of concentrate can be divided into three main phases, as follows [20]: 1st phase (0 - 2 days), 2nd phase (2 - 5 days) and 3rd phase (5 - 10 days).

```
  (1), (3)
*----------*
             (1), (2), (3)
           *----------------*
                                 (1),(2),(3),(4),(5)
                            *------------------------*


*----------*----------------*------------------------*
0          2                5         (days)         10
```

Tests were carried out with urea as the nitrogen source in the medium using a Thiobacillus ferrooxidans subcultured in the presence of this alternative nitrogen source, as is shown in figure 4 [11]. The behaviour of the extractive process resulted 25% lower in comparison to that obtained in the presence of ammonium.

Manganese dioxide bioleaching.

Manganese solubilization catalyzed by mixed cultures in the presence of different contents of pulp was accomplished by a reprecipitation in the form of manganese hydroxide after four days of biotreatment [14]. The presence of inorganic material maked unfeasible the common techniques for biomass determination such as the direct microscopic count, the viable plate counts, automated cell counting and sorting procedures dry cell weight and turbidity.

The eterotrophic nature of these microrganisms offered us an indirect method of detecting "the growth" of biomass during the process and of relating the sucrose consumption with the extractive activity as is shown in figures 6.

The isolation of the mixed cultures indicated the presence of yeasts, fungi bacteria (+)gram and (-)gram.

The active microrganisms were tested with a taxonomic analysis; one of these species was found to be Agrobacter radiobacter. Nevertheless the extractive activity resulted lower in comparison with the mixed cultures.

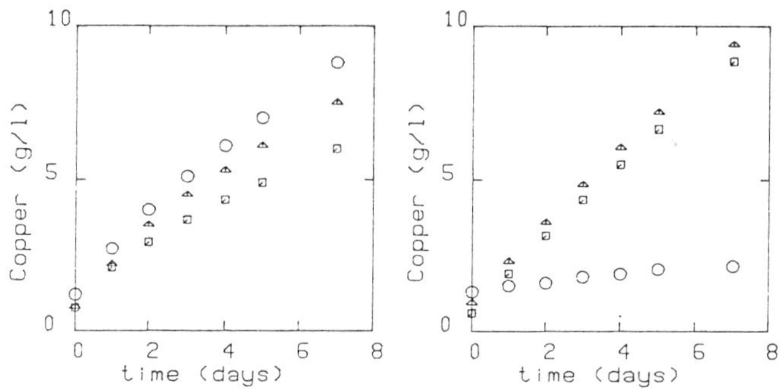

Figure 2: Copper solubilization in the presence of different contents of pulp [10% (□), 20% (Δ) and 30% (○) respectively] vs time; a) batch tests using Erlenmeyer flasks ; b) batch tests using Biomet pilot plant.

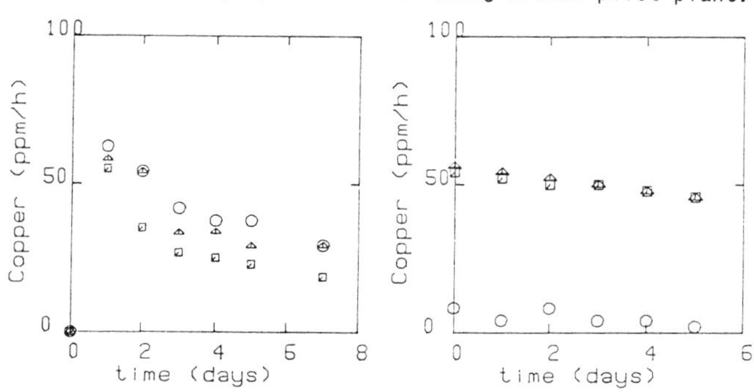

Figure 3: Rate of copper solubilization in the presence of different contents of pulp [10% (□), 20% (Δ) and 30% (○) respectively] vs time; batch tests were carried out with Erlenmeyer flasks (a) and with Biomet pilot plant (b).

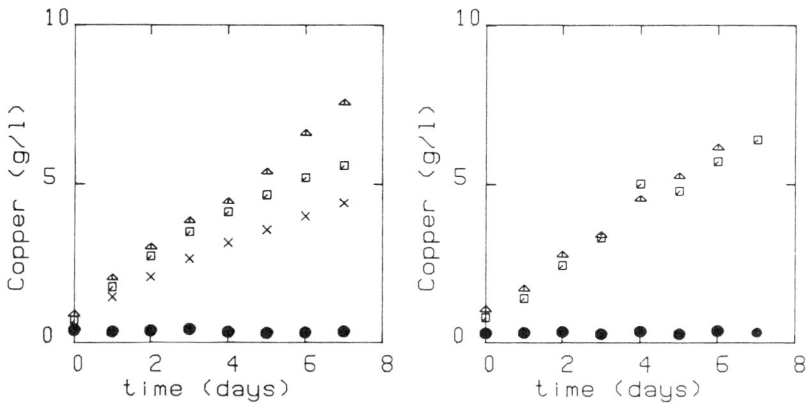

Figure 4: Copper solubilization using urea as the nitrogen source vs time; the batch tests were carried out with Erlenmeyer flasks (a) and with "Biomet" pilot plant (b) in the presence of different contents of pulp [10% (● control), 5% (×), 10% (□) and 20% (Δ) respectively].

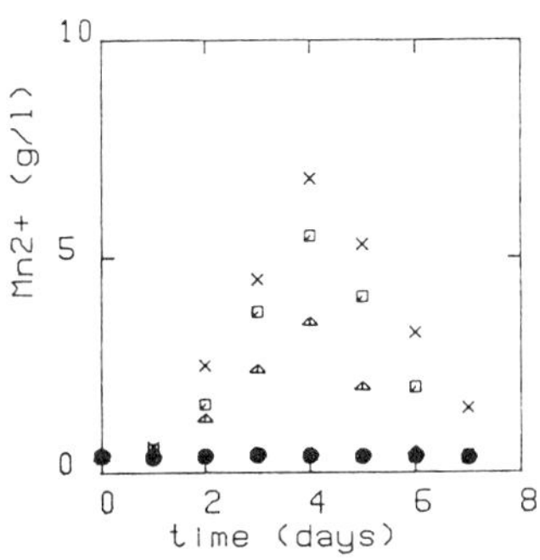

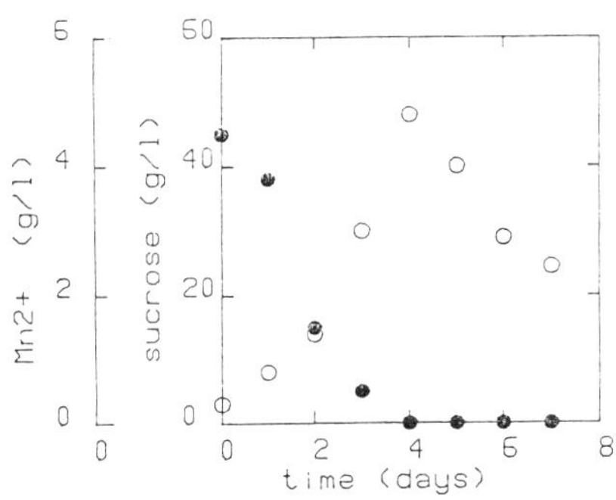

Figure 5: Manganese solubilization catalyzed by mixed cultures vs time (left); batch tests carried out with L.K.B.1601 fermenter in the presence of different contents of ore [(● control) 10% , (Δ) 10%, (□) 20%, (×) 30% respectively].
Figure 6: Manganese solubilization (○) and sucrose consumption (●) catalyzed by Agrobacter radiobacter in the presence of 10% of ore (right) vs time.

CONCLUSIONS

A new reactor was designed an built for bioleaching; it was employed in preliminary tests of chalcopyrite processing; this apparatus can produce a great quantity of information with adapt ionic electrodes.

A process scheme has been proposed on the basis of appropriate analyses in order to elucidate the bioleaching of a chalcopyrite concentrate of italian origin.

Urea resulted an interesting candidate as a nitrogen source in sulphides bioleaching.

Encouraging results in manganese processing with the quantitative reduction of MnO2 using mixed cultures were obtained.

A taxonomic analysis indicated in Agrobacter radiobacter one of the active species in mineral processing; analyses are going on for identification of other species active in mineral processing. An indirect method of detecting "the growth" of biomass during the process was set up; this last is of considerable importance bearing in mind that the presence of inorganic material in suspension is a typical condition in bioleaching experimentation; this material makes unfeasible the common techniques for biomass determination.

Chemical methods currently in use in manganese ores processing will certainly prevail in the years to come. However, biological processes can be less polluting than non biological ones; it is therefore possible that the role of biological technology in the extraction and working of minerals will become more important in the future.

ACKNOWLEDGEMENTS

The Authors would like to thank Dr. Marisa Terreri, Ing. Marco Recinella and Dr. Carla Cervelli for their helpful collaboration in this research.

REFERENCES

1. S.N. Groudev, Continuous bacterial leaching of copper sulphide concentrates, in Fundamental and Applied Biohydrometallurgy, R. W. Lawrence, R.M.R. Branion, and H.G. Ebner, Elsevier, pp. 43-50, 1986.
2. P. Bos, T.F. Huber, C.H. Kos, C. Ras and J,C. Kuenen, A dutch feasibility study on microbial coal desulphurization, in Fundamental and Applied Biohydrometallurgy, R. W. Lawrence, R.M.R. Branion, and H.G. Ebner, Elsevier, pp. 129-150, 1986.
3. P.C. Miller, R. Huberts, and E. Livesey-Goldbatt , The semicontinuous bacterial agitated leaching of nickel disulphide material, in Fundamental and Applied Biohydrometallurgy, R.W. Lawrence, R.M.R. Branion, and H. G. Ebner, Elsevier, pp. 23-42, 1986.
4. G.I. Karavaiko, L.K. Chuchalin, T.A. Pivovarova, Microbiological leaching of metals from arsenopyrite containing concentrates,in Fundamental and Applied Biohydrometallurgy, R.W. Lawrence, R.M.R. Branion, and H. G. Ebner, Elsevier, pp. 115-127, 1986.
5. S. Acevedo and G. Aroca , "Studies on the agitation and power characteristics of mineral slurries", in Fundamental and Applied Biohydrometallurgy, R. W. Lawrence, R.M.R. Branion, and H. G. Ebner, Elsevier, pp. 255-262, 1986.
6. L. Toro, G. Alberti, N. Orsi and M. C. Annesini, Kinetic Analysis of the Growth of Thiobacillus ferrooxidans in a Synthetic Medium: a Preliminary Study on Zinc Sulphide Leaching in a Batch Culture, in: G. Rossi and A.E.Torma, (Eds.), "Recent Progress in Biohydrometallurgy", A.S.M., Cagliari, pp. 317 - 324, 1983.
7. A. Bruynesteyn, R. W. Lawrence, A. Vizsolyi, "An elemental sulphur producing bioydrometallurgical process fot treating sulphide concentrates", Progress in Biohydrometallurgy, pp. 151-168, 1983.
8. H. Kandemir, "Fate of sulphides in bacterial oxidation of iron sulphides minerals", Progress in Biohydrometallurgy pp. 291-315,1983.
9. K.C. Trivedi, Microbial Leaching of Copper and Nickel Sulphides, Ph. D. dissertation, Faculty of the Graduate University of Minnesota, 1974.
10. A. Pinches , F.O. Al-Jaid , D.J.A. Williams and B. Atkinson, "Leaching of chalcopyrite concentrates with Thiobacillus ferrooxidans in batch culture", Hydrometallurgy, vol.2 pp. 87-103, 1976.

11. A. Lepidi , L. Toro , B. Paponetti and S. Di Cesare, "Urease of Thiobacillus Ferrooxidans and urea influence on chalcopyrite bioleaching", International Symposium of Warwick, July pp. 319-325, 1987.
12. D.P. Kelly & C.A. Jones ."Factors affecting metabolism and ferrous iron oxidation in suspensions and batch cultures of Thiobacillus ferrooxidans, in "Metallurgical Applications of Bacterial Leaching and Related Microbial Phenomena", pp. 20-43,1978.
13. C. Abruzzese."Valorizzazione dei minerali di Manganese Dell'Alto Lazio, mediante lisciviazione e recupero del metallo", 1986.
14. B. Paponetti, L. Toro, C. Abruzzese, A. Marabini and M. Y. Duarte, Recovery of Mn + from concentrates of MnO by means Aspergillus niger. Role of metabolic intermediates in the extractive process, 118th annual meeting of AIME, Las Vegas, February 27 - March 2, 5, 33-37, 1989.
15. A. Okuwaki, Treatment of manganese nodule II, Kagaku Kogyo, 30 (2), pp.1462-1482, 1977.
16. Kazutami Imai. On the mechanism of bacterial leaching, in Metalllurgical Applications of Bacterial Leaching and Related Microbiological Phenomena, pp 275-294, 1978.
17. J. Michael Hart and C. John Madgwick, Biodegradation of manganese dioxide tailings, Bull. Proc. Australas. Inst. Metall., Vol. 291, No 3, pp. 61-64, 1986.
18. E.M. Bierhaus, J. Perez, A.E. Torma and G. Rossi, A comparison of bacterial leachability of chalcopyrite concentrates from different origins, in Recent Progress in Biohydrometallurgy, Ed. G. Rossi and A.E. Torma, A.S.M. pp. 127-150, 1986.
19. M.P. Silverman, and D.G.L. Lundgren, Studies on the chemoautotrophic iron bacterium ferrobacillus ferrooxidans, J. Bacteriology, Vol.77, pp. 642-647, 1959.
20. L. Toro, S. Di Cesare, B. Paponetti and A. Lepidi, Biochemical and chemical Events in Copper Solubilization from a Chalcopyrite Concentrate by Thiobacillus ferrooxidans in Batch Cultures,International Journal of Mineral Processing, pp. 1-10 ,1989.
21. A.E. Torma ."Biohydrometallurgy as an Emerging Technology and Bioengineering Symposium , n.16, 149-63, 1986.
22. P.R. Norris and D.P. Kelly "Toxic metals in leaching systems", Ed. Murr E.,Torma A. and Brierly A., pp. 83-101,1978.
23. S.B. Yunker and J.M. Radovich, Enhancement of Growth and ferrous Iron Oxidation Rates of Thiobacillus by Electrochemical Reduction of Ferric iron, Biotechnology and Bioengineering, V. XXVIII, pp. 1867-1875, 1986.
24. C.A. Schnaitman , M.S. Korzynszy and D.G. Lundgren ,"Kinetic studies on iron oxidation by whole cells of Ferrobacillus Ferrooxidans", J. Bacteriol. Vol.99, pp. 552-557, 1969.
25. J. Frizt and S. Yamamura "Rapid microtitration of sulphate" Anal.Chem., Vol., 23 pp. 1461-1465, 1955.
26. D. Barham & P. Trinder , Analyst, 97, pp. 142, 1972.
27. A. Vaschetti, Procedimento per la precipitazione del ferro da soluzioni di solfato di zinco, Brevetto n. 946484, 1971.
28. H.G. Schlegel, and H. Kaltwasser, Urease, methods of enzymatic analysis. Bergmeyer H.U. Ed.,pp.1081-1085, 1974.

SILVER RECOVERY THROUGH MOLTEN SALT DESTRUCTION OF SLUDGES AND OTHER SOLIDS

S. K. JANIKOWSKI,* D. L. SMITH, G. A. REIMAN, AND R. E. McATEE
*Chemical Sciences Group, Idaho National Engineering Laboratory,
P.O. Box 1625, Idaho Falls, Idaho 83415.

ABSTRACT

Secondary recovery of precious metals from sludges is applied to minimize the wastes, reduce the waste classifications, and recover the metals. Molten salts, operating at moderate eutectic temperatures, are used to solubilize the sludges, whereupon the precious metals are reduced and gravity separated from the flux. Additional decomposition reactions occur that degrade the remaining flux contaminants. Applications are discussed for various silver bearing solids.

INTRODUCTION

Precious metal bearing sludges are generated in the metal finishing, photographic, electronic, and jewelry industries, and in salvage operations [1-8]. The majority of these sludges contain high percentages of various silver oxides and hydrated oxides. Impurities are present which may include other metals, organics and a large variety of anionic and inorganic species, i.e., halogens, cyanide, sulfur compounds, etc. Several commercial technologies have demonstrated the technical and economic ability to recover silver from these sludges [3,4,5,8]. An additional source of solid waste containing silver is generated by the nuclear industry. In this application, silver zeolites are used to capture iodine from gaseous waste streams. The zeolite is subsequently destroyed when the silver is recovered [9].

A new technology has been developed that incorporates a molten flux mixture that separates silver from solid matrices and further, makes facile the subsequent separation and quantitative recovery of silver.

EXPERIMENTAL

Silver sludges and a silver-bearing zeolite were used in this study to demonstrate the wide applicability. The sludges were generated in laboratory scale silver stripping operations using commercial processes. The sludges fit into four categories with the following characteristics: silver oxide and cyanide sludges, silver oxide and halide sludges, silver oxide and amine sludges, and silver oxide and mineral acid residues. Based on AA analyses, the weight percent of silver was greater than 85% in all of the dried (at 100^oC) sludges.

A synthetic zeolite with the nominal formula:

$$Na_{86}[(AlO_2)_{86}(SiO_2)_{106}]\cdot 276H_2O$$

was supplied by Ionex Corporation, in the form of 1.0 to 1.7 mm beads with 18% clay binder. This was converted to the silver form (AgX) by circulating

*To whom correspondence should be addressed.

Published 1990 by Elsevier Science Publishing Co., Inc.
Advances in Fine Particles Processing
John Hanna and Yosry A. Attia, Editors

an aqueous solution of silver nitrate through a column of zeolite at 90°C until replacement of the sodium was complete (99.7%). The column was rinsed with deionized water to remove fines and excess salt, then dried at 150°C for 20 hours. This left a residual 3-5 weight percent water in the zeolite. The silver content was then determined by smelting in a fire assay procedure [10] and was found to be 42% by weight of dried sample.

The zeolite sample was from a lot used in the ventilation system of a nuclear reactor and had become fouled with airborne organic contaminants, reducing its ability to capture radioiodines. The zeolite was subjected to a 1 hour heat treatment at 700°C to drive off water and organics. Assay of the resulting product indicated 38 weight percent silver.

The respective silver sludge or zeolite, (typically 1 gram of sludge or larger quantities of zeolite), was crushed and combined on a 1:1 weight ratio with the flux mixture in an alundum crucible. Table I lists the flux composition. The crucible contents were then heated in air to 1100°C in a muffle furnace, removed and allowed to cool. The contents were then removed; the top portion of solid material consisted of a slag, and the bottom portion consisted of a single silver ingot.

TABLE I. Flux composition.[a)]

Component	Amount (g)
Silicon dioxide (240 mesh)	42
Boron trioxide	13
as sodium tetraborate	18.8
as sodium tetraborate decahydrate	35.6
Sodium oxide	19
as sodium carbonate	22.6
as sodium hydroxide	17.0
Calcium oxide	19
as calcium carbonate	33.9
as calcium hydroxide	25.1
Calcium fluoride	7.0

[a)]Flux should be blended and stored in a capped bottle. Use alundum or carbon bonded silicon carbide crucibles. Use least gassy component to make up flux when possible. Components were Fisher, reagent grade.

RESULTS AND DISCUSSION

Flux technology is not new [10], and has been used extensively in extracting metals from ores, refining metals, and in salvage/recycle operations. This system is unique in its simplicity; both in terms of ease of preparation, and in its inherent reuse potential. Under the conditions employed, it is only applicable to silver bearing solids which in turn is dependent on the chemistry and physical properties of silver in the flux.

The silver assays for the ingots produced from each of the sludges and zeolite indicated 99.8+% purity. Total recovery was greater than 97% for 1050 kg of zeolite processed and greater yet for the sludges.

All of the silver in the solids evaluated existed as silver(I). The primary silver-to-oxygen linkages are ionic and not discrete in the sludges or the zeolite. The proposed model for the process depends on the decomposition of silver(I) oxide and the solubilizing effects of the flux. Silver(I) oxide decomposes to silver metal and oxygen at 160°C [11] to 230°C [12].

$$2Ag_2O \longrightarrow 4Ag + O_2 \quad (1)$$

As the sludge/flux mixture reaches 230°C, the silver(I) oxide starts to decompose. This results in minute silver particulates distributed throughout the mixture. Oxygen gas is released in the reaction and goes to the atmosphere or reacts with oxidizable components in the flux. At temperatures greater than 500°C, the flux becomes liquid. At 960°C, the silver melts and agglomerates until sufficient mass is achieved to overcome the forces keeping it in suspension. At this point the silver flows to the bottom of the crucible in nearly pure form (if present, copper, nickel, and gold may also mix with the silver). A distinct interface exists between the molten silver and slag such that, upon cooling, the silver solidifies with complete exclusion of the flux components.

Other silver(I) salts may decompose in a similar fashion to yield silver metal and the oxidized counter component. Alternatively, silver(I) salts that do not decompose via a reductive elimination type mechanism at or below 500°C, may, nonetheless, react with the liquid flux to yield silver metal. This may occur through dissolution of the silver salt in the liquid flux, whereupon, the silver(I) ions can associate with and form linkages with oxides present in the flux. This is shown in the general case in reaction (2) below:

$$Ag^{+}\ldots\text{anion (or ligand)} \xrightarrow[>500^{\circ}C]{\text{flux}} Ag^{+}\cdots\text{flux} + \text{anion}^{-}\cdots\text{flux.} \quad (2)$$

The silver-oxygen linkages are virtually the same as those which exist in the silver(I) oxide/flux melts. Electron transfer can then take place between the oxide and silver resulting in reduced silver metal and oxygen gas being liberated. The fate of the counter ions or ligands originally present in the silver sludge is dependant upon their physical and chemical properties. The ions or ligands may subsequently occupy vacant sites in the flux, be further decomposed, or liberated as a gas, respectively.

The flux is not necessarily degraded by use and may be recycled for several turnovers. The actual turnover number will be inversely proportional to the amount of nonsilver and nonoxygen impurities present in the sludge. This includes non-oxygenated silver salts as well as other salts and inert materials. The actual effect of these impurities is to lessen the availability of oxide to silver ions, and increase the eutectic melting temperature of the flux.

ACKNOWLEDGMENTS

This work was performed at the INEL in association with the University of Idaho. The industrial sludge research was supported by the United States Air Force Engineering and Services Center, ENVIRONICS Division, Tyndall Air Force Base under DOE Contract No. DE-AC07-76ID01570.

REFERENCES

1. K.J. Bush, and H. Diehl in: Recovery of Silver from Laboratory Wastes, Journal of Chemical Education, 56 No. 1, January 1979, pp. 54-55.
2. D.F. Fotus in: Recovery of Silver and Cobalt from Laboratory Wastes, Journal of Chemical Education, 61, No. 10, October 1984, p. 924.
3. E.F. Hradil and G. Hradil in: Electrolytic Recovery of Precious and Common Metals, Metal Finishing, November 1984.
4. C.A. Perman in: Recovery of Silver from Silver Chloride Residues, Talanta, 26, pp. 603-604, Great Britain, 1979.
5. Recovering Silver from Photographic Materials, Kodak Publication No. J-10, Rochester, NY, 1979.
6. S. Steed, P. Hayes, and M. Janan in: Procedure for Recovering Elemental Silver from Silver Residues, Journal of Chemical Education 49, (3), March 1972, p. 156.
7. O.L. Willbanks in: Reclaiming Silver from Silver Chloride Residues, Journal of Chemical Education, 30, (17) 347, July 1953.
8. G.I.P. Levenson in: Silver Recovery by Metal Exchange, The Journal of Photographic Science, 29, 1981.
9. G.A. Reimann in: Silver Recovery From Silver Zeolite, MS Thesis, University of Idaho, July 1986.
10. C.W. Ammen in: Recovery and Refining of Precious Metals, Van Nostrand Reinhold Publishing, NY, 1984, Chapter 12.
11. N.N. Greenwood and A. Earnshow in: Chemistry of the Elements, Pergamon Press, NY, 1986, page 1373.
12. Handbook of Chemistry and Physics, Weast, R. C., Ed., CRC Press, Cleveland, 57th ed., page B-157.

PILOT SCALE FERROUS AND SULFIDE METALS TREATMENT IN WASTEWATER CLEANUP

S. K. JANIKOWSKI,* S. N. UGAKI, P. M. WIKOFF, AND D. F. SUCIU
*Chemical Sciences Group, Idaho National Engineering Laboratory,
Idaho Falls, Idaho 83415.

ABSTRACT

Metal finishing operations generate large volumes of hazardous aqueous wastes containing metals and organics. End-of-pipe treatment usually involves coupled processes to precipitate metals followed by organic degradation. Large volumes of metal bearing sludges are generated during the treatment and are classified as RCRA regulated hazardous wastes. A two-stage treatment process was developed and piloted which reduced the volume of the waste sludge generated during wastewater cleanup. The two stages involve: 1) treatment with alkaline ferrous sulfate and sodium sulfide to reduce hexavalent chrome and precipitate all of the hazardous metals, and 2) biodegradation of the organic components. Aspects regarding the hexavalent chromium reduction, metals precipitation, metals flocculation, filtering, and the related chemistry are discussed.

INTRODUCTION

Hazardous wastewaters are generated during aircraft refurbishing and maintenance operations [1,2] during metal cleaning, stripping, plating, and in paint stripping processes. The wastes are combined to produce mixed waste solutions containing metals, other inorganics, and organic contaminants. Some of the contaminants are Cr, Cd, Cu, Pb, Ni, Zn, Fe, phosphate, carbonate, ethylenediaminetetraacetic acid (EDTA), cyanide, phenol, and 1,1,1-trichloroethane in concentrations ranging from 0.01 ppm to 200 ppm in metals, and up to 200,000 ppm in the other contaminants [3]. These aqueous mixed waste streams are typically treated in multiple stages to destroy cyanide, skim off oil, aerate to remove volatile organics, reduce chromate and precipitate metals. The effluent is passed through activated sludge to degrade all remaining organics, then filtered and readmitted to the environment.

A treatment process that decreases metal concentrations to acceptable discharge limits was developed, piloted, and implemented at Tinker Air Force Base [4]. The new process replaced an existing metals treatment process without affecting previous or subsequent steps in the total waste water treatment process. The previous metals treatment process involved acidification of the wastewater with sulfuric acid and subsequent injection of sulfur dioxide to reduce the highly soluble chromate to chromic ion according to the reaction:

$$2CrO_4^{-2} + 3SO_2 + 2H_2SO_4 \longrightarrow 2Cr^{+3} + 5SO_4^{-2} + 2H_2O. \qquad (1)$$

The resulting solutions were then made basic with caustic to precipitate all of the heavy metals as their respective hydroxides. these existed as fine,

*To whom correspondence should be addressed.

Published 1990 by Elsevier Science Publishing Co., Inc.
Advances in Fine Particles Processing
John Hanna and Yosry A. Attia, Editors

suspended particulates which were agglomerated by addition of ferric compounds. the combined reactions are:

$$3Fe^{+3} + Cr^{+3} + 12OH^- \rightarrow 3Fe(OH)_3(s) + Cr(OH)_3(s) \quad (2)$$

Other metals present underwent similar metathesis reactions.

The formerly-used process had several drawbacks. Metals removal was incomplete because migration of the metals through the gelled ferric floccules could occur and because of the varying solubility product (K_{sp}) of the respective metal hydroxides with pH. In addition, large volumes of hazardous sludge were produced.

The new metals treatment process developed and reported herein, incorporates the use of ferrous sulfate and sodium sulfide in basic solution to reduce chromate to chromic ion and precipitate all of the heavy metals. This process occurs in three steps in a continuous flow process [5]. This is achieved by first making the wastewater basic with caustic, whereupon sodium sulfide is added in proportion to the concentration of chromate to reduce the chromate with sufficient excess to precipitate the other metals (excluding chromic and ferric ions). Ferrous sulfate is then added as the solution flows through a second reaction vessel. The pH is simultaneously adjusted to between 7.2 and 7.5.

In the first step, heavy metal sulfides are precipitated. In the second step, ferrous and sulfide ions combine with chromate to form a complex through which the charge transfer occurs. The resulting chromic and ferric ions further react, as in equation (2), to form insoluble hydroxides. The ferric hydroxide formed initiates flocculation of the suspended particulates. The water then flows into a third process tank where ionic polymers are added to agglomerate all of the suspended particulates into larger floccules which are subsequently filtered from the water in a sludge filter bed [6].

EXPERIMENTAL

Laboratory tests were conducted with aqueous solutions of chromate to evaluate the effects of ferrous and sulfide ions on the chromate reduction rate, product distribution, and completeness of reaction. Initial chromate solutions were prepared in 20 ppm concentrations. This and all other concentrations expressed in ppm refer to the individual species identified. One-liter solutions were made by adding the appropriate volumes of 100 ppm stock solutions of chromate, ferrous, and sulfide ions to achieve the desired atom ratio. Chromate was added to the reaction vessel first, followed by pH adjustment with caustic to between 7.2 to 7.5. Sulfide and ferrous ions were then added sequentially and the pH was again adjusted. The resulting 1-liter solutions were then stirred for 5 minutes before aliquots were withdrawn, filtered and analyzed. Reagent grade potassium dichromate, sodium sulfide, ferrous sulfate, sodium hydroxide, and sulfuric acid from Fisher Scientific were used in the experiments. The reaction conditions and analysis results are summarized in Table I.

Duplicates of several reaction solutions were extracted with toluene immediately following the second pH adjustment. The toluene volume was then reduced with nitrogen gas sparging to a thick slurry, which was analyzed for sulfur forms using GC-MS and ion chromatography.

A pilot-scale field verification unit was constructed at the Tinker Air Force Base Industrial Waste Treatment Plant. The unit was designed as a scaled-down version of the existing Tinker Air Force Base waste treatment

TABLE I. Reaction conditions and results of jar tests with hexavalent chromium, sodium sulfide and ferrous sulfate.[a)]

Initial Conditions			Final Conditions			
Run Number	S^{-2}	Fe^{+2}	Cr^{+6} [b)]	Cr [c)]	Fe [c)]	S^{-2} [d)]
1	0	0	21	22	0.0	0
2	0	10	18	21	7.8	0
3	0	20	13	9	6.0	0
4	0	30	11	11	6.3	0
5	0	40	8	7	2.3	0
6	10	0	19	22	0.0	10
7	10	10	8	1	2.0	1
8	10	20	2	4	0.8	1
9	10	30	0	3	1.0	1
10	10	40	0	3	0.8	1
11	20	0	17	20	0.8	19
12	20	10	6	8	1.0	7
13	20	20	0	3	1.0	1
14	20	30	0	3	1.0	1
15	20	40	0	3	0.8	1
16	30	0	15	22	0.0	16
17	30	10	3	7	2.3	8
18	30	20	0	1	0.8	5
19	30	30	0	1	2.7	2
20	30	40	0	1	2.9	2
21	40	0	12	11	0.0	22
22	40	10	0	3	1.2	15
23	40	20	0	2	2.8	11
24	40	30	0	1	4.4	5
25	40	40	0	1	0.8	2
26	50	0	13	15	0.0	41
27	50	10	0	4	2.0	22
28	50	20	0	17	16.8	16
29	50	30	0	22	27.8	3
30	50	40	0	16	35.4	2

[a)] Concentrations are in ppm, initial hexavalent chromium ion concentration was 20 ppm, the pH was adjusted with sulfuric acid or sodium hydroxide to $7.2 \leq 7.5$ after the ferrous ion was added, and the final ferrous ion concentration ranged from 0 to 3 ppm.

[b)] As determined by Hach method for hexavalent chromium, total chromium, or iron.

[c)] As determined by AAS.

[d)] As determined by Hach method for sulfide ion.

facility with representative hydraulic retention times and flow velocities [5]. The physical and chemical parameters were optimized in this unit.

The influent to the pilot unit was pumped from the Industrial Waste Treatment Plant Equalization Reservoir to a 500-gallon equalization tank equipped with a pH probe and mixer so that pH could be adjusted with either caustic or sulfuric acid. Wastewater, controlled between 2 and 10-gal/min, flowed by gravity to a 50-gallon mixing tank where additional chemicals such as orthophosphate, EDTA, cyanide, and additional metals could be added. In addition, the 500-gallon Equalization Tank could be pumped through the pilot plant to provide feed of constant chemical composition during parameter tests on the effects of these additional chemicals. The wastewater flowed from the 500 gallon mixing tank into the bottom of the first in a series of three mixing tanks. Sodium sulfide was fed into this tank, designated Mixer-1. The 115 gallon Mixer-1 tank had a retention time of 23 minutes at an influent flow rate of 5 gal/min. The stream then flowed over a weir into the second mixing tank and under a second weir designed to prevent back-mixing between the two tanks. The ferrous sulfate and sulfuric acid were added in this 140-gallon tank with a 28-minute retention time. The effluent from Mixer-2 flowed over a weir into the 33-gallon Mixer-3 tank where a cationic polymer was added to aid in particle flocculation. The effluent of this tank exited from the bottom after a 6.6 minute retention time. This stream flowed into the 330-gallon solids contact clarifier (SCC) in which the treated water was filtered through a sludge bed. The SCC had a 2.75-hour retention time at a flow rate of 2-gal/min.

RESULTS AND DISCUSSION

Laboratory Studies

The predominant form of chrome(VI) in the pH range studied is as the Chromate ion [7]. The actual specie undergoing reduction is unidentified, but may be chromate, bichromate, dichromate, or any combination of these three. For simplicity, only the chromate ion is included in this discussion. The ionic reaction of interest can be written as:

$$CrO_4^{-2}(aq) + HS^{-1}(aq) + Fe^{+2}(aq) \xrightarrow[pH=7.2\ -\ 7.5]{OH^-} Cr(OH)_3(s) + Fe(OH)_3(s) + \text{sulfur species.} \qquad (3)$$

The reaction is unbalanced since it is not known how many electrons are given up by the individual reactants [1,8]. Furthermore, the chromate:monohydrogen sulfide stoichiometry changes in a nonlinear inverse fashion with the ferrous ion concentration. This is shown in Figure 1 for solutions in which the chromate concentration is reduced to less than 1 ppm.

Analyses of the reaction solutions extracted with toluene indicated that at least four sulfur species were present: elemental sulfur, sulfite, sulfate, and sulfide. The origin of these species remains unverified; however, known chemical pathways exist that can account for some of them. A 1:1:1 ion ratio of chromate:ferrous:sulfide would imply that one electron and two electrons are respectively transferred to the chromate from the ferrous and sulfide ions. Elemental sulfur would thus be formed; which is thermodynamically unstable under basic conditions and will disproportionate to form sulfide and oxy-sulfur compounds [9]. Sulfite is one potential product. Sulfite will, in addition, react with ferric ion (or oxygen) to form sulfate and, in the former case, ferrous ion. This pathway is consistent with the data in Figure 1. At low sulfide concentrations, more

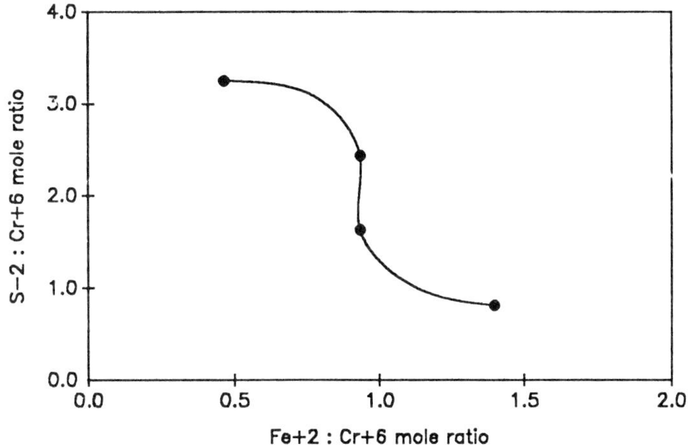

FIG. 1. Stoichiometric ratios of initial S^{-2} and Fe^{+2} to Cr^{+6} for reduction reactions where the final Cr^{+6} concentration is less than 1 ppm.

ferrous ion is needed to complete reaction (3). This can be attributed in part to the increased source of electrons in higher concentrations of ferrous ion and to the greater ability of ferrous ion to donate electrons to chromate through facile kinetic pathways.

Alternatively, at low initial ferrous ion concentrations, greater sulfide concentrations are required to complete reaction (3) in the reaction time allotted. This can be attributed in part to the relative inability of sulfide to coordinate with chromate and effect charge transfer. It may also reflect on the increased efficiency resulting from a higher steady state concentration of this proposed chromate-ferrous-sulfide intermediate [10,11]. Attempts to quantify the concentrations of the species were generally difficult due to interferences, and complete mass balances were not possible in a short time frame because of the uncertainties regarding losses through adsorption of the species onto the floccules.

Figure 2 illustrates the dependence of chromate reduction on the initial concentration of ferrous ion for solutions containing varying amounts of sulfide. It is clear that increasing concentration of sulfide and ferrous ions promote increased reduction of chromate. Figure 3 illustrates that although higher concentrations of sulfide and ferrous ions promote chromate reduction, intermediate concentrations exhibit a better overall reduction in total chrome species remaining in the effluent water. Figure 4 similarly shows that high concentrations of sulfide and ferrous ions result in high concentrations of iron species left in the effluent water. The results are verified visually by observing the formation of ferrous sulfide which produces a suspended fine particulate which is difficult to filter out and leads to the conditions known as "black water."

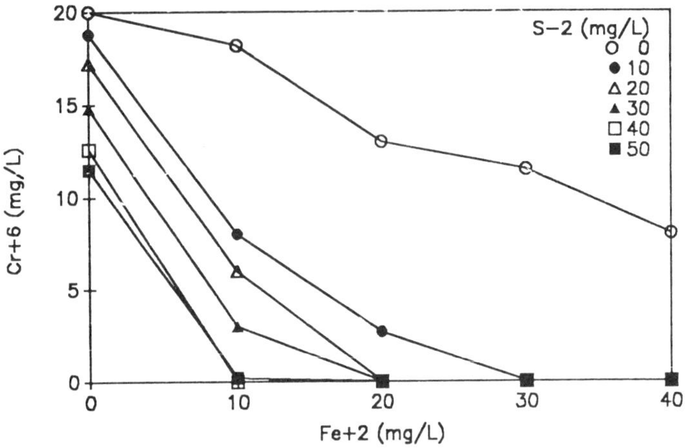

FIG. 2. Final Cr^{+6} concentration vs. the initial Fe^{+2} concentration for different sulfide concentrations.

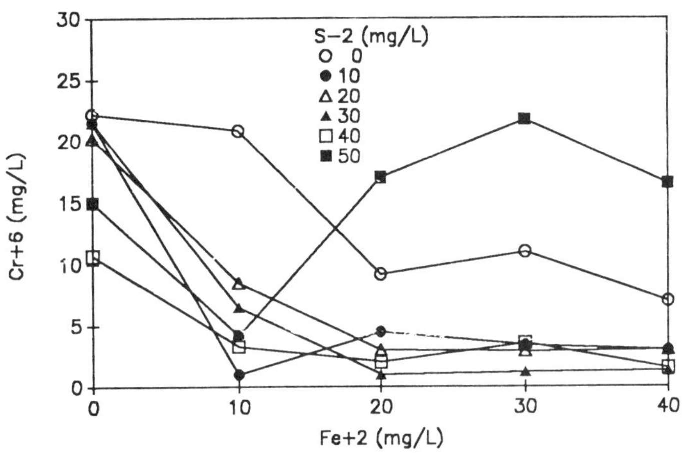

FIG. 3. Total chromium left in solution after the reaction vs. the initial Fe^{+2} concentration for different sulfide concentrations.

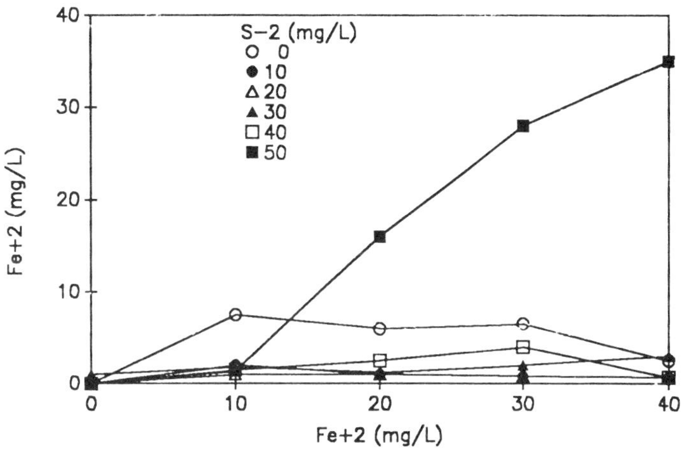

FIG. 4. Total iron remaining in the supernatant after the reaction vs. the initial Fe^{+2} concentration for different sulfide concentrations.

Pilot-Scale Studies

The results of jar and dynamic testing showed that the sodium sulfide/ferrous sulfate process is a viable method for reduction of hexavalent chrome and removing heavy metals from waste water streams while reducing the amount of chemicals and metal sludge associated with treatment.

The optimum operating conditions found in the initial jar tests provided a good starting point for pilot operations. It was found after extensive testing that an increase in the ratio of sulfide to chromate and, ferrous to chromate was required to handle fluctuations in the feed makeup. It was found that pH had a minimal effect on the process as long as it was maintained above 7.2. After the solids contact clarifier was operating at optimum conditions, 100 ppm orthophosphate, 50 ppm EDTA, temperatures from 41 to 95 ^{O}F, and mixed metals were each determined to have no effect on chromium reduction or metals removal.

Under proper operating conditions, the precipitates were effectively filtered out in the solids contact clarifier. The conditions for operating the pilot plant were at the chemical feed ratios of 2 ppm sulfide, 1.5 ppm ferrous ion, 1 ppm chromate, 20 ppm Betz 1195 cationic polymer, 0.5 ppm Betz 1120 anionic polymer, and the Mixer-2 pH at 7.2 - 7.5. Sludge production decreased approximately 90% with the sodium sulfide/ferrous sulfate process. This process has been successfully implemented at Tinker Air Force Base [4].

SUMMARY

A wastewater treatment process for hexavalent chromium reduction and metals precipitation has been developed and commercially implemented. The process uses ferrous ions and sulfide ions, added separately in the form of ferrous sulfate and sodium sulfide, to reduce the hexavalent chromium and precipitate the metals. The ferrous and sulfide ions are added separately to produce extremely fine suspended particles of ferrous sulfide, which enhance the chromium reduction kinetics beyond that observed by adding bulk ferrous sulfide. Both the ferrous ions and sulfide ions participate in the chromium reduction while excess sulfide and hydroxide serve to precipitate the hazardous metals. The process offers five advantages over other existing commercial wastewater treatment processes: 1) the process kinetics are fast enough to allow for the treatment of large volumes of wastewater, 2) the addition chemicals are inexpensive and available in technical grade, 3) the process economics are lower than for similar staged processes where ferrous ions are added first for chromium reduction followed by sulfide addition to precipitate metals, 4) the volume of generated sludge is lower than that generated from other treatment processes, and 5) no strong mineral acids or gases are required as in acid-sulfur dioxide treatment processes.

ACKNOWLEDGMENTS

This work was sponsored by the United States Air Force Engineering and Services Center, ENVIRONICS Division, Tyndall AFB under DOE Contract No. DE-AC07-76ID01570. D. Prescott, T. Harris, and R. Schober also contributed to this program and have our appreciation.

REFERENCES

1. T.E. Higgins and V.E. Slater in: Treatment of Electroplating Wastewaters by Alkaline Ferrous Reduction of Chromium and Sulfide Precipitation, ESL-TR-83-21, Engineering and Services Laboratory, Air Force Engineering and Services Center, Tyndall AFB, Florida, June 1983.
2. M.F. Herlacker in: Automated Pretreatment of Electroplating Waste at Tinker AFB, OK", Proceedings of the 5th National Conference on Hazardous Wastes and Hazardous Materials, Las Vegas, Nevada, pp. 109-113.
3. P.M. Wikoff, D.F. Suciu, P.A. Pryfogle, W.C. Schutte, G.S. Carpenter, and F.S. Lloyd in: Sodium Sulfide/Ferrous Sulfate Treatment of Hexavalent Chromium and Other Heavy Metals of Tinker AFB, ESL-TR-87-39, Engineering and Services Laboratory, Air Force Engineering and Services Center, Tyndall AFB, Florida, March, 1988.
4. P.M. Wikoff, D.F. Suciu, D.S. Prescott, R.K. Schober, T.L. Harris, and P.A. Pryfogle in: Full-Scale Implementation of the Sodium Sulfide/Ferrous Sulfate Treatment Process, Air Force Engineering and Services Center, Tyndall AFB, FL, submitted February 1989 for final review.
5. P.M. Wikoff, D.F. Suciu, D.S. Prescott, R.K. Schober, F.S. Lloyd, T.L. Harris, G.S. Carpenter, W.C. Schutte, P.A. Pryfogle, and P.L. Wichlacz in : Pilot Field Verification Studies of the Sodium Sulfide/Ferrous Sulfate Treatment Process, ESL-TR-88-13, Engineering and Services Laboratory, Air Force Engineering and Services Center, Tyndall AFB, FL, September 1988.
6. Clarifier Design, Manual of Practice FD-8, Facilities Development, Water Pollution Control Federation 1985, Figure 3.12, p. 60, Available through the WPCF, 2626 Pennsylvania Ave., N.W., Washington, DC 20037.
7. R.A. Day Jr. and A.L. Underwood in: Quantitative Analysis, Prentice Hall, Inc., NJ, 1980 4th Ed. (pK_{a2} = 6.49).

8. J.R. Aldrich in: A Better Heavy Metal Waste Treatment Method, Metal Finishing, November 1984, pages 51-55.
9. F.A. Cotton and G. Wilkinson in: Advanced Inorganic Chemistry, John Wiley and Sons, New York, 1980, Chapter 16.
10. J.H. Espensen and E.L. King in: Kinetics and Mechanisms of Reactions of Chromium (VI) and Iron (II) Species in Acidic Solution, Journal of the American Chemical Society, 85, No. 21, pp. 3328-3333, November 1963.
11. J.H. Espensen in: Rate Studies on the Primary Step of the Reduction of Chromium (IV) by Iron (II), Journal of the American Chemical Society, 92, No. 7, pp. 1880-1883, April 8, 1970.

Author Index

Subject Index